Soil Health Series: Volume 3 Soil Health and Sustainable Agriculture in Brazil

SERIES EDITOR

Douglas L. Karlen

EDITORS

Ieda Carvalho Mendes and Maurício Roberto Cherubin

CONTRIBUTORS

CHAPTER 1

Fabiane Machado Vezzani
Federal University of Rio Grande do Sul, Department of Soils, Porto Alegre - Rio Grande do Sul, Brazil

Ibanor Anghinoni
Federal University of Rio Grande do Sul, Department of Soils, Porto Alegre - Rio Grande do Sul, Brazil
Rio Grandense Rice Institute, Porto Alegre - Rio Grande do Sul, Brazil

Maurício Roberto Cherubin
"Luiz de Queiroz" College of Agriculture / University of São Paulo (ESALQ/USP), Piracicaba - São Paulo, Brazil

Ieda Carvalho Mendes
EMBRAPA Cerrados, Planaltina – Distrito Federal, Brazil

CHAPTER 2

Robélio Leandro Marchão
EMBRAPA Cerrados, Planaltina – Distrito Federal, Brazil

Ieda Carvalho Mendes
EMBRAPA Cerrados, Planaltina – Distrito Federal, Brazil

Lourival Vilela
EMBRAPA Cerrados, Planaltina – Distrito Federal, Brazil

Roberto Guimarães Júnior
EMBRAPA Cerrados, Planaltina – Distrito Federal, Brazil

Cíntia Carla Niva
EMBRAPA Cerrados, Planaltina – Distrito Federal, Brazil

Karina Pulrolnik
EMBRAPA Cerrados, Planaltina – Distrito Federal, Brazil

Kleberson Worsley Souza
EMBRAPA Cerrados, Planaltina – Distrito Federal, Brazil

Arminda Moreira de Carvalho
EMBRAPA Cerrados, Planaltina – Distrito Federal, Brazil

CHAPTER 3

João Carlos de Moraes Sá
Center for Carbon Management and Sequestration, School of Environment and Natural Resources, The Ohio State University, Columbus - Ohio, USA

Telmo Jorge Carneiro Amado
Federal University of Santa Maria, Santa Maria - Rio Grande do Sul, Brazil

Ademir de Oliveira Ferreira
Federal Rural University of Pernambuco, Recife - Pernambuco, Brazil

Rattan Lal
Center for Carbon Management and Sequestration, School of Environment and Natural Resources, The Ohio State University, Columbus - Ohio, USA

CHAPTER 4

Victória Santos Souza
"Luiz de Queiroz" College of Agriculture / University of São Paulo (ESALQ/USP), Piracicaba - São Paulo, Brazil

Beatriz da Silva Vanolli
"Luiz de Queiroz" College of Agriculture / University of São Paulo (ESALQ/USP), Piracicaba - São Paulo, Brazil

Bruna Emanuele Schiebelbein
"Luiz de Queiroz" College of Agriculture / University of São Paulo (ESALQ/USP), Piracicaba - São Paulo, Brazil

Larissa de Sousa Bortolo
"Luiz de Queiroz" College of Agriculture / University of São Paulo (ESALQ/USP), Piracicaba - São Paulo, Brazil

Martha Lustosa Carvalho
"Luiz de Queiroz" College of Agriculture / University of São Paulo (ESALQ/USP), Piracicaba - São Paulo, Brazil

Ieda Carvalho Mendes
EMBRAPA Cerrados, Planaltina - Distrito Federal, Brazil

Maurício Roberto Cherubin
"Luiz de Queiroz" College of Agriculture / University of São Paulo (ESALQ/USP), Piracicaba - São Paulo, Brazil

CHAPTER 5

Maurício Roberto Cherubin
"Luiz de Queiroz" College of Agriculture / University of São Paulo (ESALQ/USP), Piracicaba - São Paulo, Brazil

Felipe Bonini da Luz
"Luiz de Queiroz" College of Agriculture / University of São Paulo (ESALQ/USP), Piracicaba - São Paulo, Brazil
Federal University of Santa Maria, Frederico Westphalen - Rio Grande do Sul, Brazil

Renato Paiva de Lima
"Luiz de Queiroz" College of Agriculture / University of São Paulo (ESALQ/USP), Piracicaba - São Paulo, Brazil

Sarah Tenelli
Brazilian Biorenewables National Laboratory / Brazilian Center for Research in Energy and

Materials (LNBR/CNPEM), Campinas - São Paulo, Brazil.
Ricardo Oliveira Bordonal
Brazilian Biorenewables National Laboratory / Brazilian Center for Research in Energy and Materials (LNBR/CNPEM), Campinas - São Paulo, Brazil.
Bruna Gonçalves de Oliveira
"Luiz de Queiroz" College of Agriculture / University of São Paulo (ESALQ/USP), Piracicaba - São Paulo, Brazil
Leandro Carolino Gonzaga
Brazilian Biorenewables National Laboratory / Brazilian Center for Research in Energy and Materials (LNBR/CNPEM), Campinas - São Paulo, Brazil.
Carlos Eduardo Pellegrino Cerri
"Luiz de Queiroz" College of Agriculture / University of São Paulo (ESALQ/USP), Piracicaba - São Paulo, Brazil
João Luís Nunes Carvalho
Brazilian Biorenewables National Laboratory / Brazilian Center for Research in Energy and Materials (LNBR/CNPEM), Campinas - São Paulo, Brazil.

CHAPTER 6
Carla Eloize Carducci
Federal University of Grande Dourados, Dourados – Mato Grosso do Sul, Brazil
Geraldo César de Oliveira
Federal University of Lavras, Lavras – Minas Gerais, Brazil
Samara Martins Barbosa
Federal University of Lavras, Lavras – Minas Gerais, Brazil
Yuri Lopes Zinn
Federal University of Lavras, Lavras – Minas Gerais, Brazil
Daiane Pereira de Souza
Federal University of Grande Dourados, Dourados - Mato Grosso do Sul, Brazil
Clandio Favarini Ruviaro
Federal University of Grande Dourados, Dourados - Mato Grosso do Sul, Brazil
Joyce Cristina Costa
Agricultural Research Center of Piumhi, Piumhi - Minas Gerais, Brazil
Eduardo Costa Severiano
Goiás Federal Institute, Rio Verde – Goiás, Brazil

CHAPTER 7
Jucinei José Comin
Federal University of Santa Catarina, Florianópolis - Santa Catarina, Brazil
Fabiane Machado Vezzani
Federal University of Rio Grande do Sul, Porto Alegre - Rio Grande do Sul, Brazil
Monique Souza
Federal University of Santa Catarina, Florianópolis - Santa Catarina, Brazil
Claudinei Kurtz
Agricultural Research and Rural Extension Company of Santa Catarina (EPAGRI), Ituporanga – Santa Catarina, Brazil
Álvaro Luiz Mafra
Santa Catarina State University, Lages - Santa Catarina, Brazil
Paulo Emilio Lovato
Federal University of Santa Catarina, Florianópolis - Santa Catarina, Brazil
Cledimar Rogério Lourenzi
Federal University of Santa Catarina, Florianópolis - Santa Catarina, Brazil
Arcângelo Loss
Federal University of Santa Catarina, Florianópolis - Santa Catarina, Brazil

CHAPTER 8
Falberni de Souza Costa
EMBRAPA Acre, Rio Branco - Acre, Brazil
Claudenor Pinho de Sá
EMBRAPA Acre, Rio Branco - Acre, Brazil
Deborah Pinheiro Dick
Federal University of Rio Grande do Sul, Porto Alegre - Rio Grande do Sul, Brazil
Ieda Carvalho Mendes
EMBRAPA Cerrados, Planaltina – Distrito Federal, Brazil

CHAPTER 9
Elke Jurandy Bran Nogueira Cardoso
"Luiz de Queiroz" College of Agriculture / University of São Paulo (ESALQ/USP), Piracicaba - São Paulo, Brazil
José Leonardo de Moraes Gonçalves
"Luiz de Queiroz" College of Agriculture / University of São Paulo (ESALQ/USP), Piracicaba - São Paulo, Brazil
Victor Lucas Vieira Prudêncio de Araújo
"Luiz de Queiroz" College of Agriculture / University of São Paulo (ESALQ/USP), Piracicaba - São Paulo, Brazil
Antônio Marcos Miranda Silva
"Luiz de Queiroz" College of Agriculture / University of São Paulo (ESALQ/USP), Piracicaba - São Paulo, Brazil
Ademir Sérgio Ferreira de Araújo
Federal University of Piauí, Teresina - Piauí, Brazil

Arthur Prudêncio de Araújo Pereira
Federal University of Ceará, Fortaleza - Ceará, Brazil

CHAPTER 10
Ieda Carvalho Mendes
EMBRAPA Cerrados, Planaltina - Distrito Federal, Brazil
Guilherme Montandon Chaer
EMBRAPA Agrobiologia, Seropédica - Rio de Janeiro, Brazil
Fábio Bueno dos Reis Junior
EMBRAPA Cerrados, Planaltina - Distrito Federal, Brazil
Ozanival Dario Dantas
EMBRAPA Cerrados, Planaltina - Distrito Federal, Brazil
Juaci Vitoria Malaquias
EMBRAPA Cerrados, Planaltina - Distrito Federal, Brazil
Maria Inês Lopes de Oliveira
EMBRAPA Cerrados, Planaltina - Distrito Federal, Brazil
Marco Antônio Nogueira
EMBRAPA Soja, Londrina - Paraná, Brazil
Mariangela Hungria
EMBRAPA Soja, Londrina - Paraná, Brazil

CHAPTER 11
Rodrigo Estevam Munhoz de Almeida
EMBRAPA Pesca e Aquicultura, Palmas - Tocantis, Brazil
Henrique Antunes de Souza
EMBRAPA Meio-Norte, Teresina - Piauí, Brazil
Balbino Antonio Evangelista
EMBRAPA Pesca e Aquicultura, Palmas - Tocantis, Brazil
Alexandre Uhlmann
EMBRAPA Pesca e Aquicultura, Palmas - Tocantis, Brazil
Michele Ribeiro Ramos
State University of Tocantins, Palmas - Tocantis, Brazil
Edvaldo Sagrilo
EMBRAPA Meio-Norte, Teresina - Piauí, Brazil
Tais Souza dos Santos Dias
EMBRAPA Pesca e Aquicultura, Palmas - Tocantis, Brazil
Laura Resplandes de Sousa Paz Oliveira
EMBRAPA Pesca e Aquicultura, Palmas - Tocantis, Brazil
Nídia Raquel Costa
Agri Support, Botucatu - São Paulo, Brazil

CHAPTER 12
Carlos Eduardo Pellegrino Cerri
"Luiz de Queiroz" College of Agriculture / University of São Paulo (ESALQ/USP), Piracicaba - São Paulo, Brazil
Francisco Fujita de Castro Mello
Center for Knowledge Management and Horizontal Cooperation, Directorate of Technical Cooperation, Inter-American Institute for Cooperation on Agriculture (IICA), San Jose - San Jose, Costa Rica
Natália Braga Renteria
MOMBAK, São Paulo - São Paulo, Brazil
Maurício Roberto Cherubin
"Luiz de Queiroz" College of Agriculture / University of São Paulo (ESALQ/USP), Piracicaba - São Paulo, Brazil

---

EDITORIAL CORRESPONDENCE

American Society of Agronomy
Crop Science Society of America
Soil Science Society of America
5585 Guilford Road, Madison, WI 53711-58011, USA

Soil Health Series: Volume 3 Soil Health and Sustainable Agriculture in Brazil

Edited by

Ieda Carvalho Mendes and Maurício Roberto Cherubin

SSSA
Soil Science Society of America

WILEY

Editorial Correspondence:
Soil Science Society of America, Inc.
5585 Guilford Road, Madison, WI 53711-58011, USA
soils.org

Registered Offices:
John Wiley & Sons, Inc., 111 River Street, Hoboken, NJ 07030, USA

For details of our global editorial offices, customer services, and more information about Wiley products, visit us at www.wiley.com.

Wiley also publishes its books in a variety of electronic formats and by print-on-demand. Some content that appears in standard print versions of this book may not be available in other formats.

Library of Congress Cataloging-in-Publication Data Applied for

Paperback: 9780891187431

Cover Design: Wiley
Cover Image: Courtesy of Fabiano Bastos, Maurício Cherubin, Lourival Vilela and Julio Salton

Set in 9.5/12.5pt STIXTwoText by Straive, Chennai, India

Contents

Our Journey Continues

Douglas L. Karlen, Series Editor

The publication of *Soil Health in Brazil* by two outstanding Brazilian leaders for the soil health transition of the culture of agriculture brings me great joy. First as the third volume in the SSSA's (Soil Science Society of America) Soil Health series, and because Drs. Mendes and Cherubin have assembled an outstanding collection of chapters focused on multiple aspects of Brazilian agriculture. My pride stems not only from the fact that I see the next generation of soil and ecological scientists and engineers coming together to focus on sustaining our fragile soil resources, but also knowing that in less than 55 years, Brazil has been transformed from a food insecure country into one of the most important food producers and exporters in the world. Contributors to this volume recognized the advancements and transitions in agriculture led by Dr. Landers and others within my generation but were not afraid to say "we can do even better" by focusing on soil health.

Development and adoption of conservation tillage systems, an improved understanding of phosphorus chemistry in Oxisols, vast improvements in plant genetics and tolerance to various abiotic stresses, better nitrogen management, and unparalleled advancements in weed, insect, and disease control chemicals provided the backbone for productivity increases that transformed Brazil from being food dependent to being a global leader in the provision of soybean and other crops to people in countries around the world. Those were outstanding soil physical and chemical advancements, but this new generation now asks, "what about soil biology?" Thus, their focus on soil health which seeks to identify and improve the collective soil biological, chemical, and physical properties, processes, and their interactions to create an even more productive, economically efficient, and ecologically sustainable culture for 21st century agriculture.

This book is also a very important step for the SSSA because it is the first internationally focused volume for the Soil Health series. It not only establishes the role of soil health in modern Brazilian agriculture, but also links the concept to the tremendous advancements that have been made in no-tillage cropping systems and integrated (row crops, forages, and animals) agriculture within the country. Contributors then focus specifically on how soil health is useful for guiding vegetable, sugarcane, coffee, forestry, agroforestry, and family farming systems in Brazil. Finally, the editors and their team members look to the future through tools such as SoilBio, challenges in the new Brazilian agriculture, frontiers, and the public policies and initiatives that will use principles of soil health to promote carbon sequestration and help mitigate environmental and ecological effects of changing global weather patterns.

Having concluded Volumes 1 and 2 with "An Ode to Soil Health", I accepted the Editor's request to present a short sequel for which I have added the wisdom of Dr. Landers and Plato to emphasize the never-ending importance of soil health to the survival and prosperity of humanity.

An Ode to Brazilian Soil Health

Soil resources of Brazil are a tremendous wealth,
When there is a focus on soil health!

Some nay-sayers may ask why soil is important to us,
Isn't it dirt, about which only a scientist would fuss?

Past civilizations and history can advise,
Taking care of the soil is very wise.

As Plato asked more than 2,500 years ago,
Where did the forests and rich, water-filled soils go?
Like the skeleton of a sick man,
From the bare land, the rainfall and nutrients ran.
So, instead of running streams, springs, and glorious cedar trees,
His once abundant pastures now provide few plants for goats and bees.

Around the world, humankind has broken down the soil,
Using ploughs and heavy tractors, to lessen human physical toil.

The soil now says, I'll do my part,
If you care for me with science and a wise heart.

it's the fungi and bacteria that we can't see,
SoilBio elegantly shows their importance to thee.

Responding to temperature and drought,
Carbon sequestration these organisms bring about.

Although to the end of this ode, we have come,

Your journey through Volume 3 has just begun!

Foreword

In April 2022, we were thrilled to receive an invitation from Soil Science Society of America (SSSA) to show the world how Brazilian farmers and soil scientists are approaching soil health. As a result, the third volume of the Soil Health Book Series published by ASA/CSSA/SSSA is about soil health in Brazil.

In the next decades, soil scientists and agronomists from all over the world will be increasingly pressured to resolve critical issues associated with climate change and food security. By 2050, the world population is expected to reach 10 billion. This means an increase of 2 billion people within the next two decades, an amount that took almost two millennia for humankind to reach. To produce healthy food sustainably, in a sufficient amount to feed the world population, under limited resources and inputs, and despite worsening weather extremes and climate change is a tremendous challenge. Global agriculture will have to produce more with less.

In this context, Brazil stands out as one of the main players within the agricultural sector. Science-based research has enabled Brazil to evolve from a food-insecure country in early 1970s to one of the most important food producers and exporters in the world. Over the last decades, significant yield increases have been reached, establishing new records of production almost every year. Brazil is the largest country in terms of arable land, a top-five producer of 34 agricultural commodities, and the largest agricultural net exporter (Valdes, 2022).

The numbers are impressive: From 1975 to 2023, grain production increased from 38.1 million tons to 315.8 million tons (equivalent to an increase of 8.28 times), whereas in the same period the cultivated area doubled from 38 to 78.1 million ha (Conab, 2023). Crop yield increases achieved over the years have supported a land-saving effect of 274 million ha (i.e., equivalent to 11.4 times the São Paulo state area).

Meat production followed the same path, increasing from 2.9 million tons to 29 million tons from 1975 to 2020, an increase of 10 times (Aragão & Contini, 2022). The forest sector increased its productivity by more than 150%, with emphasis on *Eucalyptus* and *Pinus* species. Coffee has increased productivity by more than four times in the last 25 years (Embrapa 2020). Milk production increased significantly, from just over 14 billion liters in the early 1990s to almost 35 billion liters in 2019 (IBGE, 2020). The progress of Brazilian agriculture made it possible to regularly supply the domestic market, with about a 40% drop in the cost of the basic food basket (December 2019 compared with December 1975 in the city of São Paulo) (DIEESE, 2020), and boosted exports, rising from US$20.6 billion in 2000 to US$159 billion in 2022 (Ministerio da Agricultura e Pecuaria, 2023).

The main drivers of this profound transformation in Brazilian agriculture include the hard work and resilience of the farmers; tremendous advances in terms of research, development, and innovation; favorable environmental conditions; land availability; and increased global demand for food and animal feed, particularly over the last decade. Most importantly, Brazil's ability to harvest two to three crops a year in the same plot of land makes it unique compared with other grain- and soybean-producing countries. To put these facts into context, the Cerrado region was the first place in the world where large-scale agriculture was developed in acidic soils with low natural fertility in a region with a 6-month rainy season.

The evolution of Brazilian agriculture is closely linked with the adoption of best management practices that improve soil health (SH) and consequently crop growth and yield. In the 1970s, the No Tillage System (NTS) started in Paraná state, southern Brazil. Since then, it has spread across the country as the most important conservation practice, covering about 36.8 million ha. In 2022, Brazil celebrated 50 years since the adoption of the NTS and recently instituted the No Tillage National Day (October 23). The large-scale adoption of the NTS was an important milestone to soil and crop management, and since 2000 a new breakthrough encompassing the integration of crop, livestock, and forestry systems has become a viable option adopted successfully on 17.5 million ha throughout the country (https://redeilpf.org.br/ilpf-em-numeros/). Planting deep-rooted grasses into soybean/corn fields, either as a cover crop or pasture for cattle during the dry/winter season, increases plant residue input (i.e., carbon), provides soil protection during the dry season, and contributes to a more biologically active edaphic environment.

In this book, we present in 12 chapters, an overview of the major cropping systems and management practices that have been adopted in Brazil to improve SH and the sustainability of agricultural/forest production systems. The chapters also discuss the challenges to manage SH in the new agricultural frontiers and present the SoilBio Technology, a Brazilian pioneering initiative to evaluate and monitor SH at farm scale, based on the inclusion of soil enzymes β-glucosidase and arylsulfatase, as part of routine soil analyses. Finally, public policies and national initiatives to promote soil carbon sequestration and enhance SH are presented.

In a scenario with growing global demands for food, feed, fiber, and fuels, the intensification of Brazilian agriculture is inevitable. Soil and other natural resources will be exploited more and more. But, as the readers will be able to verify throughout this book, Brazil is working hard to be a SH ambassador worldwide through the massive adoption of sustainable agricultural practices and systems across the country. Undoubtedly, as Douglas Karlen has taught us over the past 30 years, this is a win-win situation for producers, society, and the planet!

We want to express our deepest gratitude to all the 68 authors, from 25 institutions, who joyfully embraced the challenge to show to an international audience "Soil Health in Brazil". Our special thanks go to the SSSA for this incredible opportunity and to Richard Easby for his support throughout this process.

Finally, we want to thank Douglas Karlen. Since the 1990s, his studies and passion for SH have inspired soil scientists all over the world. In 2015, we had the once-in-a-lifetime opportunity to interact with him and other US colleagues at the USDA-ARS National Laboratory for Agriculture and the Environment at Ames, IA. Maurício did part of his doctorate studies at the USDA, and Ieda spent a week visiting Doug. One year later, in 2016, we hosted Doug in Brazil when he presented the talk "Soil quality: Lessons learned and to be learned" at our Fertbio Meeting in Goiania, Goiás. Getting to know him personally provided us with the wonderful experience of finding not only a remarkable soil scientist but also an amazing human being. In his journey for SH, Doug has become a person specialized in building bridges among soil scientists worldwide. Not surprisingly, under his mentorship, SSSA invited us to edit this book. Like Doug, our wish is that as the importance of healthy soil environments becomes clear to producers, powerful transformations will take place, making reality his famous mantra: "Healthy soils, healthy landscapes, vibrant economies!!" And we agree with him, as he wrote at the end of *An Ode to Soil Health*: our quest for soil health has just begun!

Happy reading!
Sincerely,

Ieda Carvalho Mendes
Maurício Roberto Cherubin

References

Aragão, A. & Contini, E. (2022). *O agro no Brasil e no mundo um panorama de 2000 a 2021*. Embrapa. https://www.embrapa.br/documents/10180/62618376/O+AGRO+NO+BRASIL+E+NO+MUNDO.pdf/

Companhia Nacional de Abastecimento (Conab). (2023). *Produção agrícola—Safra: Série histórica dos grãos*. Conab. www.conab.gov.br/infoagro/safras/graos

Departamento Intersindical de Estatística e Estudos Socioeconômicos (DIEESE). (2020). *Cesta básica de alimentos: banco de dados*. https://www.dieese.org.br/cesta/

Embrapa. (2020). *VII Plano Diretor da Embrapa: 2020–2030*. Embrapa.

IBGE. (2020). *SIDRA: banco de tabelas estatísticas: Índice nacional de preços ao consumidos amplo – Setembro 2020*. https://sidra.ibge.gov.br/home/ipca/brasil

Ministerio da Agricultura e Pecuaria. (2023). *Exportações do agronegócio fecham 2022 com US$ 159 bilhões em vendas*. https://www.gov.br/agricultura/pt-br/assuntos/noticias/exportacoes-do-agronegocio-fecham-2022-com-us-159-bilhoes-em-vendas

Valdes, C. (2022). *Brazil's momentum as a global agricultural supplier faces headwinds*. USDA. https://www.ers.usda.gov/amber-waves/2022/september/brazil-s-momentum-as-a-global-agricultural-supplier-faces-headwinds/

Preface

John N. Landers[1]

"Soil health" (SH) is a popular term, turned scientific. It is so complex that it justifies a whole book on its multiple facets. This book demonstrates how Brazil leads the world in tropical and subtropical agriculture and in agricultural research and does justice to the world-shattering discoveries of Johanna Döbereiner (deceased) in demonstrating biological nitrogen fixation in Gramineae. Her professional example inspired the following generations of Brazilian agricultural scientists, much as the co-editors of this edition will inspire generations to come with the unveiling of key biological processes that are essential for a healthy, productive, and profitable soil. And, just as the zero-tillage revolution has inspired Brazilian farmers, researchers, and professionals in technical assistance to hone the technology to its various efficient present forms, approaching true sustainability, it goes without saying that the bedrock of all this is "plantio direto" (direct drilling, or zero tillage), now incorporated into conservation or regenerative agriculture and practiced by a band of organic pioneers. The label is unimportant; the regenerative effect on all soil processes is the reason why my poem "Throw your plough through the window" is now seen as ecological.

The perspicacious editors have assembled an impressive array of authors, all leaders in their field. They discourse on the SH context in Brazil, involving the umbrellas of SH and zero tillage for sustainability. Coupled with this are the important benefits of integrated crop–livestock and agro-forestry systems; how organic carbon levels are key to SH restoration; the multiple benefits of cover crops, especially when mixed; soil requirements for top yields of sugarcane and Arabica coffee (Cerrado region); SH in zero-tillage vegetable production; the special soil needs

[1] First zero tillage, 1976, in São Paulo; first zero tillage in the Cerrado, 1982; founder of the Farmers Association for Zero Tillage in the Cerrado 1992.

of forestry; the challenges in soil management in the new frontiers; and last, but not least, the overarching breakthrough of implanting soil enzyme bio-analysis (SoilBio) in 20 Brazilian commercial soil laboratories, a first in the world.

Finally, I congratulate all the authors for freely giving their time and knowledge to promulgate Brazil's leadership in the area of SH and in divulging such important technical information, principally for zero-tillage practitioners, be they farmers, private or public extensionists, at home or abroad, and in continuing to conquer new aspects of SH under zero tillage in conservation, regenerative, and organic agriculture.

1

Soil Health and Modern Brazilian Agriculture

Fabiane Machado Vezzani, Ibanor Anghinoni, Maurício Roberto Cherubin, and Ieda Carvalho Mendes

Chapter Overview

The relationship between the development of soil management practices and the promotion of soil health (SH) in agricultural systems of Brazil is addressed in this chapter. Agriculture in Brazil, which dates back 4000 years, underwent many changes during the 1960s. Intensive mechanization, amendments to correct soil pH (decrease acidity), increased use of fertilizers and agrochemicals, and government policies that facilitated and financed agriculture resulted in greater production of commodities (soybean [*Glycine max* (L.) Merr.], wheat [*Triticum aestivum* L.], cotton [*Gossypium hirsutum* L.], and corn [*Zea mays* L.]) in the southern states and then later in the midwestern region (Cerrado biome). In the 1970s and 1980s, several research programs proposed the adoption of conservation soil management aimed at reducing or eliminating soil tillage, planting of cover crops, rotating crops in both time and space, and appropriately managing residual plant biomass.

Development of specific agricultural management practices for complex production systems not only minimized soil erosion losses but also improved soil physical properties, enhanced fertilizer use efficiency, ameliorated acidity, enhanced residual biomass, and increased soil organic matter (SOM) accumulation. Collectively these changes stimulated biological activity and positively altered the soil structure and biota functioning. Consequently, nutrient cycling efficiency increased, and conditions for C sequestration were further improved. Simultaneously, the concept of soil quality (SQ) evolved into an improved knowledge about of SH, which embodies a more comprehensive understanding of the multifunctionality of soils and how they provide vital ecosystem services.

Soil Health Series: Volume 3 Soil Health and Sustainable Agriculture in Brazil, First Edition.
Edited by Ieda Carvalho Mendes and Maurício Roberto Cherubin.

Thus, the evolution of the concept of SQ to SH in Brazilian agriculture is closely associated with the development of agricultural management practices. This can be explained by the progressive complexity of soil structures, enabling self-organization from the simplest constitution under conventional agriculture to multiple, more complex, and biodiverse arrangements that ensure soil multifunctionality in integrated and biodiverse production systems.

Introduction

In Brazil, the concept of SH, defined as SQ in the early 1990s, was based on the perspective of an integrative assessment of soil biological, physical, and chemical properties and processes resulting from conservation practices in agricultural systems. At that time, SQ was understood as one of the three pillars, together with water quality and air quality, that constitute environmental quality, which in turn underlie agricultural sustainability. Thus, SQ is one of the fundamental concepts to guide sustainable management practices of agroecosystems, as proposed by the USDA Natural Resource Conservation Service (USDA-NRCS, 2022).

Between the 1990s and the first 10 years of the twenty-first century, there was a trend to exploit the concept SQ in studies of the soil science community of Brazil (Vezzani & Mielniczuk, 2009). Over time, efforts were focused on the search for the best SQ indicators or indices for the Brazilian edaphoclimatic conditions. A review of the papers published in the country showed that organic matter components were most frequently studied (Lopes et al., 2023). In an analysis of the studies conducted in Brazil between 2014 and the first half of 2021, Simon et al. (2022) found that more than 90% of the studies still use the terms "soil quality" or "soil health" but do not take the fundamental elements of integrative assessment and holistic interpretation embodied in the concepts into consideration.

In almost 30 years of SQ studies in Brazil, comprehension of the concept has been advanced and aligned with the international approach. Nowadays, we agree with Lehmann et al. (2020), Janzen et al. (2021), and Liptzin et al. (2022) in their understanding of the evolution of the term from "soil fertility" to "soil quality" and from there to "soil health." Soil fertility is the capacity of the soil to supply plants with nutrients in adequate amounts and proportions necessary for their development and to maintain the absence of toxic elements, especially of Al^{3+} (Bissani et al., 2004; Cantarutti et al., 2007; Lopes & Guilherme, 2007; Raij, 2011; Sousa & Lobato, 2004; Tisdale et al., 1985) to achieve high productivity. Soil fertility is associated with the soil function to support field crop production (Lehmann et al., 2020).

"Soil quality" refers to "the capacity of a specific kind of soil to function, within natural or managed ecosystem boundaries, to sustain plant and animal productivity, maintain or enhance water and air quality, and support human health and habitation" (Doran & Parkin, 1994; Karlen et al., 1997). It encompasses soil functioning related to the productivity of a plant species or community (Pankhurst et al., 1997) as well as water and air quality (Janzen et al., 2021) and reaches the level of the ecosystem and its boundaries (Lehmann et al., 2020). The scope of SH is an advance in that it aggregates the focus of soil as a living system. Most likely, Pankhurst et al. (1997) pioneered the understanding of a distinction between SQ and SH. Those authors analyzed "soil health," a concept coined by Doran and Safley (1997) as "the continued capacity of soil to function as a vital living system, within ecosystem and land-use boundaries, to sustain biological productivity, promote the quality of air and water environments, and maintain plant, animal, and human health." Based on this concept, Pankhurst et al. (1997) highlighted the inclusion of the time component contained in the expression "the continued capacity of," indicating the relevance of soil functioning over time, and the soil "as a vital living system," which relates soil functioning with the soil biota. Therefore, SH is a broader concept that recognizes the soil as an open and dynamic system in which processes are driven by life, reflected in biodiversity and in the respective functions this complex and diverse life exerts for the proper functioning of the biosphere.

The term "soil health" is becoming more popular, based on the comprehension that biological components mediate multiple processes that enable the soil to maintain full functionality in the long term (Karlen et al., 2021). From the biological perspective, soil is understood as a living and dynamic system (Pankhurst et al., 1997) and as such has to be fundamentally healthy (Liptzin et al., 2022) to promote well-being for all living beings on the planet. The path from soil fertility to SH is soil multifunctionality, where the perspective of soil functions is expanded beyond crop production to include other vital soil-related ecosystem services incorporating provision (e.g., food, feed, fiber, biofuel, water, biodiversity, raw materials), regulation (e.g., water fluxes, erosion control, climate changes), and support (e.g., biodiversity habitat, nutrient cycling) and cultural functions (e.g., aesthetic, recreation, cultural heritage, and education/research) (Guo, 2021; Janzen et al., 2021; Karlen et al., 2021; Lehmann et al., 2020; Bünemann et al., 2018). This requires more qualitative assessments and interpretations of the soil as a system and makes transdisciplinary approaches essential (Anghinoni & Vezzani, 2021; Janzen et al., 2021; Lehmann et al., 2020).

A conceptual transition of understanding about soil functioning is also taking place in Brazilian agriculture. For example, regarding revisions in the development of management practices for modern agriculture (i.e., 1960s onward) in Brazil, Anghinoni and Vezzani (2021) presented the concept of "systemic soil fertility" (SSF). This concept emphasizes that soil self-organizes into increasingly complex structures, depending on the amount of energy and organic matter deposited by plants and, along the way, awakens new system properties. Thus, SSF is the capacity of a soil to function properly, determined by the availability and stock of nutrients; fluxes of gases, water, and solutes; and biological diversity and activity. SSF is closely related to soil multifunctionality and the provision of critical ecosystem services.

The goal of this chapter is to describe the relationship between the evolution of soil management practices and promotion of SH in the agricultural systems of Brazil. To this end, we first look back on the history of Brazilian agriculture to then describe the soil conditions resulting from each management period and the SH status.

A Historical Overview of Brazilian Agriculture and Soil Management

Archeological data indicate the existence of agricultural activities in Brazil some 4000 years ago (Crestana & Sousa, 2008). At the time Brazil was discovered (1500 AC), agriculture was a common practice among native peoples. Numerous plant species were being cultivated, including cassava (*Manihot esculenta* Crantz), peanut (*Arachis hypogaea* L.), tobacco (*Nicotiana tabacum* L.), sweet potato [*Ipomoea batatas* (L.) Lam.], common bean (*Phaseolus vulgaris* L.), pumpkin (*Cucurbita* spp.), and corn; native fruits such as jaboticaba (*Myrciaria jaboticaba (Vell.) O. Berg)*, cashew (*Anacardium occidentale* L.), caja (*Spondias mombin* L.), and guava (*Psidium guajava* L.) were being harvested; and extracts from local flora such as babassu (*Ataleia* spp.) and pequi (*Caryocar brasiliense* Cambess.) were being used for food or other purposes according to local customs. This local culture influenced the food habits of newly arrived people, although the culture of the Portuguese colonizers was stronger and more dominant (Arruda, 1981).

After the arrival of the Royal Portuguese Court in Brazil (~1808), in a period called the Brazilian Empire, the most commonly grown crop was coffee (*Coffea arabica* L.), especially in the southeastern region. Other crops being cultivated during this period included sugarcane (*Saccharum officinarum* L.), tobacco, cotton (*Gossypium hirsutum* L.), and cocoa (*Theobroma cacao* L.) in the northeast and southeast and rubber [*Hevea brasiliensis* (Willd. ex A. Juss.) Müll. Arg.] in the Amazon region.

During the nineteenth century, especially in the second half, due to major socioeconomic transformations, large-scale migration from Europe began, especially into less populated continents such as the Americas and Australia. As a result of government policies, Brazil was one of the countries that received a considerable portion of these emigrants, mainly from Germany and Italy. The first immigrants were the Germans (as of 1825), who mostly occupied the lowlands of the valleys in the South, followed by Italian immigrants (as of 1875), mostly to work on sugarcane and coffee plantations in the Southeast. Newcomers from other countries arrived later and occupied the native forest areas in the South, where they developed subsistence farming systems with family labor in small areas. For food, they primarily cultivated wheat (*Triticum vulgare* Vill.), rice (*Oryza sativa* L.), beans, corn, potatoes, cassava, and vegetables and fruit species and raised various animals (dairy cattle, pigs, and poultry). As the number of families increased and land became scarce after the 1920s, descendants migrated to other areas of the southern region, always with a preference for native forests.

In the post-war period (after 1945), Brazilian agriculture was characterized by subsistence farming, and grain production was concentrated in the Southern region. Soybean cultivation was only beginning and destined for oil extraction. During that period, there was a shortage of basic foodstuffs, such as sugarcane, wheat, and common bean, which sparked a major debate about the backwardness of the agricultural sector as one of the obstacles to the development and industrialization of Brazil. Modern Brazilian agriculture was initiated in the 1960s, mostly by the industrialization policy, creating demands for food and other raw materials in the cities. The cost of labor in the countryside grew with rural-urban migration, forcing farmers to intensify production and mechanize crops. In the field of agricultural policy, three instruments played a role in that modernization: (a) subsidized credit, for the purchase of modern inputs and capital financing; (b) investments in science and technology; and (c) public and private rural extension. In addition to government policies, other factors that have contributed to the increase in the efficiency of agriculture include the abundant availability of production factors, such as mechanized land, modern inputs, and entrepreneurial people (Crestana & Sousa, 2008).

The first Agriculture Educational and Research Institutions were implemented in the country from the middle to the end of the nineteenth century. They were Agriculture School of Bahia in 1859 (Cruz das Almas, Bahia State), Agriculture Institute of Northern Rio de Janeiro in 1860 (Rio de Janeiro, Rio de Janeiro State), Agriculture School of Pelotas in 1883 (Pelotas, Rio Grande do Sul State), and Imperial Agronomy Station of Campinas (Campinas, São Paulo State), in 1887, later named Agronomic Institute of Campinas. However, the greatest scientific and technological progress occurred by implementing advanced research centers and creating graduate courses in agriculture sciences. They were a result of Federal Government negotiation under a strong influence of the United Nations, which culminated, in 1963, in an agreement between the Brazilian Ministry of Education and Culture and the United States Agency for International Development.

The main objectives were training of human resources and establishing local research in strategic areas (e.g., soil fertility and fertilizers for increased agricultural production) (Camargo, 2015; Lopes & Guilherme, 2007). Various national government programs were developed at the centers of grain production of the time. The philosophy of the programs was to achieve high yields by correcting soil acidity and fertilization with P and K (Freire et al., 2006). These programs resulted in large-scale agriculture on highly weathered, naturally acidic soils with low nutrient availability, which had been considered, until then, unsuitable for agriculture. Thus, great advances in agricultural production were achieved in Brazil, especially with regard to commodities (soybean, wheat, cotton, and corn), initially in the southern states. The driving forces of this agriculture transformation (i.e., The Green Revolution) were intensive mechanization, managing acidity, applying fertilizers and agrochemicals, and government policies that facilitated financing.

During the 1970s, the low land prices and economic incentives stimulated farmers from other regions of the country to migrate to the Cerrado region. The low soil fertility and the lack of sound technology to overcome this constraint, together with the missing agricultural infrastructure to provide for transportation, storage, agricultural inputs, and markets, were the primary factors for the extremely low agricultural activity in the Cerrado region prior to the 1970s (Mendes et al., 2018). Almost no crop could be grown for commercial purposes under the natural soil fertility conditions of the Cerrado biome. Therefore, specific programs, such as the regional development program of the Midwest of Brazil ("POLOCENTRO"), were implemented, along with financing of agricultural research and favorable public policies, which created the conditions for the movement called "march to the Cerrado." Investments in science and research—in particular the creation of the Brazilian Agricultural Research Corporation (Embrapa) in 1973—should also be mentioned as key triggers of the country's agricultural transformation (Crestana & Sousa, 2008).

Gradually, Brazil evolved from a food insecure country in the early 1970s into one of the most important food producers and exporters in the world. In the last 45 years, the area under annual crops increased 105%, while crop yields and grain production increased by 268% and 653%, respectively (Figure 1.1). In the same period, animal protein production also increased extraordinarily: 22-fold for poultry and cattle and fourfold for pork. These values are directly related to increased soybean and corn production and made Brazil the world's second largest producer of beef and chicken and the fourth largest producer of pork. Finally, the benefit that the growth of modern agriculture brought to the country is remarkable, by increasing the availability of food, especially of animal protein, contributing decisively to exports and, more recently, to renewable energy (Alves et al., 2008).

A detailed description of the events and influences that have historically shaped both Brazilian agriculture and agricultural research efforts and of the relationships of the agrarian with other sectors of the Brazilian society was given by Camargo et al. (2017).

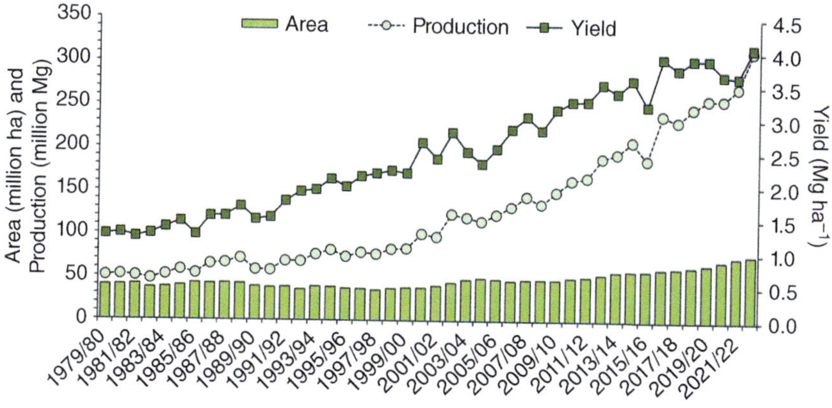

Figure 1.1 Development of cultivated area, production, and grain yield in Brazil in the past 45 years. Source: Companhia Nacional de Abastecimento (2022).

Nevertheless, the large-scale adoption of conventional agriculture with intensive soil disturbance (tillage) and monocultures or crop succession systems caused an accelerated physical degradation of Brazilian soils (Figure 1.2a). Regardless of the great gains in productivity in response to inputs and the improvement in chemical properties, more aggressive soil management, coupled with intense rainfall and burning of crop residual biomass, the efficiency of the applied inputs was low. Furthermore, soil degradation became a critical issue, threatening the sustainability of Brazilian agriculture.

In the 1970s and 1980s, rapid and wide-spread degradation of the soil physical properties was intense in crop-producing regions, especially in commodity production areas, across the entire South and Midwest (Cerrado region), where rapid agricultural expansion was taking place. In view of this, from the 1970s onward,

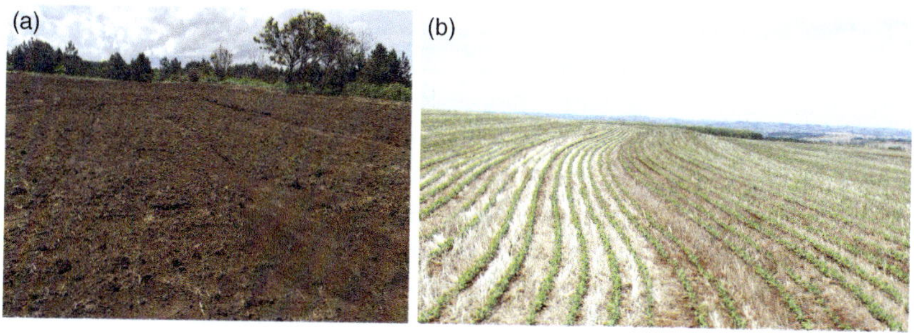

Figure 1.2 Conventional (a) and no-tillage systems (b) in the Southern region of Brazil, Pinhais, Paraná. Photos by Karina M.V. Cavalieri-Polizeli (a) and Jeferson Dieckow (b), Federal University of Paraná.

several research programs promoted by individual initiatives or groups of producers began to call for the adoption of soil conservation management. At the time, this management consisted of light harrowing, surface subsoiling, or chiseling practices, which were designated "minimum tillage." This management pursued a reduction or the elimination of soil tillage and introduced cover crops, crop rotation in time and space, and an appropriate management of residual plant biomass. These changes had positive effects in minimizing soil erosion losses, improving the soil physical properties, and increasing the efficiency of fertilizers and other additives.

Within a short time, minimum tillage was replaced by no-tillage (NT). After the 1990s, the agricultural areas under NT (Figure 1.2b) only increased significantly in the Southern states. As of the first decade of this millennium, NT was adopted in wider areas in all Brazilian states, especially those in the Cerrado region, reaching an estimated area of 33 million ha in 2017 (Fuentes-Llanillo et al., 2021). In Brazil, there is a difference between NT as a soil preparation type and NT systems (NTSs) as a management type. According to the concepts, NT is characterized by sowing a crop with minimal soil movement, restricted to the sowing row, whereas for NTSs, the soil is not only undisturbed but is also managed by "an ordered complex of interrelated and interdependent agricultural practices that must occur together; it also includes crop rotation and use of cover crops or forage grasses to form and maintain a crop residue layer on the soil surface" (Muzzili, 2000). If any of these requirements is not met, the field is considered to be only in NT and not a NTS.

The adoption of NTSs led to an accumulation of residual biomass with a subsequent increase in the soil organic matter (SOM) content in the agricultural areas of Brazil, which stimulated biological activity and altered the structure and functioning of the soil biota. Consequently, there was an increase in soil aggregation, a decrease in the phytotoxic effect of (Al^{3+}), a greater ion retention capacity, improved nutrient cycling efficiency (favoring nutrient concentration and dynamics), as well as better conditions for C sequestration. Unfortunately, a lack of diversification of agricultural systems still remains. Moreover, with the advent of precision agriculture in the early 2000s (Cherubin et al., 2022) and the preference for early sowing, basic practices of soil erosion control seem to have been forgotten because terraces were indiscriminately removed, contour sowing was abandoned, and runoff control was ignored. In addition, the formation of a compact layer in the subsurface has resulted in accelerated erosion and soil degradation, similar to the conditions in the period of conventional soil tillage (in the 1970s and 1980s), which preceded and motivated the shift to the NTS. The consolidation of NTS across the country encouraged researchers and technical assistants to recommend options of plant species to increase crop diversity (Figure 1.3a) and integrated crop–livestock (ICL) systems (Figure 1.3b) to capitalize on their biological benefits in relation to soil functioning.

Figure 1.3 Crop diversity (a) and integrated crop–livestock–forestry (ICLF) systems (b) under conservation management as factors of sustainable production in modern Brazilian agriculture. Photos by Fabiano Bastos (Embrapa Cerrados).

Biological properties, such as microbial biomass and basal respiration, and biochemical indicators, such as enzymatic activity, are increasingly valued in evaluations of agriculture soil conditions. In view of the relevance of biological activity for soil functioning under NTS, after 20 years of studies in long-term field experiments with NTS, Embrapa Cerrados has launched the Soil Bioanalysis Technology (BioAS in Portuguese, SoilBio in English) for Brazilian farmers. Briefly, SoilBio is based on the additional assessment of the soil enzymes arylsulfatase (ARYL) and β-glucosidase (GLU) in routine soil analysis, along with the calculation of soil quality indices. Using the SoilBio technology, SQ is evaluated by combining chemical and biological indicators (Mendes, Chaer et al., 2021; Mendes, Sousa et al., 2021) in a framework of three soil functions: (F1) nutrient cycling (based on ARYL and GLU activities), (F2) nutrient storage (based on SOM and cation exchange capacity), and (F3) nutrient supply (based on available Ca, Mg, K, and P and acidity indicators such as pH, H+Al, Al^{3+} and sum of bases and base saturation). The launch of SoilBio in July 2020 represented a milestone for SQ evaluation of agricultural soils in Brazil. Currently, the interpretative algorithms for SoilBio are calibrated for annual crops in soils of the Cerrado and the South (Paraná State). Ongoing research is focused on developing interpretative algorithms for other crops (e.g., sugarcane, pasture, coffee, and eucalyptus) and for other regions of the country. By combining soil enzymes and organic C, it also possible to evaluate long-term C changes in the soil (0- to 10-cm layer), using a four-quadrant model, representing high- and low-quality soils, soils undergoing biological degradation, and regenerating soils (Chaer et al., 2023). A detailed description of SoilBio can be found in Chapter 10 of this book.

In the 2020s, integrated systems of agricultural and livestock production, better known in Brazil as ICL and integrated crop–livestock–forestry (ICLF) systems, were once again used on farms (Figure 1.3). For large-scale diversified production systems, ICL and ICLF have been recognized as best options due to the associations between crop plants, animals, and the arboreal component in the same area, in a concomitant or sequential manner (see Chapter 4).

In Brazil, ICLF consists of four pillars: conservation agriculture associated with no-tillage and high biomass input, best management practices, efficient input utilization, and use of pastures at moderate grazing intensities (Anghinoni et al., 2013; Carvalho et al., 2018). The energy and nutrient fluxes and interactions that occur in natural systems are recreated in ICLF by combining herbaceous plants, shrubs, and trees with herbivore production. Diversification is ensured by agricultural rotations, interspersed with pasture/grazing periods. This synergistic arrangement stimulates the multifunctionality of the soil system, increasing the ecological processes of biological activity, decomposition of plant residues and animal excreta, soil aggregate formation and stabilization, nutrient cycling, nutrient supply to plants and organisms, self-regulation of pests and diseases, and adequate water and gas movement within the soil.

Successional or biodiverse agroforestry systems (SAFSs) constitute another type of diversified production system that has expanded in Brazil in the last decades, especially in family-based agriculture. These agroforestry systems (Figure 1.4) are combinations of tree elements with herbaceous plants and/or animals, organized in space and/or time, characterized by a high species diversity and the

Figure 1.4 Successional or biodiverse agroforestry systems (SAFSs) in modern Brazilian agriculture: combinations of the arboreal element with herbaceous plants and/or animals, organized in space and/or time, characterized by high species diversity and vertical occupation of different strata of solar radiation incidence, according to the ecological succession principles. Southern region of Brazil, Paraná. Photos by Fabiane Machado Vezzani (Federal University of Paraná).

vertical occupation of different strata of solar radiation incidence, considering the ecological succession principles. The diversified plant biomass deposited on the soil surface by pruning in the SAFSs catalyzes the ecological processes that ensure self-regulation and have a positive effect on the soil biological, physical, and chemical properties.

Integrated and biodiverse systems have greatly expanded in the agricultural areas of Brazil. More complex agricultural practices constitute recurrent themes in both research and technical assistance, with the goal to improve SH in agricultural ecosystems.

The Evolution Timeline for Brazilian Agriculture to Soil Health

The development of agricultural management practices in Brazil can be explained by the progressive complexity of soil structures, enabling self-organization from the simplest constitution under conventional agriculture, represented by the soil fertility concept, to the multiple, complex, and biodiverse arrangement in conservation agriculture, as defined for SSF and SH. The pathway to SH in agricultural soils of Brazil is presented in Figure 1.5.

Until the 1950s, large-scale crops (coffee, sugarcane, rubber, cocoa, etc.) were produced in the northern states of the country, whereas grain production was

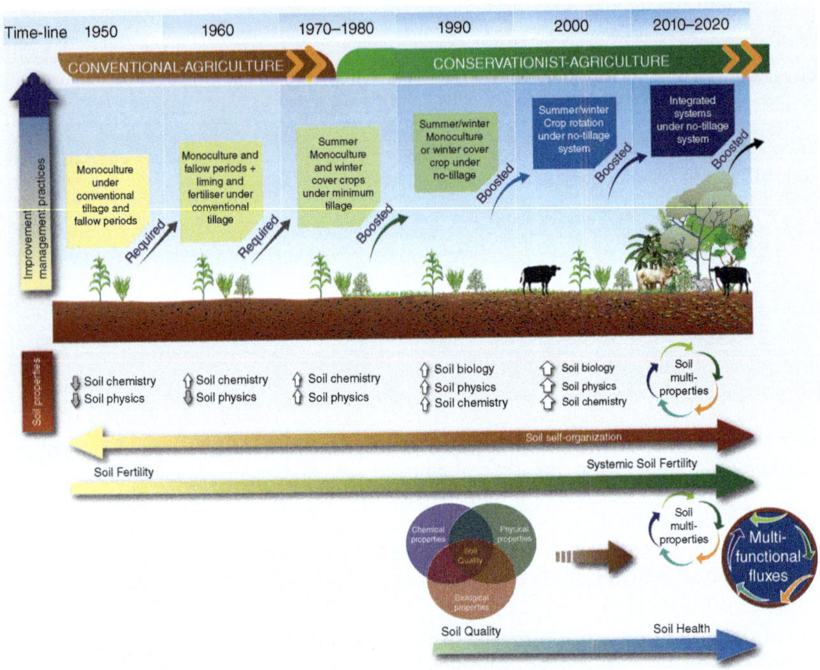

Figure 1.5 Development of management practices in Brazilian agriculture from the 1960s to the present and their relations with soil properties and the concepts of soil fertility, soil quality, systemic soil fertility, and soil health. Illustration by Wellington Cavalcanti (Embrapa Cerrados).

concentrated in southern Brazil and characterized by subsistence agriculture. During the 1950s, agricultural mechanization was introduced in that region, which increased the acreage of maize in the summer and wheat in the winter. The crop cycles were interrupted by fallow periods of 2–3 years for the recovery of plant biomass and improvement of soil physical and chemical properties. Soybean cultivation for oil production was still incipient. Acidity correctives and fertilizers were applied without research-based criteria.

From the mid-1960s onward, conventionally tilled monoculture with fallow periods or winter crops (wheat, barley, and oat) was intensified. These management practices degraded the soil chemical and physical properties (Figure 1.5, 1960s). Recommendation systems for liming and fertilization were developed by government research institutes. The concept of soil fertility best represents this period of Brazilian agriculture because it addresses acidity neutralization and a sufficient and balanced nutrient supply for plant development, which should result in increased crop productivity. In that period, tillage for crop cultivation, together with operations for soil acidity correction, caused a continuous decline in soil physical properties and the loss of organic matter, nutrients, and minerals through erosion. In addition to the harm caused by physical degradation, the agricultural areas needed continuous applications of acidity neutralizers and fertilizers to maintain crop productivity.

From 1970 to 1990, several non-governmental institutions, particularly agricultural cooperatives in the Southern region, started to develop a technical assistance network to promote winter cover crops and tillage reduction among farmers to recover soil physical properties. The combination of research and technical assistance triggered a massive shift from conventional to minimum tillage and cover crop cultivation in the winter (Figure 1.5, 1970–1980s). The increased amounts of crop residues had a positive effect on soil physical and chemical properties and motivated the change from minimum tillage to NT. No-tillage monoculture had a positive influence on the soil biological, physical, and chemical properties, which are improvements inherent to the SQ concept (Figure 1.5, 1990s). However, repeated planting of the same crop species in monoculture favors the occurrence and multiplication of weeds and soil-borne diseases.

The multiple disadvantages of plant homogeneity challenged soil science researchers to develop crop rotation options for the Brazilian agricultural systems. The reduction of tillage to minimum soil disturbance, along with crop diversification and permanent soil cover by crop residues or live mulches, became pillars of NTS management and favored the appearance of new relations between the system components (Figure 1.5, 1990–2000s). Some of these relationships resulted in an increased nutrient storage capacity and the creation of an edaphic environment more beneficial for soil life. Consequently, biological and biochemical properties emerged, resulting in new soil processes and enhanced soil and ecosystem functions, raising the conditions of agricultural soil to SH (Figure 1.5, 2000s).

The next step (from 2010 onward) was to integrate diversified annual cropping systems with livestock and forestry (ICLF) production under NT, on a large scale, and SAFSs in family-based farming. At this level of agricultural management, the greater number of productive components in the same space and time has caused a response of soil self-organization to a higher state of structural complexity, with an increased number of integrated and interconnected soil niches (Figure 1.5, 2010–2020s). In this scenario, relationships among the system components are abundant and tend toward the full functioning of the process, generating the emergence of other new properties and functions in multifunctional fluxes such as water and gas fluxes, mutualistic biological relationships, and synchrony between nutrient release and plant uptake. This condition is typical of SFF (Anghinoni & Vezzani, 2021), and SH and is characterized by multiple relationships between the components of a complex system generated by continuous fluxes of energy and organic matter driven by plants. It reflects in the capacity of the proper functioning of the soil; in the availability and stock of nutrients; in the fluxes of gases, water, and solutes; and in the biological diversity and activity.

Along this path, conventional agriculture crossed over to conservation agriculture. The greater addition of energy and organic matter in the form of plant residues, as a result of the development of agricultural practices, promotes soil self-organization into increasingly complex structures (Figure 1.5). In each state of soil organization in conservation agriculture, new properties emerge, enabling the soil to perform more complex and robust functions. In integrated and biodiverse systems, soil multi-properties positively feed back into each other in a continuously ascending, self-organizing process, until a state of multifunctionality is reached (Anghinoni & Vezzani, 2021).

The evolution from the SQ to the SH concept in Brazilian agriculture is closely associated with the evolution of farm management practices. When summer and winter monocultures or a winter cover crop under NT were planted, the soil biological, physical, and chemical properties reached a new threshold of soil functioning, allowing the soil to perform its functions in the productive ecosystem, which exemplifies how the SQ concept outperforms that of soil fertility. Moreover, when integrated and biodiverse systems were adopted, multiple properties emerged as a reflection of the fluxes of energy and matter provided by the different productive components that cause multi-functional fluxes of energy and organic matter provided by the different productive components. Integrated and biodiverse systems ensure the multifunctionality and the sustainability of vital soil-related ecosystem services, as postulated in the SH concept.

Future Trends of Soil Health in Brazilian Agriculture

The greatest legacy of the SQ and SH concepts in Brazil was possibly the challenge for scientists and technicians to reflect about the impact of conservation management practices on the soil and other ecosystem compartments. In this context, qualitative evaluations of the impact of management practices beyond the borders of agroecosystems have yet to be developed.

Today, SH has become part of the vocabulary and mindset of technicians, soil scientists, agribusiness companies, and large-scale farmers. Many of them already understand that a management to establish and consolidate SH must be preferred over one of immediate lucrativeness because long-term profitability on a more constant basis can only be obtained through a process. Small farmers in Brazil have always known and talked about the health of the soil on which they grow their crops. Perhaps we need to draw on their qualitative view to better understand the relationship of soil with environmental quality.

Currently, large-scale Brazilian agriculture is moving toward the implementation of complex systems that integrate the elements cover crops, crop rotation, livestock, forestry, and orchards and favor a more process-based rather than input-oriented agriculture. However, public policies and sectoral programs focused on SH as center of the agenda are still necessary to expand the application of conservation (or even regenerative) agriculture across the country to make the Brazilian agriculture more productive, profitable, resilient, and sustainable in the coming decades.

Acknowledgments

The spiritual influence of Professor João Mielniczuk was constantly present during the preparation of this text, as he was a great scientist who made very important contributions to the study and development of conservation practices in Brazilian agricultural systems. Above all, he was an exceptional personality who deeply influenced us in the way we understand the relationships in the living world. Therefore, we are deeply grateful for his teaching and the wealth of learning opportunities he offered us. We also thank the reviewers who kindly contributed to improving the writing of the text.

References

Alves, E. R. A., Contini, E., & Gaspes, J. G. (2008). Evolução da produção e produtividade da agricultura brasileira. In A. C. S. Albuquerque & A. G. da Silva (Eds.), *Agricultura tropical: quatro décadas de inovações tecnológicas, institucionais e políticas* (pp. 67–99). Embrapa Informação Tecnológica.

Anghinoni, I., Carvalho, P. C. F., & Costa, S. E. V. G. A. (2013). Abordagem sistêmica do solo em sistemas integrados de produção agropecuária no subtrópico brasileiro. *Tópicos em Ciência do Solo*, 8, 325–338.

Anghinoni, I., & Vezzani, F. M. (2021). Systemic soil fertility as product of system self-organization resulting from management. *Revista Brasileira de Ciência do Solo*, 45, e0210090. https://doi.org/10.36783/18069657rbcs20210090

Arruda, J. J. (1981). *História moderna e contemporânea*. Editora Ática.

Bissani, C. A., Gianello, C., Tedesco, M. . J., & Camargo, F. A. O. (2004). *Fertilidade dos solos e manejo da adubação das culturas*. Gênesis.

Bünemann, E. K., Bongiorno, G., Bai, Z., Creamer, R. E., Deyn, G. D., Goedeb, R., Fleskens, L., Geissen, V., Kuyper, T. W., Mäder, P., Pulleman, M., Sukkel, W., van Groenigen, J. W., & Brussaard, L. . L. (2018). Soil quality – A critical review. *Soil Biology and Biochemistry*, 120, 105–125.

Camargo, F. A. O. (2015). *Solo fértil e bons frutos: 50 anos da Pós-graduação em Ciência do Solo no Brasil*. UFRGS.

Camargo, F. A. O., Silva, L. S., Merten, G. H., Carlos, F. S., Baveye, P. C., & Triplett, E. W. (2017). Brazilian agriculture in perspective: Great expectations vs reality. *Advances in Agronomy*, 141, 53–114.

Cantarutti, R. B., Barros, N. F., Martinez, H. E. P., & Novais, R. F. (2007). Avaliação da fertilidade de solo e recomendação de fertilizantes. In R. F. Novais, V. H. Alvarez, N. F. Barros, R. L. F. Fontes, R. B. Cantarutti, & J. C. L. Neves (Eds.), *Fertilidade do solo* (pp. 769–850). SBCS.

Carvalho, P. C. F., Nunes, O. A. A., & Anghinoni, I. (2018). O processo de pastejo como gerador de propriedades emergentes em sistemas integrados de produção agropecuária. In E. D. Souza, F. D. Silva, T. S. Assmann, M. A. C. Carneiro, P. C. F. Carvalho, & H. B. Paulino (Eds.), *Sistemas integrados de produção agropecuária no Brasil* (pp. 39–44). Copiart.

Chaer, G. M., Mendes, I. C., Dantas, O. D., Malaquias, J. V., Reis Junior, F. B., & Oliveira, M. I. L. (2023). Evaluating C trends in clayey Cerrado Oxisols using a four-quadrant model based on specific arylsulfatase and β-glucosidase activities. *Applied Soil Ecology*, 183, 104742. https://doi.org/10.1016/j.apsoil.2022.104742

Cherubin, M. R., Damian, J. M., Tavares, T. R., Trevisan, R. G., Colaço, A. F., Eitelwein, M. T., Martello, M., Inamasu, R. Y., Pias, O. H. C., & Molin, J. P. (2022). Precision agriculture in Brazil: The trajectory of 25 years of scientific research. *Agriculture*, 12, 1882. https://doi.org/10.3390/agriculture12111882

Companhia Nacional de Abastecimento. (2022). *Produção Agrícola—Safra: Série histórica dos grãos*. Conab. www.conab.gov.br/infoagro/safras/graos

Crestana, S., & Sousa, I. S. F. (2008). Agricultura tropical no Brasil. In A. C. S. Albuquerque & A. G. da Silva (Eds.), *Agricultura tropical: Quatro décadas de inovações tecnológicas, institucionais e políticas* (pp. 41–65). Embrapa Informação Tecnológica.

Doran, J. W., & Parkin, T. B. (1994). Defining and assessing soil quality. In J. W. Doran, D. C. Coleman, D. F. Bezdicek, & B. A. Stewart (Eds.), *Defining soil quality for a sustainable environment* (pp. 1–20). SSSA.

Doran, J. W., & Safley, M. (1997). Defining and assessing soil health and sustainable productivity. In C. Pankhurst, B. M. Doube, & V. V. S. R. Gupta (Eds.), *Biological indicators of soil health* (pp. 1–28). Cab International.

Freire, J. R. J., Costa, J. A., & Stammel, J. G. (2006). Principais fatores que proporcionaram a expansão da soja no Brasil. *Revista Plantio Direto*, 92, 2006.

Fuentes-Llanillo, R. L., Telles, T., Soares Júnior, D., Melo, T., Friedrich, T., & Kassam, A. (2021). Expansion of no-tillage practice in conservation agriculture in Brazil. *Soil Tillage Research*, 208, 104877. https://doi.org/10.1016/j.still.2020.104877

Guo, M. (2021). Soil health assessment and management: Recent development in science and practices. *Soil Systems*, 5, 61. https://doi.org/10.3390/soilsystems5040061

Janzen, H. H., Janzen, D. W., & Gregorich, E. G. (2021). The 'soil health' metaphor: Illuminating or illusory? *Soil Biology and Biochemistry*, 159, 108167. https://doi.org/10.1016/j.soilbio.2021.108167

Karlen, D. L., De, M., McDaniel, M. D., & Stott, D. E. (2021). Evolution of the soil health movement. In L. K. Douglas, D. E. Stott, & M. M. Mikha (Eds.), *Approaches to soil health analysis* (Vol. 1, pp. 21–48). SSSA.

Karlen, D. L., Mausbach, M. J., Doran, J. W., Cline, R. G., Harris, R. F., & Schuman, G. E. (1997). Soil quality: A concept, definition, and framework for evaluation (a guest editorial). *Soil Science Society of America Journal*, 61, 4–10. https://doi.org/10.2136/sssaj1997.03615995006100010001x

Lehmann, J., Bossio, D. A., Kögel-Knabner, I., & Rillig, M. C. (2020). The concept and future prospects of soil health. *Nature Reviews Earth & Environment*, 1(10), 544–553. https://doi.org/10.1038/s43017-020-0080-8

Liptzin, D., Norris, C. E., Cappellazzi, S. B., Mac Bean, G., Cope, M., Greub, K. L. H., Rieke, E. L., Tracy, P. W., Aberle, E., Ashworth, A., Tavarez, O. B., Bary, A. I., Baumhardt, R. L., Gracia, A. B., Brainard, D. C., Brennan, J. R., Reyes, D. B., Bruhjell, D., Carlyle, C. N., … Honeycutt, C. W. (2022). An evaluation of carbon indicators of soil health in long-term agricultural experiments. *Soil Biology and Biochemistry*, 172, 108708. https://doi.org/10.1016/j.soilbio.2022.108708

Lopes, A. S., & Guilherme, I. R. G. (2007). Fertilidade do solo e produtividade agrícola. In R. F. Novais, V. H. Alvarez, N. F. Barros, R. L. F. Fontes, R. B. Cantarutti, & J. C. L. Neves (Eds.), *Fertilidade do solo* (pp. 1–64). Sociedade Brasileira de Ciência do Solo.

Lopes, R. D., Vezzani, F. M., & Paraguaio, E. V. (2023). Abordagem da Qualidade do Solo nos trabalhos publicados no Brasil. *Revista Brasileira de Meio Ambiente*, 11(1), 87–105.

Mendes, I. C., Chaer, G. M., Reis Junior, F. B., Sousa, D. M. G., Dantas, O. D., Oliveira, M. I. L., & Malaquias, J. V. (2021). *Tecnologia BioAS: Uma maneira simples e eficiente de avaliar a saúde do solo (Série Documentos, 369)*. Embrapa Cerrados.

Mendes, I. C., Sousa, D. M. G., Dantas, O. D., Lopes, A. A. C., Reis Junior, B., & Oliveira, M. I. D. (2021). Soil quality and grain yield: A win–win combination in clayey tropical oxisols. *Geoderma*, 388, 114880. https://doi.org/10.1016/j.geoderma.2020.114880

Mendes, I. C., Tormena, C. A., Cherubin, M. R., & Karlen, D. L. (2018). Soil health assessment and maintenance in Central and South-Central Brazil. In D. Reicosky (Ed.), *Managing soil health for sustainable agriculture: Monitoring and management* (pp. 379–415). Burleigh Dodds Science publishing. https://doi.org/10.19103/AS.2017.0033.35

Muzzili, O. (2000). *A fertilidade do solo no sistema plantio direto* [Abstract presentation]. Anais Simpósio sobre Fertilizantes de Solo e Nutrição de Plantas no Sistema Plantio Direto, Ponta Grossa, Paraná, Brasil.

Pankhurst, C. E., Doube, B. M., & Gupta, V. V. S. R. (1997). Biological indicators of soil health: Synthesis. In C. E. Pankhurst, B. M. Doube, & V. V. S. R. Gupta (Eds.), *Biological indicators of soil health* (pp. 419–435). Cab International.

Raij, B. (2011). *Fertilidade do solo e adubação*. International Plant Nutrition Institute.

Simon, C. D. P., Gomes, T. F., Pessoa, T. N., Soltangheisi, A., Bieluczyk, W., Camargo, P. B. D., & Cherubin, M. R. (2022). Literatura sobre qualidade do solo no Brasil: uma revisão sistemática. *Revista Brasileira de Ciência do Solo*, 46, e0210103. https://doi.org/10.36783/18069657rbcs20210103

Sousa, D. M. G., & Lobato, E. (2004). *Cerrado: correção do solo e adubação* (2nd ed.). Embrapa Cerrados.

Tisdale, S. L., Nelson, W. L., & Beaton, J. D. (1985). *Soil fertility and fertilizers* (4th ed.). Macmillan.

USDA Natural Resource Conservation Service(USDA-NRCS). (2022). *Soil quality for environmental health*. USDA. http://soilquality.org/basics/sustainable.html

Vezzani, F. M., & Mielniczuk, J. (2009). Uma visão sobre qualidade do solo. *Revista Brasileira de Ciência do Solo*, 33, 743–755. https://doi.org/10.1590/S0100-06832009000400001

2

Integrated Crop–Livestock–Forestry Systems for Improved Soil Health, Environmental Benefits, and Sustainable Production

Robélio Leandro Marchão, Ieda Carvalho Mendes, Lourival Vilela, Roberto Guimarães Júnior, Cíntia Carla Niva, Karina Pulrolnik, Kleberson Worsley Souza, and Arminda Moreira de Carvalho

Chapter Overview

In many parts of the world, the intensification of sustainable production on current agricultural lands has been proposed as a solution to ensure food security and environmental sustainability, reducing the conflict between expanding agricultural production and the conservation of natural ecosystems.

In Brazil, integrated crop–livestock–forestry (ICLF) production systems, in various modalities and combinations, are considered a key strategy for a sustainable agricultural development. This chapter discusses how the implementation of ICLF represents a paradigm shift and a breakthrough in tropical agriculture. The introduction of ICLF systems has a strong impact on soil health (SH) by improving soil physical quality, stimulating biological activity and C sequestration, reducing greenhouse gas (GHG) emissions, intensifying nutrient cycling, and enhancing nutrient use efficiency (NUE). The influence of integrated systems on SH translates into higher crop yields and better animal performance and welfare. In addition, ICLF systems have a "land-saving effect," in that degraded pastures are improved and wood production increased, which helps reduce the pressure on native forest areas.

Introduction

Sustainable intensification is a possible answer to one of the greatest challenges for global agriculture in that the production in existing areas must be raised while

Soil Health Series: Volume 3 Soil Health and Sustainable Agriculture in Brazil, First Edition.
Edited by Ieda Carvalho Mendes and Maurício Roberto Cherubin.

pressure on the environment should be reduced (Vilela et al., 2011). Brazil is the largest beef exporter and the second-largest beef producer in the world (Zia et al., 2019). In 2020, an estimated 162 million ha were used as pastures in Brazil, more than half (55.38%) of which is degraded to some degree (https://atlasdaspastagens.ufg.br). Pasture degradation leads to soil erosion and productivity losses and, consequently, affects animal performance (Macedo, 2009). In this context, ICLF systems, in various possible modalities or combinations, are seen as a key strategy to recovery degraded pastures considering mainly economic aspects (Cordeiro et al., 2015; Martha Jr. et al., 2011). Brazil is the only country in the world that can increase the cultivated area by 25 million ha by recovering degraded pasture areas, without felling a single tree (Vilela et al. 2012).

The strategy termed "integrated cultivation" combines different production systems—agricultural, livestock, and forestry—within the same area, in intercropping, crop rotation, or crop sequences, to exploit the synergy among the cited components. These agroecosystems can be integrated in all types of farming systems and in different combinations, such as crop–livestock (CL; agropastoral/mixed systems), crop–forestry (CF; agroforestry systems), livestock–forestry (LF; silvopastoral systems), and crop–livestock–forestry (CLF; agrosilvopastoral systems) (Figure 2.1). The synergy among the components pasture, forest, crop, and livestock is responsible for many of the benefits of ICLF.

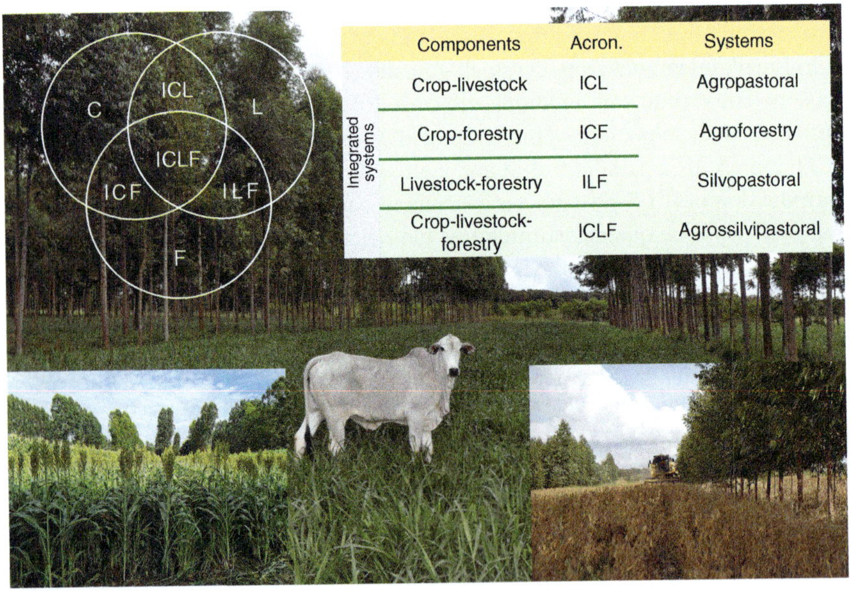

Figure 2.1 Integrated agricultural systems, in different combinations, can be implemented in all types of farming systems. Photos by Fabiano Bastos and Kleberson Souza.

The development of agroecosystem management in tropical environments is a continuous learning process. For example, in the Cerrado biome—a tropical savanna ecoregion of Brazil and the most important agricultural region of the country—the adoption of no-tillage (NT) in the early 1990s was followed by the breeding of short-cycle soybean [*Glycine max* (L.) Merr.] cultivars, which allowed for rainfed double cropping with maize (*Zea mays* L.). Until the early 2000s, these forms of crop rotation or sequence were used for grain production in the Cerrado, though still characterized by low straw yield, sparse mulch cover, and the absence of cover crops that would ensure soil protection in the dry season. For this reason, the three principles of conservation agriculture (minimum soil disturbance [i.e., tillage restricted to the planting row], permanent soil cover by crop residues or living mulch, and crop rotation/intercropping) could not be easily applied, which impaired the implementation of NT mulch-based cropping systems (Mendes et al., 2018). However, since 2000, a paradigm shift occurred in relation to the adoption of ICLF that initiated a breakthrough in tropical agriculture. The inclusion of "soil-building-plants," such as tropical deep-rooted forage grasses (mainly *Urochloa* spp. Syn. *Brachiaria* spp. and *Panicum* spp. Syn. *Megathyrsus maximus*) after intercropping with maize or sorghum [*Sorghum bicolor* (L.) Moench] or even with soybean oversowing filled this gap by providing a living soil cover or forage for livestock grazing during the dry season (winter). Figure 2.2 shows the potential yields of an integrated system with three harvests: soybean in summer, maize in winter (a typical pattern of double cropping in Brazil), and

Figure 2.2 Potential yields of an annual crop sequence in an integrated system, with three harvests in the same growing season (year): soybean in summer, maize intercropped with grasses in the winter, and off-season livestock grazing in the last months of that growing season. S, September. F, February; J, June/July; M, March; N, November. Photos by Lourival Vilela and Robélio Marchão.

dual-purpose forage, for off-season grazing and/or as cover crop in NT systems, ensuring permanent soil cover.

As will be shown in this chapter, the adoption of ICLF systems in tropical regions can raise the grain yield of cash crops; increase animal performance; improve nutrient cycling, NUE, and soil physical quality; intensify biological activity; and mitigate GHG emissions by C sequestration (Ayarza et al. 2022; Nicoloso & Rice, 2021). Data from integrated cultivation throughout Brazil allow the conclusion that these systems constitute an effective solution to improve soil chemical, physical, and biological qualities. These effects are the result of increases in organic matter content and of an improved soil structure. The synergies observed between crops and pastures in these mixed systems are responsible for gains in grain and meat production. The list of potential benefits resulting from the adoption of integrated systems in tropical regions is long (Figure 2.3).

With the inclusion of the forest (tree) component in integrated systems (livestock–forestry [ILF], crop–forestry [ICF], and crop– livestock–forestry [ICLF]), the agricultural component is usually restricted to the early implementation phase, when competition for light, water, and nutrients is still low. The livestock component is introduced in the inter-row space when the trees provide significant shading and is maintained together with the forest component until the final stage of the system (Balbino et al., 2012).

The worldwide consumption of primary processed wood products is expected to rise by 37% by 2050 (FAO, 2022). In this scenario, part of the demand of the Brazilian domestic market can be met by growing trees on low-productivity pastures in the Cerrado (more than 30 million ha). The afforestation of degraded

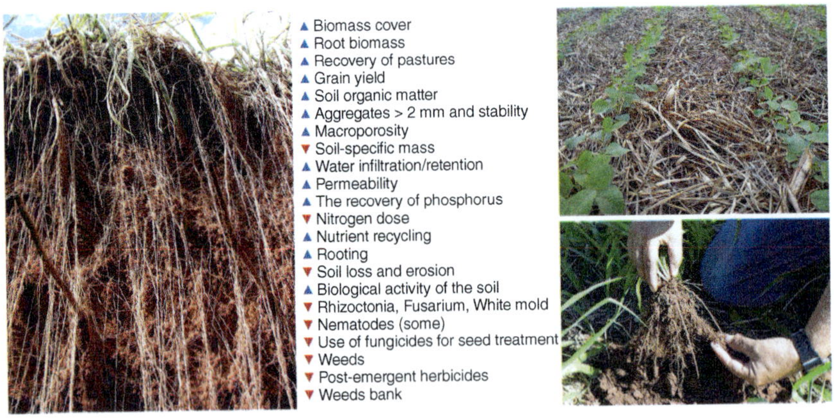

Figure 2.3 Synthesis of potential benefits from integrated cultivation systems in tropical regions. Blue triangle: increases; red triangles: reductions. Photos by Lourival Vilela, Fábio Ono and Robélio Marchão.

pastures, integrated with grain crops and the livestock component, is a possibility to reduce the production unit costs of soil fertility correction and tree planting (Vilela et al., 2012).

Integrated systems in Brazil cover a total area of 17.4 Mha (redeILPF, 2023a). Most crop farmers and ranchers have adopted the CL system (83%), followed by CLF (9%) and LF (7%). In the group of crop farmers, 99% have implemented CL, 0.4% CLF, and 0.2% CF (Skorupa & Manzatto, 2019). In this chapter, we will discuss how the progress toward sustainable intensification is profoundly changing Brazilian agriculture, not only in terms of greater crop yields and better animal performance, but also with significant improvements in SH.

Background, Conceptualization, and Combinations of Integrated Systems in Brazil

The concept of integrated systems is not new and was already mentioned in the ley farming systems of the twentieth century, which contributed to the efforts of food production in World War II. At that time, low-yielding pastures were converted into cropland by alternating a short temporary pasture with a crop phase (Stapledon, 1941). Although ley farming has been used to increase agricultural production in Europe since the seventeenth century, the possibilities of intensive fertilization and mechanization reduced the need for integration (Lemaire et al., 2014). According to the definition of André Voisin (1957), "Ley farming is the intentional introduction of an herbage crop of varying duration into a crop rotation, which should be grazed as much as possible." The reason for the last statement is that cutting and carrying forage eliminates the nutrient recycling effect, by which nutrients are returned as dung and urine to the herbage crop when animals are present. At that time, the term "herbage crop" included grasses and other forage species. In this chapter, we assume that the concept of "ley farming" was the basis for modern integrated systems and will discuss in the following how integrated systems in Brazil have been adapted in the context of modern tropical agriculture.

Currently, crop–livestock integration in Brazil is based on intercropping an annual crop with a forage species (i.e., simultaneous cultivation until the end of the annual crop cycle). Thereafter, the land use changes from agricultural to pastoral as the forage species becomes predominant (Figure 2.4). In the different modalities of integrated systems, the following planting/sowing methods are used: (a) undersowing—pasture or companion cover crop sown in the rows of the annual crop, concomitantly with a cash crop; (b) intersowing—sowing in-between crop rows, either with an additional seed hopper or an extra pass of the crop planter; (c) oversowing—cover crops and grasses, especially seeds

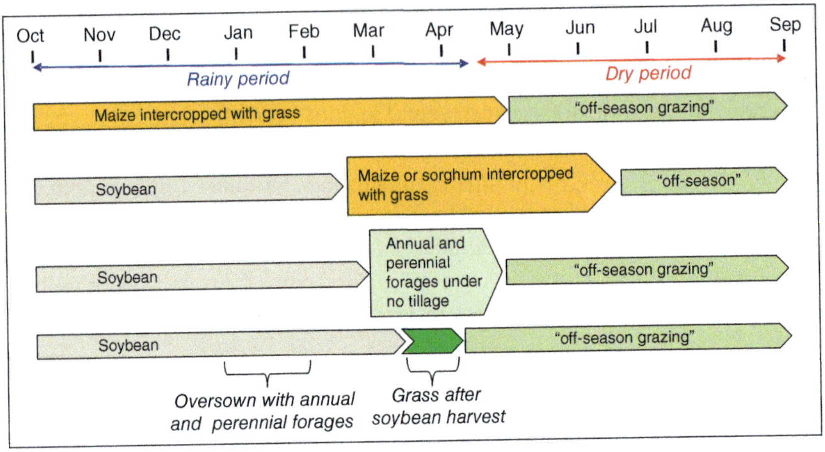

Figure 2.4 Examples of successional agricultural systems of integrated crop–livestock farming in the Cerrado region. Adapted from Vilela et al. (2018).

of *Panicum* spp., which are oversown on soybean immediately before leaf drop, either by a pendulum spreader (better distribution than a centrifugal fertilizer) or by an agricultural spray airplane. Although this system saves time, it is technically more demanding than planting directly in the soil.

Integrated systems in Brazil began with cattle grazing the stubble of irrigated rice crops, a model resulting from the influence of European immigrants in the southern region in the 1970s. In central Brazil, the "Barreirão system," a set of technologies and practices for degraded pasture recovery based on rice–forage grass intercropping, was proposed in 1991 (Kluthcouski et al., 1991). The expansion of NT management and the advent of appropriate machinery and herbicides allowed grass desiccation with subsequent soybean sowing, resulting in the development of a modality of an integrated system based on pasture–crop rotation. A common problem of forage–grain intercropping is the competition for water and nutrients. Losses in crop yield may occur, and pasture establishment may fail. Late sowing of the pasture component as well as low herbicide rates and low plant density have been identified as possibilities to minimize the competition between forage and grain crops (Kluthcouski et al., 2003).

The "Santa Fé" system (Kluthcouski et al., 2003) has been widely adopted in Brazil. It is appropriate for soils with adequate fertility levels, owing to long-term fertilization practices, in which maize is usually undersown with a forage grass species (e.g., *Brachiaria* spp. cultivars).

Throughout Brazil, integrated systems are tailored according to the profile and objectives of each farm. The different systems reflect regional and local peculiarities of farms (e.g., soil and climate conditions, infrastructure, the farmer's experience, and technology available). In the Cerrado region, for example, three modalities of integration systems are commonly used: (a) livestock farms in which grain crops (rice, maize, sorghum, soybean) are planted in pasture areas to recover pasture productivity; (b) grain-producing farms where forage grasses are planted to improve soil cover for the NT system, and, in the off-season, cattle can feed on this forage (off season grazing or "boi safrinha") (Figure 2.4); and (c) farms that systematically adopt pasture (for livestock grazing and as alternative species in the crop sequence) and crop rotation to intensify land use and benefit from the synergy between the two activities (Vilela et al., 2011). Figure 2.5 illustrates another example of crop–livestock integration: pasture implemented after soybean, with a short grazing period (three months in the dry season), in comparison with a soybean–maize double cropping system. Forage succession is an advantageous strategy of pasture diversification to feed cattle in low-rainfall periods (off-season grazing) as well as for biomass generation for a next NT crop (Dias et al., 2020; Muniz et al. 2021; Naudin et al., 2015).

For the success of the integrated system, the forages with best adaptation to the soil-climatic conditions should be chosen based on an optimized relation between forage quality and yield. In Brazil, the most widely used forage grasses in such systems are currently species of the genus *Brachiaria* spp. (Baptistella et al., 2020) and *Panicum* spp., mainly in view of the high drought tolerance, relatively low seed costs, adaptation to climate and low-fertility soils, and large annual biomass production rate. Compared with other cover and intercrop species, such

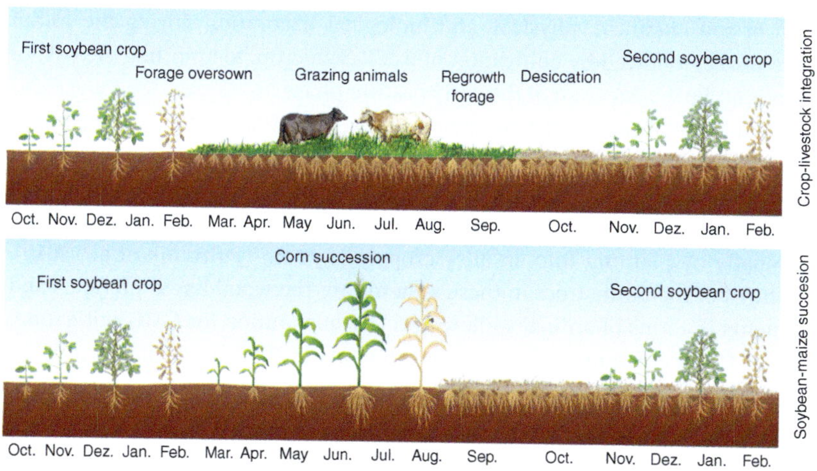

Figure 2.5 Comparative diagram of two cropping systems: an example of crop–livestock integration, of pasture implemented after soybean, with a short grazing period (three months in the dry season) and a soybean–maize double cropping system. Art by Wellington Cavalcanti; adapted from Muniz et al. (2021).

as crotalarias (*Crotalaria* spp.) and millet (*Pennisetum* spp.), *Brachiaria* spp. have a particularly good performance. On the other hand, *Panicum* spp. are the most productive seed-propagated tropical forage, of which nearly all species have a tufted growth habit (Jank et al., 2011).

Integrated crop and pasture systems have mostly been used to increase animal productivity by restoring pasture yields. One of their main advantages, regardless of the combination, is the economic aspect related to soil fertility recovery. Due to the residual fertilization effect of the previous annual crop, no additional fertilizer is needed to establish pasture in an integrated system. Moreover, the income from grain or silage sale (mainly maize and sorghum), partially or totally offsets the costs of pasture establishment (Yokoyama et al., 1999). Within 60 days after grain harvest, the pasture can already be used for animal feeding (in the same crop season).

A set of different integrated systems with regional focus has been developed, such as the systems involving leguminous species, such as pigeonpea [*Cajanus cajan* (L.) Millsp.] and cowpea [*Vigna unguiculata* (L.) Walp.], mainly to produce protein-rich forage and to increase biological N fixation, with a view to reducing mineral N fertilizer input in the successional crop (Oliveira et al., 2010). Specific integrated systems were developed for sandy or medium-textured soils. For example, the Embrapa system "São Mateus" (Salton et al., 2013), for degraded (tropical) pastures, is based on soil amendments and tillage operations to improve soil structure associated with a temporary grass crop (five to nine months), followed by soybean. This system is based on the pasture–crop system with anticipation of soil chemical, physical, and biological restoration during the initial pasture phase, followed by cultivation of a NT cash crop, to amortize the cost of fertilizers and pH correction of the early pasture phase.

Land use intensification by integrated systems, including the forest component, can provide positive results as long as the tree component is properly implemented (Pulrolnik et al., 2019). The inclusion of the forest component is an advancement in crop–livestock integration and is carried out in the initial phase of the system, usually by planting into an alley cropping system (Pulrolnik et al., 2019). Advantages of including trees in these systems are the wood-based products and the benefits in terms of animal welfare and C sequestration for GHG mitigation, mainly of CH_4 emissions from cattle (Souza et al., 2020). The flexibility of ICLF systems allows adaptations to local environmental, social, and economic conditions, with a prospect of sustainable and eco-efficient agricultural production. So far, *Eucalyptus* species (*E. cloeziana* and *E. urograndis*) are the most widely grown hardwood tree species in ICLF systems. Some reasons for the widespread use of *Eucalyptus* are the fast growth, good market price, low production costs, seedling availability, and tree architecture. Although *Eucalyptus* is an exotic forestry species in Brazil, the history of research and use for pulp and paper production traces about 60 years back.

The ICLF systems have been adopted to different degrees in Brazilian biomes in response to manifold opportunities and market demands. Aspects such as the recovery and afforestation of degraded pastures, landscape diversification, and the use of Brazilian native tree species in these systems are key to achieve sustainable agriculture at farm and regional scales. Not only is the pressure on native forests reduced, but tree cultivation in integrated systems also represents a saving in the medium to long term by supplying wood, cellulose, or charcoal, aside from other non-wood purposes, such as fruit, resin, and seed production. In summary, ICLF systems promote animal welfare and constitute an important strategy to offset GHG emissions.

In 2010, ICLF systems were included in the Brazilian Governmental ABC Plan for Low Carbon Emission Agriculture (Agricultura de Baixa Emissão de Carbono, in Portuguese) as one of the strategies to mitigate or offset GHG emissions from agriculture. In the context of the ABC plan, Embrapa developed new concepts to certify beef produced in systems integrating forestry (i.e. silvopastoral [livestock-forestry] or agrosilvopastoral [ICF] systems, respectively). Specific protocols that make the certification process possible were established to define the new parameterized and controllable concepts "Carbon Neutral Beef" and "Low Carbon Beef." The main goal of these concept brands was to reduce enteric CH_4 emissions from livestock by offering less stressful conditions in the shaded system. The certification ensures that animals are produced with welfare in mind (Alves et al., 2017).

SH as a Result of Complex Interactions in Integrated Systems

Healthy soils are biologically active and productive and are capable of storing water, sequestering C, and promoting nutrient cycling and pesticide degradation, among other important environmental services (Mendes et al., 2021b). Because ICLF systems use agricultural management approaches that mimic natural ecosystems most closely, they favor SH by positively influencing soil chemical, physical, and biological properties (Soares et al., 2019). The presence of animals and trees, along with different deep-rooted forage grasses and cash crops, raises the quantity and quality of residues entering the system, with significant impact on SH. As observed by Vezzani and Mielniczuk (2011), under integrated systems, fluxes of energy and organic matter (soil organic matter [SOM]) are high and decomposition rates are slowed, favoring a more complex structuring that resulted

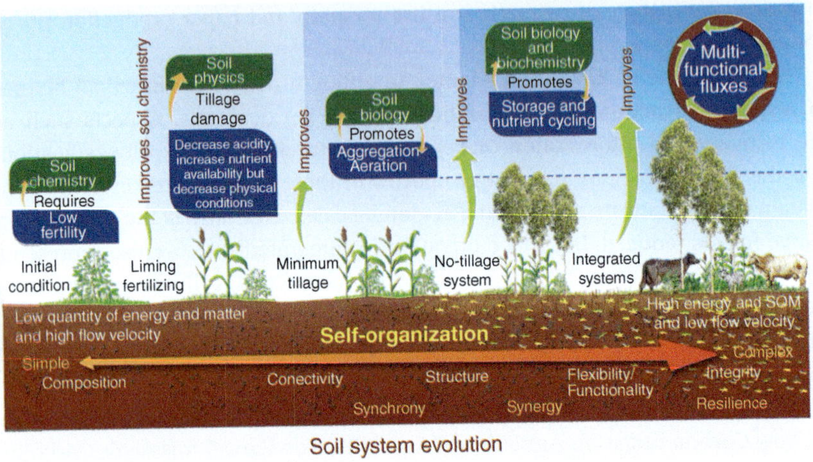

Figure 2.6 Self-organization of the soil system in the context of Systemic Soil Fertility concept in a schematic representation of the steps of sustainable intensification from conventional monocropping to integrated crop-livestock-forestry system under no-tillage. Art by Wellington Cavalcanti; adapted from Anghinoni and Vezzani (2021).

in energy and SOM retention and, consequently, in numerous nonlinear relationships. The establishment of such multifunctional flows improves the nutrient availability and stocks; fluxes of gases, water, and solutes; as well as biological diversity and activity, which together constitute Systemic Soil Fertility/Soil Health (Figure 2.6 and Chapter 1).

The complexity of interactions observed in integrated systems is uncommon in most traditional cash crop systems (Soares et al., 2019). The greater the number of components, the more numerous the connections among them. In this regard, the presence of grazing animals has a profound impact on SH in integrated systems. The essential difference between a crop–pasture intercropping under NT and an integrated system with livestock is that the latter favors more complex interactions in the soil–plant system due to residue incorporation by trampling, grazing, and the presence of animal excreta. Crop rotations will be most efficient when grazing livestock are integrated into the farming system. Livestock of all species feed on forage and/or crop residues and enhance nutrient cycling within the system (Franzluebbers, 2007). According to Bonaudo et al. (2014), livestock play the role of a catalyst by recycling and increasing the use efficiency of resources.

Grazing management is similar in integrated and non-integrated systems. However, a relevant difference is the residual effect of nutrient fertilization from the crop phase, making higher stocking rates in the pasture phase possible. The balance between compartments in integrated systems depends heavily on the grazing intensity (Carvalho et al., 2010; Soussana & Lemaire, 2014). Moderate grazing intensities favor the coupling of C and N cycles via urine and stimulate leaf area renewal. Increases in soil C stocks due to rhizodeposition have been related to grazing (Balesdent & Balabane, 1996; Carvalho et al., 2010). Among integrated systems, the technique of forage sequences is advantageous because it allows pasture establishment to feed cattle during the dry season (off-season grazing). Thereafter, biomass is formed for the following NT crop (Dias et al., 2020; Naudin et al., 2015). The conceptual diagram in Figure 2.6 explains the interrelationships among management factors in an integrated system.

SH in Integrated Systems: Effects on Nutrient Cycling and NUE

Nutrient cycling refers to the numerous transformations nutrients undergo in the soil–plant–atmosphere system. Different species have different abilities to accumulate nutrients in the biomass, which are later released into the soil through crop residue decomposition.

Brachiaria spp. in rotation with annual crops or in intercropping can accumulate between 2 and 16 Mg ha^{-1} of shoot dry weight per year, varying according to the species and management system (e.g., the fertilization regime and grazing frequency) (Bernardes et al., 2010; Dias et al., 2020; Macedo, 2009; Miguel et al., 2018). Although most studies focus on shoot dry weight production, the great potential of brachiaria as cover crop or in intercropping lies in the root system. Some reports in the literature stated that root dry weight of brachiaria can exceed 38 Mg ha^{-1} (Baptistella et al., 2020). Based on these values and the rapid root turnover of tropical grasses, the annual production of root dry weight could potentially maintain a positive C balance to reach steady-state C in the soil (Ayarza et al., 2022).

In a thorough evaluation of several mechanisms observed in *Brachiaria* spp., Baptistella et al. (2020) found an increase in N, P, and K acquisition. The authors analyzed the potential of tropical grasses in terms of nutrient cycling in more detail and reported that shoots of Brachiaria species can accumulate up to ~100 kg N ha^{-1}, 130 kg K ha^{-1}, 15 kg P ha^{-1}, 40 kg Ca ha^{-1}, and 25 kg Mg ha^{-1}. Depending on the residue quality, system management, and soil-climatic conditions, these accumulated nutrients can become available to plants by decomposition/mineralization processes or can be immobilized in the soil. The estimated decomposition and nutrient-release rates are relatively high (from three to six months) and depend on favorable rainfall and temperature conditions (Costa et al., 2014, 2016),

Equivalent amounts of fertilizer returned in the shoot biomass of *Brachiaria* spp. and *Panicum* spp. cultivars, used as mulch, were calculated by Dias et al. (2020) (Table 2.1). Significant amounts of N, P_2O_5, and K_2O were returned to the

Table 2.1 Total macronutrients in the biomass of an integrated crop–livestock system and a second-crop maize system in sequence with soybean in an Oxisol of the Cerrado region in Brazil.

Cropping systems	N	P_2O_5	K_2O
	Equivalent (kg ha^{-1})		
Winter maize	22d	19d	20c
Brachiaria brizantha 'Xaraés'	83a	60a	69a
Brachiaria ruziziensis	42c	26b	31bc
Panicum maximum 'Mombaça'	57b	41b	66a
Panicum maximum 'BRS Tamani'	70ab	35b	45b
Grass mean	63	41	52

Note. Means followed by different letters within a column differ from each other by Tukey's test at 5% probability. Adapted from Dias et al. (2020).

soil by these species, illustrating the advantages of recycling mulch in integrated cropping systems.

In addition to their great potential for nutrient cycling, the cultivation of deep-rooted forage species enhances the NUE and favors SOM accumulation and stabilization. Regarding NUE, as opposed to the more shallow and scarce roots of common crops, the vigorous, abundant, and deep root system of forage grasses (Figure 2.3) reduces nutrient losses and increases nutrient uptake by reaching nutrients beyond the rhizosphere of other crops. In an analysis of the increase of N-use efficiency in response to P fertilization of grass–legume pastures, Francisquini Junior et al. (2020) observed that the planting of legumes in *Panicum maximum* pastures may be an alternative to N fertilization when P is not applied to low-fertility soils. Broadcast and incorporated P fertilization by the time of *P. maximum* sowing not only increases forage yields but also results in higher NUE, regardless of P fertilizer solubility.

An example of the beneficial effects of the presence of brachiaria in ICL systems in terms of promoting increases in SOM and in P use efficiency is shown in Figure 2.7 and Table 2.2. These data were the result of a long-term field experiment (22 years) on a very clayey Yellow Latosol under conventional tillage at Embrapa Cerrados, conducted by Djalma Martinhão Gomes de Sousa (in memoriam). Response curves to P fertilization were evaluated in management systems with annual crops only (soybean 10 years, maize 2 years) and in an annual crop–brachiaria sequence (soybean 2 years, brachiaria pasture 9 years, soybean 2 years). Brachiaria (*Brachiaria humidicola*) was grown for cutting, without grazing (Sousa et al., 2016).

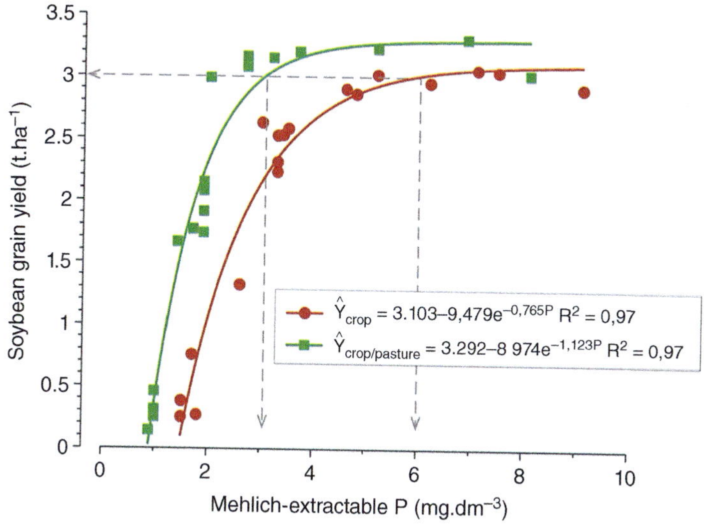

Figure 2.7 Relation between Mehlich-1 P (0- to 20-cm layer) and soybean yield in the 13th year, in cropping systems with (green) and without (red) brachiaria (Sousa et al., 1997, 2016).

Table 2.2 Soybean P use efficiency in the 13th crop season in a continuous crop system compared with crop–pasture rotation.

P fertilization[a)]		P use efficiency in the 13th crop season	
First year	Second year and thereafter	Crop[b)]	Crop–pasture[c)]
kg P_2O_5 ha^{-1}		kg grain kg^{-1} residual P_2O_5	
0	50	5.7	18.6
0	100	3.1	9.3
100	50	4.5	10.3
100	100	2.8	7.5
200	50	4.7	9.6
200	100	2.5	7.0
Mean		3.9	10.4

a) Broadcast in the first year and band-applied thereafter.
b) Soybean for 10 consecutive years, maize for 2 years, and soybean for 1 year, all under conventional tillage.
c) Soybean for two years, pasture (*Brachiaria humidicola*) for nine years, and soybean for two years. Soybean under conventional tillage.
Note. Adapted from Sousa et al. (2016).

After 13 years, the SOM contents (0- to 20-cm layer) were 28.4 and 37.3 g kg^{-1}, respectively, in the system with annual crops only and the annual crop–brachiaria sequence. In the 13th year, these differences influenced not only soybean grain yield (Figure 2.7) but also the use efficiency of residual P (Table 2.2). In the system with brachiaria (pasture), soybean produced higher yields than in the system of only annual crops, even on soils with similar extractable Mehlich-P levels. Because supply was balanced for all nutrients except P, this result indicates improved efficiency in P use after inclusion of brachiaria. For example, to produce 3.0 t ha^{-1} soybean in the system with annual crops only, 6 mg dm^{-3} of soil extractable P was required, as opposed to only 3 mg dm^{-3} in the annual/pasture system. Based on these data, the authors also calculated P use efficiency in the 13th growing season (Table 2.2), considering the soybean grain yield in relation to soil residual P. They found that the relationship between soybean yield and soil residual P in the crop–pasture rotation was on average 2.6 times greater than under continuous cropping. In the same experiment, after 22 years, the crop–pasture treatment recovered 75% of applied P, whereas the continuous cropping system recovered only 40%.

Some factors that explain greater P recovery under forage grasses should be highlighted: greater root length, density, and distribution (allowing the exploitation of a larger soil volume); higher production of root exudates and acid phosphatases (Almeida et al., 2020; Louw-Gaume, et al., 2010); enhanced mycorrhizal associations; and increased microbial activity (Rao, 2014).

Potassium is one of the macronutrients most absorbed by forage plants and has a relatively short half-life, making it readily available and the release less dependent on microbial processes (Assmann et al., 2017). Accelerated K release from biomass to soil was observed by Miguel et al. (2018), mainly in the first 30 days. In conventional systems, up to 50% of the total nutrient applied as fertilizer can be lost by leaching (Rosolem & Steiner, 2017). Studies on K cycling in cropping systems with pasture as cover crops (Benites et al., 2014; Garcia et al., 2008) have shown that *Brachiaria* spp. can extract substantial amounts of K from subsoil layers, increasing K availability in the soil surface. According to Volf et al. (2018), *Brachiaria ruziziensis* has a high K uptake capacity from the soil, including non-exchangeable forms, and can play an important role in nutrient cycling in integrated production systems. According to these authors, *B. ruziziensis* grown on low-K soils without fertilization results in a larger depletion zone of exchangeable K than in fertilized soils or soils originally high in exchangeable K, with a high K cycling potential in the system.

More recently, Hungria et al. (2016, 2021) demonstrated that it is possible to improve field performance of brachiaria, in terms of higher shoot biomass production with enhanced nutrient contents, by inoculation with elite strains of *Azospirillum brasilense* and *Pseudomonas fluorescens*. Brachiaria inoculation with these plant growth-promoting bacteria thus represented an economic and environmental feasible strategy to improve brachiaria performance, not only in integrated systems but particularly in degraded pastures. In these studies, *A. brasilense* increased brachiaria shoot biomass by a mean of 16.8% in response to both seed and leaf-spray inoculation, whereas *P. fluorescens* increased by 15.2% and 14.2%, respectively. Biomass production was always greater when inoculation was performed with an extra supply of 40 kg N ha^{-1} at sowing. In addition to increasing the brachiaria biomass production, seed and leaf-spray inoculation with *A. brasilense* also increased N contents by an average of 11.7% and 20.7% and increased K by 9.9% and 11.3%, respectively; for *P. fluorescens* mean increases were 33.3% and 36.6% for P and 10.6% and 13.6% for K, respectively. Benefits by inoculation were mainly attributed to improvements in root architecture by the synthesis of phytohormones. Biological N fixation in *A. brasilense*, P acquisition (phosphate solubilization and siderophore synthesis) and ACC-deaminase in *P. fluorescens* also contributed to plant growth and nutrient status.

Physical SH in Integrated Systems with "Soil-Building Plants": Soil Aggregation and Water Holding Capacity

There is widespread agreement that a pasture phase improves soil physical characteristics, such as bulk density, penetration resistance, infiltration rate, aggregate stability, and air and water permeability (Marchão et al., 2007).

Deep-rooted tropical grasses, in particular *Brachiaria* spp. species and Brazilian cultivars (*Brachiaria brizantha*, *Urochloa. ruzizienssis*, and *Urochloa decumbens*), are considered "soil-building plants," due to their ability to enhance soil physical quality. Greater biomass production and ability to increase microbiological activity, associated with a vigorous, fasciculate, and deep root system, leads to long and continuous biopores throughout the soil root system (Baptistella et al., 2020), which improves soil physical quality. The soil structuring caused by the fasciculate root system of grasses (Figures 2.8 and 2.9), which easily penetrates the soil, even into compacted layers (soil biological loosening), increases soil permeability (i.e., the water can move more easily through the soil) (Stone et al., 2003), favoring water infiltration and storage in the soil profile.

According to Flavio-Neto et al. (2015), the least limiting water range (LLWR) is an indicator of biological loosening effect of the root system of grasses of the genus *Brachiaria* in CL systems. These authors showed that the capacity of biological loosening of different grass species and cultivars is variable (Figure 2.10). The cultivation of *B. brizantha*, *Brachiaria decumbens*, and *B. ruziziensis* was evaluated as a management strategy to contribute to soil recovery. The cultivars Xaraes and Piatã (*B. brizantha*) increased soil loosening while increasing the LLWR

Figure 2.8 Brachiaria root system in the soil profile of crop areas, with a full root length (depth) of 3.4 m. Photos by Lourival Vilela.

Figure 2.9 Images of grass root systems in crop rotations and burrows produced by soil fauna. Photos by Fabio Ono, Júlio Salton and Patrick Lavelle.

(indicating water availability) for subsequent soybean. A high aggressiveness in the disruption of compacted layers of the root systems of these cultivars was observed.

The effects of deep-rooted pastures on physical properties and soil structure occur regardless of soil texture. This was stated by Macedo (2001) in a long-term experiment in Mato Grosso, Brazil, on a clayey Oxisol and by Vilela et al. (2001) on a sandy soil (Arenosol), who observed a similar tendency of improvement in soil physical properties.

Soil physical degradation by animal trampling is a major concern in integrated systems and has been addressed in studies. Although results indicate no harmful effect of compaction on annual crops after grazing, grain producers still use

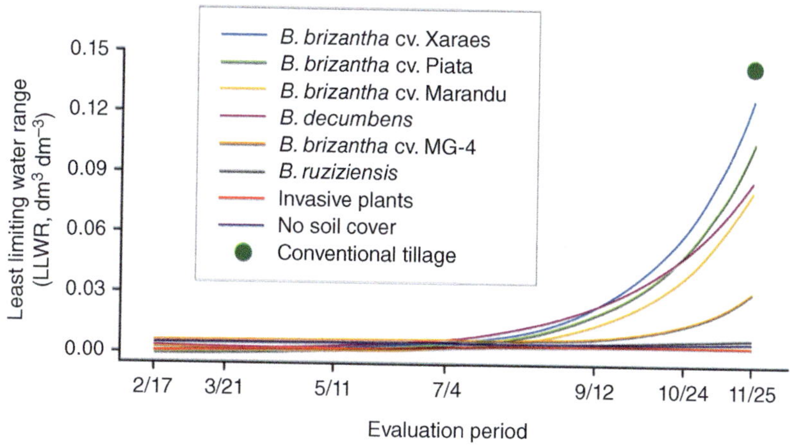

Figure 2.10 Changes in the least limiting water range (LLWR; $dm^3\ dm^{-3}$) in an Oxisol under crop–livestock integration systems in Rio Verde, Goiás, after forage cutting (first LLWR evaluation after harvest of previous maize). Art by Wellington Cavalcanti. Adapted from Flavio-Neto et al. (2015).

the argument of soil compaction for not adopting the system. In this sense, Marchão et al. (2009b) evaluated the impact of an ICL system on penetration resistance (PR) of a sandy soil in western Bahia, with low SOM and a high fine sand content. The results showed that under these conditions (fragile soil, highly susceptible to erosion and compaction due to the higher content of fine sand), the inclusion of off-season livestock grazing ("boi safrinha"), in which the animals are only maintained in the field during the off-season, did not cause drastic soil compaction. On the other hand, the results suggest that in sandy soils, the critical PR values may be lower than the threshold value of 2.5 MPa, considered critical in the literature (Taylor et al., 1966).

In southern Brazil, Lanzanova et al. (2007) pointed out that the increase in grazing frequency had a negative impact on the soil physical quality and crop yields. However, other studies have shown that, in spite of the higher soil bulk density and PR under CL, crop yields were not adversely affected (Debiasi & Franchini, 2012; Flores et al., 2007). These discrepancies regarding the effect of animal trampling on soil compaction are due the wide range of soil-climatic conditions, differences in the experimental duration and in animal stocking rate, forage mass production, and the forage species in each integrated system. Under moderate grazing, soil compaction is limited to the surface layers (usually to a depth of 0–10 cm) and may be temporary and easily reversible. Well-managed integrated systems under an appropriate stocking rate and pasture sward height (that limit forage intake to a maximum of 50%) are imperative to prevent soil physical degradation resulting from animal trampling (Fidalski, 2015; Moraes et al., 2014; Petean et al., 2010). On the other hand, overgrazing intensifies the trampling effect on the soil and can lead to detrimental soil compaction, which could negatively affect crop yield under water stress. Studies in the Cerrado region (e.g., Salton et al., 2014) and southern Brazil (Souza et al., 2010) showed that the introduction of forage grasses (brachiaria in tropical and oat + ryegrass in subtropical areas) associated with moderate grazing intensity improved soil structural and physical conditions with increased soil organic C (SOC) accumulation and biological richness and activity.

In summary, the better soil structure in integrated NT systems warrants soil physical conditions for water infiltration (Bono et al., 2012) and water availability for crops in the soil profile (Flavio-Neto et al., 2015), whereas, the effects of soil compaction are minimized so that integrated systems preserve soil physical quality in tropical and subtropical areas of Brazil.

SH in Integrated Systems: Soil Enzymes and Fauna Biodiversity

In all examples discussed so far, the potential of deep-rooted forage grasses in integrated systems as "soil-building plants" in terms of soil physical and chemical properties was clear. In the same way, under tropical conditions, the vast root systems of these plants (Figures 2.8 and 2.9) also make them an excellent biological soil conditioner because they constitute a primary source of root exudates for microorganisms. This effect is most evident when *Brachiaria* spp. are used as living cover crops during the dry season, where food resources for microbial communities are greatly restricted. In addition, the presence of brachiaria in this period results in significant reductions in soil temperature and in maintaining a more humid environment. In other words, soil microbial communities under brachiaria in the dry season, particularly in the Cerrado region, will not be stressed by lack of food, drought, and heat. As already mentioned, the greater the number of components, the higher the number of connections among them.

The synergy between the different components of the integrated systems has a direct impact on the soil biological component, resulting in multiple benefits in a win-win situation, with improved yields and greater NUEs via livestock dung/urine recycling. By providing more favorable conditions for multiplication of soil biota, integrated systems promote soil environments with intensified enzyme activities (i.e. better nutrient cycling), by which nematodes and soil diseases are suppressed and pesticide decomposition is accelerated. Integrated systems also minimize the loss in biodiversity regarding soil fauna.

Table 2.3 presents data from a long-term experiment, carried out by Embrapa Agropecuária Oeste (Embrapa Western Agriculture, Dourados, MS), on a typical

Table 2.3 Soil organic matter (SOM), pH, and biological properties (0–10 cm layer) of a clayey Oxisol under different management systems in the Cerrado, in Dourados, MS, Brazil.

Agricultural systems[a]	SOM	pH (H_2O)	MBC[b]	Acid phosphatase	βGlucosidase	Arylsulfatase
	$g\ kg^{-1}$		$mg\ kg^{-1}$	mg p-nitrophenol (PNP) kg^{-1} soil h^{-1}		
CT	19.1a[c]	5.7a	323b	419c	88b	31c
NT	19.5a	5.4a	370b	467c	138a	75b
ICLc	21.6a	5.2a	318b	739a	130a	126a
ICLp	22.4a	5.7a	530a	616b	147a	134a
PP	23.8 ± 2.2	5.9	631 ± 42	675 ± 34	170 ± 7	232 ± 27

a) CT, conventional tillage; ICLc, integrated crop–livestock under crop (soybean); ICLp, integrated crop–livestock under pasture; NTS, no-tillage system; PP: permanent pasture.
b) Microbial biomass C.
c) Values followed by a same letter in columns do not differ statistically by Duncan's test at 5%. Permanent pasture was not included in the statistical analysis.
Note. Data obtained in an experiment conducted at Embrapa Agropecuária Oeste.

dystrophic Red Latosol with a very clayey texture (Table 2.3). The plots were arranged in strips. The experiment was initiated in 1995, and evaluations of soil collected from the 0 to 10 cm layer were performed in 2015, in the 20th year. The following treatments were evaluated: (a) conventional tillage by disc harrowing (heavy + leveling), with soybean monoculture in summer and oat in autumn/winter; (b) NT with crop rotation (soybean and maize in summer and oats, turnips, and wheat in autumn/winter); (c) integrated crop–livestock (ICL) in two-year rotation cycles, with Piatã as pasture (two years) and NT soybean (2 years), constituting two subsystems: ICLc (crop phase) and ICLp (pasture phase); (d) permanent pasture (PP) of Piatã.

Twenty years after the beginning of the experiment, the SOM contents (Walkley–Black method) in the 0- to 10-cm layer did not differ between the treatments. In the treatments with soybean (CT, NT, and ICLc), microbial biomass C (MBC) was significantly lower than under ICLp. In comparison with CT, the effects of NT on the soil consisted of higher activity levels of the soil enzymes β-glucosidase and arylsulfatase, although the SOM and MBC contents observed in these treatments were similar. The ICL system also had a biological impact on the soil. In the treatments under ICL (crop and pasture phases), the activities of acid phosphatase and arylsulfatase were higher than in treatments under NT and CT. As previously discussed, the additional benefits from the presence of brachiaria in agricultural systems are mainly due to the increased input of plant residues and root exudates, in addition to soil protection during the dry season, favoring a more biologically active edaphic environment.

Arylsulfatase was the most sensitive bioindicator among the four evaluated in this study. The activity levels increased in the following order: CT < NT < ICLc = ICLp < PP. The differences between treatments with lower (CT) and higher (ICLc and ICLp) activities were 4.3-fold. Increases in arylsulfatase activity occurred regardless of MBC and SOM, which might be an indication that, over the 20-year period of this experiment, NT, ICLc, and ICLp favored the accumulation of the abiotic component of this soil enzyme (i.e. not associated with living microbial biomass). As opposed to SOM, the ability of arylsulfatase to identify changes in soil biological functioning in response to the different management systems highlights the important role of biological indicators in routine soil analysis. In fact, arylsulfatase and β-glucosidase are the pillars of the SoilBio technology (see Chapter 10). Since 2020, they have been used in large-scale on-farm SH assessments in Brazil, representing an opportunity to engage producers in soil testing beyond the standard chemical analyses.

Arylsulfatase activity was evaluated in four long-term experiments with integrated systems by Embrapa Cerrados (Planaltina, Federal District) and Embrapa Agropecuária Oeste (Naviraí, Ponta Porã, and Dourados, municipalities of South Mato Grosso) (Figure 2.11). Even though these experiments are located in different geographical regions, they had the same pattern of biological signature,

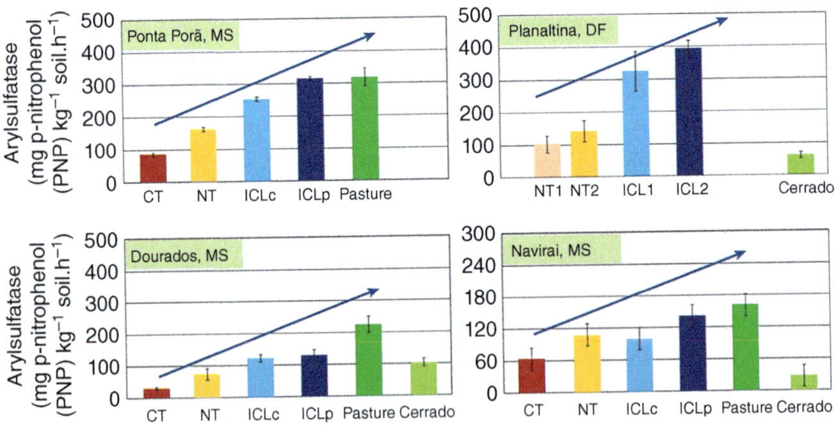

Figure 2.11 Arylsulfatase activity (mg p-nitrophenol [PNP] kg^{-1} soil $^{-1}$) in long-term integrated crop–livestock (ICL) experiments of Embrapa's Units at Planaltina, DF; Ponta Porã; Dourados and Naviraí, MS. CT: conventional tillage; ICLc: integrated crop–livestock under crop; ICLp: integrated crop–livestock under pasture; NT: no tillage. Experiments were performed under responsibility of Dr. Lourival Vilela (DF) and Julio Salton and Michelly Tomazzi (Ponta Porã, Dourados, and Naviraí).

Table 2.4 Activity of soil enzymes β-glucosidase and acid phosphatase (mean ± SE) in the 0- to 10-cm layer, and soil half-life ($T_{1/2}$) of two insecticides, under different management systems in Dourados, MS, Brazil.

Agricultural system	β-Glucosidase	Acid phosphatase	Bifenthrin	Permethrin
	–mg p-nitrophenol kg^{-1} soil h^{-1}–		–$T_{1/2}$ (days) –	
ICLc[a)]	356 ± 4	1206 ± 7	14	9
ICLp	282 ± 15	649 ± 4	25	21
NT	188 ± 3	612 ± 14	25	22
CT	99 ± 8	291 ± 14	44	47

a) CT, conventional tillage continuous crop; ICLc, integrated crop–livestock in the crop phase; ICLp, integrated crop–livestock in the pasture phase; NT, no tillage continuous crop.
Note. Adapted from Portilho et al. (2015).

represented by a higher arylsulfatase activity in response to the increase in sustainable intensification (in other words, by including better agricultural practices over a longer period of time). Increased enzyme activity levels under integrated systems in the tropics were also reported by Franco et al. (2020) and Santos et al. (2022).

Accelerated decomposition rates of pesticides in integrated systems have also been reported. In a study of Portilho et al. (2015), the persistence of the insecticides bifenthrin and permethrin was evaluated (Table 2.4). The laboratory study evaluated soil samples (0- to 10-cm layer) in four treatments of a long-term experiment at Embrapa Agropecuária Oeste (Dourados-MS). The soil samples were distributed in microcosms and incubated at 28°C and 75% field moisture capacity for 51 days. In the treatments with higher enzymatic activity (i.e. higher biological activity), the half-life values ($T_{1/2}$) of the two insecticides were significantly reduced. This result indicates another great advantage of maintaining biologically more active soils (e.g., in integrated systems) that often goes unnoticed: the ability of these soils to reduce the residence time of polluting agents in the environment.

Another advantage of the adoption of integrated systems is nematode control by rotation with crops and pastures, which can either suppress or kill nematodes. However, for better results, a careful diagnosis of nematode species and populations is required to prescribe the rotation that will minimize crop damage most efficiently. In fact, nematode susceptibility and the capacity of nematode control vary considerably between crops and cover crops and depend on the nematode species, affecting crop yield at different levels (Asmus, 2003). In the long-term experiment described in Tables 2.3 and 2.4, Salton et al. (2014) registered 3424 individuals (per 300 cm^3 of soil) of the plant parasite nematode *Rotylenchulus reniformis* under CT (soybean–fallow), compared with only four individuals in an integrated system with brachiaria grass. The use of both *Brachiaria* spp. and *Crotalaria* spp. to control nematodes has become a widespread practice in Brazil. A reduction of 99% in a *Pratylenchus brachyurus* population after three years of *B. brizantha* 'Piatã' (Palisade grass) as winter cover crop in a soybean–maize sequence was accompanied by a maize yield increase of 21.6% in comparison with the most common cover crop pearl millet [*Pennisetum glaucum* (L.) R. Br.] (Abreu & Borges, 2015). The greater the crop diversity in the rotation, the greater the control efficacy of nematodes and other pests and diseases.

Regarding soil macrofauna biodiversity, recent research has shown that (a) the introduction of rotation with pastures in agricultural areas favors certain fauna groups and (b) well-managed and vigorous pastures have good conditions for soil fauna colonization and diversity (Marchão et al., 2009a). Degraded pastures, in spite of maintaining the soil colonized, tend to increase soil bulk density and termite biomass. In the case of agricultural areas, soil disturbance and crop rotations are key drivers of qualitative and quantitative changes in soil fauna. In a study on integrated systems, Bussinger (2018) suggested correlations between *Oligochaeta*, *Coleoptera*, and *Hemiptera* with pastures and integrated systems. These findings suggested that pasture–crop rotations and permanent pastures favored fauna maintenance and tended to increase the abundance of some groups compared with the native Cerrado (reference) area.

It cannot be ignored that the conversion of native areas is expected to cause loss of biodiversity, but it is also true that optimized conservation production systems such as integrated systems are expected to reduce this loss. Studies on the introduction of rotation between annual crops and pastures have demonstrated a greater density and diversity of soil fauna species, with better results than of continuous pasture or even native Cerrado vegetation (Figure 2.12) (Marchão et al., 2009a). The presence of mulch in integrated NT systems stimulates soil fauna, root growth, and soil microflora, which can improve soil conservation (Lavelle & Spain, 2001; Santos et al., 2008). Likewise, the maintenance of a vegetation cover on the soil surface prevents a diversity loss of soil macrofauna and favors the activity of ecosystem "engineering" organisms, including the groups *Oligochaeta*, *Formicidae*, and *Isoptera*, which, in most cases, are beneficial to the ecosystem (Barros et al., 2003).

SH in Integrated Systems: C Storage and GHG Mitigation

Long-term experiments with agricultural grain crop rotations indicated that even NT systems with high amounts of crop residue input must be planned in order to balance the fast SOM decay associated with high temperature and moisture conditions in the subtropical and tropical environments of Brazil (Corbeels et al., 2016;

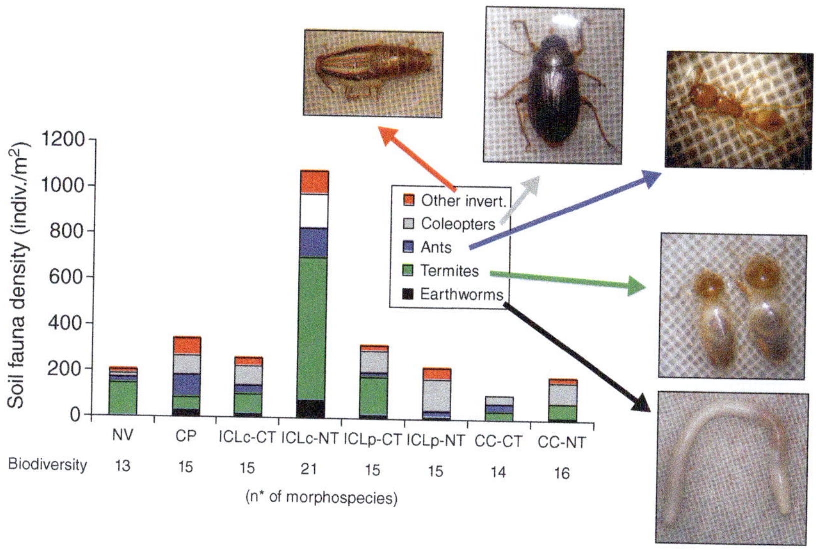

Figure 2.12 Density and diversity of soil macrofauna in different integrated and continuous systems in the Cerrado region, indicating a higher diversity in no-tillage (NT) pasture–crop rotation in the crop phase. CC-CT, continuous crop system under conventional tillage; CC-NT continuous crop system under NT; CP, continuous pasture; ICLc-CT, integrated crop–livestock system in the crop phase under conventional tillage; ICLc-NT, integrated crop–livestock system in the crop phase under NT; ICLp-CT, integrated crop–livestock system in the pasture phase under conventional tillage; ICLp-NT, integrated crop–livestock system in the pasture phase under NT; NV, native Cerrado vegetation. Adapted from Marchão et al. (2009b).

Nunes et al., 2011). Thus, in addition to equilibrated soil fertilization and the use of lime and other soil amendments (e.g., gypsum), greater crop diversity, including grasses with well-developed root systems, is very important to sustain high crop residue inputs (Dalla Nora & Amado, 2013; Sousa et al., 2016), especially on the naturally acidic and low-fertility soils of the Cerrado (Vilela et al. 2011).

Several studies in Brazil have shown that integrated systems, mainly those under NT and with crop–pasture rotations, are efficient systems that are capable of storing C in the soil profile (Ayarza et al., 2022; Batlle-Bayer et al., 2010; Sant-Anna et al., 2017). In tropical soils, the presence of forage grasses in agricultural systems is directly related to the physical protection of organic matter (Sato et al., 2019) and is the main strategy to increase C levels in deeper soil layers (Table 2.5).

Based on the oldest long-term experiment with continuous and integrated systems in Brazil, initiated in 1991 at Embrapa Cerrados, several studies related to soil C dynamics were published. In a comparison of two samplings of soil C stocks in the 0- to 100-cm layer in an 11-year period (2002–2013), the C accumulation

Table 2.5 Carbon stocks (CS) and annual accumulation rate (ΔC) of soil C in the 0- to 100-cm layer of agricultural systems in a long-term experiment with integrated crop–livestock systems, initiated in 1991, at Embrapa Cerrados, Planaltina, DF, Brazil.

Agricultural System	C Stock		ΔC
	2002[a)]	2013[b)]	
	–Mg ha^{-1}–		Mg ha^{-1} yr^{-1}
Continuous grass pasture	145	147	0.182
Continuous consortiated pasture	147	152	0.455
Crop–livestock system	143	157	1.273
Continuous crop under CT	139	148	0.818
Native Cerrado	140	–	–

a) Adapted from Sisti (2005).
b) Adapted from Sá (2011).
Note. CT, conventional tillage.

rate in the ICL system under NT was greater (1.273 Mg C ha^{-1} yr^{-1}) than under conventionally tilled continuous cropping (0.818 Mg C ha^{-1} yr^{-1}) (Table 2.5). In a study by Sant-Anna et al. (2017), changes in SOC in this area were evaluated over 22 years under pasture, crops, or ICL systems. Continuous cropping under CT promoted greater losses of SOC than under NT. The SOC stocks were greatest under ICL treatments.

Corbeels et al. (2016) estimated the total C input in the widely used soybean–maize double-cropping system in southwestern Goiás, a representative region of modern agriculture in the Brazilian Cerrado. Based on historical records of yield data and by using crop-specific harvest indices and root/shoot ratios from published biomass partitioning studies, the authors estimated that between 1995 and 2013—when NT and the second cropping season (i.e., soybean–maize double cropping in the same growing season) were implemented—the mean annual C input was 5.3 Mg C ha^{-1}, which resulted in a soil C accumulation rate of 1.61 Mg C ha^{-1} yr^{-1} in the 0- to 40-cm layer. Although the authors considered this accumulation rate relatively high for the period, a stabilization trend in C accumulation in this layer was observed, probably related to the fact that the root systems of annual crops (soybean and maize) are limited to the uppermost soil layers. These results revealed that in the widely used soybean–maize double cropping sequence system, regardless of the land use intensification in relation to soybean monoculture, no significant C amounts could be stored in the deeper soil layers (below 40 cm).

According to Faccio Carvalho et al. (2010), the conversion of well-managed pastures to crops (soybean or sorghum) caused a soil C loss of 1.44 Mg ha^{-1} yr^{-1}.

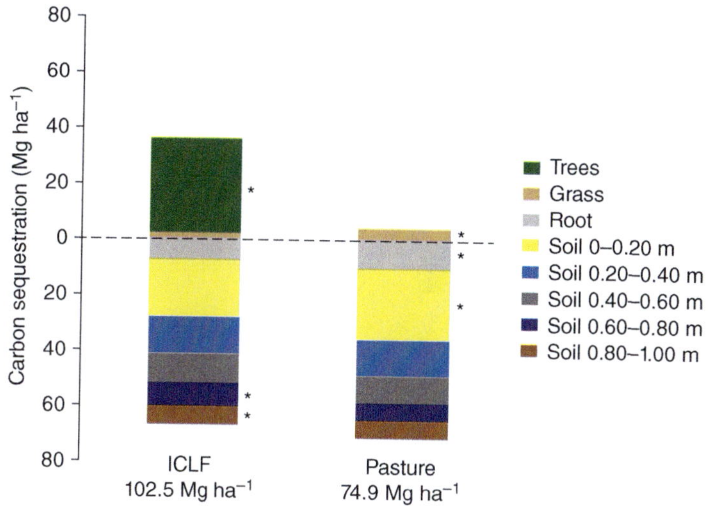

Figure 2.13 Total C sequestration above and below the soil in integrated crop–livestock system and monoculture pasture, six years after experimental setup. *Significant difference in C sequestration between integrated crop–livestock–forestry (ICLF) and monoculture pasture. Adapted from Sarto et al. (2020).

On the other hand, in the same study, the conversion to ICL systems promoted C accumulation at rates that varied between 0.82 and 2.58 Mg ha^{-1} yr^{-1}, according to the crop species, climate, location, and duration of the system. According to Sarto et al. (2020), well-managed *B. brizantha* pastures promoted increases in C stocks at a rate of 1.44 Mg ha^{-1} yr^{-1}, accompanied by an N accumulation rate of 0.33 Mg ha^{-1} yr^{-1}. They also observed that plant C sequestration in the integrated system (ICLF) was 68% greater than under monoculture pasture due to the C accumulated by aboveground tree biomass (Figure 2.13). These results demonstrate that tropical grasses are highly efficient in promoting shoot and root C input, being more efficient than leguminous species due to the multiple benefits associated to their deep root system (Rao et al., 1995; Salton et al., 2013).

Carbon stocks (0–40 cm) in a clayey Oxisol (0–40 cm layer) were evaluated in the field for four years by Coser et al. (2018) during the transition of a low-productivity pasture to an integrated system (maize + *Gliricidia sepium* + *P. maximum* 'Massai'). The integrated system accomplished the goal of building SOC, which confirmed its potential as sustainable agricultural practice with improvements in short-term C sequestration (Figure 2.14). On the other hand, C maintenance and stabilization with depth depends on several factors, including the presence of a specific biota, availability of N and other nutrients, and aggregate formation.

The intensification of land use with integrated systems can result in significant gains in terms of offsetting GHG emissions, not only by storing soil C but also

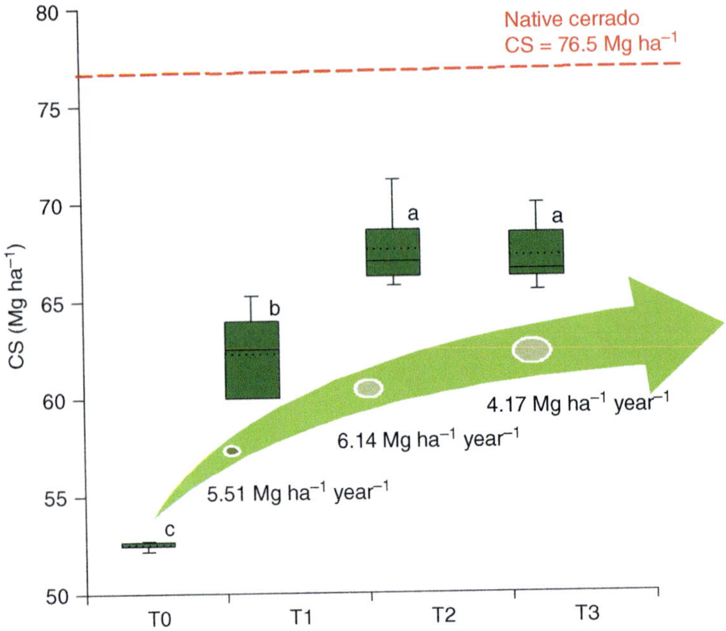

Figure 2.14 Carbon stocks (CS) in four growing seasons (T0 = November 2012; T1 = March 2014; T2 = March 2015; T3 = March 2016) in the 0.00- to 0.40-m layer and C accumulation rate between T0 (recovery of low-productivity pasture) and after three years of implementation of the integrated system. The dashed red line indicates native Cerrado (reference) CS. Different letters indicate a significant ($P < 0.05$) difference of CS between cropping seasons. Coser et al. (2018)/with permission of Elsevier.

by reducing soil emissions during cultivation. In 2016, the Brazilian agricultural sector accounted for emissions of 439,213 Gg CO_2 equivalent (CO_2eq.) (i.e., 34% of the national GHG emissions). In the context of decarbonization in tropical agriculture, integrated systems have been identified as a promising sustainable strategy (Figueiredo et al., 2017; Norse, 2012; Torres et al., 2017).

Among the GHGs, N_2O is seen as a potentially more harmful gas than CO_2 due to its 265–298 times higher global warming potential than that of CO_2 on a 100-year scale. In addition, N_2O has a longer residence time, persisting for up to 121 years in the atmosphere. In agricultural systems, N_2O emissions are influenced by soil and climate conditions, and SOM availability is a key factor in this process. Soil N_2O emissions from agricultural systems are influenced by some well-known factors, such as high water content (low aeration), N fertilization, tillage regime, and SOC availability and forms. The accumulation of different SOC fractions is influenced by soil and climate conditions and soil management (Sato et al., 2019). In tropical soils, chemical stabilization is characterized by strong organo-mineral interactions, and C accumulation in its most stable forms is associated with a higher degree of SOM stabilization (Plaza-Bonilla et al., 2014). The better the protection, the less SOM will be exposed to mineralization, hampering the access of decomposing microorganisms and, consequently, reducing SOM loss to the atmosphere in gaseous forms.

The studies of Sato et al. (2019, 2021) confirmed the hypothesis that integrated systems promote a balanced accumulation of labile and stable SOM fractions, reducing soil N_2O emissions (Figure 2.15). According to these authors, the cumulative N_2O emissions were higher in continuous crop systems than in integrated systems and smaller in the Cerrado area. Among the agricultural systems, lower cumulative N_2O emissions were observed in the NT integrated system based on pasture–crop rotation because of the greatest buildup of C in its most stable fractions and occluded in aggregates. Additionally, C fractions (labile and stable) determined in aggregate classes showed a relationship between intensified

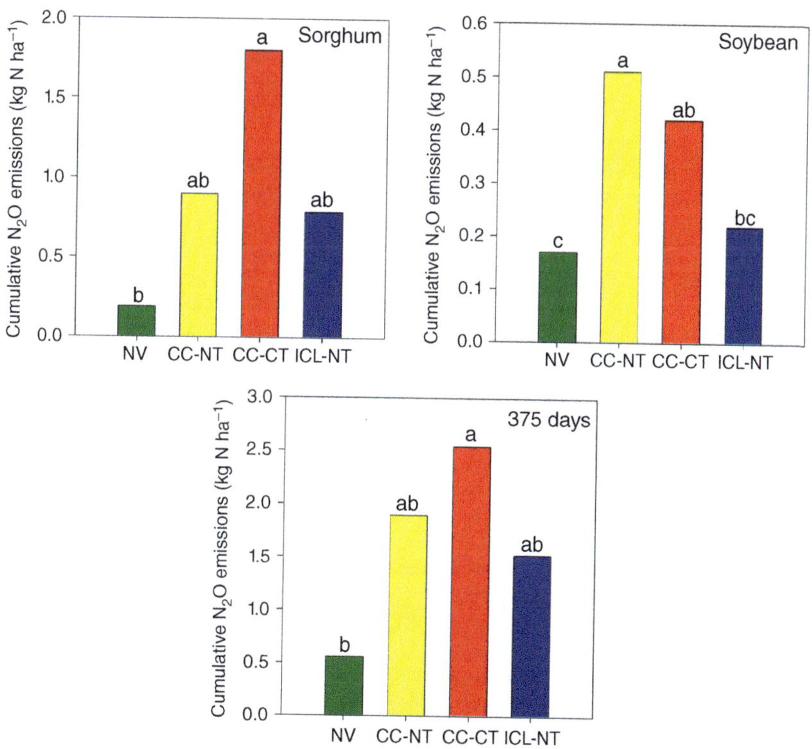

Figure 2.15 Cumulative soil N_2O emissions (kg N ha^{-1}) from continuous crop and integrated crop–livestock systems during soybean and sorghum cycles. CC-NT, continuous crop under no-tillage; CC-CT, continuous crop under conventional tillage; ICL-NT, integrated crop–livestock system under no-tillage; NV, native Cerrado vegetation. Data from Sato et al. (2021).

N_2O emissions and areas under continuous crop cultivation with soil tillage (plowing/harrowing). This result was associated with increased decomposition promoted by soil disturbance and exposure of SOM protected in aggregates. These data from the Brazilian Cerrado demonstrate a potentially positive C balance in integrated systems, even on very clayey and highly weathered soils, which reinforces the recommendation of using ICL systems to mitigate GHG emissions.

The main GHG generated in ruminant production systems is enteric CH_4. In 2016, enteric CH_4 emissions represented 56.5% of agricultural emissions (MCTIC, 2020). Emissions of this gas from integrated systems vary according to the stocking rate (number of animal units ha^{-1}) and forage quality.

Research results at Embrapa Cerrados (Souza et al., 2020) showed that an ICLF system on only 15% of the area of a farm (with a stand of 417 trees ha^{-1}) would be enough to offset all CH_4 emissions (from living cattle) and N_2O emissions (from soil and excreta together with the N_2O emissions of the initial crop phase). In other words, a property with 1000 ha of pasture must allocate 150 ha to the ICLF system (417 trees ha^{-1} with a stocking rate of 1.7 head ha^{-1}) and 850 ha to the ICL system (stocking rate 3 heads ha^{-1}) to offset GHG emissions. Carbon is also stored in the tree roots, which has not been calculated (Table 2.6). Despite the positive C

Table 2.6 Carbon balance in seven-year-old integrated crop–livestock (ICL) and integrated crop–livestock–forestry (ICLF) systems, Embrapa Cerrados, Planaltina, DF, Brazil.

System	ECH_4[a]	N_2O excreta[b]	Mean annual N_2O soil emission[c]	Stocking rate	Soil C[d]	Trunk C fixation	Cumulative mean annual emission[e]	Annual C balance[f]
	–Mg CO_{2eq} ha^{-1} yr^{-1}–			head ha^{-1}	–Mg CO_{2eq} ha^{-1} yr^{-1}–			
ICL	3.4	0.527	0.407	3.0	4.7	0	3.84	+0.86
ICLF	2.0	0.298	0.306	1.7	3.5	20.7	2.31	+21.89

a) Animal emission of CH_4 with global warming potential—PAG (100-year time horizon), related to $CO_2 = 25$ (IPCC, 2013), considering six years of the system with animals for calculation purposes (ICL = 2.91 and ICLF = 1.71)
b) Emission factor of bovine excreta (Lessa et al., 2014) in, considering six years of the system with animals for calculation
c) N_2O emitted by the soil under pasture and grain crops with global warming potential—PAG related to $CO_2 = 298$ (IPCC, 2013), mean of seven years of the system (six years under pasture + one year under grain crop)
d) Mean soil C sequestration to a depth of 100 cm
e) Mean annual emission (enteric CH_4 + soil N_2O + N_2O excreted)
f) Considering a density of 417 trees per hectare at the age of seven years, considering the annual emissions of enteric CH_4 for six years (real condition) and cumulative N_2O emissions over the seven years.

Note. Souza et al. (2020).

balance of well-managed ICL production systems, which tend to emit less C than they accumulate, this balance is based on C sequestration by the soil. This occurs, largely, due to the root system of the plant components, particularly of pasture, and the straw (litter) deposited on the soil. However, the C accumulation rate in the soil tends to stabilize over time, when the soil C stock in an agricultural system approaches the original values under native vegetation. Under these conditions, C accumulated in the soil will tend to be zero because the soil will approach the maximum C storage capacity very closely. Finally, depending on the objective, the forestry component of ICLF systems is a viable option for storing atmospheric CO_2 in the form of wood.

SH Effects on Soybean Grain Yield and Animal Production and Welfare in Tropical ICLF Systems

All effects on SH promoted by integrated systems are translated into better crop yields, animal performance, and animal welfare.

Increased soybean yields in areas where deep-rooted forage grasses had been cultivated before, either as cover crop or animal pasture, have been reported in the literature (summarized in Table 2.7). Averaged across all experiments, the gains in soybean yield after deep-rooted grasses (mainly Brachiaria) were 686 kg ha^{-1}.

Table 2.7 Soybean yield gains in areas where deep-rooted forage grasses were cultivated before for a double purpose (cover crop and animal pasture).

Reference	Soybean yield			Number of cropping cycles
	Without deep-rooted grass	With deep-rooted grass	Gain	
		–kg ha^{-1}–		
Benites et al. (2014)	3235	3806	571	3
Vilela et al. (2011)	3061	3571	510	3
Vilela et al. (2018)	3275	4049	774	1
Dias et al. (2020)	3492	4340[a)]	848	2
Muniz et al. (2021)	3409	4600[a)]	1191	2
Davi et al. (2022)	3187	3531[b)]	344	1
Chaer et al. (2023)	3821	4381	580	4
Mean	3354	4040	686	

a) Means of five and two treatments, respectively, with different species/ cultivars of deep-rooted grasses.

b) Consortia combining grass species, wheat, niger, and radish.

The off-season livestock grazing performance in integrated systems will vary according to the forage species, stocking rates, genetic potential of the herd, animal initial condition, pasture management, grazing cycles, and the farmer's objectives regarding the amount of forage biomass to be preserved as mulch. In comparison to the weight gains in a traditional system (85–110 kg ha^{-1} yr^{-1}), Martha Jr. et al. (2007) reported a threefold cow/calf weight increase (300 kg ha^{-1} yr^{-1}). A fourfold increase in animal liveweight gain (730 vs. 150–230 kg ha^{-1} yr^{-1}) in integrated systems versus traditional systems, respectively, was reported by Pedreira et al. (2018). Between 2010 and 2015, Vilela et al. (2018) observed a mean annual weight gain in finishing cattle of 45% at the Fazenda Triunfo (Formosa do Rio Preto, BA) in response to the adoption of "off-season grazing" in an ICL system.

To better illustrate the gains in animal production in pasture–crop rotation, one of the best examples is the Santa Terezinha farm in Uberlandia, MG (Vilela et al., 2008). Between 1984 and 2004, animal performance was closely evaluated. After 20 years, the original area under degraded pastures was replaced by successive cycles of crops after pastures and improved pastures after crops (Table 2.8). Despite the reduction in area, the farm carrying capacity was maintained at a high level due to the increased forage availability from improved pastures after crops.

In a recent meta-analysis, Oliveira et al. (2022) reported that in an ICLF system with *Eucalyptus* spp. and *Brachiaria* spp., with more than 28 m between tree rows and up to 99 trees ha^{-1}, the animal weight gain per area was greater than in a grass monoculture system.

In spite of these results, there are knowledge gaps about the economic, social, and environmental potential of integrated systems, and how they can be used

Table 2.8 Changes in area under different production systems and stocking rates of a crop rotation system on a sandy Oxisol of the Fazenda Santa Terezinha, Uberlandia, MG, Brazil.

	Area under different use				
Year	Degraded pasture	Crop after pastures	Pastures after crops	Herd size	Stocking rate[a)]
		–% of total–		n	head ha^{-1}
1983	100	0	0	1094	1.1
1988	29	42	29	821	1.4
1992	0	59	42	1150	2.8
1996	0	64	36	1200	3.2
2003	0	30	70	1800	2.6

a) Stocking rate determined in the wet season.
Note. Adapted from Vilela et al. (2008).

as an effective strategy to increase agricultural sustainability, mainly along the agriculture–forest frontier of Brazil. A comprehensive analysis of the economic viability of integrated systems was carried out by Reis (2021), in comparison with a "typical" continuous crop or livestock farm in the Legal Amazon region of the Brazilian state of Mato Grosso, the largest soybean and cattle producer of the country. The results of this case study approach showed higher levels of productivity, profitability and return on investment, and shorter payback periods with lower economic risk of ICLF than of continuous crop and livestock systems. However, at high crop prices, economic results of continuous cropping are better than in the integrated system. More such studies are needed to assess how far these results can be extrapolated to other parts of Brazil.

Given the multifaceted and dynamic reality of integrated systems, it is vital to assess potential environmental benefits (e.g., for animal welfare). In tropical countries such as Brazil, where cattle are frequently subjected to unfavorable thermal conditions, ICLF systems deserve special attention because they improve the microclimate and, consequently, the thermal comfort indices (Lopes et al., 2022; Magalhães et al. 2020). For example, the presence of trees in pastures of agroforestry systems in the southern Amazon improved thermal comfort in these environments and, consequently, in animal performance (Magalhães et al., 2020). The effects of shade under ICLF on thermal comfort and reproduction of lactating Gyr cows were evaluated by Reis et al. (2021), who found that trees in ICLF systems intercepted solar radiation by up to 26% and mitigated environmental heat. Animals under natural shading in the dry season produced 16% more ovarian follicles and 81% more viable oocytes than under full sun. The number of in vitro produced transferable embryos from animals under natural shading in an ICLF system was greater. Benefits of pasture afforestation on bovine immunological parameters were also described by Lopes et al. (2022). They evaluated the influence of different tree arrangements in the ICLF, in stands of single or triple rows, on gastrointestinal nematode infection and the immune response of Nellore heifers in the tropics. The shadier the studied ICLF system, the milder the parasitism and the higher the immunological lymphocyte responses in animals.

Final Considerations

The combination of crop, livestock, and forest production at the farm scale in integrated systems is seen as a promising strategy to increase agricultural sustainability. As part of the efforts toward sustainable intensification of agriculture in Brazil and in order to encourage widespread adoption of ICLF systems by rural producers, a public-private partnership, called ICLF network (redeILPF, 2023b), was created in 2012. This successful ICLF network, co-financed by private companies and Embrapa, maintains Technological Reference Units, distributed across all Brazilian biomes that is supported by 19 Embrapa Research Units (redeILPF, 2023c).

Brazilian investments in ICLF systems represent a direct contribution to advances in the pursuit of agricultural sustainability, represented by higher farmers' income, productivity gains, and optimized use of natural resources, with particular emphasis on SH. As shown in this chapter, the growing interest has boosted experimental efforts related to ICLF systems throughout the country. First studies in the Cerrado region were installed in Planaltina, DF, at Embrapa Cerrados (in 1991); in Campo Grande, MS, at Embrapa Beef Cattle (in 1993); and in Dourados (MS), at Embrapa Western Agriculture (in 1995). In the southern region, the Federal University of Rio Grande do Sul and the company Agropecuária Cerro Coroado initiated an ICL experiment in São Miguel das Missões, RS, in 2001.

Some of the results of these long-term field experiments, along with many others from across the country, were presented internationally at the 2021 United Nations Conference on Climate Change—COP 26, in Glasgow, Scotland, as a proposal of a response against climate change. In view of the extension and environmental complexity of the Brazilian territory and the high intricacy of ICLF systems, represented in all research data collected so far, these systems are undoubtedly the agricultural management forms that come nearest to substantiate the famous mantra coined by Douglas Karlen: "Healthy soils, healthy landscapes, vibrant economies!"

In the future, many variables will have to be further monitored, measured, and analyzed, such as energy use efficiency, integrated pest, disease and weed management, and new crop combinations, as well as the socioeconomic and environmental viability of the different systems. Research on new alternatives of ICLF modalities, with different arrangements and combinations of annual crops, forages, and forest species, is also needed. Studies are needed to focus on family farming, in areas where animal husbandry with low technological input is common, causing high deforestation pressure, as in the northern and northeastern regions of the country.

References

Abreu, A. B. L., & Borges, E. P. (2015). *Redução de Pratylenchus brachyurus na cultura de soja em sucessão a diferentes coberturas de solo na entressafra*. Comunicado.

Almeida, D. S., Delai, L. B., Sawaya, A. C. H. F., & Rololem, C. A. (2020). Exudation of organic acid anions by tropical grasses in response to low phosphorus availability. *Scientific Reports*, 10(1), 8. https://doi.org/10.1038/s41598-020-73398-1

Alves, F. V., Almeida, R. G., & Laura, V. A. (2017). *Carbon neutral brazilian beef: A new concept for sustainable beef production in the tropics*. Embrapa.

Anghinoni, I., & Vezzani, F. M. (2021). Systemic soil fertility as product of system self-organization resulting from management. *Revista Brasileira de Ciência do Solo*, 45, e0210090. https://doi.org/10.36783/18069657rbcs20210090

Asmus, G. L. (2003). Avaliação quali-quantitativa da ocorrência de nematóides fitoparasitos em regiões de produção de algodão de Mato Grosso do Sul. In *Boletim de Pesquisa e Desenvolvimento/ Embrapa Agropecuária Oeste*, https://core.ac.uk/download/pdf/15437479.pdf.

Assmann, J. M., Martins, A. P., Anghinoni, I., Oliveira Denardin, L. G., Holanda Nichel, G., Andrade Costa, S. E. V., Silva, R. A. P., Balerini, F., Facio Carvalho, P. C., & Franzluebbers, A. J. (2017). Phosphorus and potassium cycling in a long-term no-tillage integrated soybean-beef cattle production system under different grazing intensities in subtropics. *Nutrient Cycling in Agroecosystems*, 108(1), 21–33. https://doi.org/10.1007/s10705-016-9818-6

Ayarza, M., Rao, I., Vilela, L., Lascano, C., & Vera-Infanzón, R. (2022). Soil carbon accumulation in crop-livestock systems in acid soil savannas of South America: A review. *Advances in Agronomy*, 173, 163–226. https://doi.org/10.1016/bs.agron.2022.02.003

Balbino, L. C., Cordeiro, L. A. M., Oliveira, P., Kluthcouski, J., Galerani, P. R., & Vilela, L. (2012). Agricultura sustentável por meio da Integração Lavoura-Pecuária-Floresta (ILPF). *Informações Agronômicas International Plant Nutrition Institute*, 138, 1–18.

Balesdent, J., & Balabane, M. (1996). Major contribution of roots to soil carbon storage inferred from maize cultivated soils. *Soil Biology and Biochemistry*, 28(9), 1261–1263. https://doi.org/10.1016/0038-0717(96)00112-5

Baptistella, J. L. C., Andrade, S. A. L., Favarin, J. L., & Mazzafera, P. (2020). Urochloa in tropical agroecosystems. *Frontiers in Sustainable Food Systems*, 4, 119. https://doi.org/10.3389/fsufs.2020.00119

Barros, E., Neves, A., Blanchart, E., Fernandes, E. C., Wandelli, E., & Lavelle, P. (2003). Development of the soil macrofauna community under silvopastoral and agrosilvicultural systems in Amazonia. *Pedobiologia*, 47(3), 273–280. https://doi.org/10.1078/0031-4056-00190

Batlle-Bayer, L., Batjes, N. H., & Bindraban, P. S. (2010). Changes in organic carbon stocks upon land use conversion in the Brazilian Cerrado: A review. *Agriculture, Ecosystems & Environment*, 137(1–2), 47–58. https://doi.org/10.1016/j.agee.2010.02.003

Benites, V. M., Caetano, J. O., Ferreira Filho, W. C., Menezes, C. C. E., Polidoro, J. C., Oliveira, R. P., & Wiendl, T. (2014). Influence of brachiaria (*Urochloa brizantha*) as a winter cover crop on potassium use efficiency and soybean yield under no-tillage in the Brazilian Cerrado. *Research Findings*, 39, 24–35. https://www.ipipotash.org/uploads/udocs/e-ifc39-research-findings-41.pdf

Bernardes, T. G., da Silveira, P. M., Mesquita, M. A. M., Aguiar, R. A., & Mesquita, G. M. (2010). Decomposição da biomassa e liberação de nutrientes dos capins braquiária e mombaça, em condições de cerrado. *Pesquisa Agropecuária Tropical*, 40(3), 370–377. https://revistas.ufg.br/pat/article/view/5584

Bonaudo, T., Bendahan, A. B., Sabatier, R., Ryschawy, J., Bellon, S., Leger, F., Magda, D., & Tichit, M. (2014). Agroecological principles for the redesign of integrated crop–livestock systems. *European Journal of Agronomy*, 57, 43–51. https://doi.org/10.1016/j.eja.2013.09.010

Bono, J. A. M., Macedo, M. C. M., Tormena, C. A., Nanni, M. R., Gomes, E. P., & Müller, M. M. L. (2012). Infiltração de água no solo em um Latossolo Vermelho da região sudoeste dos cerrados com diferentes sistemas de uso e manejo. *Revista Brasileira de Ciência do Solo*, 36, 1845–1853.

Bussinger, A. P. (2018). *Efeito de diferentes usos do solo no cerrado sobre a composição da fauna edáf* [Doctoral dissertation]. Universidade de Brasília, Brasília. https://repositorio.unb.br/handle/10482/32987

Carvalho, J. L. N., Raucci, G. S., Cerri, C. E. P., Bernoux, M., Feigl, B. J., Wruck, F. J., & Cerri, C. C. (2010). Impact of pasture, agriculture and crop-livestock systems on soil C stocks in Brazil. *Soil and Tillage Research*, 110(1), 175–186. https://doi.org/10.1016/j.still.2010.07.011

Castro Dias, M. B., Pinho Costa, K. A., Costa Severiano, E., Bilego, U. O., Neto, A. E. F., Almeida, D. P., Brand, S. C., & Vilela, L. (2020). Brachiaria and *Panicum maximum* in an integrated crop–livestock system and a second-crop maize system in sequence with soybean. *The Journal of Agricultural Science*, 158(3), 206–217. https://doi.org/10.1017/S0021859620000532

Chaer, G. M., Mendes, I. C., Dantas, O. D., Malaquias, J. V., Reis-Junior, F. B., & Oliveira, M. I. L. (2023). Evaluating C trends in clayey Cerrado Oxisols using a four-quadrant model based on specific arylsulfatase and β-glucosidase activities. *Applied Soil Ecology*, 183, 104742.

Corbeels, M., Marchão, R. L., Neto, M. S., Ferreira, E. G., Madari, B. E., Scopel, E., & Brito, O. R. (2016). Evidence of limited carbon sequestration in soils under no-tillage systems in the Cerrado of Brazil. *Scientific Reports*, 6(1), 1–8. https://doi.org/10.1038/srep21450

Cordeiro, L. A. M., Vilela, L., Marchão, R. L., Kluthcouski, J., & Martha Júnior, G. B. (2015). Integração lavoura-pecuária e integração lavoura-pecuária-floresta: estratégias para intensificação sustentável do uso do solo. *Cadernos de Ciência & Tecnologia*, 32(1/2), 15–43.

Coser, T. R., Figueiredo, C. C., Jovanovic, B., Moreira, T. N., Leite, G. G., Cabral Filho, S. L. S., Kato, E., Malaquias, J. V., & Marchão, R. L. (2018). Short-term buildup of carbon from a low-productivity pastureland to an agrisilviculture system in the Brazilian savannah. *Agricultural Systems*, 166, 184–195. https://doi.org/10.1016/j.agsy.2018.01.030

Costa, C. H. M., Crusciol, C. A. C., Soratto, R. P., & Ferrari Neto, J. (2016). Taxas de decomposição e liberação de nutrientes da fitomassa de milheto, capim colonião e capim-braquiária. *Bioscience Journal*, 32, 1191–1203. https://doi.org/10.14393/BJ-v32n5a2016-32982

Costa, N. R., Andreotti, M., Fernandes, J. C., Cavasano, F. A., Ulian, N. A., Pariz, C. M., & Santos, F. G. (2014). Acúmulo de nutrientes e decomposição da palhada de braquiárias em função do manejo de corte e produção do milho em sucessão. *Revista Brasileira de Ciências Agrárias*, 9, 166–173. https://doi.org/10.5039/agraria.v9i2a3380

Davi, J. E. A., Nogueira, B. K. A., Gasques, L. R., Dalla-Côrt, A. S., Camargo, T. A., Pacheco, L. P., Silva, L. S., & Souza, E. D. (2022). Diversified production systems in sandy soils of the Brazilian Cerrado: nutrient dynamics and soybean productivity. *Journal of Plant Nutrition*, 45, 1–18.

Debiasi, H., & Franchini, J. C. (2012). Atributos físicos do solo e produtividade da soja em sistema de integração lavoura-pecuária com braquiária e soja. *Ciência Rural*, 42, 1180–1186.

FAO. 2022. *Global forest sector outlook 2050: Assessing future demand and sources of timber for a sustainable economy.* FAO Forestry Working Paper, Series 31. https://doi.org/10.4060/cc2265en

Fidalski, J. (2015). Qualidade física de Latossolo Vermelho em sistema de integração lavoura-pecuária após cultivo de soja e pastejo em braquiária. *Pesquisa Agropecuária Brasileira*, 50(11), 1097–1104. https://doi.org/10.1590/S0100-204X2015001100013

Figueiredo, E. B., Jayasundara, S., de Oliveira Bordonal, R., Berchielli, T. T., Reis, R. A., Wagner-Riddle, C., & La Scala Jr, N. (2017). Greenhouse gas balance and carbon footprint of beef cattle in three contrasting pasture-management systems in Brazil. *Journal of Cleaner Production*, 142, 420–431. https://doi.org/10.1016/j.jclepro.2016.03.132

Flávio Neto, J., Severiano, E. D. C., Costa, K. A. D. P., Junnyor, W. S. G., Gonçalves, W. G., & Andrade, R. (2015). Biological soil loosening by grasses from genus Brachiaria in crop-livestock integration. *Acta Scientiarum Agronomy*, 37, 375–383. https://doi.org/10.4025/actasciagron.v37i3.19392

Flores, J. P. C., Anghinoni, I., Cassol, L. C., Carvalho, P. C. D. F., Leite, J. G. D. B., & Fraga, T. I. (2007). Atributos físicos do solo e rendimento de soja em sistema plantio direto em integração lavoura-pecuária com diferentes pressões de pastejo. *Revista Brasileira de Ciência do Solo*, 31(4), 771–780. https://doi.org/10.1590/S0100-06832007000400017

Francisquini Junior, A., Calonego, J. C., Rosolem, C. A., Santos, C. H., & Tiritan, C. S. (2020). Increase of nitrogen-use efficiency by phosphorus fertilization in grass–legume pastures. *Nutrient Cycling in Agroecosystems*, 118, 165–175. https://doi.org/10.1007/s10705-020-10091-9

Franco, A. D. J., Silva, A. P. V. D., Souza, A. B. S., Oliveira, R. L., Batista, É. R., Souza, E. D. D., … Carneiro, M. A. C. (2020). Plant diversity in integrated crop-livestock systems increases the soil enzymatic activity in the short term. *Pesquisa Agropecuária Tropical*, 50. https://doi.org/10.1590/1983-40632020v5064026

Franzluebbers, A. J. (2007). Integrated crop-livestock systems in the southeastern USA. *Agronomy Journal*, 99, 361–372. https://doi.org/10.2134/agronj2006.0076

Garcia, R. A., Crusciol, C. A. C., Calonego, J. C., & Rosolem, C. A. (2008). Potassium cycling in a corn-brachiaria cropping system. *European Journal of Agronomy*, 28(4), 579–585. http://dx.doi.org/10.1016/j.eja.2008.01.002

Hungria, M., Nogueira, M. A., & Araujo, R. S. (2016). Inoculation of *Brachiaria* spp. with the plant growth-promoting bacterium *Azospirillum brasilense*: an environment-friendly component in the reclamation of degraded pastures in the tropics. *Agriculture, Ecosystems & Environment*, 221, 125–131. https://doi.org/10.1016/j.agee.2016.01.024

Hungria, M., Rondina, A. B. L., Nunes, A. L. P., Araujo, R. S., & Nogueira, M. A. (2021). Seed and leaf-spray inoculation of PGPR in brachiarias (*Urochloa* spp.) as an economic and environmental opportunity to improve plant growth forage yield and nutrient status. *Plant and Soil*, 463(1), 171–186. http://dx.doi.org/10.1007/s11104-021-04908-x

IPCC. (2013). Climate change 2013: the physical science basis. In T. F. Stocker, D. Qin, G. K. Plattner, M. Tignor, S. K. Allen, J. Boschung, A. Nauels, Y. Xia, V. Bex, & P. M. Midgley (Eds.), *Contribution of working group 1 to the fifth assessment report of the intergovernmental panel on climate change*. Cambridge University. http://www.ipcc.ch/report/ar5/wg1/

Jank, L., Valle, C. D., & Resende, R. M. S. (2011). Breeding tropical forages. *Crop Breeding and Applied Biotechnology*, 11, 27–34. https://doi.org/10.1590/S1984-70332011000500005

Kluthcouski, J., Pacheco, A. R., Teixeira, S. M., & Oliveira, E. T. (1991). *Renovação de pastagens de cerrado com arroz: I. Sistema Barreirão* [Documentos, 33]. Embrapa-CNPAF. https://ainfo.cnptia.embrapa.br/digital/bitstream/CNPAF/1907/1/doc_33.pdf

Kluthcouski, J., Stone, L. F., & Aidar, H. (2003). *Integração lavoura-pecuária*. Embrapa Arroz e Feijão. http://livimagens.sct.embrapa.br/amostras/00075060.pdf

Lanzanova, M. E., Nicoloso, R. D. S., Lovato, T., Eltz, F. L. F., Amado, T. J. C., & Reinert, D. J. (2007). Atributos físicos do solo em sistema de integração lavoura-pecuária sob plantio direto. *Revista Brasileira de Ciência do Solo*, 31, 1131–1140. https://doi.org/10.1590/S0100-06832007000500028

Lavelle, P., & Spain, A. V. (2001). Soil organisms. *Soil ecology*, 201–356). Springer. https://doi.org/10.1007/978-94-017-5279-4

Lemaire, G., Franzluebbers, A., Faccio Carvalho, P. C., & Dedieu, B. (2014). Integrated crop–livestock systems: Strategies to achieve synergy between agricultural production and environmental quality. *Agriculture, Ecosystems and Environment*, 190, 4–8. https://doi.org/10.1016/j.agee.2013.08.009

Lessa, A. C. R., Madari, B. E., Paredes, D. S., Boddey, R. M., Urquiaga, S., Jantalia, C. P., & Alves, B. J. (2014). Bovine urine and dung deposited on Brazilian savannah pastures contribute differently to direct and indirect soil nitrous oxide emissions. *Agriculture, Ecosystems & Environment*, 190, 104–111. https://doi.org/10.1016/j.agee.2014.01.010

Lopes, L. B., Kamchen, S. G., Gomes, F. J., Natividade, U., Magalhães, L. M. D., de Paula Pimenta, A., & Araujo, R. N. (2022). Influence of silvopastoral systems on gastrointestinal nematode infection and immune response of Nellore heifers under tropical conditions. *Veterinary Parasitology*, 309, 109765. https://doi.org/10.1016/j.vetpar.2022.109765

Louw-Gaume, A. E., Rao, I. M., Gaume, A. J., & Frossard, E. (2010). A comparative study on plant growth and root plasticity responses of two *Brachiaria* forage grasses grown in nutrient solution at low and high phosphorus supply. *Plant and Soil*, 328(1), 155–164. https://doi.org/10.1007/s11104-009-0093-z

Macedo, M. C. M. (2001). Integração lavoura e pecuária: Alternativa para sustentabilidade da produção animal. In P. G. da Cruz, J. R. Alves, & R. A. de Lima (Eds.), *Anais do Simpósio Sobre Manejo da Pastagem—Planejamento de sistemas de produção em pastagens* (pp. 257–283). FEALQ.

Macedo, M. C. M. (2009). Integração lavoura e pecuária: o estado da arte e inovações tecnológicas. *Revista Brasileira de Zootecnia*, 38, 133–146. https://doi.org/10.1590/S1516-35982009001300015

Magalhaes, C. A. S., Zolin, C. A., Lulu, J., Lopes, L. B., Furtini, I. V., Vendrusculo, L. G., Zaiatz, A. P. S., Pedreira, B. C., & Pezzopane, J. R. M. (2020). Improvement of thermal comfort indices in agroforestry systems in the southern Brazilian Amazon. *Journal of Thermal Biology*, 91, 102636. https://doi.org/10.1016/j.jtherbio.2020.102636

Marchão, R. L., Balbino, L. C., Silva, E. M. D., Santos Junior, J. D. D. G. D., Sá, M. A. C. D., Vilela, L., & Becquer, T. (2007). Qualidade física de um Latossolo Vermelho sob sistemas de integração lavoura-pecuária no Cerrado. *Pesquisa Agropecuária Brasileira*, 42, 873–882. https://doi.org/10.1590/S0100-204X2007000600015

Marchão, R. L., Lavelle, P., Celini, L., Balbino, L. C., Vilela, L., & Becquer, T. (2009a). Soil macrofauna under integrated crop-livestock systems in a Brazilian Cerrado Ferralsol. *Pesquisa Agropecuária Brasileira*, 44, 1011–1020. https://doi.org/10.1590/S0100-204X2009000800033

Marchão, R. L., Vilela, L., Paludo, A. L., & Guimarães Jr, R. (2009b). *Impacto do pisoteio animal na compactação do solo sob integração lavoura-pecuária no Oeste Baiano* [Comunicado Técnico, 163]. Embrapa Cerrados. https://ainfo.cnptia.embrapa.br/digital/bitstream/CPAC-2010/31546/1/comtec-163.pdf

Martha Jr, G. B., Alves, E., & Contini, E. (2011). *Pecuária brasileira e a economia de recursos naturais* [Perspectiva Pesquisa Agropecuária, 1]. Embrapa, Estudos e Capacitação. https://www.infoteca.cnptia.embrapa.br/infoteca/bitstream/doc/908616/1/Pecuariabrasileira.pdf

Martha, G. B., Jr., Vilela, L., & Sousa, D. M. G. (2007). *Cerrado: Uso eficiente de corretivos e fertilizantes em pastagens*. Embrapa Cerrados. https://ainfo.cnptia.embrapa.br/digital/bitstream/item/203801/1/Livro.pdf

Mendes, I. C., Sousa, D. M. G., Dantas, O. D., Lopes, A. A. C., Reis Junior, F. B., Oliveira, M. I. L., & Chaer, G. M. (2021b). Soil quality and grain yield: a win-win combination in clayey tropical Oxisols. *Geoderma*, 388, 114880. https://doi.org/10.1016/j.geoderma.2020.114880

Mendes, I. C., Tormena, C. A., Cherubin, M. R., & Karlen, D. L. (2018). Soil health assessment and maintenance in central and south-Central Brazil. In D. Reicosky (Ed.), *Managing soil health for sustainable agriculture: Monitoring and management* (pp. 379–415). Burleigh Dodds. http://dx.doi.org/10.19103/AS.2017.0033.35

Miguel, A. S. D. C. S., Pacheco, L. P., Carvalho, Í. C. D., Souza, E. D. D., Feitosa, P. B., & Petter, F. A. (2018). Phytomass and nutrient release in soybean cultivation systems under no-tillage. *Pesquisa Agropecuária Brasileira*, 53(10), 1119–1131. https://doi.org/10.1590/S0100-204X2018001000005

Ministério da Ciência, Tecnologia e Inovação (MCTI). (2020). *Fourth national communication of Brazil to the UNFCCC*. https://unfccc.int/sites/default/files/resource/4a%20Comunicacao%20Nacional.pdf

Ministry of Agriculture, Livestock and Food Supply (Ministério da Agricultura, Pecuária e Abastecimento—MAPA). (2022). BrazilianAgricultural Policy for Climate Adaptation and Low Carbon Emission: Executive Summary 2020–2030 (pp. 1–36). https://www.gov.br/mre/pt-br/delbraspar/seminario-brasil-portugal/abcsumarioexecutivo2022ingles_final.pdf.

Moraes, A., de Faccio Carvalho, P. C., Anghinoni, I., Lustosa, S. B. C., de Andrade, S. E. V. G., & Kunrath, T. R. (2014). Integrated crop–livestock systems in the Brazilian subtropics. *European Journal of Agronomy*, 57, 4–9. https://doi.org/10.1016/j.eja.2013.10.004

Moraes Sá, J. C., Séguy, L., Tivet, F., Lal, R., Bouzinac, S., Borszowskei, P. R., Briedis, C., Santos, J. B., Cruz Hartman, D., Bertoloni, C. G., Rosa, J., & Friedrich, T. (2015). Carbon depletion by plowing and its restoration by no-tillage cropping systems in Oxisols of subtropical and tropical agro-ecoregions in Brazil. *Land Degradation and Development*, 26, 531–543. https://doi.org/10.1002/ldr.2218

Muniz, M. P., Pinho Costa, K. A., Costa Severiano, E., Bilego, U. O., Almeida, D. P., Neto, A. E. F., … Castro Dias, M. B. (2021). Soybean yield in integrated crop–livestock system in comparison to soybean–maize sequence system. *The Journal of Agricultural Science*, 159(3–4), 188–198. https://doi.org/10.1017/s0021859621000393

Naudin, K., Bruelle, G., Salgado, P., Penot, E., Scopel, E., Lubbers, M., Ridder, N., & Giller, K. E. (2015). Trade-offs around the use of biomass for livestock feed and soil cover in dairy farms in the Alaotra lake region of Madagascar. *Agricultural Systems*, 134, 36–47. https://doi.org/10.1016/j.agsy.2014.03.003

Nicoloso, R. S., & Rice, C. W. (2021). Intensification of no-till agricultural systems: An opportunity for carbon sequestration. *Soil Science Society of America Journal*, 85(5), 1395–1409. https://doi.org/10.1002/saj2.20260

Nora, D. D., & Amado, T. J. C. (2013). Improvement in chemical attributes of Oxisol subsoil and crop yields under no-till. *Agronomy Journal*, 105(5), 1393–1403. https://doi.org/10.2134/agronj2013.0031

Norse, D. (2012). Low carbon agriculture: Objectives and policy pathways. *Environment and Development*, 1(1), 25–39. https://doi.org/10.1016/j.envdev.2011.12.004

Nunes, R. S., Lopes, A. A. C., Sousa, D. M. G., & Mendes, I. C. (2011). Management systems and the carbon and nitrogen stocks of cerrado oxisol under soybean-maize sequence. *Revista Brasileira de Ciência do Solo*, 35(4), 1407–1419. https://doi.org/10.1590/S0100-06832011000400035

Oliveira, A. F., Menezes, G. L., Gonçalves, L. C., de Araújo, V. E., Ramirez, M. A., Júnior, R. G., & Lana, Â. M. Q. (2022). Pasture traits and cattle performance in silvopastoral systems with Eucalyptus and Urochloa: Systematic review and meta-analysis. *Livestock Science*, 262, 104973. https://doi.org/10.1016/j.livsci.2022.104973

Oliveira, P. D., Kluthcouski, J., Favarin, J. L., & Santos, D. D. C. (2010). *Sistema Santa Brígida-Tecnologia Embrapa: Consorciação de milho com leguminosas* [Circular Técnica, 88]. Santo Antônio de Goiás. https://ainfo.cnptia.embrapa.br/digital/bitstream/item/33775/1/circ-88.pdf

Pedreira, B. C., Domiciano, L. F., & Vilela, L. (2018). O estado da arte e estudos de caso em sistemas integrados de produção agropecuária no Centro-Oeste do Brasil. In E. D. de Souza, F. D. da Silva, & T. S. Assmann (Eds.), *Sistemas Integrados de Produção Agropecuária no Brasil* (1st ed pp. 277–300) Copiart, Tubarão.

Petean, L. P., Tormena, C. A., & Alves, S. J. (2010). Intervalo hídrico ótimo de um Latossolo Vermelho distroférrico sob plantio direto em sistema de integração lavoura-pecuária. *Revista Brasileira de Ciência do Solo*, 34, 1515–1526.

Plaza-Bonilla, D., Álvaro-Fuentes, J., & Cantero-Martínez, C. (2014). Identifying soil organic carbon fractions sensitive to agricultural management practices. *Soil and Tillage Research*, 139, 19–22. https://doi.org/10.1016/j.still.2014.01.006

Portilho, I. I. R., Scorza Jr, R. P., Salton, J. C., Mendes, I. D. C., & Mercante, F. M. (2015). Persistência de inseticidas e parâmetros microbiológicos em solo sob sistemas de manejo. *Ciência Rural*, 45, 22–28. https://doi.org/10.1590/0103-8478cr20131328

Pulrolnik, K., Marchão, R., Vilela, L., Guimarães Junior, R., Souza, K. W., & Moraes Neto, S. P. (2019). *Recomendações para inserção do componente arbóreo em sistemas integrados lavoura-pecuária-floresta*. [Comunicado Técnico, 182] Embrapa Cerrados (Infoteca-e). https://ainfo.cnptia.embrapa.br/digital/bitstream/item/199881/1/ComTec-182.pdf

Rao, I. M. (2014). Advances in improving adaptation of common bean and Brachiaria forage grasses to abiotic stresses in the tropics. In M. Pessarakli (Ed.), *Handbook of plant and crop physiology* (pp. 847–889). CRC Press/Taylor and Francis Group.

Rao, I. M., Ayarza, M., & Garcia, R. (1995). Adaptive attributes of tropical forage species to acid soils I. differences in plant growth, nutrient acquisition and nutrient utilization among C_4 grasses and C_3 legumes. *Journal of Plant Nutrition*, 18, 2135–2155. https://doi.org/10.1080/01904169509365052

redeILPF. (2023a). *ILPF in numbers*. https://redeilpf.org.br/ilpf-em-numeros

redeILPF. (2023b). *Crop-livestock-forest integration*. https://redeilpf.org.br

redeILPF. (2023c). *ICLF in numbers*. https://redeilpf.org.br/images/ICLF_in_Numbers-Harvest.pdf

Reis, J. C. (2021). *Integrated crop-livestock-forest systems: a Brazilian alternative for agriculture sustainability*. [Doctoral dissertation]. University of Brasília.

Reis, N. S., Ferreira, I. C., Mazocco, L. A., Souza, A. C. B., Pinho, G. A. S., Fonseca Neto, Á. M., Malaquias, J. V., Macena, F. A., Muller, A. G., Martins, C. F., Balbino, L. C., & McManus, C. M. (2021). Shade modifies behavioral and physiological responses of low to medium production dairy cows at pasture in an integrated crop-livestock-Forest system. *Animals*, 11, 2411. https://doi.org/10.3390/ani11082411

Rosolem, C. A., & Steiner, F. (2017). Effects of soil texture and rates of K input on potassium balance in tropical soil. *European. Journal of Soil Science*, 68(5), 658–666. http://dx.doi.org/10.1111/ejss.12460

Sá, J. M. (2011). *Dinâmica de matéria orgânica do solo e eficiência energética em Latossolo Vermelho do Cerrado* [Doctoral dissertation]. Universidade Federal Rural do Rio de Janeiro, Seropédica.

Salton, J. C., Kichel, A. N., Arantes, M., Kruker, J. M., Zimmer, A. H., Mercante, F. M., & Almeida, R. G. (2013). *Sistema São Mateus-Sistema de integração lavoura-pecuária para a região do Bolsão Sul-Mato-Grossense* [Comunicado Técnico, 186]. Embrapa Agropecuária Oeste. https://www.infoteca.cnptia.embrapa.br/infoteca/bitstream/doc/960712/1/COT2013186.pdf

Salton, J. C., Mercante, F. M., Tomazi, M., Zanatta, J. A., Concenço, G., Silva, W. M., & Retore, M. (2014). Integrated crop-livestock system in tropical Brazil: toward a sustainable production system. *Agriculture, Ecosystems and Environment*, 190, 70–79. https://doi.org/10.1016/j.agee.2013.09.023

Sant-Anna, S. A. C., Jantalia, C. P., Sa, J. M., Vilela, L., Marchão, R. L., Alves, B. J. R., Urquiaga, S., & Boddey, R. M. (2017). Changes in soil organic carbon during 22 years of pastures, cropping or integrated crop/livestock systems in the Brazilian Cerrado. *Nutrient Cycling in Agroecosystems*, 108(1), 101–120. https://doi.org/10.1007/s10705-016-9812-z

Santos, G. G., Silveira, P. M. D., Marchão, R. L., Becquer, T., & Balbino, L. C. (2008). Macrofauna edáfica associada a plantas de cobertura em plantio direto em um Latossolo Vermelho do Cerrado. *Pesquisa Agropecuária Brasileira*, 43, 115–122. https://doi.org/10.1590/S0100-204X2008000100015

Santos, J. V., Raimundo, B. L., Dias, J. B., Corso, M. M., Perondi, P. A. O., Pezzopane, J. R. M., Bernardi, A. C. C., Mendes, I. C., & Martin-Neto, L. (2022). The long-term effects of intensive grazing and silvopastoral systems on soil physicochemical properties, enzymatic activity, and microbial biomass. *Catena*, 219, 106619. https://doi.org/10.1016/j.catena.2022.106619

Sarto, M. V., Borges, W. L., Sarto, J. R., Rice, C. W., & Rosolem, C. A. (2020). Deep soil carbon stock, origin, and root interaction in a tropical integrated crop–livestock system. *Agroforestry Systems*, 94(5), 1865–1877. http://dx.doi.org/10.1007/s10457-020-00505-6

Sato, J. H., Figueiredo, C. C., Marchão, R. L., Oliveira, A. D., Vilela, L., Delvico, F. M., Alves, B. J. R., & Carvalho, A. M. (2019). Understanding the relations between soil organic matter fractions and N_2O emissions in a long-term integrated crop–livestock system. *European Journal of Soil Science*, 70(6), 1183–1196. https://doi.org/10.1111/ejss.12819

Sato, J. H., Marchão, R. L., Figueiredo, C. C. D., Vilela, L., Oliveira, A. D., & Carvalho, A. M. (2021). How soil organic carbon fractions affect N_2O emissions in a long-term integrated crop-livestock system: a case study. In H. Datta & R. S. Meena (Eds.), *Soil carbon stabilization to mitigate climate change* (pp. 307–332). Springer. https://doi.org/10.1007/978-981-33-6765-4_9

Sisti, C. (2005) *Estudo de sistemas de uso do solo e rotações de culturas em sistemas agrícolas brasileiros: Dinâmica de nitrogênio e carbono no sistema solo–planta-atmosfera*. [Doctoral dissertation]. Universidade Federal Rural do Rio de Janeiro.

Skorupa, L. A., & Manzatto, C. V. (2019). Avaliação da adoção de sistemas de integração lavoura-pecuária-floresta (ILPF) no Brasil. In L. A. Skorupa & C. V. Manzatto (Eds.), *Sistemas de integração lavoura-pecuária-floresta no Brasil: estratégias regionais de transferência de tecnologia, avaliação da adoção e de impactos* (pp. 340–379)). Embrapa.

Soares, D. S., Ramos, M. L. G., Marchão, R. L., Maciel, G. A., Oliveira, A. D., Malaquias, J. V., & Carvalho, A. M. (2019). How diversity of crop residues in long-term no-tillage systems affect chemical and microbiological soil properties. *Soil and Tillage Research*, 194, 104316. https://doi.org/10.1016/j.still.2019.104316

Sousa, D. M. G., Nunes, R. S., Rein, T. A., & Santos Junior, J. D. G. (2016). Manejo do Fósforo na Região de Cerrado. In R. F. Flores & P. P. Da Cunha (Eds.), *Práticas de manejo do solo para adequada nutrição de plantas no cerrado* (pp. 291–357). Gráfica UFG.

Sousa, D. M. G., Vilela, L., Rein, T. A., & Lobato, E. (1997). Eficiência da adubação fosfatada em dois sistemas de cultivo em um Latossolo de Cerrado. Embrapa. https://ainfo.cnptia.embrapa.br/digital/bitstream/item/102888/1/pesq-10.pdf

Soussana, J. F., & Lemaire, G. (2014). Coupling carbon and nitrogen cycles for environmentally sustainable intensification of grasslands and crop-livestock systems. *Agriculture, Ecosystems and Environment*, 190, 9–17. https://doi.org/10.1016/j.agee.2013.10.012

Souza, E. D. D., Costa, S. E. V. G. D. A., Anghinoni, I., Lima, C. V. S. D., Carvalho, P. C. D. F., & Martins, A. P. (2010). Biomassa microbiana do solo em sistema de integração lavoura-pecuária em plantio direto, submetido a intensidades de pastejo. *Revista Brasileira de Ciência do Solo*, 34, 79–88.

Souza, K. W., Pulrolnik, K., Júnior, R. G., Marchão, R. L., Vilela, L., Carvalho, A. M., ... de Oliveira, A. D. (2020). *Offsetting greenhouse gas (GHG) emissions through crop-livestock-forest integration* [Circular Técnica, 43]. Embrapa Cerrados (INFOTECA-E). https://ainfo.cnptia.embrapa.br/digital/bitstream/item/215587/1/CT-43-Kleberson-Worslley.pdf

Stapledon, R. G. (1941). Make fruitful the land! A policy for agriculture (the democratic order). K. Paul, Trench, Trubner & Co., Ltd.

Stone, L. F., Moreira, J. A. A., & Kluthcouski, J. (2003). Influência das pastagens na melhoria dos atributos físico-hídricos do solo. In J. Kluthcouski, L. F. Stone, & H. Aidar (Eds.), *Integração lavoura-pecuária* (pp. 171–181). Embrapa Arroz e Feijão.

Taylor, H. M., Roberson, G. M., & Parker, J. J., Jr., (1966). Soil strength-root penetration relations for medium-to coarse-textured soil materials. *Soil Science*, 102(1), 18–22.

Torres, C. M. M. E., Jacovine, L. A. G., de Olivera, N., Neto, S., Fraisse, C. W., Soares, C. P. B., de Castro Neto, F., ... Lemes, P. G. (2017). Greenhouse gas emissions and carbon sequestration by agroforestry systems in southeastern Brazil. *Scientific Reports*, 7(1), 1–7.

Vezzani, F. M., & Mielniczuk, J. (2011). Agregação e estoque de carbono em Argissolo submetido a diferentes práticas de manejo agrícola. *Revista Brasileira de Ciência do Solo*, 35, 213–223. https://doi.org/10.1590/S0100-06832011000100020

Vilela, L., Barcellos, A. D. O., & Sousa, D. M. G. (2001). *Benefícios da integração entre lavoura e pecuária* [Documentos, 42]. Embrapa Cerrados. https://ainfo.cnptia.embrapa.br/digital/bitstream/CPAC-2010/23704/1/doc-42.pdf

Vilela, L., Manjabosco, E. A., Marchão, R. L., & Guimarães Jr, R. (2018). *Integrated crop-livestock in western Bahia state: The off-season cattle model* [Circular Técnica 37] Embrapa Cerrados (Infoteca-e). https://ainfo.cnptia.embrapa.br/digital/bitstream/item/178318/1/CircTec-37-Vilela.pdf

Vilela, L., Martha Junior, G. B., Macedo, M. C. M., Marchão, R. L., Guimarães Jr, R., Pulrolnik, K., & Maciel, G. A. (2011). Sistemas de integração lavoura-pecuária na região do Cerrado. *Pesquisa Agropecuária Brasileira*, 46, 1127–1138.

Vilela, L., Martha Junior, G. B., & Marchão, R. L. (2012). *Integração lavoura-pecuária-floresta: alternativa para intensificação do uso da terra*. Revista UFG. https://ainfo.cnptia.embrapa.br/digital/bitstream/item/94789/1/33779.pdfo

Vilela, L., Martha Júnior, G. B., Marchão, R. L., Guimarães Júnior, R., Barioni, L. G., & Barcellos, A. D. O. (2008). Integração lavoura-pecuária. In F. G. Faleiro & A. L. F. Neto (Eds.), *Savanas: Desafios e estratégias para o equilíbrio entre sociedade, agronegócio e recursos naturais* (pp. 933–962). Embrapa Cerrados.

Voisin, A. (1957). *Productivité de L'herbe*. Flammarion.

Volf, M. R., Guimarães, T. M., Scudeletti, D., Cruz, I. V., & Rosolem, C. A. (2018). Potassium dynamics in ruzigrass rhizosphere. *Revista Brasileira de Ciência do Solo*, 42, e0170370. https://doi.org/2018;42:e0170370

Yokoyama, L. P., Viana Filho, A., Balbino, L. C., Oliveira, I. P. D., & Barcellos, A. D. O. (1999). Avaliação econômica de técnicas de recuperação de pastagens. *Pesquisa Agropecuária Brasileira*, 34, 1335–1345. https://doi.org/10.1590/S0100-204X1999000800003

Zia, M., Hansen, J., Hjort, K., & Valdes, C. (2019). Brazil once again becomes the world's largest beef exporter. *Amber waves: The economics of food, farming, natural resources, and RURAL America*. USDA–ERS. https://doi.org/10.22004/ag.econ.302722.

3

Soil Organic Carbon Restoration as the Key Driver to Promote Soil Health in No-till Systems of the Tropics

João Carlos de Moraes Sá, Telmo Jorge Carneiro Amado, Ademir de Oliveira Ferreira, and Rattan Lal

Chapter Overview

Biomass-C agriculture is the primary pathway to develop resilient, profitable, and low-environmental-impact production systems. The legacy that the no-till system (NTS) has promoted since about 1970 has been to minimize soil loss by erosion of 7.1–21.7 billion t and to protect 2.8–8.4 million ha. The NTS on-farm scale is based on three principles: lack of soil disturbance, permanent soil cover, and diverse crop systems. It has been consolidated as the soil management system that produces food in harmony with "Mother Nature." The NTS evolution scale shows clearly that pedological processes and the attendant changes are guided by interconnected steps acting as a system, and each phase implies some challenges that must be addressed. The establishment of the minimum amount of biomass-C to reach the soil C dynamic equilibrium (3.2 and 5.1 Mg ha^{-1} yr^{-1} for subtropical and tropical environments, respectively) was fundamental to understand that production systems that add biomass-C amounts lower than the dynamic equilibrium requirement will have a negative C balance and render soil vulnerable to weather conditions. Therefore, the challenge is to develop production systems with a high biomass-C input based on quantity, quality, and frequency that also lead to low CO_2 emissions. Examples of long-term on-farm systems in tropical regions adding 7.4–8.38 Mg C ha^{-1} yr^{-1} illustrate how NTS can increase soil organic C (SOC) stocks and generate healthy edaphic environments. In addition, NTS can restore C from natural capital (NC) and recover it in a shorter period than originally thought. Data from several NTS-based studies show that it is possible to recover the historic C lost upon the conversion of soil under native vegetation (NV) into agriculture between 49 and 77 years. The restorative path

Soil Health Series: Volume 3 Soil Health and Sustainable Agriculture in Brazil, First Edition.
Edited by Ieda Carvalho Mendes and Maurício Roberto Cherubin.

starts with recovering and protecting aggregates, increasing biological activity, and activating the flow of C and N to be stored in the soil. There is a strong nexus among soil biota, soil health (SH), plant vigor, productivity, healthy food production, human health, and environmental quality. Activities of soil enzymes, such as β-glucosidase (GLU) and arylsulfatase (ARYL), which constitute the pillars of Soil Bioanalysis Technology (SoilBio), have proven to be an efficient tool for understanding field variability in crop yields and for supporting management decisions. A large proportion of the dataset investigated revealed the depleted SOC and total N (TN) stocks that were associated with low enzyme activity and limited microbial biodiversity. These results reinforce the need to fully integrate the three principles of NTS that operate synergistically in order to build and sustain SH in agricultural production systems. Furthermore, the restoration of SOC and TN stocks showed a positive linear relationship with the enzymatic activity that expresses the biological activity. The physical improvement of the soil through aggregation, aeration, and water storage also stimulated the presence of plant growth–promoting soil microorganisms.

Introduction

Soil is an essential component to sustain life and human basic needs such as food, energy, clean water and air, and biodiversity (Keesstra et al., 2016). However, soil is a finite resource and is vulnerable to degradation by human mismanagement (Lal, 2015). The world terrestrial land is 13.00 billion ha (FAO, 2013), and the Global Assessment of Soil Degradation study estimated that nearly 2 billion ha of agricultural land, pasture, forest, and woodland have been degraded (i.e., 15.4%) at some level since the twentieth century, of which 0.140 billion ha (7%) are in Brazil (Oldeman et al., 2014). In this tropical and subtropical region, the continuous use of intensive plow and long-term monocropping grain crops by replacing NV (forest or grassland) has led to the depletion of SOC, a decline in SH, and a reduction in the provision of essential environmental services (Sá et al., 2015, 2017; Wingeyer et al., 2015).

The no-till system (NTS) was adopted 50 years ago in Brazil. The Paraná State, followed by Rio Grande do Sul (Southern Region), were the ecoregions where the NTS was initially adopted to control soil erosion. Over this time, the NTS proved to be one of the most important technologies that supported upscaling of Brazilian agriculture. The NTS contributes to the net removal of C from the atmosphere (Bayer et al., 2000; Sá et al., 2017, 2022), restoration of SOC (Amado et al., 2006; De Oliveira Ferreira et al., 2021a, 2021b; Sá et al., 2006, 2015, 2018, 2022), improved SH, and increased crop yield (Bolliger et al., 2006; Sá et al., 2014, 2015). Application of NTS with minimal soil disturbance, permanent soil biomass mulch cover, diversified crops, and improved plant nutrient availability, is essential to the sustainability of soil productivity (Kassam et al., 2009).

Currently, Brazil has 36.8 million ha under NTS (Sá & De Oliveira Ferreira, 2018), corresponding to around 55.8% of total grain production area. Paraná and Rio Grande do Sul States were the pioneers in South America NTS adoption in the early 1970s, but the system began upscaling during the mid-1990s. In 1990, the land area under NTS growth was 0.9 million ha, with an annual rate of increase of 0.081 million ha yr^{-1}. By the year 2000, there was a significant upscaling, reaching 14.0 million ha, with an annual rate of 1.64 million ha yr^{-1}. The NTS adoption area has doubled from 2010 to 30.3 million ha (1.51 million ha yr^{-1}) and in 2016/2017 reached the current area of 36.8 million ha (Sá & De Oliveira Ferreira, 2018) (Figure 3.1).

The Brazilian Midwest region called "Cerrado," with a potential agricultural area of 70 million ha and one of the biggest world agricultural soil reserves, together with the South are the focus of the present chapter, being relevant agricultural regions in the national and international scenario. The Cerrado has more than 24.84 million ha cultivated under annual crops, of which more than 13.7 million ha are cultivated with NTS (Fuentes et al., 2021). In the recent past, however, these regions of Brazil (South and Midwest) have experienced severe soil degradation and rapid SOC depletion associated with intensive tillage, low crop residue input, and a high rate of SOC mineralization (Bayer et al., 2009;

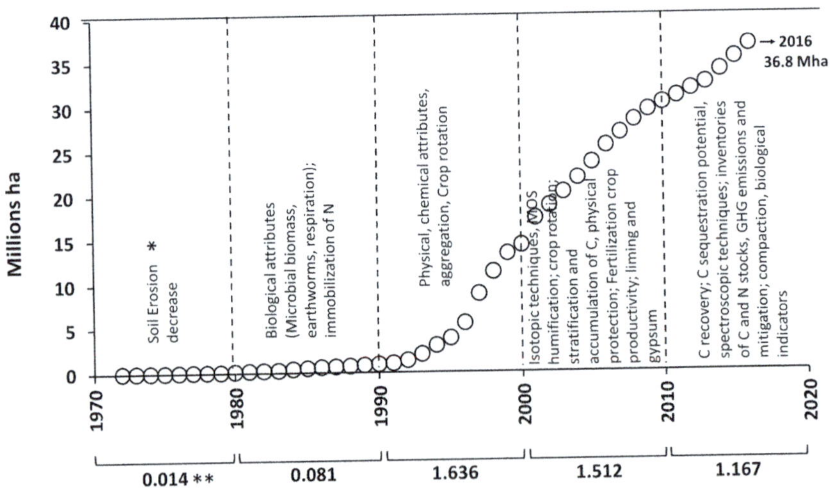

Figure 3.1 Timeline of No-till System (NTS) in Brazil: annual expansion of the area in millions of ha (Mha $year^{-1}$), most relevant topics discussed in each decade (*) and annual rate of expansion of the surface under NTS per decade (**) . Source. Sá and De Oliveira Ferreira (2018).

Dieckow et al., 2009; Sá et al., 2001). The main challenge under a warm and wet climate is the fast turnover of soil organic matter (SOM) and nutrient leaching, which necessitate the adoption of cropping systems with characteristics of high biomass-C inputs that could be reached with complex crop rotation that provide closed nutrient cycling combined with minimal soil disturbance by NTS adoption (Sá et al., 2015; Tivet et al., 2013).

Of the 36.8 million ha managed under NT in the country currently (Sá & De Oliveira Ferreira, 2018), only 10–15% have successfully adopted NTS based on the three pillars (lack soil disturbance, permanent soil cover, and diverse crop systems). About 85%–90% of the area, which today is under NTS, has failed in the use of one or two pillars. Therefore, significant portions of farmers are not applying the three fundamental principles properly, resulting in the occurrence of old problems, such as soil compaction, soil erosion, and low soil C restoration common in the conventional tillage system. However, the mission of adopting the NTS since the 1970s has been to avoid the loss of 7.1–21.7 million t of soil (Baptista & Levien, 2010; Cogo et al., 1984, 2003; Sá et al., 2022), which is equivalent to protecting 2.8–8.4 million ha of land (Sá et al., 2022).

De Oliveira Ferreira et al. (2012) reported that a minimum amount of biomass-C input of 3.21 Mg ha^{-1} yr^{-1} (~7.13 Mg ha^{-1} yr^{-1} of plant biomass) is required to maintain SOC stock. Subsequently, Sá et al. (2015) reported that biomass-C input of 4.15 Mg ha^{-1} yr^{-1} (~9.23 Mg ha^{-1} yr^{-1} of plant biomass) under a long-term (20 years) NTS was necessary to maintain the steady state of SOC stocks for subtropical ecoregions. However, for a tropical climate, Sá et al. (2015, 2022) estimated a minimum rate of biomass-C input of 5.1–5.8 Mg ha^{-1} yr^{-1} to achieve a dynamic C equilibrium according to the specific soil characteristics (Sá et al., 2022), which is equivalent to 11.7–13.3 Mg ha^{-1} yr^{-1} of crop residues (shoot + roots) input. The cropping systems should be redesigned to attain values above the minimum amount of crop residues required to maintain the steady state of SOC and enhance SH. In tropical and subtropical regions, De Oliveira Ferreira et al. (2021a) estimated that it would be necessary to add 50%–75% more crop residues in order to harness the benefits of long-term NTS (the maintenance phase), which is characterized by high nutrient cycling, plant photosynthesis rate, and crop residue input; increased faunal activity and soil microbial biomass C (SMBC); and SOC stabilization through the formation of macroaggregates with a direct impact on water infiltration (Sá, 2004). Long-term experiments (Dalla Nora et al., 2014) show that chemical subsoil amelioration (increased Ca^{2+} and reduced Al^{3+} contents) improves access to plant available water in deep layers and maintains high plant photosynthesis rates even during drought spells (Arachchige et al., 2018; Bossolani et al., 2022). Other management strategies that can stimulate the development of deep root systems in compacted soils are combining mechanical, chemical, and biological strategies (e.g., chisel tillage, lime/gypsum applications, and use of a cover crop [CC] mix) with a 3-month period free of machinery or livestock traffic to restore soil structure (Pott et al., 2021). In a subtropical environment, Inagaki et al. (2016) demonstrated that the use of gypsum after lime can enhance C sequestration by 0.26–0.33 Mg C ha^{-1} yr^{-1}. The deep P fertilization input in tropical soils through chisel tillage was also efficient in stimulating deep root systems (Hansel et al., 2017). The deep root growth creates bio-pores that communicate shallow and deep layers in NTS, allowing C translocation and its accumulation in subsoil (De Oliveira Ferreira et al., 2021b; Nicoloso et al., 2018; Sá et al., 2022). Among different CCs, De Oliveira Ferreira et al. (2021b) highlighted the role of oilseed radish (*Raphanus sativus* L. ssp.) in creating stable bio-pores that are preferential pathways of water infiltration and root growth of cash crops. Promoting the deep root system is a key strategy to enhance biological activity and to improve SH in acid tropical soils. Poffenbarger et al. (2022) estimated that more than 50% of total SOC is stored below 0.20 m, and in the US Midwest it could be a sink of 0.31 Mg C ha^{-1} yr^{-1}. For subtropical and tropical agro-ecoregions, approximately 45%–52% of SOC stock occurs below the 0.40-m layer (Sá et al., 2013, 2022).

Adoption of crop rotation and mix of CCs to fill up short time windows is one strategy that enhances NTS performance in tropical environments because of its positive impacts on SOC, nutrient loss control, soil physical attributes, soil temperature and moisture regulation, and soil biota (Amado et al., 2006; De Oliveira Ferreira et al., 2013, 2018a, 2018b; Sá et al., 2014, 2015, 2022). These positive soil effects lead to increase in crop productivity (De Oliveira Ferreira et al., 2009; Sá et al., 2014). However, such benefits (e.g., improved SH, improved productivity, and increased SOC stock) are realized gradually over a long time period.

The temporal evolution of NTS established in a previous conventionally tilled field can be divided into four distinctive evolution phases (Sá, 2004) (Figure 3.2). The initial phase (0–5 years) is characterized by the low accumulation and high decomposition rate of straw, low SOC content, build-up of aggregates and rearrangement of soil structure, and restoration of fauna and microbial biomass activity as a result of lack of soil disturbance, along with an increase in the demand of N to build up SOC. The transition phase (6–10 years) is characterized by a notable increase in SOC content and soil bulk density (BD), an increase in the size and water stability of aggregates, high water infiltration, and positive effects on faunal and microbial biomass activity in the soil. The consolidation phase (11–20 years) encompasses the accumulation of a mulch layer on the soil surface with crop residues of previous years because of decreased decomposition rates of biomass. The presence of a permanent living mulch layer protects soil aggregates against the deleterious effects of direct raindrop impact. It also protects soil biota against direct sunlight. Therefore, this mulch layer creates a continuous soil C flux input that increases SOC stock and enhances a diversified biota with high SMBC. In this phase, there is a restoration in TN stock, which

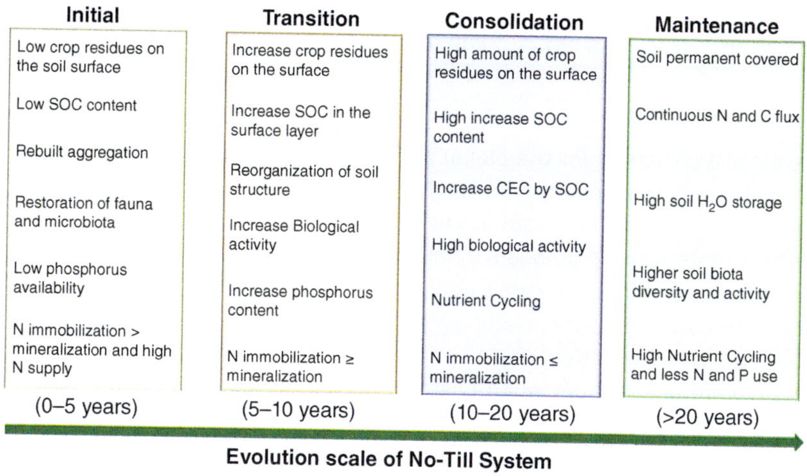

Figure 3.2 No-till system evolution scale. CEC, cation exchange capacity; SOC, soil organic C. Source: Sá (2004).

results in a high plant N supply and supports a diversified and active biota community. It is important to restore the TN stock of soils of the tropics in order to improve soil productivity. In this context, the use of leguminous CCs, such as vetch species (*Vicia villosa* Roth and *Vicia* sativa L.), and crotalaria (*Crotalaria spectabilis* Roth and *Crotalaria juncea* L.) in conjunction with black oats (*Avena sativa* L.), oilseed radish (*R. sativus* L.), and pearl millet [*Pennisetum americanum* (L.) R. Br.], are key strategies. The importance of legume CCs to restore SOC and TN stocks in soils of the tropics has been widely reported by various sequential and complementary studies (Amado et al., 2006; Bayer et al., 2006, 2009; Boddey et al., 2010; De Oliveira Ferreira et al., 2018b; Leal et al., 2020). The nutrient cycling has high efficiency in the third phase (Sá et al., 2009), with low nutrient leaching of key nutrients (K, N, and S). The maintenance phase (>20 years) is characterized by a high accumulation of plant residues of different growth seasons and high SOC stock, stabilization of soil BD, a high amount of macroaggregates, and increased faunal and SMBC (Figure 3.2). In summary, the increase in SOC content, soil temperature and moisture regulation, improvement of soil structure, biological diversity and activity, and high nutrient plant availability stand out as the main components for the maintenance of long-term SH under NTS.

In the next sections, two case studies are presented that document the ability of NTS to restore the C budget of the NC in tropical and subtropical regions of Brazil and that promote the soil environment to be biologically more active and biodiverse with a direct impact on crop yields and soil physical properties.

Case Study 3.1 No-till Systems and the Environmental Value

Context

The biggest challenge for the global agricultural sector in the coming years will be to upscale cropping systems that have the capacity to produce large amounts of biomass-C for input in soil (Sá et al., 2022) with diversity and frequency to improve crop productivity, mitigate climate change, and restore NC. The NC is an asset that supports a flow of benefits to society and encompasses natural resources (i.e., soil, water, air, biodiversity) and represents all the C accumulated in vegetation, organisms, and soil serving the basis for ecosystems to produce food, fiber, wood, energy, and ecosystem services (Mace, 2019; Manning et al., 2018). The NC is one of the pillars on which the national or regional economy is based, and its depletion can limit economic growth and the capacity to deal with global challenges of food insecurity, population growth, finite agricultural land, and climate change adaptation and mitigation (Lal, 2015, 2019; Smith et al., 2016; Wackernagel & Rees, 1997).

Soil, being the foundation of natural resources and the key component for food security, is also critical to environmental sustainability (Amelung et al., 2020; Lal, 2014). Soil can be defined as an "organic C-mediated realm in which solid, liquid, gaseous, and biological components interact from nanometer to landscape scale to make production and ecosystem services essential for all terrestrial life" (Lal, 2014). Soil is the main capital component of agricultural, livestock, forestry, and environmental services production and is also an essential NC whose value depends on its status of use or degradation level (Lal et al., 2020).

The SOC content directly or indirectly controls chemical, physical, and biological attributes and is the main indicator for SH and productive capacity (Conceição et al., 2005; Sá et al., 2009) and must be considered in the valuation of NC (Crossman & Bryan, 2009). Carbon accumulation in the soil occurs only when the biomass-C input is greater than the C losses by erosion, biological decomposition, and leaching and C exportation in grain harvest (Lal, 2014; Lal et al., 2018; Sá et al., 2015). The challenge is to manage the biomass-C input in order to exceed the minimum amount of C to achieve the dynamic equilibrium level according to the site-specific soil characteristics. For the tropical region, the minimum amount of C input has been estimated to be in the range of 5.1–5.8 Mg C ha^{-1} yr^{-1} (Sá et al., 2015, 2022) in order to compensate for the fast biological decomposition associated with precipitation and high temperatures, whereas for the subtropical region C input could range between 3.2 and 4.0 Mg C ha^{-1} yr^{-1}

(De Oliveira Ferreira et al., 2012, 2021a; Sá et al., 2014). Cropping systems involving tropical pasture with well-developed root systems that could reach 2 m depth (e.g., *Brachiaria* spp.) are very efficient in building up aggregates that protect the new C inputs (see Chapter 2). The input of C of *Brachiaria* may be 3–3.5 Mg C ha^{-1} for aboveground (Boddey et al., 2010) and 5–15 Mg C ha^{-1} for the root biomass (de Oliveira et al., 2004), leading to C sequestration rates of >1 Mg ha^{-1} yr^{-1}. Sá et al. (2022) reported that production systems involving quantity, quality, and frequency of C input can regenerate and restore degraded soils and the ecosystem over a 77-year period.

The SOC accumulation and the C restoration of partial or total NC is closely associated with the capacity of the cropping systems to intensify the annual input of biomass-C (Briedis et al., 2016, 2018; De Oliveira Ferreira et al., 2016, 2018a, 2021a, 2021b; Sá et al., 2014, 2015, 2022). The level and the turnover time of C recovery depends on the biome through its attributes comprising of climate, altitude, land scape position, parent material, soil texture, and soil disturbance and the intensity of crop rotation (Amundson & Biardeau, 2018; Paustian et al., 2016; Sá et al., 2022; Smith et al., 2016).

Some questions to be addressed for advancing this scenario of NC restoration include (a) assessing the economic value of total C and C pools in soil and their losses upon conversion of natural vegetation to agricultural land use and (b) evaluating the benefits to the society with the adoption of NT cropping systems (NTCS). Therefore, the study by Sá et al. (2022) is based on the hypothesis that NTCS—with permanent soil cover, crop diversification, and the addition of amount of biomass above the C dynamic equilibrium level—can create positive C balance in soil, stimulate a diversified biota, promote soil aggregation, and recover the C stock either partially or totally to the level of the NV.

This case study builds on Sá et al. (2022) and is aimed at evaluating the NC for (a) continuous plow-based tillage (SCT) with degraded and C-depleted soil, (b) continuous NTCS soil with C-restored soil, and (c) the capacity of NTCS through intensive cropping systems via C enhancement restoring NC and the environmental value through the monetization of each incremental increase of Mg C ha^{-1} using the land market price.

Experimental Procedures

The present study was based on a set of two tillage experiments conducted in the subtropical and tropical climate of Brazil. The first experiment, at the Ponta Grossa (PG) site represents the subtropical environment in southern Brazil, and the second experiment, at the Lucas do Rio Verde (LRV) site, represents the humid tropical environment in Midwest located in Central Brazil.

(Continued)

Case Study 3.1 (Continued)

Geographic coordinates, location, characteristics related to soil and climate, and experimental details of these two sites are presented in Table 3.1 and in Sá et al. (2015).

Table 3.1 Geographic location, characteristics related to soil and climate, and experimental details of the two studied sites.

Characteristic	Subtropical (Ponta Grossa site)	Tropical (Lucas do Rio Verde site)
Research institute	Agronomic Institute of Paraná	Rio Verde Foundation
Geographic coordinates	25°09′S, 50°09′W	13°00′S, 55°58′W
Altitude, m asl	865	380
Soil		
Brazilian Soil Classification System	Latossolo Vermelho distrófico	Latossolo Vermelho-Amarelo
FAO classification	Ferralsol	Ferralsol
USDA soil taxonomy	Oxisol, Rhodic Hapludox	Oxisol, Typic Haplustox
Parent material	Shale	Shale and sandstone
Clay, silt, sand, $g\ kg^{-1}$	650, 240, 110	402, 106, 492
Clay type	Kaolinite, hematite, gibbsite	Kaolinite, gibbsite, hematite
Climate		
Classification[a)]	Subtropical, Cfb	Tropical, Aw
Mean annual temperature, °C[b)]	18.5	25.7
Mean annual precipitation, mm	1545	1950
Mean annual potential evapotranspiration, mm	900–1000	1300
Native vegetation	Herbaceous species, with shrubs and woody plants	Cerrado, tropical savannah dominated by arboreal species

a) Aw, tropical savanna climate, summer hot and very wet, winter hot and dry; Cfb, humid subtropical climate, summer and winter wet and warm summer.

b) Values of mean annual temperature, precipitation, and potential evapotranspiration refer to the period of 1954–2001 for the Ponta Grossa site and 1990–2009 for the Lucas do Rio Verde site.

Note. Data from Sá et al. (2022).

The experiments comprised two tillage systems: (a) soil under continuous conventional plow-based tillage (CT), including disc plowing to a depth of 0.18–0.20 m, followed by two narrow-disk plowing events, and (b) soil under continuous NTCS, according to the principles of the NT system, which consists of no soil plowing (any disturbance is restricted to the seeding row), permanent soil cover, and diversification of crop rotation. The two sets of experiments were conducted next to one another and comprised of an undisturbed soil under NV for a reference of non-anthropized land use that represents the NC. At both the PG and LRV sites, treatments were established as whole plots, which were divided into six (PG site) and three (LRV site) subplots for the purpose of soil sampling (Figure 3.3).

The land area for the PG site experiment was converted from NV to pasture in 1967 and continued for 10 years (Figure 3.3). In 1978, the land was converted to annual crops and cultivated in CT for 3 years until the tillage experiment began in 1981. The long-term tillage experiment includes (a) CT, consisting of disc plowing to a depth of 20 cm twice a year (autumn and spring), followed by two narrow-disk plowing events to a depth of 10 cm, and (b) NT, consisting of direct seeding through the previous crop residues with minimum soil disturbance. The experiment was set up as a whole plot, where dimensions were 50 m by 140 m for the CT and 100 m by 100 m for the NT treatment. Six sub-plots were identified in CT (25 m by 46 m) and NT (33 m by 50 m) treatments for soil sampling (Figure 3.3). The cropping sequence for both tillage systems comprised of two annual crops, including soybean [*Glycine max* (L.) Merr.] and maize (*Zea mays* L.) in the summer, alternating with oats (*Avena strigosa* Schreb), wheat (*Triticum aestivum* L.), and vetch in the winter (Table 3.2).

The area from the LRV site experiment was converted from NV to cropland in 1986 (Figure 3.3) and managed under CT for 15 years until the experiment was begun in 2001, which also includes CT and NT. The experiment was set up as whole plots, with the plot dimensions of 216 m by 252 m for CT and 216 m by 42 m for NT, and later divided into three sub-plots (72 m by 42 m for NT and 216 m by 84 m for CT) for soil sampling. Because the original goal of the experiment was to compare the standard tillage management used in the region (soybean–cotton [*Gossypium hirsutum* L.] sequence under CT) and the NT system with different cropping systems (i.e., diverse biomass-C inputs), the cropping sequence between tillage systems was not the same (Table 3.2). Thus, the cropping sequence for CT was soybean–cotton, and that for NT involved soybean as the first crop in the summer followed by a mix of sorghum [*Sorghum bicolor* (L.) Moench] and brachiaria (*Brachiaria ruziziensis* Germ. & Evrard) as the second crop.

(Continued)

Case Study 3.1 (Continued)

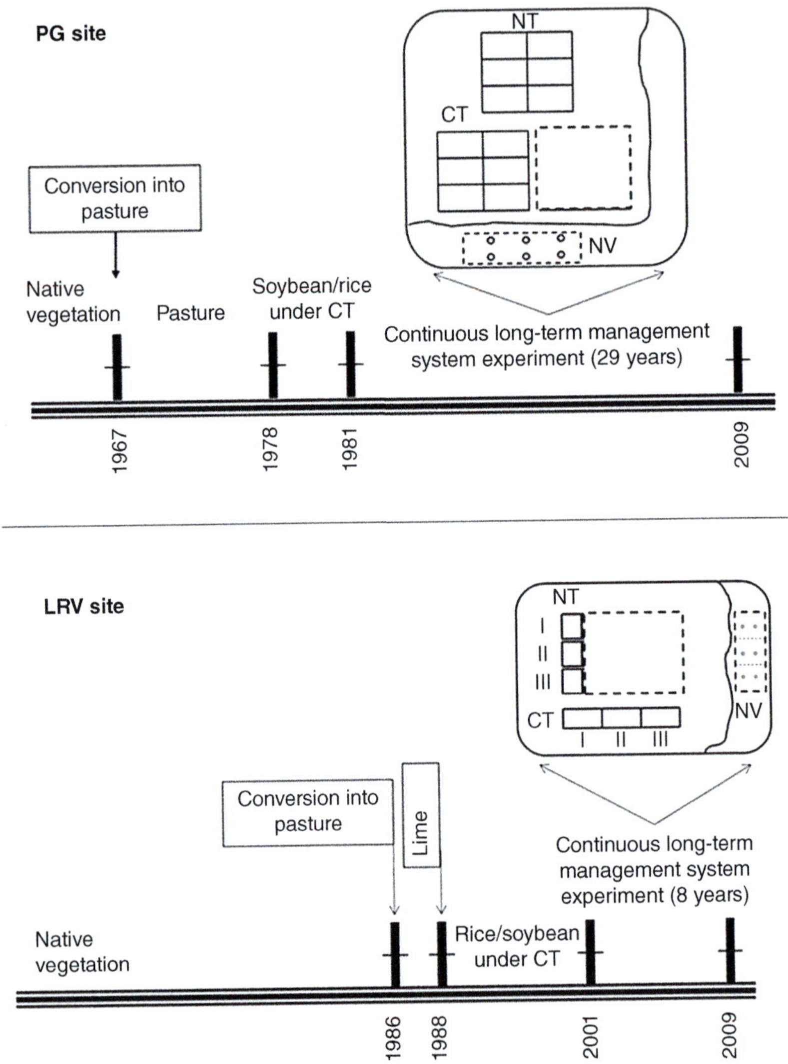

Figure 3.3 Chronology of land use and experimental design at the long-term tillage experiments at the Ponta Grossa (PG) site and the Lucas do Rio Verde (LRV) site. Sá et al. (2022).

Table 3.2 Crop sequence and carbon input in the 29-year experiment period at the Ponta Grossa site and in the 8-year experiment period at the Lucas do Rio Verde site.

Tillage systems[a)]	Crop sequence[b)]	C input[c)] Cumulative	Annual
		–Mg ha^{-1}–	
	Ponta Grossa site		
CT	S/O – M/O – S/W – S/O + V – S	86.1	3.07
NT	S/O – M/O – S/W – S/O + V – S	116.1	4.15
	Lucas do Rio Verde site		
CT	S/M – S/M – S/C – S/C – S/C – S/C – S/C – S/C	32.1	4.01
NT	S/Sg + B – Rc – M – S/Sg + B – S/Sg + B – S/Sg + B – M – S/Sg + B	67.0	8.38

a) CT, conventional tillage; NT, no-till.
b) B, brachiaria (*Brachiaria ruziziensis* Germ. & Evrard); C, cotton (*Gossypium hirsutum* L.); L, white lupin (*Lupinus albus* L.); M, maize (*Zea mays* L.); O, black oats (*Avena strigosa* Schreb); S, soybean [*Glycine max* (L.) Merr.]; V, vetch (*Vicia sativa* L.); W, wheat (*Triticum aestivum* L.).
c) Carbon input represents aboveground (shoots) and belowground (roots).

Soil Sampling and Biomass-C Input Estimation Soil samples to a depth of 1 m were obtained after 29 and 8 years of experimentation at the PG and LRV sites, respectively. Soil bulk samples for depth intervals of 0–0.05, 0.05–0.10, 0.10–0.20, 0.20–0.40, 0.40–0.60, 0.60–0.80, and 0.60–1.00 m were collected at four random points per plot and composited. Soil BD was determined by the core method (Blake & Hartge, 1986), and cores were obtained by using steel cylinders of known volume (0.05 m in height and diameter) in the middle of each depth. At each site, soil samples were also obtained for the same depth intervals from an area under NV in close proximity to the experiment to determine the baseline SOC and vegetation C stock. Inputs of aboveground and belowground C under crops were calculated by using the harvest index, the root-to-shoot ratio, and the C concentration in biomass of each crop. Estimates of the aboveground, belowground, and litter biomass under NV were made based on the site measurements according to the regression equation proposed by Brown et al. (1989).

(Continued)

Case Study 3.1 (Continued)

Soil Fractionation, Carbon Analyses, and Calculations Soil fractionation was done following a process described by Briedis et al. (2018). Briefly, the first fractionation step involved a partial dispersion and physical sieving of the soil (<2 mm) to obtain three size fractions: (a) coarse particulate organic C (cPOC; 250–2000 μm), (b) microaggregate fraction (μAgg; 53–250 μm), and (c) easily dispersed mineral-associated organic C (dMAOC; <53 μm). The second step involved a density fractionation of the μAgg fraction (obtained in the first step) to obtain the light POC (LPOC) fraction and a heavy fraction, which was fully dispersed and sieved to obtain two fractions: (a) intra-microaggregate POC (iPOC; >53 μm) and (b) MAOC derived from microaggregates (μMAOC; <53 μm). The third and last step involved acid hydrolysis of the dMAOC and μMAOC fractions (obtained in the previous steps) to obtain nonhydrolyzable (i.e., NH-dMAOC and NH-MAOC) and hydrolyzable fractions of C (i.e., H-dMAOC and H-μMAOC). To realize the objectives of this study, some fractions were merged and considered the hydrolyzable fractions only as MAOC, nonhydrolyzable fractions as recalcitrant MAOC (R-MAOC), and all particulate C fractions (i.e., cPOC, LPOC, and iPOC) as POC.

The total organic C content of bulk soils and fractions, herein referred to as SOC, was determined for the finely ground samples (<150 μm) by the dry combustion method using a CN elemental analyzer (TruSpec CN, LECO). The soil type of this study (Red Latosol, equivalent to Oxisol, FAO soil classification 2008) has less than 0.2% of inorganic C (Sá et al., 2013). The C stocks of bulk soils and fractions were calculated on equivalent soil mass–depth basis (Ellert & Bettany, 1995), where soil under NV was used as reference.

Calculations

(1) C losses (Mg ha^{-1}) = SCT SOC stock – SNV SOC stock
(2) SOC recovered by NTCS (Mg ha^{-1}) = SOC stock NTCS – SOC stock SCT
(3) Conversion rate of biomass-C into SOC by NTCS (%) = Cumulative biomass-C input/SOC restored by NTCS
(4) C balance: C losses – C Recovered by NTCS (Mg ha^{-1})
(5) Soil C sequestration rate by NTCS (Mg ha^{-1} yr^{-1}) = C recovered by NTCS/years of experiment
(6) Turnover time by NTCS = C balance/C sequestration rate (years).

Statistical Analyses

Differences among treatments for SOC stock were tested through ANOVA. Mean values were compared using the LSD at the 5% probability level

(Webster, 2007). Analysis was performed by soil depth, and results were considered statistically significant at $P < 0.05$. All statistical calculations were carried out using R version 2.11.1 (Team, 2006), package *aov*. SigmaPlot 12.0 was used for graphic representation.

C-budget and Turnover Time Perspective

The recovery of C lost from NC is closely related to the strategy of C input via biomass (i.e., aboveground and belowground of crop residues) through adoption of innovative production systems (Sá et al., 2022). In the subtropical region where winter has lower temperatures and frequent frosts, soil C decomposition rates are several times lower than those for the tropical region (Bayer et al., 2006; Sá et al., 2006, 2014, 2015). In this study, the annual rate of C loss due to intensive soil tillage (for the 0- to 10-m layer) for the subtropical altitude region (PG site) was 0.29 Mg C ha^{-1} yr^{-1} and was approximately 10 times lower than that of 3.0 Mg C ha^{-1} yr^{-1} for the tropical region (LRV site). Based on the SOC losses from the 0- to 20-m layer (Figure 3.4a), the rate of C loss for the PG site was almost three times lower (0.85 Mg C ha^{-1} yr^{-1}) than the rate observed for the LRV site (1.83 Mg ha^{-1} yr^{-1}) (Figure 3.4b). These data highlight that one of the biggest world agricultural soil resources has a rapid rate of C loss that puts its sustainability and the NC at high risk.

The minimum amount of C to be added annually to maintain the dynamic equilibrium level for the PG site was estimated at 3.2 Mg ha^{-1} yr^{-1} (De Oliveira Ferreira et al., 2012, 2021a). The annual biomass-C input through NTCS in this study was 4.0 Mg C ha^{-1} yr^{-1}, representing 25% more C above the equilibrium level, and with a C sequestration rate of 0.59 Mg C ha^{-1} yr^{-1} that effectively offset the soil C losses. However, the SOC sequestration rate of this study is far from the rates of 1.30 (Sisti et al., 2004), 1.61 (Sá et al., 2014), and 1.95 Mg C ha^{-1} yr^{-1} (Diekow et al., 2005) reported for the subtropical region, indicating that the production system can be vastly improved to add more biomass-C and trigger an increase in the rate of C sequestration. At the PG site, the crop rotations with maize and black oats (S/O – M/O – S/W – S/O + V – S) led to a total biomass-C input of 116.1 Mg ha^{-1} over 29 years (annual biomass-C rate of 4.15 Mg ha^{-1}). Such a high input of biomass-C resulted in a rapid POC recovery. Based on another study conducted in close proximity to the PG site, De Oliveira Ferreira et al. (2018b) reported that the contribution of biomass C of 5.9 Mg ha^{-1} yr^{-1} in a sandy clay Oxisol and 7.0 Mg ha^{-1} yr^{-1} in a clayey Oxisol resulted in a similar level of POC recovery as observed in the present study. However, for the tropical zone (LRV site), even the high annual C input

(Continued)

Case Study 3.1 (Continued)

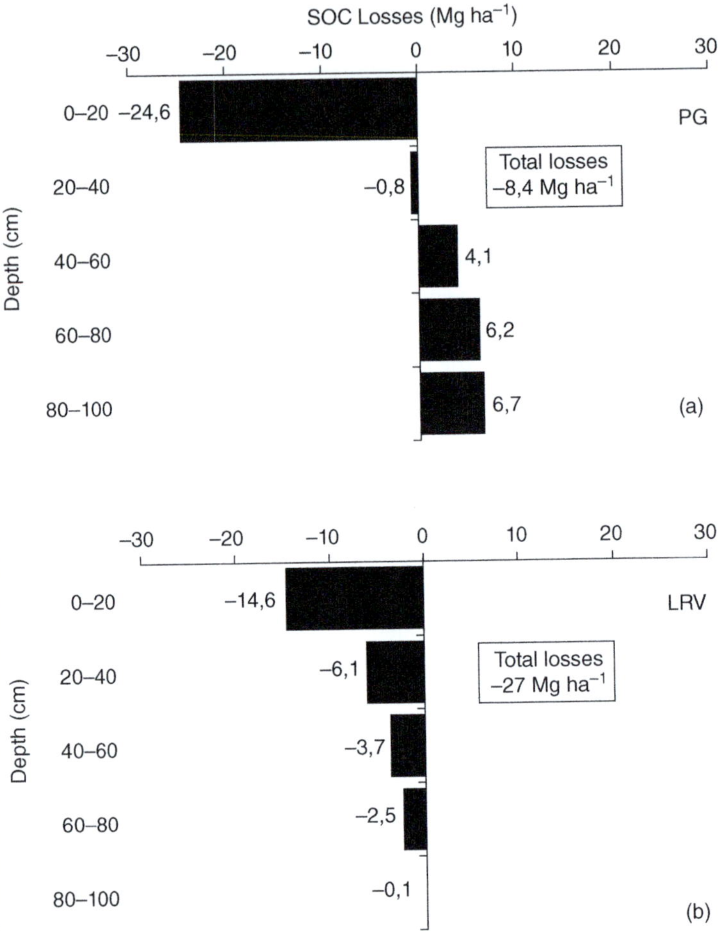

Figure 3.4 Soil organic C (SOC) losses at the 0.00- to 1.00-m layer due continuous tillage: (a) subtropical region (Ponta Grossa [PG] site) and (b) tropical region (Lucas do Rio Verde [LRV] site). Sá et al. (2022).

of 8.4 Mg C ha^{-1} yr^{-1} (i.e., 18.6 Mg ha^{-1} yr^{-1} of crop residues) was not high enough to recover the POC over the 8-year period. The data from the present study indicated the addition of 98.4% (18.6 Mg ha^{-1} yr^{-1}) more crop residues than that required to attain the steady-state. Furthermore, POC fraction was found to be the most management sensitive C fraction (Sá et al., 2014).

The results presented above support the hypothesis that the formation of extra-large macroaggregates and the C protection within intra-macroaggregate may be the primary mechanisms to enhance the stability of SOC. Therefore, by managing the quantity (amount), quality (diversity of crop residues), and annual frequency of addition (number of annual additions), the turnover time of C in the soil can be shortened from 35 to 23 years. Thus, the intensification of biomass-C input leads to two distinct pathways to the stabilization of SOC: (a) protection of C within intra-macroaggegates and reducing its oxidation (Briedis et al., 2016, 2018; De Oliveira Ferreira et al., 2018b; Tivet et al., 2013) and (b) maintenance of the flow of labile C and N forms that stimulate the transformation of these organic compounds into more stable levels and leading to the formation of new macroaggregates (De Oliveira Ferreira et al., 2018b; Sá et al., 2014). At the LRV site, an increase in the rate of decomposition tripled the rate of the observed loss of SOC because of the distribution of rainfall throughout the summer combined with the high mean annual temperature and high soil fertility. At the PG site, however, greater C protection occurred within intra-aggregates due to the high clay content and lower annual average temperature. The rapid decomposition throughout the year due to the more uniform rainfall distribution at the PG site may have generated a greater amount of free labile C compounds, weakly adsorbed on dispersed clay and silt particles and prone to translocation into the 0.40–1.00 m layer with the percolating water. Indeed, the leaching caused by intense rainfall induced translocation of 11.0 Mg C ha^{-1} into the 0.40- to 1.00-m layer (Figure 3.4a).

Furthermore, the input of biomass C is not enough to offset the losses of C by decomposition and the erosional impact of heavy rainfalls. Translocation of C into the subsoil may be due to the displacement of decomposable labile compounds (i.e., low-molecular-weight organic compounds) and C associated with dispersed clay particles in the surface layer (Bayer et al., 2000; Tivet et al., 2013). The inversion of layers by soil tillage and the downward percolation of water to the deeper layers of the soil profile carrying displaced clay particles attached to organic compounds is an important pathway (Figure 3.4a) (Tivet et al., 2013).

The recovery of C occurs through the POC fraction (De Oliveira Ferreira et al., 2018b) that acts as an indicator of changes in soil C stock and even of SH, as reported for the mid-Atlantic United States by Weil et al. (2003). Furthermore, it is the pathway to enhance C fractions associated with minerals. Therefore, the management of the trinity (i.e., quantity, diversification, and frequency of addition of biomass C) is the main strategy to maximize the input of biomass-C input for enhanced soil C sequestration. Thus, the rate of C loss can be curtailed by reducing the rate of oxidation and enhancing the

(Continued)

Case Study 3.1 (Continued)

aggregation of dispersed particles into stable macroaggregates (De Oliveira Ferreira et al., 2018b) and even in microaggregates (Arachchige et al., 2018), increasing input of biomass-C into the system. Cover crops with a prolific and robust root systems plus POC could be a key component of the re-aggregation process because it tends to form macroaggregates (De Oliveira Ferreira et al., 2018b) that increase intra-aggregate protection of encapsulated C (Six et al., 2000). In addition, the soil surface covered with crop residues or a live CC leads to the protection of more labile forms of C among aggregates (De Oliveira Ferreira et al., 2018b; Tivet et al., 2013). A diversified and active soil biota boosted by crop diversification is relevant to increase the soil C retention efficiency. The C losses at the LRV site in the 0.40- to 1.00-m layers of ~40% may have been caused by the upward movement of water (capillarity) from the subsoil layers below 1 m (Figure 3.4b), stimulating the oxidation of the C of the soil between the 0.40- and 1.00-m layers during the dry period. The bare soil surface without vegetation cover is prone to the upward flow that may cause the emission of the labile and more volatile C compounds out of the soil profile. The proposed trinity may stabilize more labile forms of C over time increase soil resilience (Sá et al., 2014). Formation of a stronger bond between C compounds with clay minerals over time would lead to the creation of a more humified organo–mineral complex and resilient macroaggregates with a longer half-life of C and more stable soil structure. Therefore, cropping systems based on input of biomass-C with quantity, quality, and frequency will create a continuous C and N flow that stimulates the interactions among soil attributes (Figure 3.5), increasing C translocation, restoring SH, and enhancing the NC.

Natural capital balance

The conversion of NV to agricultural land use had a strong environmental impact caused by the loss of 100% of C stored in the aboveground vegetation through the burning of biomass. The C losses due to the continuous soil tillage, despite being proportionally smaller compared with those caused by the burning of vegetation, have led to severe changes in the physical, chemical, and biological attributes affecting soil functionality and health. The economic value of C losses in tropical regions due to frequent soil tillage was equivalent to 21% of the NC (Sá et al., 2022).

Considering the economic investment made in the purchase of land and its improvement to achieve a high grain yield, the loss of NC limits the return capacity of the economic capital invested.

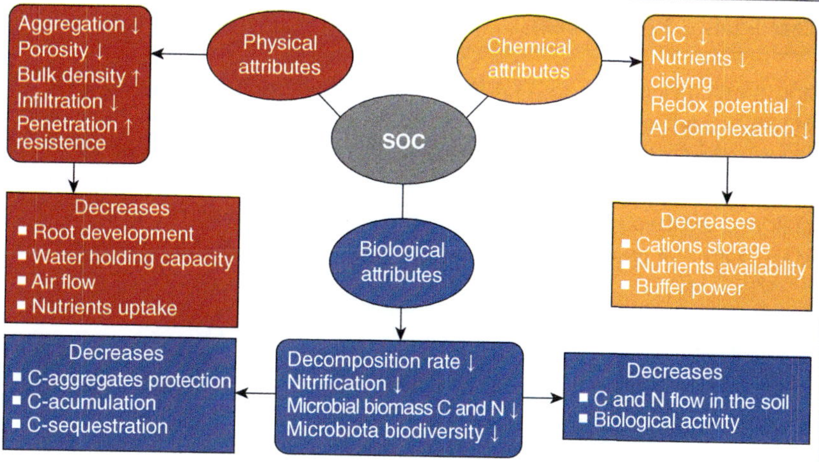

Figure 3.5 The impacts of the decline in soil organic C (SOC) on soil physical, chemical, and biological attributes. Up and down arrows next to each physical, chemical, or biological attribute indicate an increase or a decrease, respectively. Sá et al. (2022).

The drastic loss of the terrestrial C stock can collapse the soil functionality chain, which is set in motion by changing the interactions that occur between the SOC content and the attributes (physical, chemical, and biological) that drive the capacity of the soil to provide ecosystem services (Sá et al., 2009). The trigger for recovering the NC is set in motion by the intensification of input of biomass-C–creating production systems that include CCs with robust and abundant root systems and aboveground biomass, providing a continuous soil cover and protecting the stability of aggregates and continuous exudates as C input to feed a diverse biotic community. The annual rate of C sequestration for the PG site at 0.59 Mg C ha^{-1} yr^{-1}, based on 4.0 Mg C ha^{-1} yr^{-1}, was 25% more than the minimum amount of C required to attain the dynamic equilibrium level for that region (i.e., 3.2 Mg C ha^{-1} yr^{-1}) (De Oliveira Ferreira et al., 2012, 2021a). However, the rate of soil C sequestration can be increased by 2.2–3.3 times (Diekow et al., 2005; Sá et al., 2015; Sisti et al., 2004), and the turnover time of NC can be reduced to <25 years (Sá et al., 2022). For the tropical region, the minimum C required to attain the dynamic equilibrium level is 5.1 Mg ha^{-1} yr^{-1} (Sá et al., 2015), and the annual biomass-C input was 8.38 Mg C ha^{-1} yr^{-1}, in which the C sequestration rate was 2.15 Mg C ha^{-1} yr^{-1}. Even the high current rate of C sequestration can be increased to 2.6 Mg C ha^{-1} yr^{-1} (Sá et al., 2006) by enhancing the crop production system based on a mixture of leguminous

(Continued)

Case Study 3.1 (Continued)

and grass species (i.e., crotalaria + brachiaria; pigeonpea [*Cajanus cajan* (L.) Millsp.] + corn + brachiaria; pearl millet + corn), which can add high amounts of biomass-C to the system. Increased longevity of C in soil can be achieved by minimizing the SOC oxidation level and re-aggregating the dispersed clay and silt fractions to create new macro-aggregates by cementation and stabilization of micro-aggregates in soil (Tisdall & Oades, 1982) (Figure 3.6).

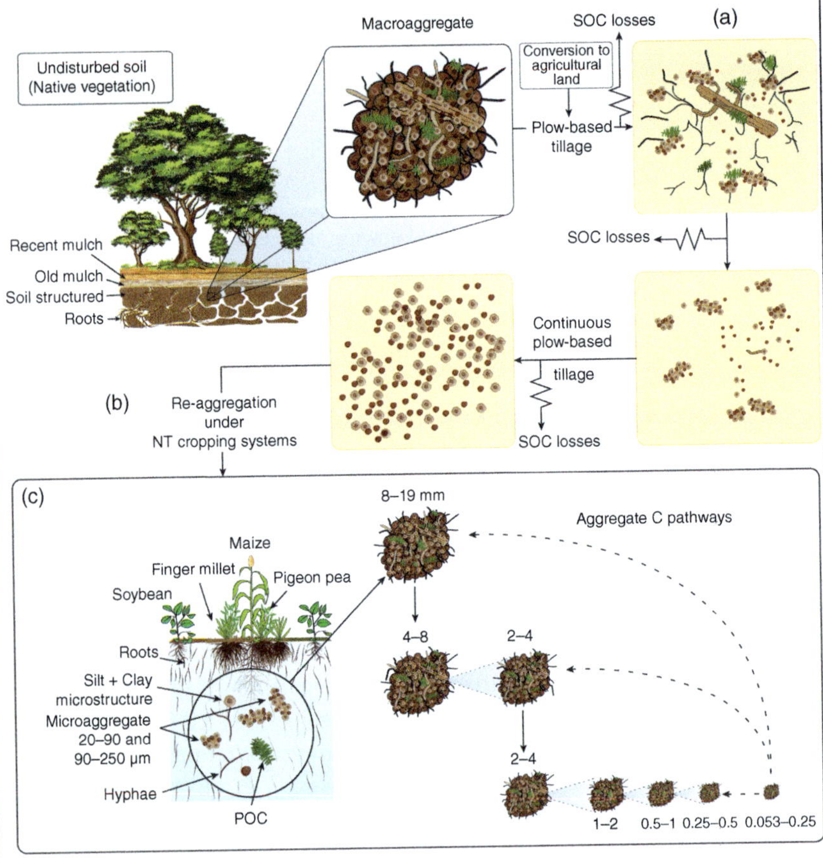

Figure 3.6 Schematic diagram illustrating the effect of land use conversion on soil aggregation and soil organic C (SOC) dynamics. (a) Conversion of native vegetation to conventional plow-based tillage (CT) disrupts soil aggregates, promotes the dispersion of clay particles and silt + clay microaggregates, and diminishes the formation of new aggregates. (b) Build-up of aggregates by diverse biomass-C inputs under no-tillage (NT). (c) Processes involved in accumulation and redistribution of SOC between aggregate size classes. Aggregate C pathway for soils under NT cropping systems. Tivet et al. (2013)/with permission of Elsevier.

Carbon Balance in a Production System with High Carbon Input

Making Brazilian agriculture emission neutral or negative can be achieved by maximizing the soil C sink by increasing the CO_2 fixed by crops during photosynthesis and its accumulation in soil profile upon land application of biomass-C. Thus, C accumulation in soil occurs only when the biomass-C input is greater than the losses of C by erosion, decomposition, and leaching (Lal et al., 2018; Sá et al., 2015). The challenge will be to manage the biomass-C input in a way that exceeds the amount of C to reach the potentially minimum level of C in the soil. For the tropical environment, the amount of C input is estimated at 5.1–5.8 Mg ha^{-1} yr^{-1} (Sá et al., 2015). An example of a farming system in Lucas do Rio Verde–MT documents that achieving a positive C balance is feasible (Table 3.3), which can be calculated by using the following expression:

$$\begin{aligned}\text{Soil C balance (net C sequestration rate)}\\ &= (\text{Biomass-C input} \times \text{Humification rate})\\ &\quad - (\text{Soil C decomposition rate} \times \text{Soil C stock})\end{aligned}$$

where the C is the balance is represented by the net C sequestration rate considering the biomass-C input as the amount via biomass (shoots + roots), the humification rate represented by the % of C conversion from the plant biomass to soil C at the LRV site (19% or 0.19 annually, respectively); the soil C stock was calculated from the data obtained from soil analyses (soil C content and soil BD) for a specific layer.

The positive C balance in Lucas do Rio Verde-MT, Capuaba Farm (Table 3.3) is the result of the combination of quantity of biomass added + quality of biomass + frequency of addition (Sá et al., 2022). The positive balance can be achieved by adopting the site-specific NTS (i.e., permanently covered soil and with diverse crop rotation).

Highlights

Restoring and sustaining agricultural SH is a complex challenge in a tropical environment and is achievable by adoption of the continuous NT through the input of biomass-C based on the high quantity, quality, and frequency to offset the C losses.

The SH evaluated by the C pathway and restoration from cultivated soils involves a crop production system that adds a high input of biomass-C in conjunction with the use of NT that reduces the rate of oxidation, promotes a continuous C and N flow, stimulates the C protection

(Continued)

Case Study 3.1 (Continued)

within intra-macroaggregates, and leads to SOC sequestration and restoration of soil structure.

Table 3.3 Carbon balance in a crop production system under no-till system in Lucas do Rio Verde-MT (Capuaba Farm)

C balance components	C balance period; rotation				
	Year 1; soybean–corn + Brq Rz	Year 2; soybean–corn + Crt Ochrol	Year 3; soybean–mix of CC	Year 4; rice + radish	Year 5; soybean–corn + Brq Rz
			–Mg ha^{-1}–		
Annual DM input (shoots)	15.94	14.21	9.03	5.84	17.69
Annual DM input (root)	3.94	3.19	2.54	1.49	3.92
Total annual DM input	19.88	17.40	11.56	7.32	21.61
Total C input by biomass	7.02	6.12	5.01	3.14	9.59
Gross C sequestration rate	1.33	1.16	0.95	0.60	1.82
Soil C stock	40.50	41.35	42.02	42.46	42.55
Annual C decomposition rate (losses)	0.49	0.50	0.50	0.51	0.51
Net C sequestration rate = C balance	**0.84**	**0.66**	**0.45**	**0.09**	**1.31**
Average 5-year net sequestration rate					0.66

Abbreviations: Brq Rz, *Brachiaria ruziziensis*; Crt Ochrol, *Crotalaria ochroleuca*; CC, cover crops, DM, dry matter.
Note. Data from J. C. M. Sa (personal communication, 2022).

The turnover time based on the C sequestration rates for subtropical and tropical sites led to recovery of the C lost from NV and soil in 77 and 49 years, respectively. These results indicate that achieving C recovery is feasible in one or two generations (25–50 years). Further, the key strategy to restore SH evaluated by NC is to manage the C sequestration rate of the production systems through high input of biomass-C and low CO_2 emissions.

In summary, the restoration of SH evaluated by NC must be based on the following strategies: (a) increasing the biomass-C input according to the three pillars that support conservation agriculture (CA): adding the amount at the rate above that of the dynamic equilibrium, guaranteeing a controlled soil C oxidation rate, and promoting the continuous C and N flow to reorganize the aggregates and the new structure; (b) stimulating macroaggregation and enhancing C protection within intra-aggregate and stabilizing it over time and restoring soil functionality by improving soil's chemical, physical, and biological attributes; and (c) managing the rate of input of biomass-C as a key strategy to decrease C turnover time, leading to the restoration of NC and ecosystem services to society.

Case Study 3.2 Soil Health Check-up Based on Enzyme Activity and DNA Analysis in Brazilian No-Till Fields with Different Yield Environments

Context

The current global challenges are driven by the population growth and the attendant high demands for food, fiber, biofuel, energy, water, and overall agricultural products at low price in order to be accessible to the whole society. Such ever-increasing human demands exert an unprecedented pressure on both natural and anthropogenic (managed) ecosystems.

The agroecosystems are already faced with numerous challenges of soil degradation, loss of SOC stock, soil compaction, water and wind erosion, temporal yield instability, lack of crop rotation, pressure from pests, and excessive reliance on agrochemical industry inputs. Over and above these, new challenges include climate change, loss of biodiversity, weed resistance to herbicides, harnessing land-based C sink, decreasing the environmental footprint, adopting greenhouse gas emission–neutral or –negative agriculture, increasing use efficiency of inputs, improving the shortage of rural labor, conciliating aspiration of urban and rural societies, and producing more with less. Together, these complex challenges could be summarized by being sustainable without impairing the farmer's conditions to keep farm activity economically viable. Therefore, soil and crop management should be aimed at decreasing environmental, economic, and social costs of business-as-usual in production agriculture.

(Continued)

Case Study 3.2 (Continued)

Soil health is defined as “the continued capacity of the soil to function as a vital living ecosystem that sustains plants, animals and humans subsistence” (Lehmann et al., 2020; USDA-NRCS, 2023). Understanding soil as a vital living system implies accepting that the most important soil processes are mediated by a diverse biotic community that is dynamic but vulnerable to farming operations and therefore must be carefully managed to support soil’s ability to enhance chemical, physical, and biological attributes to sustain life (Moebius-Clune et al., 2016).

There is a strong interaction among soil biota, SH, plant vigor and crop yield, healthy and nutritious food production, human health, and environmental quality that goes from the microbial scale to the landscape scale and beyond. Soil health is a critical subcomponent of ecosystem health (Passinato et al., 2021). Thus, soil should be managed to create a favorable habitat for the soil microbiome in order to stabilize the temporal community that can stimulate continuous energy and C fluxes into soil. This would make soil a C sink, increase nutrient cycling, improve soil aggregation, enhance water infiltration and storage, decrease plant natural pest control, strengthen resilience to abiotic stresses, promote plant vigor, and increase crop productivity.

The microbiome, associated with the rhizosphere, must be efficiently managed by farmers so that microbiota can play their critical role as plant growth promoters and regulators. Thus, farmers must have tools to evaluate SH at the farm level. In addition, there is a strong need to redesign cropping systems that are beneficial to the whole society and to future generations.

This case study reports results from a recent investigation of GLU and ARYL activity as bioindicators to access SH at the field level in croplands located at main Brazilian grain-producing states (Passinato et al., 2021). These soil enzymes constitute the pillars of SoilBio technology launched in July 2020 (see Mendes et al., 2019, 2021, and Chapter 10).

Experimental Procedures

This case study involved assessment of seven farms distributed over four key agricultural states and that follow management practices such as continuous NTS (range, 10–20 years) and that have applied precision farming to generate a soil and crop dataset that allowed to establish a relationship between crop performance (averaged over a 3-year period) and soil biological activity accessed by SoilBio technology (last year period). In each farm the field selected was split into three crop yield environments (high, medium, and low crop performance) based on a dataset composed of yield maps, satellite images, and farmer and consultant experience regarding the crop yield and within-field spatial variability. In general, these fields had soybean yields ranging from 3.6 to 5.6 Mg ha^{-1}, being classified above the national average yield of 3.5 Mg ha^{-1} in 2020/2021. The Brazilian agricultural ecoregions selected for this study were the Southern, Central-West, and Northeastern regions (Figure 3.7). Additional details about these studies are presented by Passinato et al. (2021). In each study environment, three random soil samples were collected, with a total of 63 sampling points.

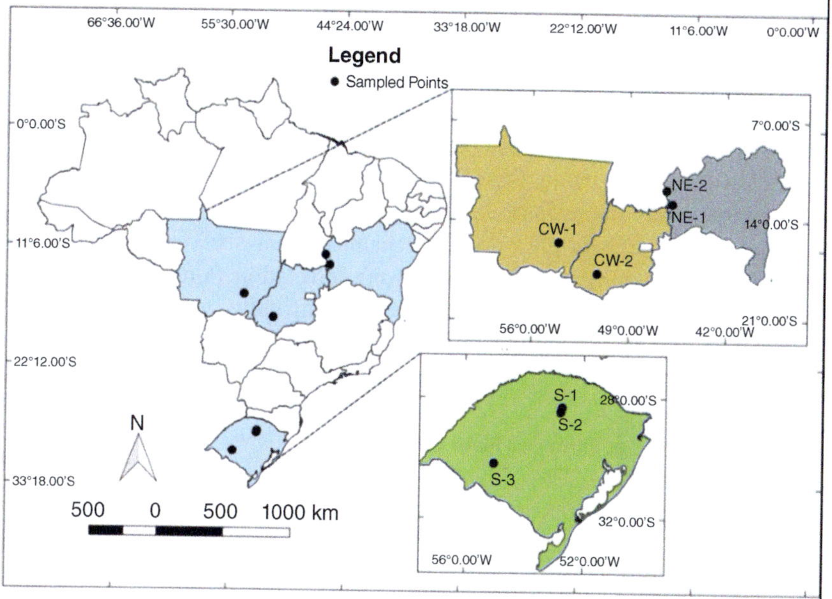

Figure 3.7 Fields sampled in the main Brazilian agro-ecoregions. Agro-ecoregions are differentiated by colors: green = Southern (municipalities S-1 = Carazinho, S-2 = Não-Me-Toque and S-3 = Rosário do Sul, in Rio Grande do Sul state); brown = Central West (municipalities CW-1 = Primavera do Leste, in Mato Grosso state and CW-2 = Rio Verde, in Goiás state); gray = Northeastern (municipalities NE-1 = Luis Eduardo Magalhães and NE-2 = Placas, in Bahia state). Passinato et al. (2021)/MDPI/CC BY 4.0.

Soil was sampled at the 0- to 0.10-m depth for chemical analysis. Soil samples for bio-analysis were collected at the same depth during the soybean reproductive crop stage. The methodology proposed by Tabatabai (1994) and Mendes et al. (2019) was followed to evaluate GLU and ARYL enzyme activity. The SOC and TN stocks, at the same soil depth of previous soil analyses,

(Continued)

Case Study 3.2 (Continued)

were determined by the dry combustion method through elemental analysis (FlashEA 1112, Thermo Fisher Scientific), taking into account the soil bulk density to express the results in equivalent mass method (Ellert & Bettany, 1995). One of the fields selected for this study was a pioneer project on Precision Agriculture–Aquarius Project (Amado & Santi, 2011) with a 22-year database. This field is located in Não Me Toque, Rio Grande do Sul. In this field, in addition to the soil enzyme activity, soil DNA characterization was performed in three crop yield environments. The DNA soil was analyzed by Biome Markers (https://biomemakers.com) in California. The DNeasy 420 PowerLyzer PowerSoil Kit (Qiagen) was used for DNA extraction. Bacterial and fungal microbial communities were identified according the 16S rRNA and ITS marker regions. DNA sequencing was carried out by the Illumina MiSeq instrument using pair-end sequencing. The soil of Aquarius project is a clay Rhodic Hapludox. More details are given by Passinato et al. (2021) and Shaffer et al. (2022).

The enzyme activity data and soil chemical analysis were submitted to ANOVA ($p < 0.01$ and $p < 0.05$) and Pearson's correlation. Principal component analysis was used to identify the soil chemical attributes that were drivers of SoilBio. The Tukey test and other statistics were performed using the R Studio statistical package.

Soil Health Status at Farm and Field Levels

Even though average SOM contents in the seven fields (Figure 3.7) were similar in the high-, medium-, and low-crop-yield environments, the activity of GLU and ARYL captured the differences in low- and high-yield environments. This trend indicated that the soil enzyme activity is a more sensitive indicator of yield environment than SOM content (Figure 3.8). In spite of similar SOM levels, in general, lower-yield environments were associated with low enzyme activity, whereas higher-yield environments had high enzyme activity.

The SOC stocks in the 0.0- to 0.10-m soil depth ranged from 8.5 to 45.0 Mg ha^{-1}, depending on the ecoregion, soil texture and type, and management. Despite the broad range, SOC stock is associated with differences among ecoregions (Southern is subtropical, and Central-West and Northeast are tropical regions). In addition, the SOC stock was affected by within-field variability, defined in this study as crop yield environments. The TN stocks, for the same depth, ranged from 0.7 to 3.2 Mg ha^{-1}. GLU enzyme activity ranged from 90.8 to 256.0 mg p-nitrophenol (PNP) kg^{-1} soil h^{-1}, and ARYL enzyme activity ranged from 32.3 to 287.3 mg PNP kg^{-1} soil h^{-1} (Figure 3.9). This broad range of data creates an opportunity to explore the relationships of soil attributes and managements with enzyme activity.

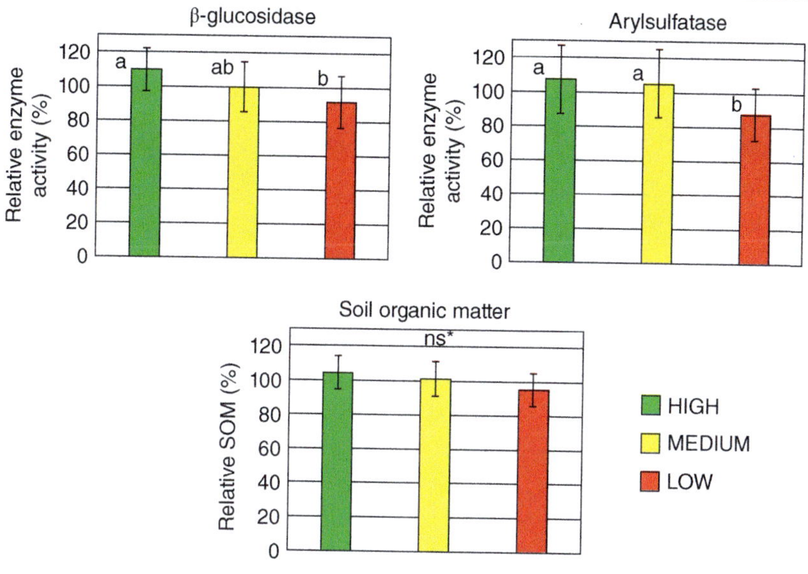

Figure 3.8 β-Glucosidase and arylsulfatase activities and soil organic matter (SOM) content relative to the average of the field in different yield potential environments (high, medium, and low) in seven Brazilian fields from main agro-ecoregions. Values followed by the same letter do not differ statistically by Tukey test ($p < 0.05$); *ns = not significant ($n = 63$). Error bars correspond to standard deviation. Passinato et al. (2021)/MDPI/CC BY 4.0.

The GLU and ARYL activity presented linear relationships, with SOC and r^2 values of 0.53 ($p < 0.001$) and 0.48 ($p < 0.001$), respectively (Figure 3.9). The higher enzyme (GLU) activity in this study was obtained with SOM in the range of 4%–5%, whereas the maximum ARYL activity was observed with SOM content of 3.5% (Passinato et al., 2021). Therefore, these data suggest that SOM content of about 3% may be a reference value to have enzyme activity that supports satisfactory soil bioactivity. In addition, the quality of SOM, mainly the labile fractions, should also be taken into account to deeply characterize the relationship of SOM and soil enzyme activity. In this case study, the SOM level prior to the adoption of CA was usually in the range of 2%–3%, suggesting that fields with 4%–5% SOM have young C (labile), which originated from inputs of biomass by cash crops and CCs.

The GLU and ARYL enzyme activity relationships with TN stock were positive and significant at R^2 of 0.65 ($p < 0.001$) and 0.59 ($p < 0.001$), respectively. The maximum enzyme activity was observed with TN stock of >2.5 Mg ha^{-1} in the 0.0- to 0.10-m soil depth. In this study, the TN had a linear positive relationship with ARYL and GLU, suggesting that legume CCs and organic fertilizer

(Continued)

Case Study 3.2 (Continued)

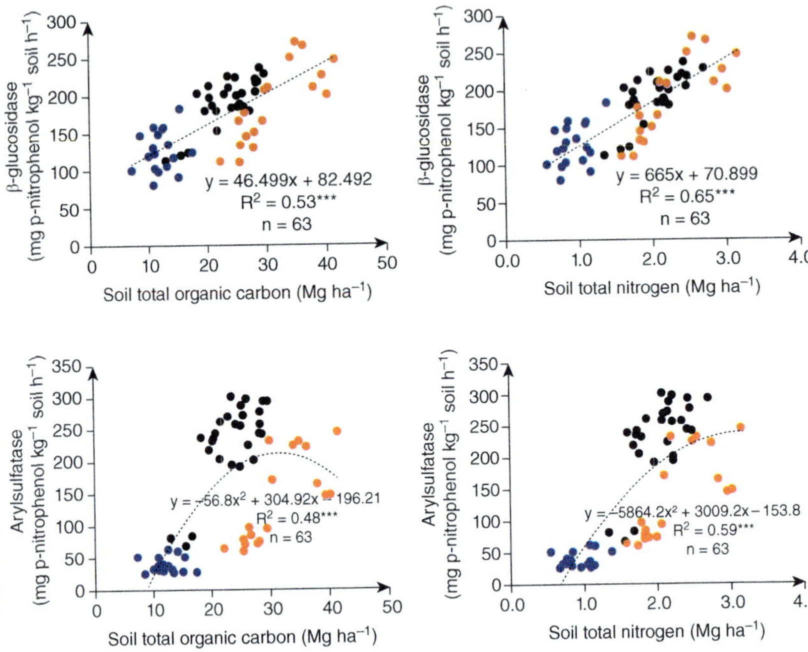

Figure 3.9 Relationships between the activities of β-glucosidase and arylsulfatase with total organic C (TOC) and total N (TN) at the 0- to 0.10-m soil depth, determined by dry combustion. $^{***}p < 0.001$. Orange, blue, and black circles represent Central-West, Northeast, and South regions, respectively. Passinato et al. (2021)/MDPI/CC BY 4.0.

rich in N may be the key strategy to restore TN stocks and boost soil biological activity under NTS. The biological inputs for replacing some agrochemicals to suppress root and foliar crop diseases and promote plant growth are growing in Brazilian agriculture. These biological inputs have numerous advantages, such as decreasing the cost and environmental footprint. In addition, they play a complementary role of enhancing soil biology.

The GLU and ARYL activity relationships with SOC and TN have been previously reported by Mankolo et al. (2012) in Alabama, with *r* values of 0.58 and 0.66 for GLU and ARYL, respectively. Xu et al. (2015) reported that SOM and TN had an *r* value of 0.83 with GLU and ARYL enzyme activity. These values are similar to those observed in the present study. Such strong relationships may be attributed to the uptake of nutrients from labile fractions of SOM by microorganisms for use as energy and nutrient sources. In addition, an increase in SOM content retains soil moisture, enhances cation exchange capacity (CEC), reduces the xenobiotic effect on soil biota, and promotes the buildup of aggregates that strengthen a diverse microbial community (Passinato et al., 2021).

Soil texture determines soil aggregation, CEC, SOM content, and water holding capacity, which affects soil biological activity. In the present study, as expected, the increase of sand content in soil texture was associated with a decrease in activity of both enzymes investigated. This case study showed that, despite remarkable differences in soil types, soil texture, elevation, and climate, GLU and ARYL activity was an efficient indicator of SH, with high levels where appropriate soil management was adopted.

Examining chemical soil data from Southern Brazil laboratories involving 35,362 soil analyses, Tiecher et al. (2016) reported that 55% had a SOM content <2.5%. In the present case study, around 40% of the sampling points of the dataset had low a SM content of <2.5%. These numbers suggest that a significant percentage of farmers, including NTS farmers, who represent 90% of soybean growers in Southern Brazil, are not fully integrating the three principles of NTS in order to enhance SH. Therefore, there is a need to redesign cropping systems with increased crop rotation and increased C and N inputs and to enhance the use of CCs in short windows between cash crops to restore SOC and TN stocks and support a diversified soil biota (Amado et al., 2006; Leal et al., 2020).

On the other hand, around 15% of the farms evaluated in the dataset were successful in restoring SOC and TN stocks and had high enzyme activity. In Southern Brazil, the use of a winter cocktail of CCs (black oat + hairy vetch + rye + oilseed radish adagio + oilseed radish common) supported the greatest levels of enzyme activity (210 and 287 mg PNP kg^{-1} soil h^{-1}, respectively). In the Central-West ecoregion, the use of *Brachiaria* spp. as pasture/CC and the cocktail of summer CCs had similar effects on enzyme activity (233 and 199 mg PNP kg^{-1} soil h^{-1}, respectively) (Passinato et al., 2021). These farmers, regardless of being located in the Southern or in Central-West agroecoregions, were able to accomplish continuous input of 10–15 Mg ha^{-1} yr^{-1} of aboveground biomass in NTS. This biomass, as a primary soil source of C and N input, decreased soil temperature and increased soil moisture via the mulching effect and enhanced the associated microbial activity throughout the season. The fungal community is one of the most sensitive to soil disturbance by frequent tillage and to the quality of biomass input, mainly by the use of leguminous CCs. Therefore, the cocktail of CCs with a mixture of legumes and grasses is a key strategy to increase beneficial fungal communities.

(Continued)

Case Study 3.2 (Continued)

In the field located in Não Me Toque (Rio Grande do Sul), Aquarius Project, the relationships between GLU and ARYL activity with soil DNA analysis were investigated (Figure 3.10). There were positive and significant relationships between GLU and ARYL and soil diversity, based on DNA analysis, with R^2 of 0.85 ($p < 0.01$) and 0.79 ($p < 0.05$), respectively. These results reinforce the role of these two soil enzymes as efficient bioindicators of soil biology and SH. Moreover, there exists a knowledge gap in the literature regarding this relationship. The data from this case study show a strong connection between soil enzyme activity and microbial diversity in agricultural soils.

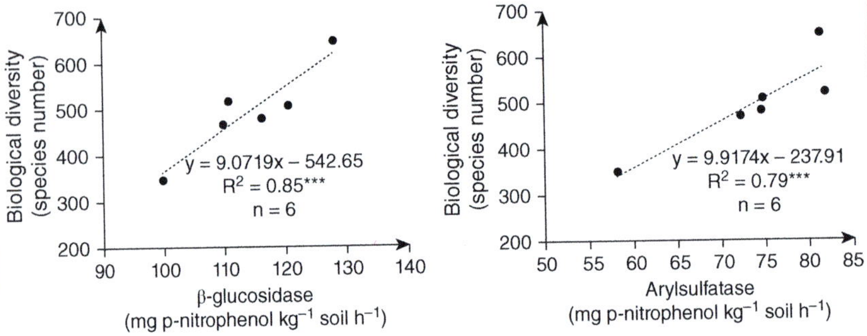

Figure 3.10 Relation of β-glucosidase and arylsulfatase activity under no-till system and biodiversity assessed by DNA characterization. Não Me Toque, Rio Grande do Sul. $^*p < 0.05$; $^{**}p < 0.01$. Passinato et al. (2021)/MDPI/CC BY 4.0.

The nexus between crop yield environment and the biota diversity was also evaluated in this study through visual inspection (Figure 3.11). In general, the high-crop-yield environment had a better balance among the species of microorganisms that comprise the community. On the contrary, the low-crop-yield environments had a lower microbial diversity, with remarkable supremacy of some species, in detriment of community diversity, which was generally associated with a negative impact on plant vigor and grain yield (Passinato et al., 2021).

Plant–microbe interactions in the rhizosphere are key drivers of plant health and crop yield. The mechanisms by which plant vigor is enhanced by microbe activity in the rhizosphere are highly complex and diverse, but some important ones related to plant nutrition include bioavailability of soil nutrients, such as biological N fixation and P and K solubilization, and biotic (insect and diseases) and abiotic (temperature and drought) stress alleviation (Passinato et al., 2021). Among the important genuses of plant

Figure 3.11 Biota diversity evaluated by DNA analysis in two distinct crop yield environments. Não Me Toque, Rio Grande do Sul. Colored circle represents different types of microorganisms, and the size of circle is related to its representativeness in the community.

growth–promoting organisms are *Bacillus* spp. and *Pseudomonas* spp., known as plant growth–promoting bacteria. On the other hand, the *Penicillium* and *Trichoderma* genuses are reported as plant growth–promoting fungi by providing soil pathogenic biocontrol. Biological N fixation can be provided by *Bradyrhizobium* and *Rhizobium* genus, with important roles in restoration of TN stock.

Both soil aggregation and the plant efficiency in soil water and nutrient uptake are increased by arbuscular mycorrhizal fungi. Plant root growth is stimulated by *Azospirillum* spp., and recently Brazilian farmers have been evaluating the use of *Bacillus aryabhattai* as a tool to deal with short-term drought during grain the cash crops season. The use of beneficial biological organisms in Brazilian agriculture was estimated at approximately 25 million ha^{-1}, with a sharp increasing trend in its adoption over the last 5 years. Typically, farmers are moving away from inputs of single organisms, with a specific goal of using a community or blend of organisms with more broad-based goals of improving soil and plant health. In addition, there is a trend toward a one-time biological input rather than several inputs throughout the year, compromising CC and cash crop inputs (T. Amado, personal observation by following innovate farmers in biological inputs). In the present case study, the plant-beneficial organisms were more abundant in the high-crop-yield environment compared with the medium- and low-crop-yield environments (Figure 3.12). These results suggest that farmers and advisors should be aware of the relevance of soil biotic diversity restoration in low-yield environments.

(Continued)

Case Study 3.2 (Continued)

This goal could be achieved by the intensification of CCs and biological inputs in that environment to restore its productivity potential.

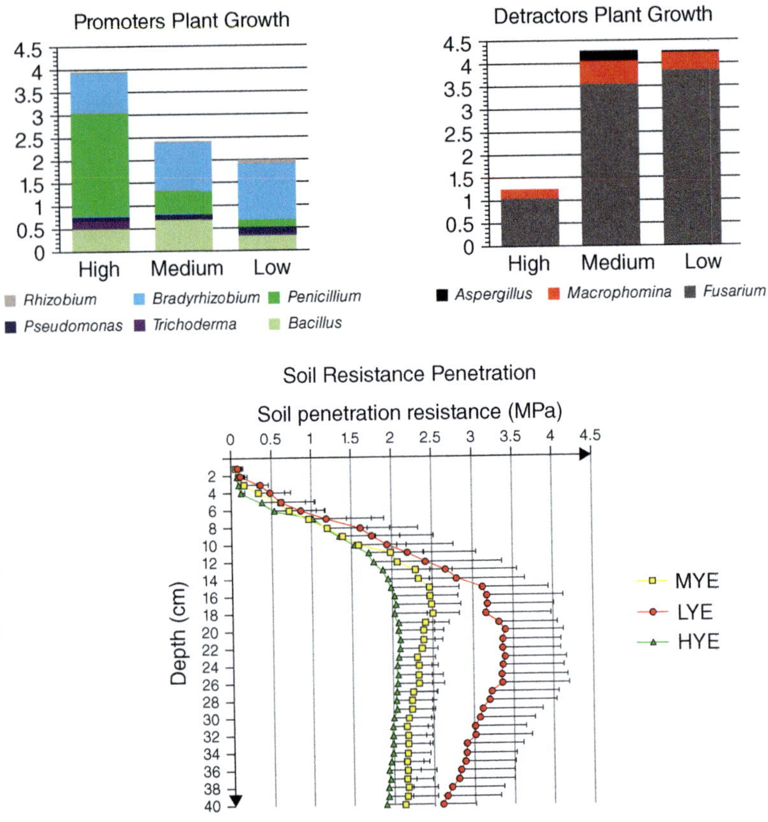

Figure 3.12 Accumulated proportion of most abundant genera of promoters' plant growth and detractors' plant growth by crop yield environment and soil resistance penetration by crop yield environment. Não Me Toque, Rio Grande do Sul. Passinato et al. (2021)/MDPI/CC BY 4.0.

The most prevalent and important plant pathogenic microorganism genuses observed in this case study were *Fusarium*, *Macrophomina*, and *Aspergillus*; these organisms were more prevalent in medium- and low-yield environments compared with a high-yield environment. Therefore, the plants growing in medium- and low-yield environments were under higher plant disease pressure in comparison to that for the high-yield environment. This could be an important factor that helps to explain the lower crop performance in these environments compared with a high-yield environment. Fields managed under soybean monocropping and without crop diversification with CCs and crop rotation are more prone to diseases caused by *Fusarium* and *Macrophomina* (Pérez-Brandán et al., 2013).

Finally, the relationship between soil penetration resistance (PR) and crop yield environment was also investigated in this Aquarius Project field. A PR >2.5 MPa is assumed to be a critical value for soybean root growth, impairing grain yield. The lower-yield environment had PR above the critical value because of poor soil physical attributes that determine water infiltration rates and soil aeration. The lack of soil aeration aggravates soil disease biota, such as *Fusarium* spp., and creates a hostile environment to plant-beneficial organisms.

Highlights

This case study added more scientific evidence supporting GLU and ARYL enzyme activity as an efficient indicator of soil biodiversity under NTS. In addition, these soil enzymes serve as an efficient tool to distinguish low-yield environments from high-yield environments and help to understand within-field crop variability, which is important in precision agriculture.

The dataset obtained in this study show that a large proportion of sites had depleted SOC and TN stocks, which were associated with low enzyme activity, limited biodiversity, and high soil penetration resistance, which are restrictive to root growth. These results reinforce the need to fully integrate the three principles of NTS that operate synergistically to build a friendly environment for soil biota and sustain SH in agroecosystems.

The soil microbiome characterization was an efficient tool to establish the connection between crop yield performance and the rhizosphere-associated microbiome. The high-crop-yield environment had a more diverse soil microbial community associated with a higher presence of plant growth-biota promoters (e.g., *Bacillus* spp., *Penicillium* spp., *Trichoderma* spp., *Pseudomonas* spp., *Bradyrhizobium* spp., and *Rhizobium* spp.). On the other hand, in the low-crop-yield environment there was a higher abundance of *Fusarium* spp. and *Macrophomina* spp., which limit plant growth. On the contrary, in this low-yield environment there was a lower frequency of plant growth-biota promoters. The frequency of plant pathogenic organisms was linked to poor soil physical attributes, such as high soil penetration resistance, low infiltration rate, and poor soil aeration attributes associated with high soil compaction.

(Continued)

Case Study 3.2 (Continued)

This case study was based on the research by Passinato et al. (2021), which was recognized at the eighth World Congress on Conservation Agriculture. This work evidenced that applying the three integrated principles of CA with a focus on crop rotation and CCs in the cropping system results in enhancement of soil biota, plant vigor, and crop yield. The key drivers of the SH regeneration process are the restoration of SOM and TN contents that could be achieved through crop diversification, the use of CCs, control of soil compaction, improvement of soil aggregate build-up, and stimulation of diverse and active plant root growth that will support plant-growth–promoting microorganisms and a diverse soil biota community. Results from this case study support Brazilian farmers' recent efforts to work with biological products and CCs as a way to decrease the dependence on agrochemicals. This approach allows increasing the efficiency of CA with positive impacts for farmers and society. This case study also supports the use of the SoilBio technology as a tool for SH assessments improving farmers' knowledge about the status of SH and providing support for future management adjustments.

Future studies

Brazil has one of the largest NTS areas in the world at ~38 million ha. The benefits of NTS include soil erosion control, saving time in crop establishment, and saving diesel and labor. NTS also provides the basis of agricultural intensification and allowing two or even three crops per year despite the scarcity of rural labor. It enhances SOC sequestration and natural ecosystem preservation by increasing crop yields and water use efficiency. However, the level of success in achieving these goals is highly variable, depending on a complex interaction of management practices, soil type, climate conditions, and farmer's knowledge and perseverance in keeping the systems free of soil disturbance for long term. The emerging production systems with a high biomass-C input based on quantity, quality, and frequency that promote high SOC sequestration rates are the key drivers to achieve SH, plant productivity, food health, and environment preservation. In this context, innovative Brazilian farmers are using a mixture of three to seven different species of CCs (polycultive) combined with inputs of biological organisms throughout the year (CCs and cash crops inputs), gradually replacing agrochemical inputs. There is a need of on-farm research to track the progress in reducing chemical inputs, improving plant vigor, increasing yield, and enhancing SH. The SoilBio and temporal evaluations of SOC stocks are important tools for characterizing the success of NTS adoption and function as warning indicators to redesign cropping systems to restore SH. Understanding spatial crop performance and variability of soil attributes creates an opportunity to develop variable application rates of biological inputs that currently are not available in the digital platforms, allowing better integration among precision agriculture, conservation agriculture, and regenerative agriculture. The high volume of data generated by precision/digital agriculture will support important management decisions in terms of regenerative and conservation agriculture. This integration will help to overcome the challenges of upscaling processes from the micro- or plot scale to the landscape, national, and continental scale.

References

Amado, T. J. C., Bayer, C., Conceição, P. C., Spagnollo, E., De Campos, B.-H. C., Veiga, d., & M. (2006). Potential of carbon accumulation in no-till soils with intensive use and cover crops in southern Brazil. *Journal of Environmental Quality*, 35, 1599–1607. https://doi.org/10.2134/jeq2005.0233

Amado, T. J. C., & Santi, A. L. (2011). Using precision farming to overcome yield-limiting factors in southern Brazil Oxisols: A case study. In D. Clay, J. Shanahan, & J. Francis Pierce (Eds.), *GIS applications in agriculture—Nutrient management for improved energy efficiency* (Vol. 3, pp. 31–60). CRC.

Amelung, W., Bossio, D., de Vries, W., Kögel-Knabner, I., Lehmann, J., Amundson, R., Bol, R., Collins, C., Lal, R., Leifeld, J., Minasny, B., Pan, G., Paustian, K., Rumpel, C., Sanderman, J., van Groenigen, J. W., Mooney, S., van Wesemael, B., Wander, M., & Chabbi, A. (2020). Towards a global-scale soil climate mitigation strategy. *Nature Communications*, 11, 5427.

Amundson, R., & Biardeau, L. (2018). Opinion: Soil carbon sequestration is an elusive climate mitigation tool. *Proceedings of the National Academy of Sciences*, 115, 11652–11656.

Arachchige, P. S. P., Hettiarachchi, G. M., Rice, C. W., Dynes, J. J., Maumann, L., Wang, J., Karunakaran, C., Kilcoyne, A. L. D., Attanayake, C. P., Amado, T. J. C., & Fiorin, J. E. (2018). Sub-micron level investigation reveals the inaccessibility of stabilized carbon in soil microaggregates. *Scientific Reports*, 8, 16810.

Baptista, J., & Levien, R. (2010). Methods of soil tillage and its influence in water erosion and initial development of *Eucalyptus Saligna* on the inceptisoil of the central depression of the State of Rio Grande do Sul. *Revista Arvore*, 34, 567–575.

Bayer, C., Dieckow, J., Amado, T. J. C., Eltz, F. L. F., & Vieira, F. C. B. (2009). Cover crop effects increasing carbon storage in a subtropical no-till sandy Acrisol. *Communications in Soil Science and Plant Analysis*, 40, 1499–1511.

Bayer, C., Martin-Neto, L., Mielniczuk, J., Pavinato, A., & Dieckow, J. (2006). Carbon sequestration in two Brazilian Cerrado soils under no-till. *Soil and Tillage Research*, 86, 237–245.

Bayer, C., Mielniczuk, J., Amado, T. J. C., Martin-Neto, L., & Fernandes, S. V. (2000). Organic matter storage in a sandy clay loam Acrisol affected by tillage and cropping systems in southern Brazil. *Soil and Tillage Research*, 54, 101–109.

Blake, G. R., & Hartge, K. H. (1986). Bulk density. In A. Klute (Ed.), *Methods of soil analysis. Part I: Physical and mineralogical methods* (pp. 363–375). ASA and SSSA.

Boddey, R. M., Jantalia, C. P., Conceição, P. C., Zanatta, J. A., Bayer, C., Mielniczuk, J., Dieckow, J., Dos Santos, H. P., Denardin, J., Aita, C., Giacomini, S. J., Alves, B. J. R., & Urquiaga, S. (2010). Carbon accumulation at depth in Ferralsols under zero-till subtropical agriculture. *Global Change Biology*, 16, 784–795.

Bolliger, A., Magid, J., Amado, J. C. T., Skora, N. F., Ribeir, M., Calegari, A., Ralisch, R., de Neergaard, A., & Donald, L. S. (2006). Taking stock of the Brazilian 'Zero Till Revolution': A review of landmark research and farmers' practice. *Advances in Agriculture*, 91, 47–110.

Bossolani, J. W., Crusciol, C. A. C., Moretti, L. G., Garcia, A., Portugal, J. R., Bernart, L., Vilela, R. G., Caires, E. F., Amado, T. J. C., Calonego, J. C., & Reis, A. R. (2022). Improving soil fertility with lime and phosphogypsum enhances soybean yield and physiological characteristics. *Agronomy for Sustainable Development*, 42, 26.

Briedis, C., Sá, J. C. M., Lal, R., Tivet, F., De Oliveira Ferreira, A., Franchini, J. C., Schimiguel, R., Hartman, D. C., & Santos, J. Z. (2016). Can highly weathered soils under conservation agriculture be C saturated? *Catena*, 147, 638–649.

Briedis, C., Sá, J. C. M., Lal, R., Tivet, F., Franchini, J. C., De Oliveira Ferreira, A., Hartman, D. C., Shimiguel, R., Bressan, P. T., Inagaki, T. M., Romaniw, J., & Gonçalves, D. R. P. (2018). How does no-till deliver carbon stabilization and saturation in highly weathered soils? *Catena*, 163, 13–23.

Brown, S., Gillespie, A. J. R., & Lugo, A. E. (1989). Biomass estimation methods for tropical forests with applications to forest inventory data. *Forest Science*, 35, 881–902.

Cogo, N. P., Levien, R., & Schwarz, R. A. (2003). Soil and water losses by rainfall erosion influenced by tillage methods, slope-steepness classes, and soil fertility levels. *Revista Brasileira de Ciência do Solo*, 27, 743–753.

Cogo, N. P., Moldenhauer, W. C., & Foster, G. R. (1984). Soil loss reductions from conservation tillage practices. *Soil Science Society of America Journal*, 48, 368–373.

Conceição, P. C., Amado, T. J. C., Mielniczuk, J., & Spagnollo, E. (2005). Soil organic matter and other attributes as indicators to evaluate soil quality in conservation systems. *Revista Brasileira de Ciência do Solo*, 29, 777–788.

Crossman, N. D., & Bryan, B. A. (2009). Identifying cost-effective hotspots for restoring natural capital and enhancing landscape multifunctionality. *Ecological Economics*, 68, 654–668.

Dalla Nora, D., Amado, T. J. C., Bortolotto, R. P., Ferreira, A. O., Reichardt, K., & Santi, A. L. (2014). Subsoil chemical amelioration and crop yields under continuous long-term no-till in a subtropical Oxisol. *African Journal of Agricultural Research*, 9, 3338–3349.

De Oliveira Ferreira, A., Amado, T., Rice, C. W., Diaz, D. R., Keller, C., & Inagaki, T. M. (2016). Can no-till grain production restore soil organic carbon to levels natural grass in a subtropical Oxisol? *Agriculture, Ecosystems & Environment*, 229, 13–20.

De Oliveira Ferreira, A., Amado, T. J. C., Nicoloso, R. S., Sá, J. C. M., Fiorin, J. E., Hansel, D. S. S., & Menefee, D. (2013). Soil carbon stratification affected by long-term tillage and cropping systems in southern Brazil. *Soil and Tillage Research*, 133, 65–74.

De Oliveira Ferreira, A., Amado, T. J. C., Rice, C. W., Diaz, D. A. R., Briedis, C., Inagaki, T. M., & Goncalves, D. R. P. (2018a). Driving factors of soil carbon accumulation in Oxisols in long-term no-till systems of South Brazil. *Science of the Total Environment*, 622–623, 735–742.

De Oliveira Ferreira, A., Amado, T. J. C., Rice, C. W., Goncalves, D. R. P., & Diaz, D. A. R. (2021b). Comparing on-farm and long-term research experiments for soil carbon recovery of conservation agriculture in southern Brazil. *Land Degradation & Development*, 32, 3365–3376.

De Oliveira Ferreira, A., Sá, J. C. M., Briedis, C., & Figueiredo, A. G. (2009). Desempenho de geno'tipos de milho cultivados com diferentes quantidades de palha de aveia-preta e doses de nitrogênio. *Pesquisa Agropecuária Brasileira*, 44, 73–179.

De Oliveira Ferreira, A., Sá, J. C. M., Harms, M. G., Miara, S., Briedis, C., Quadros Netto, C., Santos, J. B., & Canalli, L. B. (2012). Carbon balance and crop residue management in dynamic equilibrium under a no-till system in Campos Gerais. *Revista Brasileira de Ciência do Solo*, 36, 1583–1590.

De Oliveira Ferreira, A., Sá, J. C. M., Lal, R., Amado, T. J. C., Inagaki, T. M., Briedis, C., & Tivet, F. (2021a). Can no-till restore soil organic carbon to levels under natural vegetation in a subtropical and tropical typic quartzipisamment? *Land Degradation & Development*, 32, 1742–1750.

De Oliveira Ferreira, A., Sá, J. C. M., Lal, R., Tivet, F., Briedis, C., Inagaki, T. M., Gonçalves, D. R. P., & Romaniw, J. (2018b). Macroaggregation and soil organic carbon restoration in a highly weathered Brazilian Oxisol after two decades under no-till. *Science of the Total Environment*, 621, 1559–1567.

De Oliveira, O. C., de Oliveira, I. P., Urquiaga, S., Alves, B. J. R., & Boddey, R. M. (2004). Chemical and biological indicators of decline/degradation of *Brachiaria* pastures in the Brazilian Cerrado. *Agriculture, Ecosystems and Environment*, 103, 289–300.

Dieckow, J., Bayer, C., Conceição, P. C., Zanatta, J. A., Martin-Neto, L., Milori, D. B. M., Salton, J. C., Macedo, M. M., Mielniczuk, J., & Hernani, L. C. (2009). Land use, tillage, texture and organic matter stock and composition in tropical and subtropical Brazilian soils. *European Journal of Soil Science*, 60, 240–249.

Diekow, J., Mielniczuk, J., Knicker, H., Bayer, C., Dick, D., & Kogel-Knabner, I. (2005). Carbon and nitrogen stocks in physical fractions of a subtropical Acrisol as influenced by long-term no-till cropping systems and N fertilisation. *Plant and Soil*, 268, 319–328.

Ellert, B. H., & Bettany, J. R. (1995). Calculation of organic matter and nutrients stored in soils under contrasting management regimes. *Canadian Journal of Soil Science*, 75, 529–538.

FAO. (2013). *FAO statistical yearbook 2013: World food and agriculture*. https://www.fao.org/3/i3107e/i3107e.pdf

Fuentes, R. L., Telles, T., Soares Júnior, D., Melo, T., Friedrich, T., & Kassam, A. (2021). Expansion of no-tillage practice in conservation agriculture in Brazil. *Soil and Tillage Research*, 208, 104877.

Hansel, F. D., Telmo, J. C., Amado, D. A., Diaz, R., Rosso, L. H. M., Nicoloso, F. T., & Schorr, M. (2017). Phosphorus fertilizer placement and tillage affect soybean root growth and drought tolerance. *Agronomy Journal*, 109, 2936–2944.

Inagaki, T. M., Sá, J. C. M., Caires, E. F., & Gonçalves, D. R. P. (2016). Lime and gypsum application increases biological activity, carbon pools, and agronomic productivity in highly weathered soil. *Agriculture, Ecosystems and Environment*, 231, 156–165.

Kassam, A., Friedrich, T., Shaxson, F., & Pretty, J. (2009). The spread of conservation agriculture: Justification, sustainability and uptake. *International Journal of Agricultural Sustainability*, 7, 292–320.

Keesstra, S. D., Bouma, J., Wallinga, J., Tittonell, P., Smith, P., Cerdà, A., Montanarella, L., Quinton, J. N., Pachepsky, Y., van der Putten, W. H., Bardgett, R. D., Moolenaar, S., Mol, G., Jansen, B., & Fresco, L. O. (2016). The significance of soils and soil science towards realization of the UN sustainable development goals. *The Soil*, 2, 111–128.

Lal, R. (2014). Societal value of soil carbon. *Journal of Soil and Water Conservation*, 69, 186A–192A.

Lal, R. (2015). Sequestering carbon and increasing productivity by conservation agriculture. *Journal of Soil and Water Conservation*, 70, 55A–62A.

Lal, R. (2019). Rights-of-soil. *Journal of Soil and Water Conservation*, 74, 81A–86A.

Lal, R., Brevik, E. C., Dawson, L., Field, D., Glaser, B., Hartemin, A. E., Hatano, R., Lascelles, B., Monger, C., Scholten, T., Singh, B. R., Spiegel, H., Terribile, F., Basile, A., Zhang, Y., Horn, R., Kosaki, T., & Sánchez, L. B. R. (2020). Managing soils for recovering from the COVID-19 pandemic. *Soil Systems*, 4, 46.

Lal, R., Smith, P., Jungkunst, H. F., Mitsch, W. J., Lehmann, J., Ramachandran Nair, P. K., McBratney, A. B., de Moraes Sá, J. C., Schneider, J., Zinn, Y. L., Skorupa, A. L. A., Zhang, H.-L., Minasny, B., Srinivasrao, C., & Ravindranath, N. H. (2018). The carbon sequestration potential of terrestrial ecosystems. *Journal of Soil and Water Conservation*, 73, 145A–152A.

Leal, O. A., Amado, T. J. C., Fiorin, J. E., Keller, C., Reimche, G. B., Rice, C. W., Nicoloso, R. S., Bortolotto, R. P., & Schwalbert, R. (2020). Linking cover crop residue quality and tillage system to CO_2-C emission, soil C and N stocks and crop yield based on a long-term experiment. *Agronomy*, 10, 1848.

Lehmann, J., Bossio, D. A., Kögel-Knabner, I., & Rillig, M. C. (2020). The concept and future prospects of soil health. *Nature Reviews Earth & Environment*, 1, 544–553.

Mace, G. M. (2019). The ecology of natural capital accounting. *Oxford Review of Economic Policy*, 35, 54–67.

Mankolo, R., Reddy, C., Senwo, Z., Nyakatawa, E., & Sajjala, S. (2012). Soil biochemical changes induced by poultry litter application and conservation tillage under cotton production systems. *Agronomy*, 2, 187–198.

Manning, P., van der Plas, F., Soliveres, S., Allan, E., Maestre, F. T., Mace, G., Whittingham, M. J., & Fischer, M. (2018). Redefining ecosystem multifunctionality. *Nature Ecology & Evolution*, 2, 427–436.

Mendes, I. C., de Souza, L. M., de Sousa, D. M. G., de Castro Lopes, A. A., dos Reis Junior, F. B., Lacerda, M. P. C., & Malaquias, J. V. (2019). Critical limits for microbial indicators in tropical oxisols at post-harvest: The FERTBIO soil sample concept. *Applied Soil Ecology*, 139, 85–93.

Mendes, I. C., Sousa, D. M. G., Dantas, O. D., Lopes, A. A. C., Junior, F. B. R., Oliveira, M. I., & Chaer, G. M. (2021). Soil quality and grain yield: A win–win combination in clayey tropical Oxisols. *Geoderma*, 388, 114880.

Moebius-Clune, B. N., Moebius-Clune, D. J., Gugino, B. K., Idowu, O. J., Schindelbeck, R. R., Ristow, A. J., van Es, H. M., Thies, J. E., Shayler, H. A., McBride, M. B., Kurtz, K. S. M., Wolfe, D. W., & Abawi, G. S. (2016). *Comprehensive assessment of soil health, Part 1: Soil health concepts*. Cornell University. http://www.css.cornell.edu/extension/soil-health/1concepts.pdf

Nicoloso, R. S., Rice, C. W., Amado, T. J. C., Costa, C. N., & Akley, E. K. (2018). Carbon saturation and translocation in a no-till soil under organic amendment. *Agriculture, Ecosystems & Environment*, 264, 73–84.

Oldeman, L. R., Hakkeling, R. T. A., Sombroek, W. G., & Batjes, N. (2014). *Global assessment of human-induced soil degradation (GLASOD). World map of the status of human induced soil degradation*. Winand Staring Centre–ISSSFAO–ITC.

Passinato, J. H., Amado, T. J. C., Kassam, A., Acosta, J. A. A., & de Amaral, L. P. (2021). Soil health check-up of conservation agriculture farming systems in Brazil. *Agronomy*, 11, 2410.

Paustian, K., Lehmann, J., Ogle, S., Reay, D., Robertson, G. P., & Smith, P. (2016). Climate-smart soils. *Nature*, 532, 49–57.

Pérez-Brandán, C., Huidobro, J., Grümberg, B., Scandiani, M. M., Luque, A. G., Meriles, J. M., & Vargas-Gil, S. (2013). Soybean fungal soil-borne diseases:

A parameter for measuring the effect of agricultural intensification on soil health. *Canadian Journal of Microbiology*, 60, 73–84.

Poffenbarger, H., Archontoulis, S., Burras, C. L., Danalatos, G., Dixon, P., Russell, A., Six, J., Takle, E., Veenstra, J., & Castellano, M. (2022). A large subsoil carbon sink in the United States Corn Belt. *ESS Open Archive.*

Pott, L. P., Amado, T. J. C., Schwalbert, R. A., Gebert, F. H., Reimche, G. B., Pes, L. Z., Reimche, G. B., Pes, L. Z., & Ciampitti, I. A. (2021). Effect of hairy vetch cover crop on maize nitrogen supply and productivity at varying yield environments in southern Brazil. *Science of the Total Environment*, 759, 144313.

Sá, J. C. M. (2004). Adubação fosfatada no sistema plantio direto. In T. Yamada & R. S. Abdalla (Eds.), *Simpósio sobre fósforo na agricultura Brasileira* (pp. 201–222). Potafós.

Sá, J. C. M., Cerri, C. C., Dick, W. A., Lal, R., Solismar Filho, P. V., Piccolo, M. C., & Feigl, B. E. (2001). Organic matter dynamics and carbon sequestration rates for a tillage chronosequence in a Brazilian Oxisol. *Soil Science Society of America Journal*, 65, 1486–1499.

Sá, J. C. M., Cerri, C. C., Lal, R., Dick, W. A., De Cassia Piccolo, M., & Feigl, B. E. (2009). Soil organic carbon and fertility interactions affected by a tillage chronosequence in a Brazilian Oxisol. *Soil and Tillage Research*, 104, 56–64.

Sá, J. C. M., & De Oliveira Ferreira, A. (2018). The soil science in the evolution of no-till system in Brazil. *SBCS Bulletim (Brazilian Society of Soil Science)*, 44, 54–57.

Sá, J. C. M., Lal, R., Briedis, C., De Oliveira Ferreira, A., Tivet, F., Inagaki, T. M., Gonçalves, D. R. P., Canalli, L. B., Santos, J. B., & Romaniw, J. (2022). Can C-budget of natural capital be restored through conservation agriculture in a tropical and subtropical environment? *Environmental Pollution*, 298, 118817.

Sá, J. C. M., Lal, R., Cerri, C. C., Lorenz, K., Hungria, M., & de Faccio Carvalho, P. C. (2017). Low-carbon agriculture in South America to mitigate global climate change and advance food security. *Environment International*, 98, 102–112.

Sá, J. C. M., Potma Gonçalves, D. R., Ferreira, L. A., Mishra, U., Inagaki, T. M., Furlan, F. J. F., Moro, R. S., Floriani, N., Briedis, C., & de Oliveira Ferreira, A. (2018). Soil carbon fractions and biological activity based indices can be used to study the impact of land management and ecological successions. *Ecological Indicators*, 84, 96–105.

Sá, J. C. M., Santos, J. B., Lal, R., de Moraes, A., Tivet, F., Sá, M. F. M., Briedis, C., de Oliveira Ferreira, A., Eurich, G., Farias, A., & Friedrich, T. (2013). Soil-specific inventories of landscape carbon and nitrogen stocks under no-till and native vegetation to estimate carbon offset in a subtropical ecosystem. *Soil Science Society of America Journal*, 77, 2094–2110.

Sá, J. C. M., Séguy, L., Gozé, E., Bouzina, S., Husson, O., Boulaki, S., Tivet, F., Forest, F., & Burkner dos Santos, J. (2006). Carbon sequestration rates in no-tillage soils under intensive cropping systems in tropical agroecozones. *Edafologia*, 13, 139–150.

Sá, J. C. M., Séguy, L., Tivet, F., Lal, R., Bouzinac, S., Borszowskei, P. R., Briedis, C., dos Santos, J. B., da Cruz Hartman, D., Bertoloni, C. G., Rosa, J., & Friedrich, T. (2015). Carbon depletion by plowing and its restoration by no-till cropping systems in Oxisols of subtropical and tropical agro-ecoregions in Brazil. *Land Degradation & Development*, 26, 531–543.

Sá, J. C. M., Tivet, F., Lal, R., Briedis, C., Hartman, D. C., Santos, J. Z., & Santos, J. B. (2014). Long-term tillage systems impacts on soil C dynamics, soil resilience and agronomic productivity of a Brazilian Oxisol. *Soil and Tillage Research*, 136, 38–50.

Shaffer, J. P., Carpenter, C. S., Martino, C., Salido, R. A., Minich, J. J., Bryant, M., Sanders, K., Schwartz, T., Humphrey, G., Swafford, A. D., & Knight, R. (2022). A comparison of six DNA extraction protocols for 16S, ITS and shotgun metagenomic sequencing of microbial communities. *BioTechniques*, 73, 34–46.

Sisti, C. P. J., Santos, H. P., Kohhann, R., Alves, B. J. R., Urquiaga, S., & Boddey, R. M. (2004). Change in carbon and nitrogen stocks in soil under 13 years of conventional or zero tillage in southern Brazil. *Soil and Tillage Research*, 76, 39–58.

Six, J., Elliott, E. T., & Paustian, K. (2000). Soil macroaggregate turnover and microaggregate formation: A mechanism for C sequestration under no-tillage agriculture. *Soil Biology and Biochemistry*, 32, 2099–2103.

Smith, P., House, J. I., Bustamante, M., Sobocká, J., Harper, R., Pan, G., West, P. C., Clark, J. M., Adhya, T., Rumpel, C., Paustian, K., Kuikman, P., Cotrufo, M. F., Elliott, J. A., McDowell, R., Griffiths, R. I., Asakawa, S., Bondeau, A., Jain, A. K., … Pugh, T. A. M. (2016). Global change pressures on soils from land use and management. *Global Change Biology*, 22, 1008–1028.

Tabatabai, M. A. (1994). Soil enzymes. In P. J. Bottomley, J. S. Angle, & R. W. Weaver (Eds.), *Methods of soil analysis. Part 2. Microbiological and biochemical properties* (pp. 775–833). SSSA.

Team, R. D. C. (2006). *R: A language and environment for statistical computing*. R Foundation for Statistical Computing.

Tiecher, T., Martins, A. P., Peretto, E. J. S., Fink, J. R., Santos, L. S., Denardin, L. G. O., & Tiecher, T. L. (2016). *Evolução e Estado da fertilidade do solo no norte do Rio Grande Do Sul e sudoeste de Santa Catarina*. UFRGS.

Tisdall, J. M., & Oades, J. M. (1982). Organic matter and water-stable aggregates in soils. *Journal of Soil Science*, 33, 141–163.

Tivet, F., Sá, J. C. M., Lal, R., Briedis, C., Borszowskei, P. R., dos Santos, J. B., Farias, A., Eurich, G., da Cruz Hartman, D., Nadolny, M., Jr., Bouzinac, S., & Séguy, L. (2013). Aggregate C depletion by plowing and its restoration by diverse biomass-C inputs under no-till in sub-tropical and tropical regions of Brazil. *Soil and Tillage Research*, 126, 203–218.

USDA-NRCS. (2023). *Soil health*. https://www.nrcs.usda.gov/conservation-basics/natural-resource-concerns/soils/soil-health

Wackernagel, M., & Rees, W. E. (1997). Perceptual and structural barriers to investing in natural capital: Economics from an ecological footprint perspective. *Ecological Economics*, 20, 3–24.

Webster, R. (2007). Analysis of variance, inference, multiple comparisons and sampling effects in soil research. *European Journal of Soil Science*, 58, 74–82.

Weil, R. R., Islam, K. R., Stine, M. A., Gruver, J. B., & Samson-Liebig, S. E. (2003). Estimating active carbon for soil quality assessment: A simplified method for laboratory and field use. *American Journal of Alternative Agriculture*, 18, 3–17.

Wingeyer, A. B., Amado, T. J. C., Pérez-Bidegain, M., Studdert, G. A., Varela, C. H. P., Garcia, F. O., & Karlen, D. L. (2015). Soil quality impacts of current south American agricultural practices. *Sustainability*, 7, 2213–2242.

Xu, Z., Yu, G., Zhang, X., Ge, J., He, N., Wang, Q., & Wang, D. (2015). The variations in soil microbial communities, enzyme activities and their relationships with soil organic matter decomposition along the northern slope of Changbai Mountain. *Applied Soil Ecology*, 86, 19–29.

4

Cover Crops and Soil Health in Brazilian Agricultural Systems

Victória Santos Souza, Beatriz da Silva Vanolli, Bruna Emanuele Schiebelbein, Larissa de Souza Bortolo, Martha Lustosa Carvalho, Ieda Carvalho Mendes, and Maurício Roberto Cherubin

Chapter Overview

Monoculture of soybean [*Glycine max* (L.) Merr.] that utilized annual tillage-based cropping was a prevalent practice in the Brazilian farmlands systems before the 1970s. However, currently, the adoption of no-tillage (NT) is widespread in Brazilian farmlands for soybean cultivation. Although this change was motivated by a need to stop rampant soil erosion, it brought about other benefits and led to the establishment of the no-till system (NTS). The NTS is characterized by minimal soil disturbance and higher plant biodiversity. Cover crops (CCs) are increasingly used by Brazilian farmers to solve problems that inadvertently arise from removing tillage from the soil management strategies, such as soil compaction and soil-borne diseases. The main CC species used vary according to the regional climate and planting window. In subtropical Brazil, a region with high water availability, CCs are usually planted in the fall or winter, and the main species used are ryegrass (*Lolium multiflorum* L.), black oat (*Avena strigosa* Schreb.), and forage radish (*Raphanus sativus* L.). For the Cerrado region, with a warm and dry climate, farmers use different species of sunn hemp (*Crotalaria* spp.) brachiaria grass (*Urochloa* spp.), pearl millet [*Pennisetum glaucum* (L.) R. Br.], and sorghum [*Sorghum bicolor* (L.) Moench]. Overall, using CCs helps maintain soil health (SH) by improving soil structure and water flow, increasing soil organic matter (SOM), favoring nutrient cycling, and establishing a biologically active and diverse soil environment. All these factors result in better crop production. More recently, Brazilian farmers have started using CC multispecies seed mixes to amplify these benefits. Despite the advantages of diversified crop

Soil Health Series: Volume 3 Soil Health and Sustainable Agriculture in Brazil, First Edition.
Edited by Ieda Carvalho Mendes and Maurício Roberto Cherubin.

production systems for modern and sustainable tropical agriculture, operational issues still hamper the large-scale adoption of CCs. Therefore, further research and technology transfer efforts are necessary to motivate more farmers toward the adoption of crop diversification strategies.

Introduction

One stormy night in the spring of 1972, Herbert Bartz stood in his freshly tilled, freshly sown fields at the Rhenânia Farm in the southern Brazilian state of Paraná. Streams of water mixed with sediment carried away downhill his hard work and his investment, throwing his seeds and fertile topsoil into the river. As retold by Bartz in his memoirs (Santin, 2018), on that rainy night in 1972 he made a decision that would spark a revolution in agricultural soil management in Brazil. He decided to find a solution to the erosion problem destroying farms and waterways. He traveled across Europe and the United States, talking to researchers and farmers. Finally, while in Kentucky, Bartz met Harry Young Jr., the first US farmer to work commercially with what he called "zero-tillage." Inspired, Bartz imported to Brazil the specialized seeder necessary to sow without removing plant residue. That same year, his farm would become home to the first commercial annual crop field sown without previous soil mobilization in Brazil. Herbert Bartz, together with his fellow farmers and friends Franke Dijkstra and Manuel "Nonô" Pereira, are recognized as visionaries and pioneers whose inventiveness, resilience, and enthusiasm were central to the "Plantio Direto" (no-tillage) revolution in Brazilian agriculture (Figure 4.1).

Historically, the occupation of Brazilian territory was preceded by deforestation. The agricultural model in these deforested areas did not adopt adequate practices to preserve the soil. Most of the technologies and cultivation methods, including agricultural machinery and implements, were developed in European countries with entirely different climates and hydrological regimes and imported to Brazil. With the heavy rains that are characteristic of tropical and subtropical regions, Brazilian agricultural soils were intensely eroded because production systems of annual crops were conducted with little to no soil cover. The loss of the most fertile surface layer and the productive potential of the soils due to severe erosion stimulated the migration of farmers to new areas. However, with little concern for the sustainability of soil and other natural resources, the same cycle of destruction happened in the new areas as well. By the 1960s, severe erosion was recognized as one of the main challenges faced by Brazilian farmers. The combination of highly erosive rainfall, deep tillage, and downhill sowing created a perfect "storm" with severe agricultural and environmental consequences. Brazil needed a new model of agricultural production, adapted to the edaphoclimatic conditions of the main agricultural regions of the country.

Figure 4.1 Pioneer farmers Manoel Pereira, Herbert Bartz, and Franke Djikstra (from left to right) on the adoption and diffusion of the no-till system in Brazil. Archives of the Federação Brasileira do Sistema Plantio Direto. Marie Bartz.

The NT revolution in Brazil that began in the 1970s was the result of intense collaboration between farmers and researchers from Brazil and the United States. This revolution started with the introduction and development of specialized machinery to educational tours to promote the transition. In the 50 years following the initiative of the three farmers in Paraná state that pioneered the implementation of the new technique, the practice of sowing without previous soil mobilization has been successfully implemented in multiple regions of annual crops in Brazil where diverse edaphoclimatic conditions prevail. This success has been recorded for the past 50 years following the initiative of the three farmers in Paraná State who pioneered the implementation of the new technique. The area of annual crop production under NT in Brazil reached 33 million ha in 2017, representing about 61% of the total area under annual crops in that year (Fuentes-Llanillo et al., 2021). From the initial NT, farmers and researchers developed the NTS, a system of soil management with three basic principles: minimal soil disturbance, permanent soil cover, and crop diversification. The correct implementation of the NTS mitigates or corrects problems that arise from monocultures and continuous field traffic, such as the degradation of the soil's physical quality by compaction, loss of soil biodiversity, and decrease of soil C stocks. Many Brazilian farmers who have adopted NT do not implement the NTS correctly, neglecting mainly the crop diversification principle. This principle is responsible for maintaining SH in the NTS; improving chemical, physical, and biological soil properties; and reducing erosion, which is the main purpose of the other two principles. In recent years, research and educational initiatives to promote the wider adoption of diversification by Brazilian farmers have supported the use of CCs as an alternative to enhance NT in annual cropping systems.

Cover crops keep the soil surface covered, which can improve or maintain SH by improving soil aggregation and water flow, increasing SOM, favoring nutrient cycling, and promoting the establishment of a biologically more active and diverse soil environment. The addition of plant residues provided by CCs is a source of nutrients and energy for soil organisms, from the macrofauna to the microbiome. These organisms are responsible for most of the processes related to nutrient cycling in the soil and suppressing pathogens. The root systems of CCs also play an important role in maintaining or restoring several soil functions as compared to commercial crops, which were selectively bred to express desired characteristics at the expense of the roots. Both plants with fasciculate (e.g., *Urochloa* spp., widely used in the Brazilian Cerrado both as forage and as a CC) and pivoting root systems (e.g., *R. sativus*, traditionally used in the Brazilian subtropical region) can recycle nutrients. This is achieved by scavenging for them in deeper soil layers and improving soil structural quality through aggregation and the formation of biopores.

These biopores function as channels for the movement of gasses, water, root growth, mitigation of soil compaction, and improving water flow regulation. With greater water, air, and nutrient availability and lower pathogen population pressure, the cash crops introduced in the system have a more conducive production environment, resulting in higher yields.

Main Cover Crops and Their Introduction into the Annual Cropping Systems in Brazil

The choice of CCs in the diversification system, whether single or multispecies, is linked to several characteristics: plant architecture, root morphology, plant physiology, and biotic interactions. The architecture determines the spatial configuration of the plant's root system, which includes depth, length, density, and branching of the roots (Lesmes-Vesga et al., 2023). Morphology refers to the individual characteristics of the roots, such as specific root diameter, root length, tissue density, and dry mass content of the roots. This has a strong relationship with soil physical quality due to the formation of biopores, regulation of water flow, maintenance of soil structure, and nutrient cycling (Schiebelbein et al., 2022). Physiology refers to root characteristics related to the kinetics of nutrient uptake, respiration, and exudate release that will be essential to reestablish soil biodiversity and nutrient cycling. Biotic interactions refer to interactions between roots and soil biota that affect nutrient uptake, such as associations with mycorrhizal fungi and N-fixing bacteria (e.g., *Rhizobium* spp. in legumes), as well as interactions with pathogens. As roots die and decompose, they incorporate C deeper, favoring soil aggregation. In addition, the biopores formed by the roots contribute to greater infiltration of water into the soil and provide a habitat for organisms (Fischler et al., 1999).

Ultimately, CCs must be practical. Farmers may choose a CC species considering the cost of the seeds, minimal need for management, efficient use of water, fast growth and short cycle, and the low possibility of diseases and pest carryover from the CC to the main cash crops (Baligar & Fageria, 2007).

Main Cover Crop Species Cultivated in Brazil

The choice of best CCs differs according to the Brazilian agricultural region, climate, soil, and management system. Table 4.1 shows the most common species cultivated and their appropriate planting season. Farmers from southern Brazil adopt CCs preferentially in the fall/winter season, after the main season (summer season). Farmers from the Cerrado, whose main concern is water availability, adopt CCs in the spring/summer season and off-season. In this region, it has become increasingly common to sow CCs before or even alongside the cash crop (e.g., the Santa Fé management strategy, which allows the producer to intercrop maize with *Urochloa* spp.).

Table 4.1 Main cover crop species cultivated in the different seasons in Brazil.

Season	Cover crop species
Spring-summer (October–March)	Sunn hemp (*Crotalaria spectabilis* Roth, *Crotalaria juncea* L., *Crotalaria ochroleuca* G. Don), pearl millet [*Pennisetum glaucum* (L.) R. Br.], pigeon pea [*Cajanus cajan* (L.) Millsp.], sorghum [*Sorghum bicolor* (L.) Moench], sunflower (*Helianthus annuus* L.), brachiaria [*Urochloa brizantha* (Hochst. ex A. Rich.) R. Webster, *Urochloa decumbens* (Stapf) R. Webster, and *Urochloa ruziziensis* (Germ. & Evrard) Crins], stylosanthes (*Stylosanthes capitata* Vogel and *Stylosanthes macrocephala*)
Fall-winter (April–September)	Black oat (*Avena strigosa* Schreb.), vetch (*Vicia sativa* L.), forage radish (*Raphanus sativus* L.), forage ryegrass (*Lolium multiflorum* L.), wheat (*Triticum aestivum* L.), white lupine (*Lupinus albus* L.), buckwheat (*Fagopyrum esculentum* Moench)

Spring/Summer Cover Crops

The spring/summer species are cultivated between the harvest of the summer crop and the sowing of the winter crop, thus preventing fallow during this period. They can be divided into families such as cereals, legumes, grasses, and others. Commonly called "brachiaria" (syn. *Urochloa*) (Figure 4.2a), the grasses of the genus *Urochloa* are tropical grasses, and their main representatives are palisade grass [*Urochloa brizantha* (Hochst. ex A. Rich.) R. Webster], spreading liverseed grass [*Urochloa decumbens* (Stapf) R. Webster], and Congo grass [*Urochloa ruziziensis* (Germ. & Evrard) Crins]. The three species are good alternatives for crop rotation, but they have some differences between them. *Urochloa brizantha* and *U. ruziziensis* show medium to moderate tolerance to water deficit, whereas *U. decumbens* does not tolerate water deficit or low temperatures (Carvalho et al., 2022). Both have a high potential for plant production, which is reflected in a greater ability to sequester C (Ayarza et al., 2022; Salton et al., 2014).

The management of grasses in a NTS, directly and indirectly, influences SH (see Chapter 2). In addition to promoting soil cover and weed suppression

Figure 4.2 Examples of spring/summer cover crops sown in Brazil. (a) *Urochloa ruziziensis* (Germ. & Evrard) Crins; (b) *Pennisetum glaucum* (L.) R. Br.; (c) *Sorghum bicolor* (L.) Moench; (d) *Crotalaria spectabilis* Roth. (a) Mariana Teixeira; (b) Jaqueline Gomes; (c) Maurício Roberto Cherubin; (d) Bruna E. Schiebelbein.

(van Eekeren et al., 2010), their vast root system "explores" a larger soil volume to obtain water and nutrients (Crush et al., 2005). Even after decomposition, they leave biopores in the soil, facilitating water infiltration and the subsequent development of plant roots. In addition, grasses are complete foods rich in organic and mineral compounds, and their leaves and stems are easily digested by animals (Borowik et al., 2020), making them an excellent ally in the crop–livestock integration system.

Pearl millet (Figure 4.2b) and sorghum (Figure 4.2c) are the most important cereals grown in tropical regions due to their adaptability to abiotic stresses and their commercial importance as animal feed (Baligar & Fageria, 2007; de Carvalho et al., 2011). Pearl millet is an annual cereal with C_4 metabolism and erect growth. Its main feature is its low need for water. Thus, it is widely cultivated in drier regions (Fageria, 1992), such as the Brazilian Cerrado (Carvalho & Amabile, 2006; de Carvalho et al., 2014). Pearl millet has the potential to be used as a CC due to its high root density, high resistance to penetration, vegetative vigor, and ability to accumulate macronutrients (Carvalho & Amabile, 2006; de Carvalho et al., 2014; Rosolem et al., 2002), which allows for greater soil exploration and consequent cycling of nutrients, mainly those present in the subsurface layers. It is also considered a "bad host" for plant-parasitic nematodes, thus reducing the population (Carvalho et al., 2022). Sorghum has similar characteristics to maize and provides the same amount of metabolizable energy but differs in its higher protein content and lower digestibility (Dowling et al., 2002). It is a drought-tolerant crop and can provide enough biomass to cover the soil surface (Baligar & Fageria, 2007; de Carvalho et al., 2012). In addition to its capacity to grow on a wide variety of soils, sorghum is not nutritionally demanding. In Brazil it is cultivated after the harvest of the cash crops and develops relatively well with residual fertilization from the previous crop (Baligar & Fageria, 2007). Its development is fast and promotes total soil coverage in the shortest possible time, even in compacted areas (Rosolem et al., 2002).

One of the main advantages of using legume plants as CCs is their ability to fix atmospheric N through symbiosis with *Rhizobium* bacteria located in the root nodules (Duchene et al., 2017). These plants have a low C/N ratio, decomposing more quickly and releasing more NO_3^- compared with nonlegumes with a high C/N ratio. To obtain the benefits of the N supply and at the same time reduce its leaching loss, it is recommended to intercrop legumes with other CCs, such as grasses and crucifers (Daryanto et al., 2018).

Tropical legumes such as sunn hemp (particularly *Crotalaria juncea* L. and *Crotalaria spectabilis* Roth) (Figure 4.2d), besides promoting biological N fixation, have nematicidal action that affects larval development, exude nematotoxic compounds, and stimulates antagonistic microflora during biomass decomposition (Ratnadass et al., 2012). More recently, *Sthylosanthes* spp. also has become an attractive option as a CC (e.g., the "Mirai system"). Due to their pivoting root system, these legumes also promote the formation of biopores in the soil, which in turn allow water infiltration and reduce soil density, thus contributing to the physical health of the soil (Fischler et al., 1999). Pigeonpea [*Cajanus cajan* (L.) Millsp.] is another important legume that is used as a CC. Its vigorous and vast root system is adapted to low-fertility soils and acts as a "biological subsoiler" by breaking up the most compacted soil layers. Another important function of this legume is the allelopathic effect on weeds. Long-term NT systems with the presence of legume CC and N fertilization improve soil C and N stores, thereby improving SH and promoting atmospheric C sequestration (Diekow et al., 2005). For degraded soils, the recommendation would be black mucuna [*Mucuna aterrima* (L.) DC. var. *utilis* (Wall. ex Wight) Baker ex Burck], which is a rustic legume with an aggressive root system (Carvalho et al., 2022). In addition to high yields of seeds, leaves, and stems, it also has a rapid initial development (Sathiyanarayanan & Arulmozhi, 2007).

Sunflower (*Helianthus annuus* L.), from the Asteraceae botanical family, is a second crop option for regions with uneven annual rainfall distribution due to its resistance to drought (Zegada-Lizarazu & Monti, 2011). However, it is sensitive to acidic soils and those with a high Al content. Under ideal conditions, the sunflower presents good biomass production and a taproot system that is capable of absorbing nutrients in deeper soil layers.

Fall/Winter Cover Crops

Fall/winter CCs are sown from the end of the Brazilian summer (March) until the beginning of the Brazilian winter (June), providing an alternative for colder regions (de Lima Filho et al., 2014). In addition, these plants promote soil cover, reduce weed incidence, and allow the increase of biomass in the production system. In the winter period, wheat (*Triticum aestivum* L.) is one of the main income generation alternatives for farmers. The grain is widely used to produce flour and animal feed (Carvalho et al., 2022). It is a grass that has a deep but not vigorous root system; therefore, it requires soils that do not offer physical barriers to the development of its roots. Despite this limitation, wheat roots can associate with arbuscular mycorrhizal fungi and promote better development of subsequent crops, even after wheat senescence (Dabney et al., 2001). Forage ryegrass is widely used in Brazil as a winter forage for cattle feeding due to its high nutritional value. Initial development is slow, but it outperforms other winter CCs in biomass production throughout its development (Carvalho et al., 2022). Although its root system is not deep, which impairs its development under conditions of water stress, it has a fasciculate and abundant development, which positively interferes with the formation of biopores and soil aggregation. Black oat (Figure 4.3a) is another widely cultivated grass in the colder regions of Brazil. It is used for grazing, due to its high protein content and high digestibility, and as a CC. Despite not having the capacity

Figure 4.3 Examples of fall/winter cover crops present in Brazil. (a) *Avena strigosa* Schreb.; (b) *Raphanus sativus* L.; (c) *Fagopyrum esculentum* Moench Source: Pixabay; (d) *Vicia sativa* L. (a) Beatriz da Silva Vanolli; (b) Maurício Roberto Cherubin; (c) Beatriz da Silva Vanolli; (d) Mauricio Roberto Cherubin.

to fix N, black oat has a high potential for N accumulation and cycling due to its deep root system (Carvalho et al., 2022). It is considered a rustic plant and easy to adapt to different soils but does not tolerate waterlogged soils or soils with a high Al content (Burle et al., 2006).

Radish (Figure 4.2b) is the most common species of the Brassicaceae family used as a CC in Brazil, particularly in southern Brazil. It has been sown as a single CC, but it is also used in crop mixes. Radish has a root system capable of exploiting compacted and dense soils without affecting its development, in addition to having a rapid crown development that favors weed suppression (Burle et al., 2006; Lawley et al., 2012). More recently, the cultivation of buckwheat (*Fagopyrum esculentum* Moench) (Figure 4.2c), a species of the Polygonaceae family, has been expanded in different regions of Brazil. It is sown in acid soils with low fertility (e.g., low P content). Buckwheat has been used to diversify CC mixes.

Vetch (*Vicia sativa* L.) is an alternative leguminous plant (Figure 4.2d) to diversify winter CCs because, in addition to having deep and branched roots, which help in structuring the soil, it can fix N, promoting the increase of this nutrient for the subsequent crop. Vetch does not tolerate very humid and acidic soils; however, it is an excellent weed suppressor due to its allelopathic effect and rapid development (Carvalho et al., 2022). On the other hand, an option for drought tolerance is white lupine (*Lupinus albus* L.), which has a deep root system (Dabney et al., 2001). In addition, its proteinoid roots promote strong acidification of the rhizosphere and high absorption of cations/anions due to the excretion of citric acid, which leads to intensive chemical extraction of a limited volume of soil (Dinkelaker et al., 1989).

For more information about the main CCs used in Brazil, including agronomical and technical recommendations, we suggest consulting the *Practical Guide of Cover Crops* published by Carvalho et al. (2022).

Introducing Cover Crops in Rotation Annual Cropping Systems

Cover crops can be used for several purposes, depending on the objective and the region. For instance, in the South and Central-west regions of Brazil (de Carvalho et al., 2014), intercropped or multispecies mixes are examples of crop rotation plans. Crop rotation consists of the planned and previously arranged alternation of the cultivation of different species in a planned area and/or region that helps maintain SH and reduces pests and diseases (Ambrosini et al., 2022; Rigon et al., 2022).

Adopting crop rotation practices can contribute to the conservation and protection of soils and promote a greater diversity of species (i.e., a more sustainable agricultural production) (Cordeiro et al., 2021). For maximum efficiency in increasing the productive capacity of the soil, crop rotation planning should consider the use of crops with different root systems (e.g., grasses and legumes), each of which has a positive residual impact on the soil and for the crop planted in succession (Rosolem et al., 2002). The choice of CCs that produce large amounts of biomass and grow quickly alone or with cash crops is essential for a good rotation system. Crop rotation plans differ mainly according to region, climate, main crop, and rainfall distribution. Considering the specificities of each region, two examples were presented to illustrate the adoption of CCs in the South (temperate climate) and Cerrado (tropical climate) regions of Brazil.

In the southern region for example, it is necessary to adopt low-temperature–tolerant CCs such as oats, mainly black oat, radish, vetch, and others (Figure 4.4) because these species are tolerant to low temperatures (Krenchinski et al., 2018). Black oat and vetch recycle large amounts of K, while vetch, black velvet bean, lupin, and pigeonpea recycle large amounts of N (Borkert et al., 2003). These plants can be grown as a winter crop after the harvest of the summer crop (cash crop) or between summer and winter crops in the shorter fall window. Because

Figure 4.4 Example of cover crop grown in winter in southern Brazil. Black oat (grass) + rye (grass) + vetch (legume) in Capinzal, Santa Catarina state, Brazil. Edson M.T. da Rocha.

soybean is of undisputed importance for economic sustainability in Brazil, the example of rotation plans provided have soybean as the main crop.

To promote diversification during the main cropping season, it is encouraged that the the farmland be divided to enable the farmer to produce different crops simultaneously in the same season per year. For example, in summer, one of the divided farmlands could be used for soybeans while the other part could be used maize (Table 4.2). Dividing the area also applies to winter, resulting in better diversification of the area over time where the CC will be implemented in the southern region of Brazil. In winter one can adopt half of the productive area for the cultivation of wheat as a cash crop and 25% of the area as a CC mixture (e.g., black oat + radish + vetch) that precedes the next crop's maize and 25% of the area destined to pasture with the use of black oat, for example.

Another strategy suggested to increase crop diversification and optimize the benefits of CCs is to fit CCs in the window between the summer and winter crops, called the shortfall window (period from March to June). This pre-seeding winter window can be filled with a mixture of CCs to generate a larger biomass input and increase soil protection, decomposition, and soil nutrient recycling for the subsequent cash crop (de Moraes et al., 2016; Piva et al., 2021). Maintaining

Table 4.2 Example of a classic 4-year crop rotation plan including cover crops in the winter season, highlighting the opportunity to use cover crops in the shortfall window.

	Year 1		Year 2		Year 3		Year 4	
Area	**Summer**	**Winter**	**Summer**	**Winter**	**Summer**	**Winter**	**Summer**	**Winter**
25%	Soybean	Wheat	Soybean	Cover crop mix	Maize	Wheat	Soybean	Oats (grazing)
25%	Soybean	Cover crop mix	Maize	Wheat	Soybean	Oats (grazing)	Soybean	Wheat
25%	Maize	Wheat	Soybean	Oats (grazing)	Soybean	Wheat	Soybean	Cover crop mix
25%	Soybean	Oats (grazing)	Soybean	Wheat	Soybean	Cover crop mix	Maize	Wheat

Note: Example of cover crop mix: oats+radish+vetch.

or improving the productive capacity of the soil requires crop alternatives that can be adopted in rotation with cash crops to improve the management of production systems. The development of these CCs depends on soil conditions, climate, and growing season.

The main crops targeted by producers in the Cerrado region are usually soybean and maize (Abrahão & Costa, 2018). Thus, the ideal when planning a rotation plan is to divide the area into plots that allow the cultivation of both crops in the same year and the insertion of CCs (de Carvalho et al., 2014) to provide ecosystem services. After the maize harvest, there is room for another CC, which can be *Urochloa* spp. as a CC or grazing, so that it keeps the soil covered in winter and forms biomass for the subsequent crop (Table 4.3) (Rossi et al., 2013). Summer grass species, such as pearl millet, in addition to producing large amounts of biomass, can provide significant improvements to the soil through the action of the fasciculated roots, especially due to the structure and accumulation of organic matter (OM) (Nascimento et al., 2022).

The improvements in soil properties benefit the subsequent crop in this region because the root systems of CCs have better development and explore a greater volume of soil in the profile, improving the availability of nutrients and water, which is essential for crops to cope with droughts. Summer legume species, such as *C. spectabilis*, and *Crotalaria breviflora* are alternatives also used in the Cerrado of Brazil with a similar window to pearl millet (Figure 4.5) (Berriel et al., 2020; Cruz et al., 2020). In addition to the potential as biomass and soil cover, some species of pearl millet and sunn hemp are examples of CCs that help in the management of the main nematodes (*Meloydogine incognita*, *Rotylenchulus reniformis*, and *Pratylenchus brachyurus*) that cause economic damage to the species grown for grain and fiber production (Debiasi et al., 2016; Silva et al., 2022a,b).

Table 4.3 Example of a classic 4-year crop rotation plan, including cover crops in summer and highlighting the opportunity to use cover crops in the winter dry season, in the Cerrado.

Area	1st Harvest	2nd Harvest		1st Harvest	2nd Harvest		1st Harvest	2nd Harvest		1st Harvest	2nd Harvest	
25%	Soybean	Maize	Urochloa (cover crop)	Soybean	Cotton		Millet (cover crop)	Maize	Urochloa (cover crop)	Soybean	Cotton	
25%	Soybean	Cotton		Soybean	Maize	Urochloa (cover crop)	Soybean	Cotton		Millet (cover crop)	Maize	Urochloa (cover crop)
25%	Millet (cover crop)	Maize	Urochloa (cover crop)	Soybean	Cotton		Soybean	Maize	Urochloa (cover crop)	Soybean	Cotton	
25%	Soybean	Cotton		Millet (cover crop)	Maize	Urochloa (cover crop)	Soybean	Cotton		Soybean	Maize	Urochloa (cover crop)

Figure 4.5 *Crotalaria ochroleuca* G. Don cultivated as a cover crop before soybean [*Glycine max* (L.) Merr.] cultivation in Campinorte/Goiás, Brazil. Maria Emilia Moreira Marostica.

In summary, CCs improve soil physical, chemical, and biological properties. They also play a critical role in increasing SOM, which is key to SH (Locatelli et al., 2022; Morais et al., 2020). Organic matter is highly correlated with soil aggregate stability, water infiltration and retention, erosion resistance, biological activity, cation exchange capacity, and plant nutrient availability (da Silva Neto et al., 2016; dos Reis Ferreira et al., 2020). Normally, these species produce large amounts of biomass and can be grown alone or with other crops. Thus, they offer excellent conditions for soil maintenance, high productivity, and recovery of degraded areas (Chapagain et al., 2020).

Impact of Cover Crops on Soil Health

Cover crops enhance SH, sustaining its multi-functionality (Schipanski et al., 2014). In the following subsections, we presented the direct and indirect benefits of CCs to soil functioning, focusing specifically on water flow regulation, maintenance of soil structure, cycling of OM, nutrient cycling, habitat for soil organisms, and crop production (Figure 4.6).

Regulating Water Flow

Agricultural production is affected by soil moisture content, water infiltration, and drainage quality. Soil hydrological processes are influenced by multiple factors, such as land cover management, climate change, vegetation cover, and agriculture. The dynamics of soil hydraulic processes are determined primarily as a function of the physical properties of the soil (da Luz et al., 2020; Robinson et al., 2022). Aggregates, for example, are components of soil structure and therefore are of paramount importance in maintaining soil porosity and aeration, microbial and plant growth, water infiltration, and erosion control (Sekaran et al., 2021). In addition, other factors, such as granulometry, density, and porosity, are also fundamental for the regulation of water in the soil (Libardi, 2005). Thus, soil management and SH are directly linked to ecosystem services related to soil water dynamics, especially for soil physical properties (Kassam et al., 2014; Wagenet & Hutson, 1997).

Soil physical degradation processes, such as erosion and compaction, begin with land use change, where soil disturbance causes depletion of soil Cs stocks (Guo & Gifford, 2002) and reduction of porosity (Balbino et al., 2002). Consequently, there is a loss in ecosystem service provision, especially water availability, which depends on the physical health of the soil (Lal, 2015). The delivery of the ecosystem service of water cycling and availability by soil is therefore conditioned by both physical qualities (i.e., the permanent attributes) and the soil aggregate itself (sensitive to management).

Soil physical multifunctionality can be altered by processes such as soil management, plant root growth, soil properties, mineral composition, texture, SOM, pedogenic processes, microbial activities, exchangeable ions, and soil compaction (Adetunji et al., 2020). Soil covered with living plants retains more water than bare soils, which results in better infiltration rates and reduced evaporation (Guswa et al., 2002; Zhang & Schilling, 2006). Agricultural management practices, including CCs and crop residues, can improve soil water flow, availability, and retention primarily as a function of biopore formation (De Moraes et al., 2018). The mechanisms by which CCs affect soil water regulation can be direct and indirect and is addressed in the next sections.

Direct Mechanisms

The canopy and mulch of CCs protect the soil surface by absorbing the kinetic energy of raindrops and reducing the velocity and volume of runoff water (Figure 4.7). Thus, CCs prevent the shedding of particles by the direct impact of raindrops on the soil surface and reduce the amount and intensity of soil runoff

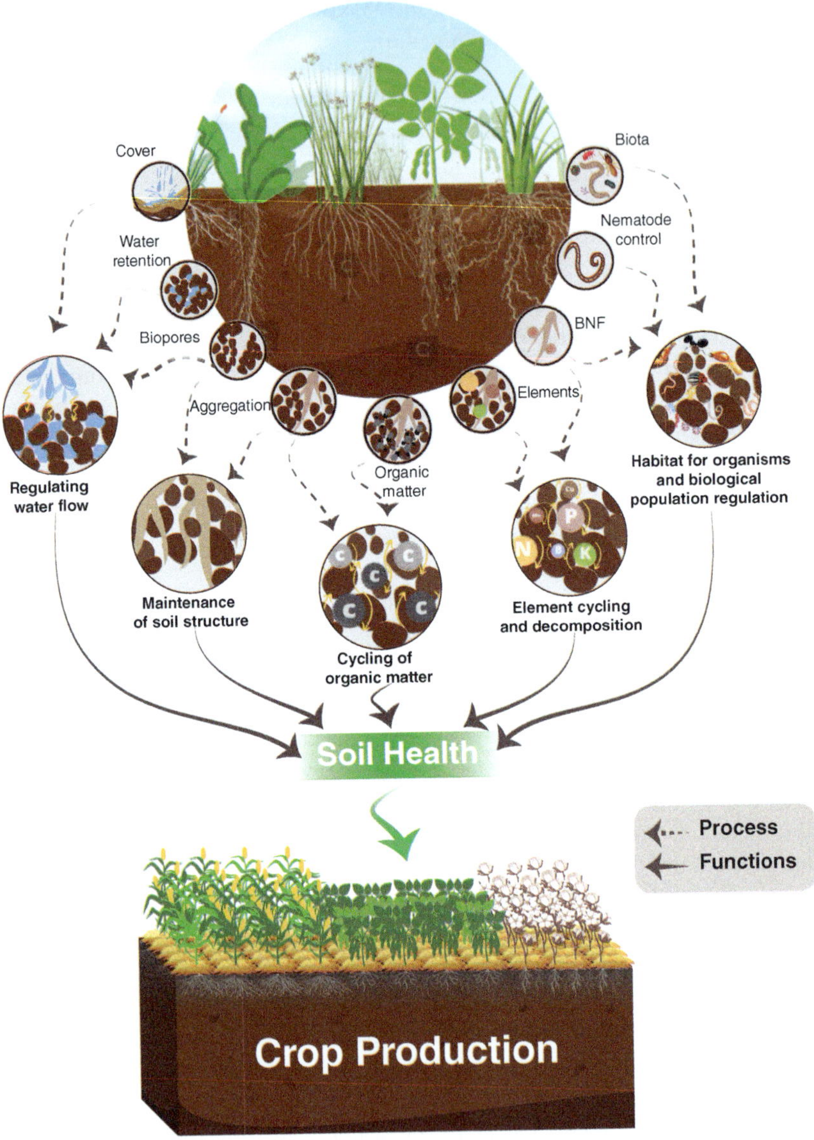

Figure 4.6 Schematic representation of the soil functions that are potentially affected by the inclusion of cover crops in the production system and their direct and indirect relationships.

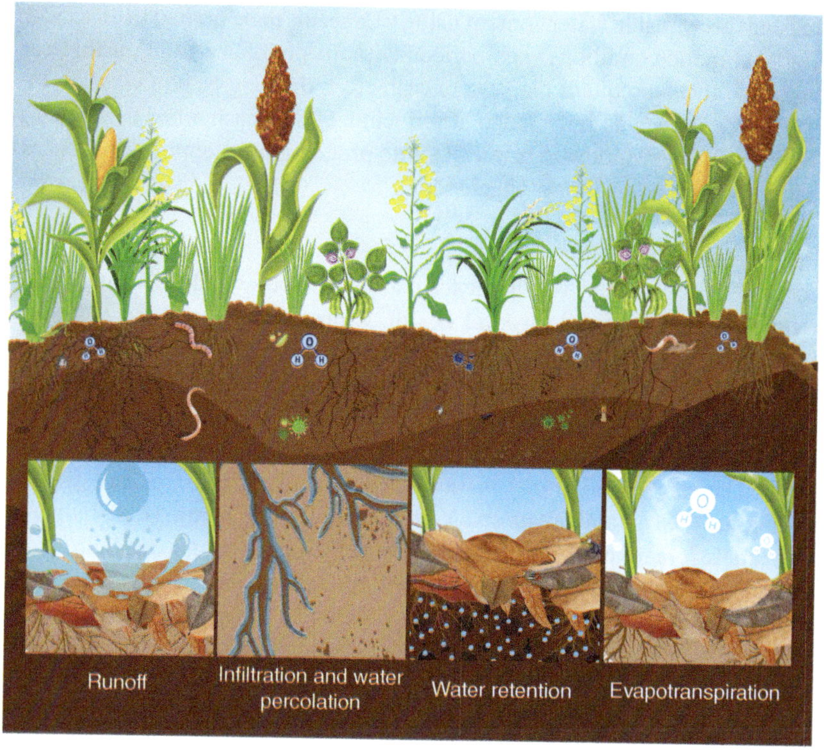

Figure 4.7 Representation of the effects of cover crops on soil water regulation.

that would result in erosion. Surface-related factors interfere with the movement of water across the air–soil interface. Cover crop residues on the soil surface increase soil roughness and consequently reduce the effect of surface runoff, increasing water infiltration and reducing the effects of temperature on soil water evaporation, thus optimizing soil water storage (Figure 4.7) (de Almeida et al., 2018). The use of CCs is one of the most widely applied soil and water conservation methods to control soil moisture and soil temperature variations. Cover cropping systems can reduce soil temperature during the hottest period of the day and increase soil temperature a night (de Lima et al., 2021).

Several factors interfere with water infiltration in the same soil, such as the volume of rainwater or irrigation, total porosity, pore size distribution, and the presence or absence of compacted layers or surface sealing (Helalia et al., 1988). The presence of living plants and biomass covering the soil reduces soil temperature fluctuations, especially in the surface layers. Slower increases in soil temperature are also reflected in lower rates of water loss through evaporation and higher water availability compared with bare soils (Ward et al., 2012). However, increasing the available water content using CCs seems to be limited to areas with favorable climatic conditions (i.e., periods without prolonged drought that can compromise the subsequent planting of cash crops due to a lack of water in the soil). Therefore, specific knowledge of soil, climate, and cash crops is necessary when planning the window to insert CCs into production systems. In this regard, irrigation systems allow more "wiggle room."

Indirect Mechanisms

Cover crop roots play a key role in the formation of biopores and aggregate, which consequently improves soil physical health. Crops with vigorous root systems can modify soil aggregation (Zhang & Peng, 2021) and mitigate the compaction and stratification effects that are commonly observed in agricultural soils. They form deep and abundant biopores that improve moisture flow and increase water percolation. The more water that infiltrates, the more it will be retained in the deeper soil layers (Yu et al., 2016). Plants with abundant and vigorous root systems can penetrate the soil and create stable and continuous biopores. These biopores and channels remain after decomposition of the crop roots and can be used as preferential growth paths for subsequent crop roots, especially in compacted soils. Biopores also become preferential flow paths for air and water, which can increase the overall saturated hydraulic conductivity of the soil by several orders of magnitude, thereby increasing the availability of water and oxygen for plant roots (Bertollo et al., 2021). Positive effects of CC use to improve soil aggregation were evidenced by Bertollo et al. (2021), who evaluated soil physical attributes after subsequent soybean cultivation after four pre-sowings (*U. ruziziensis*, oats, wheat, or maize) under four levels of soil compaction. *Urochloa ruziziensis* increased the root biomass and length density of soybean, especially in the deeper soil layers of the compacted treatments.

Compacted soils can be recovered by mechanical and/or biological methods, using plants with vigorous root systems, thereby increasing infiltration and retention of water in the soil. In Botucatu, state of São Paulo, Brazil, an experiment was conducted from 2003 to 2006 by Calonego and Rosolem (2010). They evaluated the immediate and medium-term effects of mechanical scarification and crop rotation on root growth and soybean yield for 4 years. Mechanical scarification resulted in lower soil resistance to penetration and higher soybean yields in the first year. However, in subsequent years, soybean root growth in depth was increased in rotation with triticale (X *Triticosecale* Wittmack) and pearl millet due to the presence of biopores and decreased soil resistance to penetration.

The increase in soil water content is highly dependent on the biomass of the CC and soil characteristics. Available water storage is the most important soil water characteristic for plants because the root system of plants has little to no access to the water from the other characteristics. Higher water availability in systems with CCs is attributed to the accumulation of soil organic C (SOC) and other beneficial changes in soil properties that increase soil water retention (Figure 4.7). Thus, CCs can enhance soil water availability indirectly by increasing SOC through the decomposition of roots and stem and the release of organic exudates (Carneiro et al., 2004).

Maintenance of Soil Structure

Soil aggregation is a vital metric for assessing SH and crop productivity (Bronick & Lal, 2005; Peng et al., 2015; Rabot et al., 2018). The structural organization of soil contains various aggregate size classes, ranging from micro to macro scale. Traditionally, soil aggregates are divided into two categories: macroaggregates (>0.25 mm) and microaggregates (<0.25 mm) (Topps et al., 2021). These different aggregate size classes have physical, chemical, and biological properties, such as organic C content, and are considered the result of ongoing agricultural and environmental processes that determine the ability to sustain terrestrial life (Díaz-Zorita et al., 2002; Rabot et al., 2018).

The adoption of CCs into the system can improve soil aggregation and reduce decomposition (Adetunji et al., 2020). The extensive fibrous roots of CCs tend to increase aggregation by producing stable macroaggregates (Bronick & Lal, 2005). Specifically, plant roots destroy aggregates and provide root fragments, mucin, and exudates (Six et al., 2000) that act as nucleation points and binders for the initial stages of aggregate formation (Rabot et al., 2018; Six et al., 2004; Tisdall & Oades, 1982), thus contributing to the evolution of organomineral associations into macroaggregates (Blanco-Canqui & Ruis, 2020; Qi et al., 2022; Zhang & Peng, 2021).

The root system architecture holds a significant position in shaping soil aggregation. Plants that generate substantial quantities of fine and clustered roots seem to be more efficient in aggregate formation as compared to taproots, which facilitate soil decompaction and increase soil porosity through the creation of biopores (Adetunji et al., 2020). Biopores, being low in mechanical resistance and high in O_2 concentrations and thereby providing quick transport pathways for solutes and nutrients, are deemed preferential growth channels for the new roots and offer a superior physical environment for root growth (Ogilvie et al., 2021). Thus, root presence in soil enhances aggregate stability, permeability, porosity, and pore connectivity and impacts the structure of the microbial community in terms of richness and diversity (Galdos et al., 2020).

Paschoal et al. (2020) investigated the effect of intercropping maize and forage grasses on improving the structural quality of the soil through visual evaluation of soil structure. Their results showed that the intercropping system improved the quality of the soil structure, increasing the porosity and reducing soil density in relation to cultivation of maize only. Similar results were obtained by Rigon et al. (2020), in a long-term experiment, in which there was an increase in the stability and mean weight diameter of aggregates in the surface layer of the soil under pearl millet as a winter CC in relation to non-covered soil, improving the physical quality of the soil. This improvement may be linked to an increase in binding agents (Six et al., 2004), which is accompanied by a nearly significant increase in root biomass compared with non-covered soil. The findings of this study suggest that CCs enhance or maintain soil aggregate stability in grain-based cropping systems.

Moreover, the maintenance of soil cover, particularly by CCs such as grasses, offers numerous benefits in terms of reducing soil erosion. This is due to the reduction in soil density and the increase in aggregate stability resulting from the fasciculated root systems of CCs (Blanco-Canqui & Ruis, 2020; Koudahe et al., 2022). A study by Galdos et al. (2020) showed that species with deep roots and vigorous growth, such as black oat, were able to maintain soil porosity and pore connectivity. On the other hand, species like radish and white clover did not have a significant impact on soil pore connectivity. The distinct root architecture of these plant species influences soil pore characteristics differently, based on the plant's growth strategy.

Furthermore, the use of CCs improves soil aggregation at deeper layers compared with mechanical tillage. In studies conducted by Inagaki et al. (2021), under long-term NT planting areas, biological tillage performed by the forage radish resulted in significant increases in soil C storage, demonstrating the potential of incorporating the radish in crop rotation to replace mechanical tillage in long-term NTSs. Additionally, there was a significant increase in bacterial population and diversity through the stimulation of microbial activity favored by the root system (Balota et al., 2014). Therefore, CCs have a positive effect on soil macroporosity. However, CCs have little effect on mesoporosity and microporosity because these are scale levels dependent on the intrinsic properties of each soil (Six et al., 2004). The increase in macroporosity can be partially attributed to the increase and stability of soil macroaggregates. The presence of ecosystem engineers (e.g., earthworms) and biopores formed by decomposing roots can increase macroporosity in soils with CC cultivation (Blanco-Canqui & Ruis, 2020).

In summary, CCs increase the stability of aggregates, protecting the soil surface from the impact of raindrops, providing additional aboveground and belowground biomass input, and increasing the concentration of SOC and microbial activity. An increase in SOC concentration was positively correlated with increased soil aggregate stability (Inagaki et al., 2021). Cover crop residues can generate temporary, transient, and permanent organic binding agents to promote soil aggregation (Calegari et al., 2013; Six et al., 2000) and help to enhance SH and sustainable agriculture in the long term.

Cycling of Organic Matter

Plants, through photosynthesis, assimilate atmospheric CO_2 and incorporate it into their tissues. In the process of decomposition, plants release C into the soil; of this C about two-thirds returns to the atmosphere in the form of gasses, and one-third remains in the soil. Thus, depending on the species used and the prevailing climatic conditions, biomass production by CCs 12.5 Mg ha^1 of dry matter is an important strategy for increasing soil C stocks (Sá et al., 2013). Several studies have shown that simplified cropping systems, such as soybean–fallow, that add low amounts of biomass (<5 Mg ha^{-1} of crop residue) resulted in higher CO_2 emissions, about 1.4 t C-CO_2 ha^{-1} higher than the average of spring CCs. In the soybean season, CO_2 emissions were 23% higher in the fallow compared with the average CC species (Rigon et al., 2020). Comparing conventional systems with the use of CCs (e.g., *U. ruziziensis* and *Crotalaria spectabilis* Roth), an increase of 0.35 Mg C ha^{-1} yr^{-1} in the upper soil layer (0–20 cm) was observed in the system with CCs in the Cerrado region (Anghinoni et al., 2021). This might be attributed to the large amount of crop residues that are incorporated into the soil and the absence of soil disturbance. Hence, this increase could be equivalent to a humification rate of 10.5% (Metay et al., 2007).

In this scenario, conventional management that induces soil disturbance (e.g., plowing) contributes to exposing OM, accelerating the rate of oxidation and CO_2 emissions (Bayer et al., 2016). In conventional systems (e.g., soybean followed by fallow), there are losses of 80% of soil C at 0–15 cm after 5 years of cultivation in soils of the Cerrados region of western Bahia, Brazil (da Silva et al., 1994). Similarly, the low biomass input (<10 Mg ha^{-1} of dry matter) in the Cerrado in areas of winter fallow with low crop diversification contribute to lower C addition and stabilization in soil, which reduces the potential for C sequestration in face of greenhouse gas emissions (Bayer & Dieckow, 2020).

On the other hand, conservation management of the soil, with a special emphasis on plant species diversity in the production systems and CCs with a high C/N ratio in legume–grass mixtures, can increase its ability to sequester C (Silva et al., 2022a,b). Metay et al. (2007) analyzed a 5-year system with CCs in the Brazilian Cerrado and observed that C enrichment affected the smaller soil fractions (mineral-associated fraction), which are responsible for long-term C storage. Soil organic C plays an important role in physical, chemical, and biological processes, and increasing SOC content improves SH and the provision of ecosystem services (Figure 4.8). It is important to note that all soil C pools play critical functions in the soil. Particulate OM provides food for soil life, and OM occluded within the aggregates helps in greater soil aggregation, while mineral-associated OM is more stable and provides greater C sequestration (Cotrufo & Lavalle, 2022). The transitioning

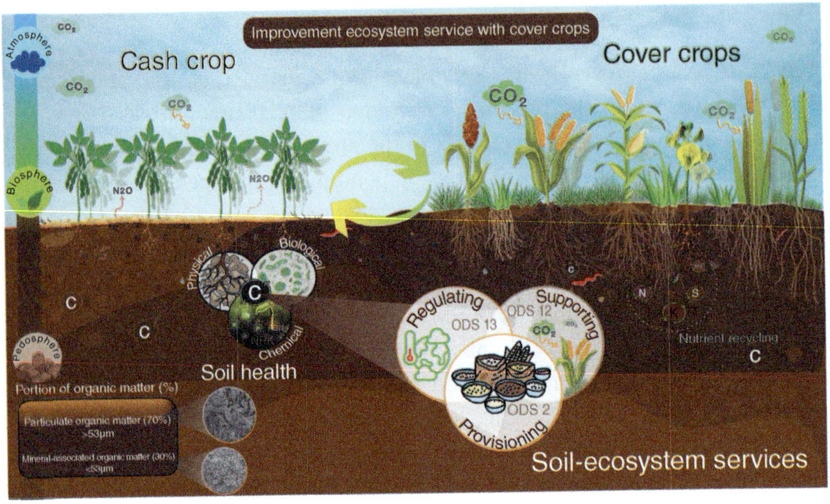

Figure 4.8 Representation of the effects of more biodiverse cropping system in Brazilian Cerrado, introducing cover crops in succession to soybean [*Glycine max* (L.) Merr.].

soybean production from soybean/maize double crop to CCs might contribute to achieving sustainability in Brazilian farmlands in full (Anghinoni et al., 2021).

The quantitative effect of increased biomass addition to the soil by the CCs will directly influence other functions, such as (a) cover crop species diversity (mix) increases the biochemical diversity of biomass, which increases the activity of organisms and soil functionality, favoring biomass transformation (Almeida et al., 2016); (b) higher quality biomass increases the efficiency of the transformation promoted by the microorganisms, producing more chemically stable compounds (Ribeiro et al., 2011); and (c) maintaining the C of the mineral-associated fraction is seen as one of the main climate change mitigation strategies because this fraction has a longer residence time and lower turnover in the soil (Bayer et al., 2003) (Figure 4.9).

Nutrient Cycling

The cycling of elements and the decomposition of organic material are very important for the functioning of the soil in both natural and agricultural environments (Carvalho et al., 2023). Both are directly linked to the dynamics of nutrients in the ecosystem, influencing stability and the ability to maintain other functions (Cerri et al., 2004). The cycling of elements can be understood as the natural process of continuous transfer of elements to different compartments of the planet (e.g., air, water, soil, organisms), which allows its recovery and reuse in the system, including between soil and plant and vice versa (Weiler et al. 2019).

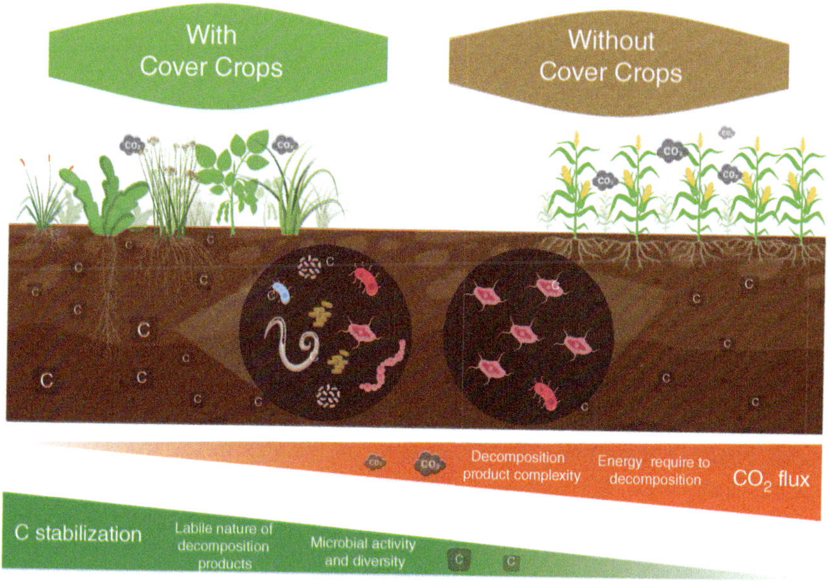

Figure 4.9 A representation of the effect of including a more diversified system with cover crops (mix) compared with conventional farming in Brazil.

Cover crops can be inserted in the production system in rotation, succession, and consortia with the cash crop. The use of CCs promotes nutrient cycling and recycling, preventing nutrient loss through leaching and runoff processes, which are generally intensified when the soil is bare and fallow, especially in tropical regions (Sharma et al., 2018). Knowing the rate of decomposition and nutrient release from the biomass of CCs is important to choose plants that have compatible cycles and take advantage of the release of nutrients with the demand of the cash crop (Doneda et al., 2012; Gonçalves et al., 2007). Decomposition rate will depend on biotic and abiotic factors, such as plant composition (C/N ratio, cellulose, lignin, and hemicellulose content), soil biological activity, and edaphoclimatic conditions (mainly temperature and rainfall) (Carvalho et al., 2010). These factors will determine the residence time of residues in soil and the release of nutrients present in their tissues (Leite et al., 2010).

The low C/N ratio of legume crops (usually <20) favors rapid decomposition compared with grass crops (>25) (Weiler et al., 2019). Thus, due to the higher decomposition of the legume residues, the soil will be unprotected at the beginning of the commercial crop development, but the nutrients will be quickly released and become available. On the other hand, because grasses have slow decomposition rates, the residues remain longer in the soil, with a slower release of nutrients, which occurs over time (Pereira et al., 2016).

Canalli et al. (2020), in Southern Brazil, observed that in a grass intercrop (black oat + ryegrass) the biomass stays longer in the soil (139 days to decompose half of the biomass). However, when there is a mixture of grasses and legumes (black oat + blue lupine), this period drops to 69 days. A possible alternative is to increase the diversification of plants (CCs mix), benefiting the soil with vegetal cover for a longer period, which is associated with the supply of nutrients according to the demand of successive crops. In addition to the greater speed of decomposition, there is a higher biomass production in triple consortia between grasses, brassicas, and legumes (6.265 kg ha^{-1}) and the double consortium between grasses and legumes (5.625 kg ha^{-1}) than in the legume treatments alone (3.388 kg ha^{-1}) (Paye et al., 2022). Therefore, the presence of a suitable complementary species enhances biomass production in the system.

In addition, CCs influence the reduction of N leaching, which accounts for 10–30% losses in conventional grain production systems in the tropical region (Sharma et al., 2012). Maltas et al. (2009), in the Brazilian Cerrado, showed that growing CCs (second crop) (e.g., sorghum and pigeon pea) in succession to cash crops (first crop) induces a positive effect on N uptake and yield of cash crops. The use of CCs provides different effects on the cycling of elements such as N, P, K, Ca, and Mg (Pacheco et al., 2017. As observed by Soratto et al. (2012), dry matter production is higher in pearl millet, which also accumulates more N, P, K, Mg, S, C, and Si. On the other hand, *C. juncea* accumulates a greater amount of Ca. Mean values of the amount of dry matter produced and the nutrient cycled of the main cover crops in Brazil are summarized in Table 4.4 (Carvalho et al., 2022).

Habitat for Organisms and Biological Population Regulation

The multifunctionality of soil is directly associated with its heterogeneous and dynamic characteristics, which are ensured especially by its biological component. Therefore, among all soil functions, provision of habitat for organisms can be considered as one of the major modulators of the soil–atmosphere system. The habitat is defined as the environment for the establishment and maintenance of communities of organisms, which can establish ecological relationships of cooperation, antagonism, commensalism, and others, thus ensuring the regulation of the biological population. Thus, soils are known to host a vast diversity of organisms, surpassing even several other environmental compartments (Andreote et al., 2014).

Plants are the main modulators of soil organism communities, being responsible for suppressing or attracting communities of associated microorganisms (Mendes et al., 2013). Figure 4.10 presents a summary of the impacts of CC species on weed, pests, and some diseases. It makes clear the necessity of local recommendations to adopt CCs to help solve specific problems in each farm.

Table 4.4 Mean values of biomass production and nutrient extraction of the main cover crops used in Brazil.

Cover crops	Biomass production	Nutrient extraction (N–P–K)
	t ha^{-1}	kg ha^{-1}
Pearl millet [*Pennisetum glaucum*(L.) R. Br.]	8.0–15.0	113–3.9–93
Sorghum [*Sorghum bicolor* (L.) Moench]	9.0–12.0	125–17–132
Congo grass [*Urochloa ruziziensis* (Germ. & Evrard) Crins]	12.0–16.0	200–18–372
Crotalária spectabilis	3.0–6.0	220–24–220
Black velvet bean [*Mucuna pruriens* (L.) DC.]	6.0–9.0	185–14–125
Lab lab (*Dolichos lablab* L.)	3.0–9.0	60–20–178
Pigeon pea [*Cajanus cajan* (L.) Millsp.]	3.0–12.0	150–29–74
Sunflower (*Helianthus annuus* L.)	7.0–15.0	31–2–150
Buckwheat (*Fagopyrum esculentum* Moench)	3.0–6.0	46–11–88
Black oat (*Avena strigosa* Schreb.)	5.0–10.0	80–12–200
White lupin (*Lupinus albus* L.)	2.0–3.0	186–11–130
Canola (*Brassica napus* L.)	2.0–3.0	40–19–105
Radish (*Raphanus sativus* L.)	3.0–9.0	63–4.5–78

Note. Carvalho et al. (2022).

Grasses such as *Urochloa* have a high potential as bioactivator plants of the soil biological machinery by increasing enzyme activity (Mendes et al., 2021). The extensive root system is an important source of root exudates for microorganisms, especially when *Urochloa* is grown during the dry season when microbial communities have few food options. In addition, the presence of *Urochloa* in this period results in reductions in soil temperature and maintenance of soil moisture. In the Brazilian Cerrado, in an experiment conducted by Fundação MT, they observed greater soybean (cash crop) productivity and biological activity when *U. ruziziensis* was used as a CC, especially during dry periods (Mendes et al., 2021). The soybean/*U. ruziziensis* treatment showed 3.6 and 8 times more β-glucosidase and arylsulfatase activity, respectively (Mendes et al., 2021).

In the southern region of Brazil, Almeida et al. (2016), in a long-term experiment, observed a 44% increase in β-glucosidase enzyme activity, higher numbers of Hymenoptera, a 38% increase in organic C, a 21% increase in organic N, and a 25% increase in urease enzyme activity when a legume species (e.g., vetch, lablab [*Lablab* Adans.], and cowpea [*Vigna* Savi]) was inserted into the system.

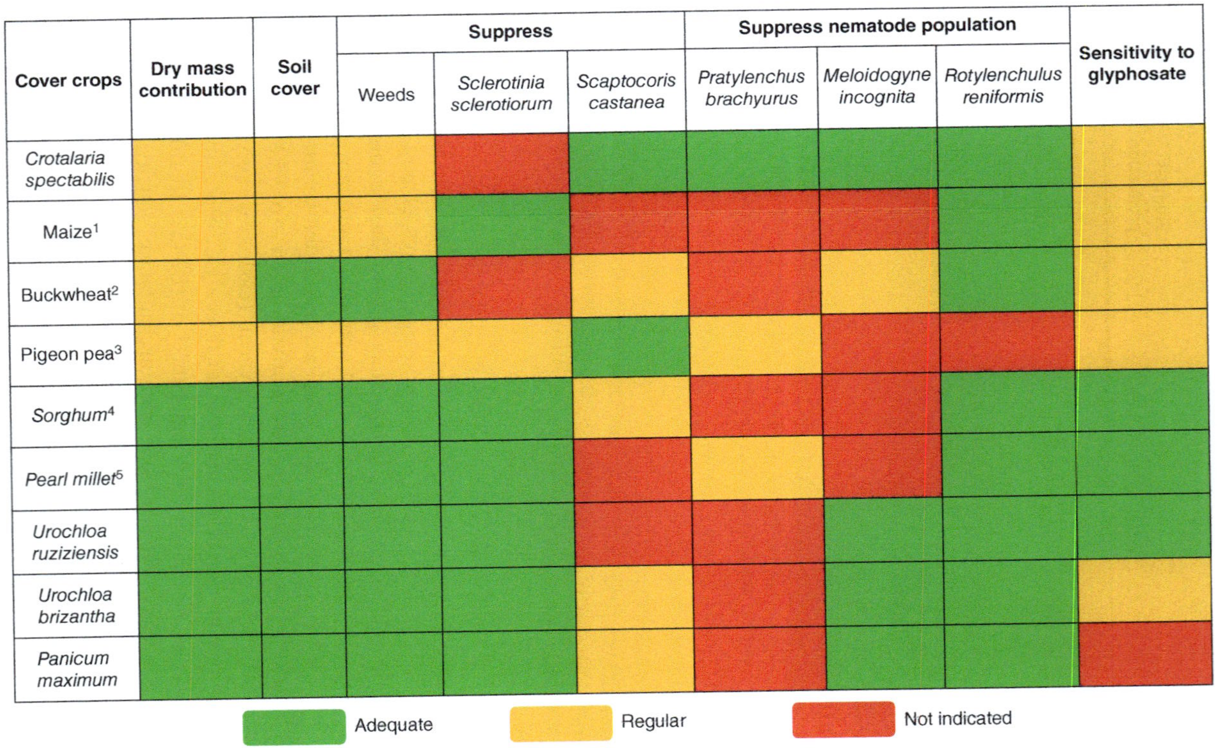

Figure 4.10 Main cover crops used in Brazil and their potential to suppress communities of organisms.

The greater diversity of CC species causes the C and N of the biomass to be found in different proportions; this greater diversity favors the occurrence of edaphic fauna organisms and microbial activity, which promotes the cycling of N and C in the soil.

In addition, CCs (e.g., *C. spectabilis*) reduce the nematode population (Debiasi et al., 2016) and break the cycle of pathogen and disease development (Aguiar-menezes, 2004) (Figure 4.11). Souza (2024) evaluated the nematode population in naturally infested areas in the Brazilian Cerrado, after 3 years of a system with second-crop CCs and first-crop soybean. The CCs that reduced the population of *Pratylenchus* in the soil and *Helicotylenchus* in the root were *C. spectabilis*, pearl millet, and the mix of CCs (*C. spectabilis*, pearl millet, and *U. ruziziensis*); these treatments also resulted in higher soybean yields on average over the 3 years. Debiasi et al. (2016) found that the cultivation of *C. spectabilis* in the off-season, in monoculture or intercropped with pearl millet cultivar ADR

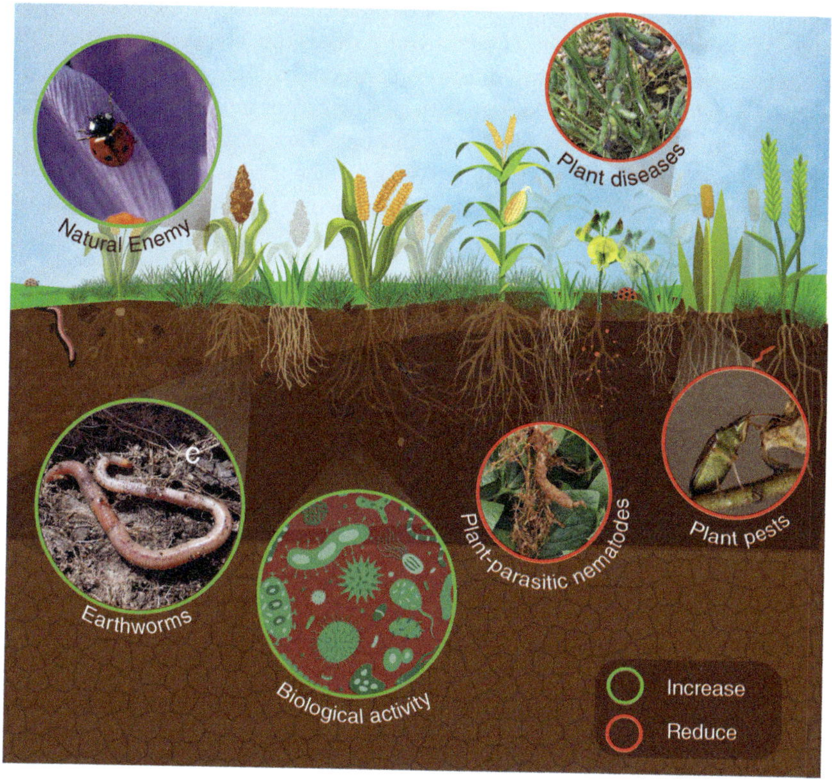

Figure 4.11 Effects promoted by the presence of cover crops on the biological indicators. Source: Souza, 2022/https://repositorio.ifgoiano.edu.br/handle/prefix/2358/last accessed 4 July, 2023.

300, reduced the population of *P. brachyurus* by about 59.7% compared with the treatment without CCs. It also reduced damage to the soybean crop, increasing productivity by approximately 58.6%.

Crop Production

Agricultural sustainability depends on the economic, social, and environmental components of agricultural production systems. Cover crops are an essential component of sustainable agriculture in Brazil (Carvalho et al., 2022), providing a range of benefits for the soil and supporting the production of food, fiber, and fuel. Garbelini et al. (2020) evaluated soybean production systems in the southern region of Brazil and observed a 6% increase in soybean yield when *U. ruziziensis* was used as CC compared with the conventional soybean–maize cropping system.

The difference in soybean yield is greater when there is water deficit during the soybean growth cycle because the roots of *U. ruziziensis* enhance the soil aggregation and the plant residue from *U. ruziziensis* reduces the temperature of the soil surface, resulting in less water loss due to evaporation (Balbinot Junior et al., 2017). A more diversified production system with CCs makes for more stable crop yields, better nutrient and water use efficiency, and increased profit margin compared with that for the less diverse cropping systems (Liebig et al., 2014). Anghinoni et al. (2021) reinforces in their work the importance of a more diversified system with crop rotation to increase soybean yield (28.2% higher) compared with the traditional soybean–maize cropping system.

However, when considering an economic analysis of the production system, the production systems that proved most profitable for the producer in Southern Brazil were a crop rotation between soybean (cash crops) and *U. ruziziensis* intercropped with maize (CCs), which showed an operating profit that was \$333.98 ha^{-1} (11% higher) higher than the soybean–maize rotation. If we consider a 45-ha farm, production systems that involve planting *U. ruziziensis* intercropped with maize during the winter season every year or rotating monoculture of *U. ruziziensis* and maize every 3 years can lead to an average increase in the cumulative operating profit (over 6 years) of the production system, ranging from \$9,929.32 to \$15,029.02, respectively (Garbelini et al., 2020).

According to Anghinoni et al. (2021), in the Cerrado region, a more diversified rotation system with soybeans (first crop) and CCs (second crop) for 2 years and 1 year of *U. ruziziensis* resulted in a statistically identical profit compared with the one obtained through the cultivation of soybean for 3 years in less diverse cropping sequences (such as soybean–*U. ruziziensis* and soybean–maize). These findings not only indicate the saved efficiency for soybean cultivation under a diversified sequence but also that crop rotations potentially reduce inherent risks associated with production under stressful or uncertain conditions, contributing to the stability of food production. This is because the same or higher profit could be achieved through fewer cropping seasons, which involve production uncertainty to some extent (Weil & Kremen, 2007) due to the negative effects of droughts, pests, or other adverse events during the growing season.

Even though the benefits of using CCs are known, the cultivation of CCs must be done in place of off-season maize, a highly profitable crop, so farmers give priority to maize in their cultivation. However, in the Brazilian Cerrado there is a well-established rainy period (November–February), and therefore maize becomes at risk. This risk makes the crop unviable when delaying the sowing of soybean or if long-cycle soybean cultivars are used. A mix of CCs (*P. glaucum*, *C. spectabilis*, and *U. ruziziensis*) increase plant diversification and keep the soil covered during periods of low rainfall CCs. As remarked by Souza et al. (2024), the use of a more diversified system such as mix of CCs provided greater soil coverage (9.269 kg ha^{-1}), suppression of nematodes of the genus *Helicotylenchus*, greater stability of soybean yields (5.635 kg ha^{-1} average during three years), and the cycling of nutrients during the soybean cycle (30 kg ha^{-1} N, 2.50 kg ha^{-1} P, and 50 kg ha^{-1} K) (Figure 4.12).

Overall, CCs play an important role in supporting the production of food, fiber, and fuel in Brazil. By improving SH and reducing erosion, CCs can increase agricultural productivity, support sustainable land use, and contribute to a more resilient agricultural system. Promoting the implementation of cover cropping

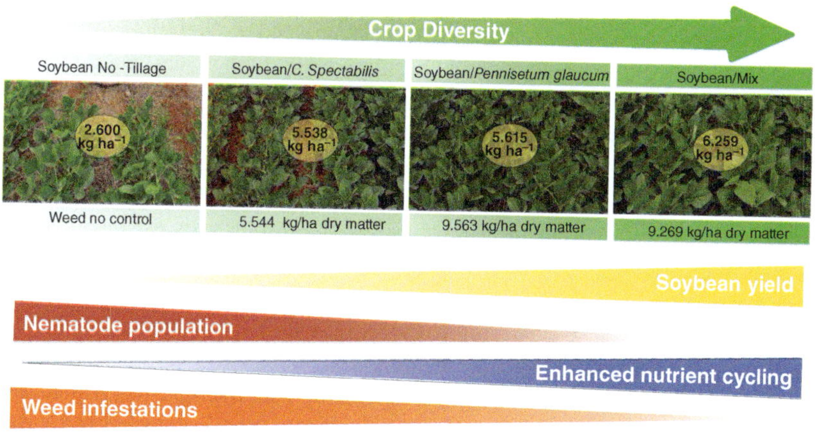

Figure 4.12 Effect of higher diversity of cover crops on sustainability aspects of soybean production in the Brazilian Cerrado. Mix of cover crops: *Pennisetum glaucum* (L.) R. Br., *Crotalaria spectabilis* Roth, *Urochloa ruziziensis* (Germ. & Evrard) Crins. Souza et al. (2024).

techniques, in conjunction with other sustainable soil management methods, can safeguard the enduring sustainability of Brazil's agricultural sector.

Final Remarks

The interest of Brazilian farmers in CCs has been growing rapidly in recent years, mainly as a response to reduced inputs, pests, increased sustainability, and profitability of the cash crop. Soil compaction and the spread of pathogenic nematodes are common in NT farms in Brazil, especially in farms that do not follow the plant diversification principle of NTS. The need for sustainable solutions to these problems is currently the main driver for research and the adoption of CCs by farmers. As the knowledge of the importance of C for the maintenance of soil health and the provision of soil-related ecosystem services spreads, it is expected that farmers' demand for CCs will be increasingly tied to C accumulation in soils, reaching production systems that do not traditionally use CCs, such as horticulture (see Chapter 7). Therefore, the demand for CC seed and information on how to use them correctly tends to increase among Brazilian farmers in the next few years.

For CCs to become standard practice in Brazilian farms, some challenges remain to be addressed. The production of commercial seeds for the main CCs is not enough to meet the current demand, in addition to the high cost and access to seed being restricted to regions close to the few companies producing and selling them due to the logistical costs inevitable in a country with continental dimensions. Seed of less established CCs that may be suited for specific conditions is even more scantily produced. In some cases, farmers resort to saving seed or buying uncertified seed, which may lead to yield problems such as weed proliferation and higher disease incidence. More knowledge of CC mixes containing several plant species is needed; very few studies have evaluated their impact on soil properties when compared with single species or more traditional blends. Additionally, species population in mixes is still defined by each manufacturer without clear information on their calculation methods. Finally, a large portion of Brazilian farmers lack the capital or the access to credit necessary to buy the implements for termination CCs, and many still must resort to improvised equipment or are unable to adopt this practice.

Overall, CCs can be a powerful management tool for maintaining SH in Brazilian agricultural systems, and farmers are increasingly recognizing their importance and potential. Despite the current challenges, the topic has piqued the interest of researchers and businesses. Thus, a reasonable expectation is for the available knowledge to grow exponentially with the joined efforts of academics, businesses, consultants, and farmers. With information and education, we hope that the use of CCs may yet be a standard management practice in Brazilian farms.

Acknowledgments

We thank the Coordenação de Aperfeiçoamento de Pessoal de Nível Superior—Brasil (CAPES, Finance Code 001), Conselho Nacional de Desenvolvimento Científico e Tecnológico (CNPq) and São Paulo Research Foundation (FAPESP # 2022/16368-6, 2022/13531-3; 2022/14212-9) for the scholarships. We are also grateful for the support of the CCARBON—Center for Carbon Research in Tropical Agriculture (2021/10573-4); RCGI—Research Centre for Greenhouse Gas Innovation (2020/15230-5), and Bayer S/A (Project —"Balanço de carbono em sistemas agrícolas: revelando o impacto da adoção de práticas de manejo sustentáveis nos estoques de carbono do solo e nas emissões de gases de efeito estufa"). Maurício Roberto Cherubin thanks the CNPq for his Research Productivity Fellowship (311787/2021-5).

References

Abrahão, G. M., & Costa, M. H. (2018). Evolution of rain and photoperiod limitations on the soybean growing season in Brazil: The rise (and possible fall) of double-cropping systems. *Agricultural and Forest Meteorology*, 32–45. https://doi.org/10.1016/j.agrformet.2018.02.031

Adetunji, A. T., Ncube, B., Mulidzi, R., & Lewu, F. B. (2020). Management impact and benefit of cover crops on soil quality: A review. *Soil and Tillage Research*, 204(June), 104717. https://doi.org/10.1016/j.still.2020.104717

Aguiar-Menezes, E. L. (2004). *Diversidade vegetal: uma estratégia para o manejo de pragas em sistemas sustentáveis de produção agrícola*. Embrapa Agrobiologia.

Almeida, D. O., Bayer, C., & Almeida, H. C. (2016). Fauna and microbial attributes of an Ultisol under cover crop systems in the south of Brazil. *Pesquisa Agropecuária Brasileira*, 51, 1140–1147.

Ambrosini, V. G., de Almeida, J. L., de Araujo, E. A., Alves, L. A., Filippi, D., Flores, J. P. M., Fostim, M. L., Fontoura, S. M. V., Bortoluzzi, E. C., Bayer, C., & Tiecher, T. (2022). Effect of diversified cropping systems on crop yield, legacy, and budget of potassium in a subtropical Oxisol. *Field Crops Research*, 275, 108342. https://doi.org/10.1016/j.fcr.2021.108342

Ambrosini, V. G., de Almeida, J. L., de Araujo, E. A., Alves, L. A., Filippi, D., Flores, J. P. M., Fostim, M. L., Fontoura, S. M. V., Bortoluzzi, E. C., Bayer, C., & Tiecher, T. (2022). Effect of diversified cropping systems on crop yield, legacy, and budget of potassium in a subtropical Oxisol. *Field Crops Research*, 275, 108342. https://doi.org/10.1016/j.fcr.2021.108342

Andreote, F. D., Gumiere, T., & Durrer, A. (2014). Exploring interactions of plant microbiomes. *Scientia Agricola*, 71(6), 528–539. https://doi.org/10.1590/0103-9016-2014-0195

Andreote, F. D., Gumiere, T., & Durrer, A. (2014). Exploring interactions of plant microbiomes. *Scientia Agricola*, 71(6), 528–539. https://doi.org/10.1590/0103-9016-2014-0195

Anghinoni, G., Anghinoni, F. B. G., Tormena, C. A., Braccini, A. L., Mendes, I. C., Zancanaro, L., & Lal, R. (2021). Conservation agriculture strengthens sustainability of Brazilian grain production and food security. *Land Use Policy*, 108, 105591.

Ayarza, M., Rao, I., Vilela, L., Lascano, C., & Vera-Infanzón, R. (2022). Chapter Three - Soil carbon accumulation in crop-livestock systems in acid soil savannas of South America: A review. *Advances in Agronomy*, 173, 163–226. https://doi.org/10.1016/bs.agron.2022.02.003

Anghinoni, G., Anghinoni, F. B. G., Tormena, C. A., Braccini, A. L., Mendes, I. C., Zancanaro, L., & Lal, R. (2021). Conservation agriculture strengthens sustainability of Brazilian grain production and food security. *Land Use Policy*, 108, 105591.

Balbino, L. C., Bruand, A., Brossard, M., Grimaldi, M., Hajnos, M., & Guimarães, M. D. F. (2002). Changes in porosity and microaggregation in clayey Ferralsols of the Brazilian Cerrado on clearing for pasture. *European Journal of Soil Science*, 53(2), 219–230.

Balbinot Junior, A. A., Santos, J. C. F., Debiasi, H., & Yokoyama, A. H. (2017). Contribution of roots and shoots of Brachiaria species to soybean performance in succession. *Pesquisa Agropecuária Brasileira*, 52, 592–598. https://doi.org/10.1590/s0100-204x2017000800004

Baligar, V. C., & Fageria, N. K. (2007). Agronomy and physiology of tropical cover crops. *Journal of Plant Nutrition*, 30(8), 1287–1339. https://doi.org/10.1080/01904160701554997

Balota, E. L., Calegari, A., Nakatani, A. S., & Coyne, M. S. (2014). Benefits of winter cover crops and no-tillage for microbial parameters in a Brazilian Oxisol: A long-term study. *Agriculture, Ecosystems & Environment*, 197, 31–40. https://doi.org/10.1016/j.agee.2014.07.010

Bayer, C., & Dieckow, J. (2020). Lessons learnt from long-term no-till systems regarding soil management in humid tropical and subtropical regions. In P. D. Yash, R. C. Dalal, & N. W. Menzies (Eds.), *No-till farming systems for sustainable agriculture: Challenges and opportunities* (pp. 437–457). Springer.

Bayer, C., Gomes, J., Zanatta, J. A., Vieira, F. C. B., & Dieckow, J. (2016). Mitigating greenhouse gas emissions from a subtropical Ultisol by using long-term no-tillage in combination with legume cover crops. *Soil and Tillage Research*, 161, 86–94. https://doi.org/10.1016/j.still.2016.03.011

Bayer, C., Spagnollo, E., Wildner, L. P., Ernani, P. R., & Albuquerque, J. A. (2003). Increase of carbon and nitrogen in a south Brazilian oxisol by using of summer cover-crops. *Ciência Rural*, 33, 469–475.

Berriel, V., Monza, J., & Perdomo, C. H. (2020). Cover crop selection by jointly optimizing biomass productivity, biological nitrogen fixation, and transpiration efficiency: Application to two crotalaria species. *Agronomy*, 10(8), 1116. https://doi.org/10.3390/agronomy10081116

Bertollo, A. M., de Moraes, M. T., Franchini, J. C., Soltangheisi, A., Junior, A. A. B., Levien, R., & Debiasi, H. (2021). Precrops alleviate soil physical limitations for soybean root growth in an Oxisol from southern Brazil.

Blanco-Canqui, H., & Ruis, S. J. (2020). Cover crop impacts on soil physical properties: A review. *Soil Science Society of America Journal*, 84(5), 1527–1576. https://doi.org/10.1002/saj2.20129

Borkert, C. M., Gaudêncio, C. D. A., Pereira, J. E., Pereira, L. R., & Oliveira Junior, A. D. (2003). Mineral nutrients in the shoot biomass of soil cover crops. *Pesquisa Agropecuária Brasileira*, 38, 143–153. https://doi.org/10.1590/S0100-204X2003000100019

Borowik, A., Wyszkowska, J., & Kucharski, J. (2020). Impact of various grass species on soil Bacteriobiome. *Diversity*, 12(212), 20. https://doi.org/10.3390/d12060212

Bronick, C. J., & Lal, R. (2005). Soil structure and management: A review. *Geoderma*, 124(1–2), 3–22. https://doi.org/10.1016/j.geoderma.2004.03.005

Burle, M. L., de Carvalho, A. M., Amabile, R. F., & Pereira, J. (2006). Caracterização das espécies de adubo verde. In A. M. de Carvalho & R. F. Amabile (Eds.), *Cerrado: adubação verde* (pp. 71–142). Embrapa Cerrados.

Canalli, L. B. S., Santos, J. B., Silva, A. A., Muchinski, A., Feliciano, E., & Penteado, D. (2020). Residues Decomposition in Crop Rotations under No-till System. *Brazilian Archives of Biology and Technology*, 63. http://dx.doi.org/10.1590/1678-4324-solo-2020190609

Calegari, A., Tiecher, T., Hargrove, W. L., Ralisch, R., Tessier, D., Tourdonnet, S., Guimarães, M. F., & Santos, D. R. (2013). Long-term effect of different soil management systems and winter crops on soil acidity and vertical distribution of nutrients in a Brazilian Oxisol. *Soil Tillage Research*, 133, 32–39. https://doi.org/10.1016/j.still.2013.05.009

Calonego, J. C., & Rosolem, C. A. (2010). Soybean root growth and yield in rotation with cover crops under chiseling and no-till. *European Journal of Agronomy*, 33(3), 242–249.

Carneiro, R. G., de Mendes, I. C., Lovato, P. E., de Carvalho, A. M., & Vivaldi, L. J. (2004). Indicadores biológicos associados ao ciclo do fósforo em solo de Cerrado sob plantio direto e plantio convencional. *Pesquisa Agropecuária Brasileira*, 39, 661–669.

Carvalho, A. M., & Amabile, R. F. (2006). Plantas condicionadoras de solo: Interações edafoclimáticas, uso e manejo. In A. M. Carvalho & R. F. Amabile (Eds.), *Cerrado: Adubação verde* (pp. 143–170). Embrapa Cerrados.

Carvalho, A. M., Dantas, R. A., Coelho, M. C., Lima, W. M., Souza, J. P. S. P., Fonseca, O. P., & Guimarães Júnior, R. (2010). *Hemicellulosis, cellulose and lignin contents in cover crops with potential for no-tillage system in Cerrado. Brazil.*

Carvalho, M. L., da Vanolli, B. S., Schiebelbein, B. E., de Borba, D. A., da Luz, F. B., Cardoso, G. M., de Bortolo, L. S., Marostica, M. E. M., Souza, V. S., & Cherubin, M. R. (2022). Guia prático de plantas de cobertura: Aspectos filotécnicos e impactos sobre a saúde do solo. In *Guia prático de plantas de cobertura: Aspectos filotécnicos e impactos sobre a saúde do solo*. ESALQ-USP. https://www.livrosabertos.sibi.usp.br/portaldelivrosUSP/catalog/book/782

Cerri, C. C., Bernoux, M., Cerri, C. E. P., & Feller, C. (2004). Carbon cycling and sequestration opportunities opportunities in South America: The case of Brazil. *Soil Use and Management*, 20, 248–254.

Chapagain, T., Lee, E. A., & Raizada, M. N. (2020). The potential of multi-species mixtures to diversify cover crop benefits. *Sustainability*, 12(5), 2058. https://doi.org/10.3390/su12052058

Cordeiro, C. F. S., Echer, F. R., & Araujo, F. F. (2021). Cover crops impact crop yields by improving microbiological activity and fertility in sandy soil. *Journal of Soil Science and Plant Nutrition*, 21(3), 1968–1977. https://doi.org/10.1007/s42729-021-00494-0

Cotrufo, F. M., & Lavalle, J. M. (2022). Soil organic matter formation, persistence, and functioning: A synthesis of current understanding to inform its conservation and regeneration. *Advances in Agronomy*, 172, 1–66. https://doi.org/10.1016/bs.agron.2021.11.002

Crush, J. R., Waller, J. E., & Care, D. A. (2005). Root distribution and nitrate interception in eleven temperate forage grasses. *Grass and Forage Science*, 60(4), 385–392. https://doi.org/10.1111/J.1365-2494.2005.00488.X

Cruz, T. T., Asmus, G. L., & Garcia, R. A. (2020). Crotalaria species in succession to soybean for the management of *Pratylenchus brachyurus*. *Ciência Rural*, 50. https://doi.org/10.1590/0103-8478cr20190645

da Luz, F. B., Carvalho, M. L., Aquino de Borba, D., Schiebelbein, B. E., Paiva de Lima, R., & Cherubin, M. R. (2020). Linking soil water changes to soil physical quality in sugarcane expansion areas in Brazil. *Water*, 12(11), 3156.

da Silva, J. E., Lemainski, J., & Resck, D. V. S. (1994). Perdas de matéria orgânica e suas relações com a capacidade de troca catiônica em solos da região de Cerrados do oeste baiano. *Revista Brasileira de Ciência do Solo*, 18, 541–547.

da Silva Neto, E. C., Pereira, M. G., Fernandes, J. C. F., & Neto, T. D. A. C. (2016). Aggregate formation and soil organic matter under different vegetation types in

Atlantic Forest from southeastern Brazil. *Semina: Ciências Agrárias*, 37(6), 3927–3939.

Dabney, S. M., Delgado, J. A., & Reeves, D. W. (2001). Using winter cover crops to improve soil and water quality. *Communications in Soil Science and Plant Analysis*, 32, 1221–1250. https://doi.org/10.1081/CSS-100104110

Daryanto, S., Fu, B., Wang, L., Jacinthe, P. A., & Zhao, W. (2018). Quantitative synthesis on the ecosystem services of cover crops. *Earth-Science Reviews*, 185, 357–373. https://doi.org/10.1016/J.EARSCIREV.2018.06.013

Debiasi, H., Franchini, J. C., Dias, W. P., Junior Ramos, E. U., & Junior Balbinot, A. A. (2016). Cultural practices during the soybean off-season for the control of Pratylenchus brachyurus (in Portuguese, with English abstract). *Pesquisa Agropecuária Brasileira*, 51, 1720–1728. https://doi.org/10.1590/S0100-204X2016001000003

de Almeida, W. S., Panachuki, E., de Oliveira, P. T. S., da Silva Menezes, R., Sobrinho, T. A., & de Carvalho, D. F. (2018). Effect of soil tillage and vegetal cover on soil water infiltration. *Soil and Tillage Research*, 175, 130–138.

de Carvalho, A. M., Coelho, M. C., Dantas, R. A., Fonseca, O. P., Júnior, R. G., & Figueiredo, C. C. (2012). Chemical composition of cover plants and its effect on maize yield in no-tillage systems in the Brazilian savanna. *Crop & Pasture Science*, 63, 1075–1081.

de Carvalho, A. M., Miranda, J. C. C., Gerosa, M. L., & Ribeiro Junior, W. Q. (2014). Adubação verde e plantas de cobertura no cerrado. In O. F. de Lima Filho, E. J. Ambrosano, F. Rossi, & J. A. D. Carlos (Eds.), *Adubação verde e plantas de cobertura no Brasil (fundamentos e prática)* (Vol. 2, pp. 343–372). Embrapa.

de Carvalho, A. M., Pereira, L. L., Alves, P. C. A. C. A., Junior Guimaraes, R., & Vivaldi, L. J. (2011). Cover plants with potential use for crop-livestock integrated systems in the Cerrado region. *Pesquisa Agropecuária Brasileir*, 46, 1200–1205.

de Carvalho, A. M., Quintana, L. G., Souza, A. M., Coser, T. R., Sousa, T. R., de Oliveira, A. D., & Gerosa, M. L. (2023). Decomposition, biochemical composition and nutrient cycling of macauba palm in a Cerrado Entisol. *Scientia Agricola*.

de Lima, J. L., da Silva, J. R., Montenegro, A. A., Silva, V. P., & Abrantes, J. R. (2021). The effect of vegetal mulching on soil surface temperature in semiarid Brazil. *Die Bodenkultur: Journal of Land Management, Food and Environment*, 71(4), 185–195.

de Lima Filho, O. F., Edmilson, J. A., Fabrício, R., & José, A. D. C. (2014). *Adubação verde e plantas de cobertura no Brasil (fundamentos e prática)* (1st ed., Vol. 2).

de Moraes, M. T., Bengough, A. G., Debiasi, H., Franchini, J. C., Levien, R., Schnepf, A., & Leitner, D. (2018). Mechanistic framework to link root growth models with weather and soil physical properties, including example applications to soybean growth in Brazil. *Plant and Soil*, 428, 67–92.

de Moraes, M. T., Debiasi, H., Carlesso, R., Franchini, J. C., da Silva, V. R., & da Luz, F. B. (2016). Soil physical quality on tillage and cropping systems after two decades in the subtropical region of Brazil. *Soil and Tillage Research*, 155, 351–362. https://doi.org/10.1016/j.still.2015.07.015

Pereira, F. C. B. L., de Mello, L. M. M., Pariz, C. M., de Mendonça, V. Z., Yano, É. H., de Miranda, E. E. V., & Crusciol, C. A. C. (2016). Autumn maize intercropped with tropical forages: Crop residues, nutrient cycling, subsequent soybean and soil quality. *Revista Brasileira de Ciência do Solo*, 40, 1–20. https://doi.org/10.1590/18069657rbcs20150003

Díaz-Zorita, M., Perfect, E., & Grove, J. H. (2002). Disruptive methods for assessing soil structure. *Soil and Tillage Research*, 64(1–2), 3–22. https://doi.org/10.1016/S0167-1987(01)00254-9

Diekow, J., Mielniczuk, J., Knicker, H., Bayer, C., Dick, D. P., & Kögel-Knabner, I. (2005). Soil C and N stocks as affected by cropping systems and nitrogen fertilisation in a southern Brazil Acrisol managed under no-tillage for 17 years. *Soil and Tillage Research*, 81(1), 87–95. https://doi.org/10.1016/J.STILL.2004.05.003

Dinkelaker, B., Römheld, V., & Marschner, H. (1989). Citric-acid excretion and precipitation of calcium citrate in the rhizosphere of white lupin (*Lupinus albus* L). *Plant, Cell & Environment*, 12, 285–292. https://doi.org/10.1111/j.1365-3040.1989.tb01942.x

Doneda, A., Aita, C., Giacomini, S. J., Miola, E. C. C., Giacomini, D. A., Schirmann, J., & Gonzatto, R. (2012). Fitomassa e decomposição de resíduos de plantas de cobertura puras e consorciadas [Phytomass and decomposition of cover crops residues in monoculture and intercropping]. *Revista Brasileira de Ciência do Solo*, 36(6), 1714–1723. https://doi.org/10.1590/S0100-06832012000600005

dos Reis Ferreira, C., da Silva Neto, E. C., Pereira, M. G., do Nascimento Guedes, J., Rosset, J. S., & dos Anjos, L. H. C. (2020). Dinâmica da agregação do solo e frações de carbono orgânico ao longo de 23 anos de manejo do plantio direto. *Soil and Tillage Research*, 198, 104533.

Dowling, L. F., Arndt, C., & Hamaker, B. R. (2002). Economic viability of high digestibility sorghum as feed for market broilers. *Agronomy Journal*, 94, 1050–1058. https://doi.org/10.2134/agronj2002.1050

Duchene, O., Vian, J. F., & Celette, F. (2017). Intercropping with legume for agroecological cropping systems: Complementarity and facilitation processes and the importance of soil microorganisms. A review. *Agriculture, Ecosystems & Environment*, 240, 148–161. https://doi.org/10.1016/J.AGEE.2017.02.019

Fageria, N. (1992). *Maximizing crop yields*, 1. https://books.google.com.br/books?hl=en&lr=&id=NbPMD99XIkEC&oi=fnd&pg=PR5&ots=fas-nXDxEQ&sig=L5b-2bQb2jOiNG0ca2kmPmr1NvU&redir_esc=y#v=onepage&q&f=false

Fischler, M., Wortmann, C. S., & Feil, B. (1999). Crotalaria (*C. ochroleuca* G. Don.) as a green manure in maize-bean cropping systems in Uganda. *Field Crops Research*, 61(2), 97–107. https://doi.org/10.1016/S0378-4290(98)00150-6

Fuentes-Llanillo, R., Telles, T. S., Junior, D. S., de Melo, T. R., Friedrich, T., & Kassam, A. (2021). Expansion of no-tillage practice in conservation agriculture in Brazil. *Soil and Tillage Research*, 208, 104877. https://doi.org/10.1016/j.still.2020.104877

Galdos, M. V., Brown, E., Rosolem, C. A., Pires, L. F., Hallett, P. D., & Mooney, S. J. (2020). Brachiaria species influence nitrate transport in soil by modifying soil structure with their root system. *Scientific Reports*, 10(1), 1–11. https://doi.org/10.1038/s41598-020-61986-0

Garbelini, L. G., Franchini, J. C., Debiasi, H., Balbinot, J. A., Betioli, J. E., & Telles, T. S. (2020). Profitability of soybean production models with diversified crops in the autumn–winter. *Agronomy Journal*, 112, 4092–4103. https://doi.org/10.1002/agj2.20308

Gonçalves, S. L., Gaudencio, C. A., Franchini, J. C., Galerani, P. R., & Gracia, A. (2007). *Rotação de culturas [Crops rotation]* (45th ed.). Embrapa.

Guo, L. B., & Gifford, R. M. (2002). Soil carbon stocks and land use change: A meta analysis. *Global Change Biology*, 8(4), 345–360.

Guswa, A. J., Celia, M. A., & Rodriguez-Iturbe, I. (2002). Models of soil moisture dynamics in ecohydrology: A comparative study. *Water Resources Research*, 38(9), 5–1.

Helalia, A. M., Letey, J., & Graham, R. C. (1988). Crust formation and clay migration effects on infiltration rate. *Soil Science Society of America Journal*, 52(1), 251–255. https://doi.org/10.1016/j.agsy.2013.11.004

Inagaki, T. M., de Sá, J. C. M., Tormena, C. A., Dranski, A., Muchalak, A., Briedis, C., de Oliveira Ferreira, A., Giarola, N. F. B., & da Silva, Á. P. (2021). Mechanical and biological chiseling impacts on soil organic C stocks, root growth, and crop yield in a long-term no-till system. *Soil and Tillage Research*, 211, 104993. https://doi.org/10.1016/j.still.2021.104993

Kassam, A., Derpsch, R., & Friedrich, T. (2014). Global achievements in soil and water conservation: The case of conservation agriculture. *International Soil and Water Conservation Research*, 2(1), 5–13.

Koudahe, K., Allen, S. C., & Djaman, K. (2022). Critical review of the impact of cover crops on soil properties. *International Soil and Water Conservation Research*, 10, 343–354. https://doi.org/10.1016/j.iswcr.2022.03.003

Krenchinski, F. H., Cesco, V. J. S., Rodrigues, D. M., Albrecht, L. P., Wobeto, K. S., & Albrecht, A. J. P. (2018). Agronomic performance of soybean grown in succession to winter cover crops. *Pesquisa Agropecuária Brasileira*, 53, 909–917. https://doi.org/10.1590/S0100-204X2018000800005

Lal, R. (2015). Managing carbon for restoring degraded soils. *Sustainability*, 7, 5875–5895.

Lawley, Y. E., Teasdale, J. R., & Weil, R. R. (2012). The mechanism for weed suppression by a forage radish cover crop. *Agronomy Journal*, 104, 205–214. https://doi.org/10.2134/agronj2011.0128

Leite, L. F. C., Freitas, R. C. A., Sagrilo, E., & Galvão, S. R. S. (2010). Decomposition and nutrients release from crop residues on a Yellow Latosol in the savanna, Manharão state, Brazil. *Revista Ciência Agronômica*, 41, 29–35.

Lesmes-Vesga, R. A., Cano, L. M., Ritenour, M. A., Sarkhosh, A., Chaparro, J. X., & Rossi, L. (2023). Variation in the root system architecture of peach × (peach × almond) backcrosses. *Plants*, 12, 1874. https://doi.org/10.3390/plants12091874

Libardi, P. L. (2005). *Dinâmica da água no solo* (Vol. 61). Edusp.

Liebig, M. A., Archer, D. W., & Tanaka, D. L. (2014). Crop diversity effects on near-surface soil condition under dryland agriculture. *Applied and Environmental Soil Science*, 2014, 703460. https://doi.org/10.1155/2014/703460

Locatelli, J. L., Santos, R. S., Cherubin, M. R., & Cerri, C. E. (2022). Changes in soil organic matter fractions induced by cropland and pasture expansion in Brazil's new agricultural frontier. *Geoderma Regional*, 28, e00474.

Maltas, A., Corbeels, M., Scopel, E., Wery, J., & Silva da Macena, F. A. (2009). Cover crop and nitrogen effects on maize productivity in no-tillage systems of the Brazilian Cerrados. *Agronomy Journal*, 101, 1036–1046. https://doi.org/10.2134/agronj2009.0055

Mendes, I. C., Chaer, G. M., Reis Junior, F. B., Sousa, D. M., Dantas, O. D., Oliveira, M. I. L., & Malaquias, J. V. (2021). *Tecnologia BioAS uma maneira simples e eficiente de avaliar a saúde do solo*. Embrapa Cerrados.

Mendes, R., Garbeva, P., & Raaijmakers, J. M. (2013). The rhizosphere microbiome: Significance of plant beneficial, plant pathogenic, and human pathogenic microorganisms. *FEMS Microbiology Reviews*, 37, 634–663.

Metay, A., Oliver, R., Scopel, E., Douzet, J. M., Moreira, J. A. A., Maraux, F., Feigl, B. J, & Feller, C. (2007). N2O and CH4 emissions from soils under conventional and no-till management practices in Goiânia (Cerrados, Brazil). *Geoderma*, 88, 141:78. https://doi.org/10.1016/j.geoderma.2007.05.010.

Morais, M. C., Siqueira-Neto, M., Guerra, H. P., Satiro, L. S., Soltangheisi, A., Cerri, C. E., Feigl, B. J., & Cherubin, M. R. (2020). Trade-offs between sugarcane straw removal and soil organic matter in Brazil. *Sustainability*, 12(22), 9363.

Nascimento, V. D., Arf, O., Alves, M. C., Souza, E. J. D., Silva, P. R. T. D., Kaneko, F. H., & Teixeira Filho, M. C. M. (2022). Mechanical chiseling and the cover crop effect on the common bean yield in the Brazilian Cerrado. *Agriculture*, 12(5), 616. https://doi.org/10.3390/agriculture12050616

Ogilvie, C. M., Ashiq, W., Vasava, H. B., & Biswas, A. (2021). Quantifying root-soil interactions in cover crop systems: A review. *Agriculture*, 11(3), 218. https://doi.org/10.3390/agriculture11030218

Paschoal, M. C. G., Cagna, C. P., Filho, O. G., & Mazzini-Guedes, R. B. (2020). Visual evaluation of soil structure in maize and forage grasses intercropping under no-tillage. *Brazilian Archives of Biology and Technology*, 63, 1–8. https://doi.org/10.1590/1678-4324-solo-2020190498

Pacheco, L. P., Monteiro, M. M. S., Petter, F. A., Nóbrega, J. C. A., & Santos, A. S. (2017). Biomass and nutrient cycling by cover crops in Brazilian cerrado in the state of Piaui. *Revista Caatinga*, 30, 13–23. https://doi.org/10.1590/1983-21252017v30n102rc

Paye, W. S., Acharya, P., & Ghimire, R. (2022). Water productivity of forage sorghum in response to winter cover crops in semi-arid irrigated conditions. *Field Crops Research*, 283, 108552. https://doi.org/10.1016/j.fcr.2022.108552

Peng, X., Horn, R., & Hallett, P. (2015). Soil structure and its functions in ecosystems: Phase matter & scale matter. *Soil and Tillage Research*, 146(part A), 1–3. https://doi.org/10.1016/j.still.2014.10.017

Piva, J. T., Bratti, F., Locatelli, J. L., Ribeiro, R. H., Besen, M. R., Brancaleoni, E., & Schmitt, D. E. (2021). Use of winter cover crops improves maize productivity under reduced nitrogen fertilization: A long-term study. *Bragantia*, 80. https://doi.org/10.1590/1678-4499.20200165

Qi, J., Jensen, J. L., Christensen, B. T., & Munkholm, L. J. (2022). Soil structural stability following decades of straw incorporation and use of ryegrass cover crops. *Geoderma*, 406, 115463. https://doi.org/10.1016/j.geoderma.2021.115463

Rabot, E., Wiesmeier, M., Schlüter, S., & Vogel, H. J. (2018). Soil structure as an indicator of soil functions: A review. *Geoderma*, 314, 122–137. https://doi.org/10.1016/j.geoderma.2017.11.009

Ratnadass, A., Fernandes, P., Avelino, J., & Habib, R. (2012). Plant species diversity for sustainable management of crop pests and diseases in agroecosystems: A review. *Agronomy for Sustainable Development*, 32, 273–303. https://doi.org/10.1007/s13593-011-0022-4

Ribeiro, P. H., Santos, J. V. V. M., Coser, S. M., Nogueira, N. O., & Martins, C. A. S. (2011). Green manure, carbon and nitrogen stocks and soil organic matter quality. *Revista Verde*, 6, 43–50.

Rigon, J. P. G., Crusciol, C. A. C., Calonego, J. C., Pavinato, P. S., Azevedo, A. C., & Rosolem, C. A. (2022). Intensive crop rotations and residue quality increase soil phosphorus lability under long-term no-till in tropical soils. *Soil and Tillage Research*, 223, 105446. https://doi.org/10.1016/j.still.2022.105446

Rigon, J. P. G., Franzluebbers, J. C., & Calonego, J. C. (2020). Soil aggregation and potential carbon and nitrogen mineralization with cover crops under tropical no-till. *Journal of Soil and Water Conservation*, 75(5), 601–609. https://doi.org/10.2489/jswc.2020.0018

Robinson, D. A., Nemes, A., Reinsch, S., Radbourne, A., Bentley, L., & Keith, A. M. (2022). Global meta-analysis of soil hydraulic properties on the same soils with differing land use. *Science of the Total Environment*, 852, 158506.

Rosolem, C. A., Foloni, J. S. S., & Tiritan, C. S. (2002). Root growth and nutrient accumulation in cover crops as affected by soil compaction. *Soil and Tillage Research*, 65(1), 109–115. https://doi.org/10.1016/S0167-1987(01)00286-0

Rossi, C. Q., Pereira, M. G., Giacomo, S. G., Betta, M., & Polidoro, J. C. (2013). Litter decomposition and nutrient release from Brachiaria, sorghum and soybean in no-tillage areas in the Cerrado region, Goiás. *Semina Ciências Agrárias*, 34(4), 1523–1533. https://doi.org/10.5433/1679-0359.2013v34n4p1523

de Sá, J. C. M., Séguy, L., Tivet, F., Lal, R., Bouzinac, S., Borszowskei, P. R., Briedis, C., Santos, J. B., da Hartaman, D. C., Bertoloni, C. G., Rosa, J., & Riedrich, T. (2013). Carbon depletion by plowing and its restoration by no-till cropping systems in oxisols of subtropical and tropicalagro-ecoregions in Brazil. *Land degradation & development*, 26, 531–543. https://doi.org/10.1002/ldr.2218

Salton, J. C., Mercante, F. M., Tomazi, M., Zanatta, J. A., Concenço, G., Silva, W. M., & Retore, M. (2014). Integrated crop-livestock system in tropical Brazil: Toward a sustainable production system. *Agriculture, Ecosystems & Environment*, 190, 70–79. https://doi.org/10.1016/J.AGEE.2013.09.023

Santin, W. (2018). *O Brasil Possível: A Biografia de Herbert Bartz* (1st ed.). Londrina.

Sathiyanarayanan, L., & Arulmozhi, S. (2007). *Mucuna pruriens* Linn.: A comprehensive review. *Pharmacognosy Reviews*, 1, 157–162.

Sharma, P., Singh, A., Kahlon, C. S., Brar, A. S., Grover, K. K., Dia, M., & Steiner, R. L. (2018). The role of cover crops towards sustainable soil health and agriculture – a review paper. *American Journal of Plant Sciences*, 9, 1935–1951. https://doi.org/10.4236/ajps.2018.99140

Schiebelbein, B. E., Carvalho, M. L., Locatelli, J. L., Mendonça, R. S., Rodrigues, Y. F., & Cherubin, M. R. (2022). *Plantas de cobertura e as inter-relações com a saúde do solo*. SolloAgro/ESALQ/USP.

Schipanski, M. E., Barbercheck, M., Douglas, M. R., Finney, D. M., Haider, K., Kaye, J. P., Kermanian, A. R., Mortensen, D. A., Ryan, M. R., Tooker, J., & White, C. (2014). A framework for evaluating ecosystem services provided by cover crops in agroecosystems. *Agriculture Systems*, 125, 12–22.

Sekaran, U., Sagar, K. L., & Kumar, S. (2021). Soil aggregates, aggregate-associated carbon and nitrogen, and water retention as influenced by short and long-term no-till systems. *Soil and Tillage Research*, 208, 104885.

Sharma, P., Shukla, M. K., Sammis, T. W., Steiner, R. L., & Mexal, J. G. (2012). Nitrate-nitrogen leaching from three specialty crops of New Mexico under furrow irrigation system. *Agricultural Water Management*, 109, 71–80.

Silva, L. S., dos Laroca, J. V. S., Coelho, A. P., Gonçalves, E. C., Gomes, R. P., Pacheco, L. P., de Carvalho, P. C. F., Pires, G. C., Oliveira, R. L., de Souza, J. M. A., Freitas, C. M., Cabral, C. E. A., Wruck, F. J., & de Souza, E. D. (2022a). Does grass-legume intercropping change soil quality and grain yield in integrated crop-livestock systems? *Applied Soil Ecology*, 170. https://doi.org/10.1016/j.apsoil.2021.104257

Silva, B. N. P. D., Masetto, T. E., Garcia, R. A., & Toledo, M. Z. (2022b). Row spacing and seed physiological quality of Crotalaria species. *Pesquisa Agropecuária Tropical*, 52. https://doi.org/10.1590/1983-40632022v5272674

Six, J., Bossuyt, H., Degryze, S., & Denef, K. (2004). A history of research on the link between (micro)aggregates, soil biota, and soil organic matter dynamics. *Soil and Tillage Research*, 79(1), 7–31. https://doi.org/10.1016/j.still.2004.03.008

Six, J., Paustian, K., Elliott, E. T., & Combrink, C. (2000). Soil structure and organic matter I. Distribution of aggregate-size classes and aggregate-associated carbon. *Soil Science Society of America Journal*, 64(2), 681–689. https://doi.org/10.2136/sssaj2000.642681x

Soratto, R. P., Crusciol, C. A. C., Costa, C. H. M., Ferrari Neto, J., & Castro, G. S. A. (2012). Production, decomposition and nutrient cycling in residues of sunnhemp and pearl millet in monocropped and intercropped systems. *Pesquisa Agropecuário Brasileira*, 47, 1462–1470.

Souza, V. S., Santos, D. C., Ferreira, J. G., Souza, S. O., Gonçalo, T. P., Sousa, J. V. A., Cruvinel, A. G., Vilela, L., Paim, T. P., Almeida, R. E. M., Canisares, L. P., Cherubin, M. R. (2024). Cover crop diversity for sustainable agriculture: Insights from the Cerrado biome. *Soil Use and Management*, (40), early view. https://doi.org/10.1111/sum.13014.

Tisdall, J. M., & Oades, J. M. (1982). Organic matter and water-stable aggregates in soils. *European Journal of Soil Science*, 33, 141–163.

Topps, D., Khabir, M. I. u., Abdelmagid, H., Jackson, T., Iqbal, J., Robertson, B. K., Pervaiz, Z. H., & Saleem, M. (2021). Impact of cover crop monocultures and mixtures on organic carbon contents of soil aggregates. *Soil Systems*, 5(3), 43. http://dx.doi.org/10.3390/soilsystems5030043

van Eekeren, N., Bos, M., de Wit, J., Keidel, H., & Bloem, J. (2010). Effect of individual grass species and grass species mixtures on soil quality as related to root biomass and grass yield. *Applied Soil Ecology*, 45(3), 275–283. https://doi.org/10.1016/J.APSOIL.2010.05.003

Wagenet, R. J., & Hutson, J. L. (1997). Soil quality and its dependence on dynamic physical processes. *Journal of Environmental Quality*, 26(1), 41–48.

Ward, P. R., Flower, K. C., Cordingley, N., Weeks, C., & Micin, S. F. (2012). Soil water balance with cover crops and conservation agriculture in a Mediterranean climate. *Field Crops Research*, 132, 33–39.

Weil, R., & Kremen, A. (2007). Thinking across and beyond disciplines to make cover crops pay. *Science of Food and Agriculture*, 87, 551–557.

Weiler, D. A., Giacomini, S. J., Aita, C., Schmatz, R., Pilecco, G. E., Chaves, B., & Bastos, L. M. (2019). Summer cover crops shoot decomposition and nitrogen release in a no-tilled sandy soil. *Revista Brasileira de Ciência do Solo*, 43. https://doi.org/10.1590/18069657rbcs20190027

Yu, Y., Loiskandl, W., Kaul, H. P., Himmelbauer, M., Wei, W., Chen, L., & Bodner, G. (2016). Estimation of runoff mitigation by morphologically different cover crop root systems. *Journal of Hydrology*, 538, 667–676.

Zegada-Lizarazu, W., & Monti, A. (2011). Energy crops in rotation: A review. *Biomass and Bioenergy*, 35(1), 12–25. https://doi.org/10.1016/J.BIOMBIOE.2010.08.001

Zhang, Y. K., & Schilling, K. E. (2006). Effects of land cover on water table, soil moisture, evapotranspiration, and groundwater recharge: A field observation and analysis. *Journal of Hydrology*, 319(1–4), 328–338.

Zhang, Z., & Peng, X. (2021). Bio-tillage: A new perspective for sustainable agriculture. *Soil and Tillage Research*, 206(71), 104844. https://doi.org/10.1016/j.still.2020.104844

5

Soil Health in Sugarcane Production Systems

Maurício Roberto Cherubin, Felipe Bonini da Luz, Renato Paiva de Lima, Sarah Tenelli, Ricardo Oliveira Bordonal, Bruna Gonçalves de Oliveira, Leandro Carolino Gonzaga, Carlos Eduardo Pellegrino Cerri, and João Luís Nunes Carvalho

Chapter Overview

Sugarcane (*Saccharum* spp.) is a semi-perennial crop cultivated in approximately 10 million ha in Brazil. The large-scale cultivation of sugarcane makes Brazil the world's largest producer, responsible for 40% of global sugarcane production. Soil health (SH) is one of the main pillars of the sustainability of sugarcane production. Therefore, in this chapter, we present the science-based evidence of the impact of land use change for sugarcane expansion and adoption of management practices (e.g., harvesting system, soil tillage, crop rotation, organic amendments, and traffic control) on SH and sustainability of the production systems. In summary, this chapter reveals the profound transformation that the Brazilian sugarcane sector has experienced in the last decades. This transformation has resulted in important advances in SH and land productivity, although some challenges (soil compaction, erosion, and biodiversity loss) still need to be overcome. Investments in research and innovation coupled with sectoral public policies need to be promoted and strengthened in Brazil toward enhancing SH, sugarcane production, and sustainability of the production system.

Introduction

Sugarcane has had an important and strategic socio-economic and environmental role in Brazil since the sixteenth century. Brazil is the world's leading sugarcane

Soil Health Series: Volume 3 Soil Health and Sustainable Agriculture in Brazil, First Edition.
Edited by Ieda Carvalho Mendes and Maurício Roberto Cherubin.

producer and is responsible for about 40% of global production (FAO, 2023). In the last four decades, the Brazilian sugarcane sector has undergone a profound transformation, resulting in a 400% increase in national production (Figure 5.1), driven by the increment of cultivated area but also by the substantial increase in

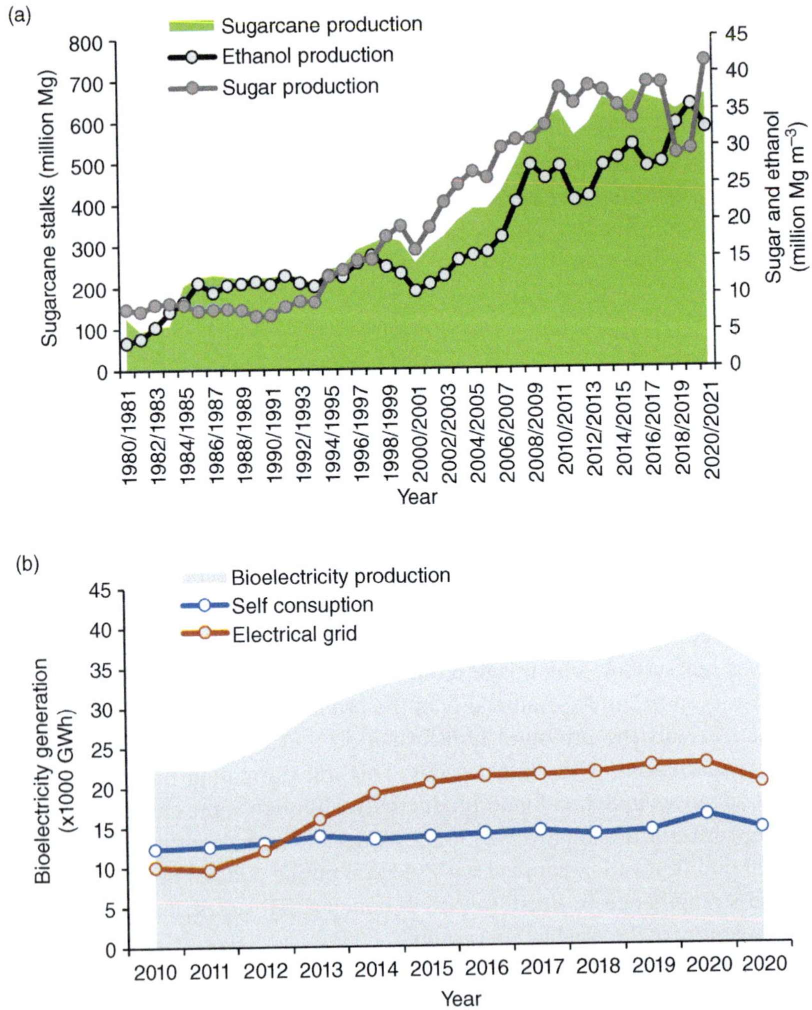

Figure 5.1 Evolution of sugarcane production in Brazil and main sugarcane-derived products. (a) Sugar and ethanol; (b) Bioelectricity (used for mill's self-consumption and delivered to the national electric energy grid). UNICA (2022).

the efficiency of field and industrial operation (Cherubin et al., 2021b). Therefore, Brazil became the largest producer of sugar and ethanol (i.e., 41 million tons and 32 billion liters in the 2020/2021 season, respectively) (Figure 5.1). The sugarcane sector also produces about 6% of national electricity (34 TWh in 2021), using bagasse and straw to feed integrated cogeneration power plants. Currently, 40% of produced bioelectricity is for the mill's consumption, and 60% is delivered to the national grid (Figure 5.1). More recently, the sugarcane sector has evolved to produce other environmentally friendly options of bioenergy, such as cellulosic ethanol produced from bagasse and straw, biomethane from vinasse biodigestion, and green hydrogen from ethanol.

Despite the Brazilian know-how for cultivating sugarcane, the complexity of the production system imposes several challenges and opportunities to preserve and enhance SH and crop yields. The concept of SH (i.e., "the continued capacity of a soil to function as a vital living ecosystem that sustains plants, animals, and humans" [Chessman et al., 2019]), should be used to support sustainable land use and soil management decisions in sugarcane fields. Recently, SH has been considered as the frontier of soil science (Shen & Teng, 2022). Healthy soils provide a suitable growing environment for sugarcane as well as other vital soil-related ecosystem services, such as erosion control, soil C sequestration, water regulation, nutrient cycling, and biodiversity promotion (Carvalho et al., 2022; Cherubin et al., 2021a; Oliveira et al., 2019). In addition, healthy soils are more resistant and resilient to biotic and abiotic stresses (e.g., climate changes), increasing crop yields and yield stability over time (Qiao et al., 2022). Healthy soil is particularly important for sugarcane, a semi-perennial crop that is exposed to long periods (six months) of water stress in the producing regions of Brazil.

Sugarcane stands out as a promising alternative to meet the growing global demand for bioenergy options. Therefore, it requires sustainable expansion and intensification of large-scale production in Brazil (Cherubin et al., 2021b; Rossetto et al., 2022). In that direction, substantial advances in SH management have been achieved in the last decades through the adoption of several management practices, such as transition from burned harvest system to green mechanized system, adoption of reduced tillage or no-tillage (NT) systems, introduction of cover crops during sugarcane field renovation, recycling and reuse of organic industrial by-products such as vinasse and filter cake, straw management, and use of machinery traffic control technologies. Therefore, in this chapter, we present and discuss scientific evidence of how sugarcane expansion and management practices affect SH and related ecosystem services in sugarcane production systems in Brazil.

Soil Health Responses to Land Use Change for Sugarcane Expansion in Brazil

Introduced in Brazil in the sixteenth century, sugarcane cultivation grew rapidly in the Northeast region. Later, the crop migrated to the Central-South region, particularly to São Paulo state, driven by the more suitable edaphoclimatic conditions for plant growth and the socio-economic development of that region. In the last four decades, the sugarcane cultivation area tripled in the country, reaching about 10 million ha (Figure 5.2). Currently, the Central-South region comprises 90% of the sugarcane cultivation area (9 million ha); about 10% remains in the Northeast/North region. The majority of sugarcane expansion has occurred in areas previously occupied with poor-managed pasture (Cherubin et al., 2021b) in São Paulo (SP) and surrounding states, such as Goiás (GO), Minas Gerais (MG), Mato Grosso do Sul (MS), and Paraná (PR) (Figure 5.2). Extensive pasture areas were and still are the main opportunities for sugarcane expansion in Brazil (Oliveira et al., 2019). According to Hernandes et al. (2021), even excluding

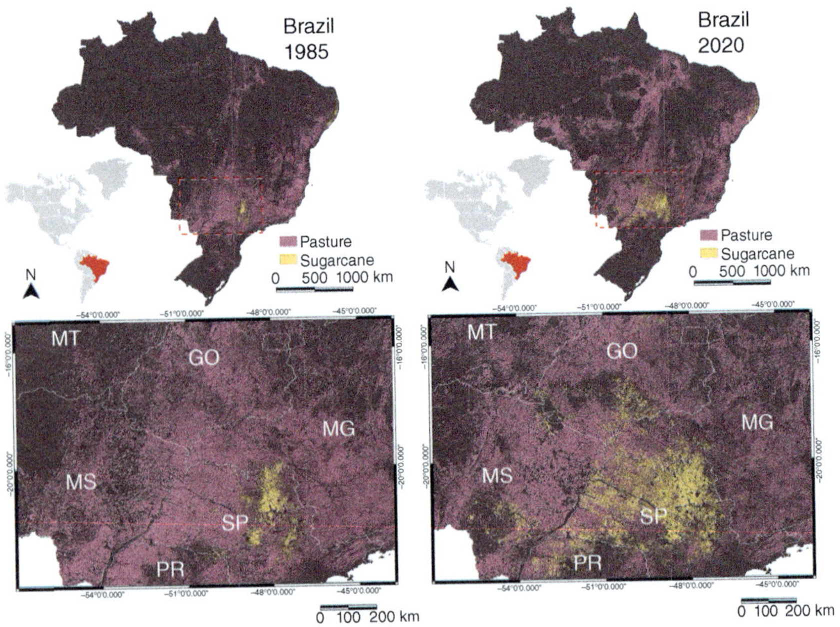

Figure 5.2 Historic evolution of sugarcane cultivation area in Brazil. (a) Land use change for sugarcane expansion over extensive pasture areas in Brazil, with emphasis in the Central-South region (bottom panels); (b) evolution of sugarcane cultivation area from 1980 to 2022 in two main production regions in Brazil. (a) MapBiomas (2022)/ https://plataforma.mapbiomas.org/last accessed June 09, 2022; (b) IBGE and CONAB (2023).

environmentally relevant areas, 20 million ha of pasture are suitable for sugarcane expansion in Brazil.

The land use change for large-scale sugarcane expansion has raised concerns about the impacts on SH and other soil-related ecosystem services. To address such concerns, several studies have been conducted across the country in the last decade, as recently synthesized by Cherubin et al. (2021b). Overall, the main promising land transition scenario is extensive pasture to sugarcane. Pasture represents the second largest land use in Brazil (behind only native vegetation), covering about 160 million ha. It is estimated that at least a half of this pasture area is under some stage of degradation (LAPIG, 2022). The continuing SH degradation in pasture areas of the Central-South region was detected by Cherubin et al. (2016a) by integrating soil chemical, physical, and biological indicators into an overall SH index (SHI). They concluded that, on average, pasture soils were functioning at 70% of their potential capacity, whereas soil in native vegetation soils was functioning at 87% (Figure 5.3). Soil health degradation in long-term extensive pasture is driven by soil C depletion (Franco et al., 2015; Oliveira et al., 2016), chemical impoverishment (acidification and nutrient depletion [Cherubin et al., 2015]), losses of soil biodiversity (Franco et al., 2016; Vanolli et al., 2021), and physical degradation and erosion (Cherubin et al., 2016b). Finally, unhealthy soils and poor management/investments have led to low pasture productivity and loss of social-economic benefits.

Based on this scenario, converting pasture areas into sugarcane production areas enhances SH (SHI, 74%) (Cherubin et al., 2016a). It is a sustainable option because it not only prevents further soil degradation but also can restore specific soil functions (i.e., improvement in soil chemical properties) in the short term (Cherubin et al., 2015) and land productivity (Oliveira et al., 2019; Cherubin et al., 2021b). In addition, sugarcane cultivation can rapidly restore (in two to three years) soil C stocks lost during land conversion from pasture (Mello et al., 2014). Therefore, soil C restoration coupled with low-C bioenergy production makes the sugarcane production system a powerful tool for mitigating climate change. However, sugarcane cultivation presents several challenges. If not properly managed, high mechanization can lead to serious soil compaction problems in sugarcane cultivation, reducing water infiltration, increasing runoff and erosion risks, and causing losses of habitat for soil biota and crop yield decline (Cherubin et al., 2016b; Cavalcanti et al., 2020; Carvalho et al., 2022; Luz et al., 2022a; Thomaz et al., 2022; Youlton et al., 2016). Soil tillage increases C losses (Tenelli et al., 2019) and greenhouse gas (GHG) emissions. Excessive use of pesticides and mineral fertilizers can impair soil biodiversity (Franco et al., 2016; Vanolli et al., 2021) and reduce downstream water quality (Gunkel et al., 2007; Taniwaki et al., 2017). Therefore, the adoption of best management practices is crucial to reduce or even eliminate these tradeoffs of sugarcane cultivation.

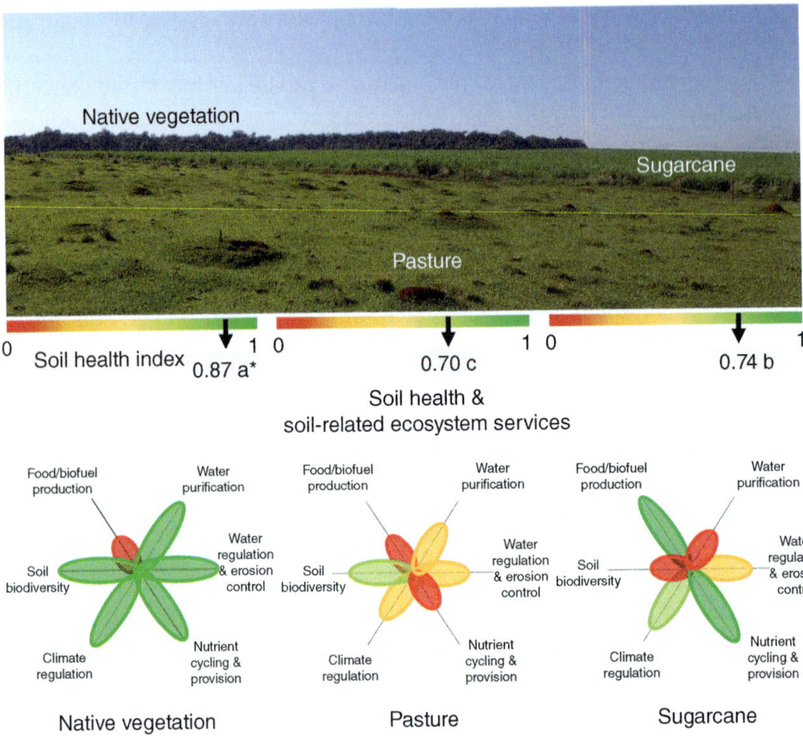

Figure 5.3 Effects of land use change for sugarcane expansion on soil health and on the soil-related ecosystem service in Brazil. *Mean SHI scores followed by the same uppercase letter do not differ according to Tukey's test ($p < 0.05$). Soil health index scores are based on Cherubin et al. (2016a, 2016b). The provision of soil-related ecosystem services of each land use was ranked qualitatively: poor (red petal), moderate (orange petal), good (light green petal), and optimum (dark green petal). The criteria used for ranking were the frequency of studies reporting positive or negative land use change effects on each soil-related ecosystem service and the statistical significance of those changes. Cherubin et al. (2021b)/MDPI/CC BY 4.0.

Soil Health Responses to Management Practices Adopted in Sugarcane Cultivation

Soil Tillage Systems

Conventional tillage (CT; i.e., intensive topsoil mobilization) is traditionally used in Brazilian sugarcane fields. The mechanical implements used in this system are heavy harrows, subsoilers, disc harrows, and moldboard plows. Soil compaction, weeds, and pest infestations are the main reasons for using CT; however, soil mobilization leads to negative impacts on SH, such as degradation of soil structure (Cavalcanti et al., 2020), aggregate slaking, and reduction in water availability to plants and soil biota. Soil mobilization also increases soil erosion (Thomaz et al., 2022) and oxidation of soil organic C (SOC) (Franco et al., 2015). In addition, recent studies have indicated that the benefits of soil tillage are short lasting and that the soil returns to the same or even a worse degree of compaction after one or two harvesting events (Barbosa et al., 2019) due to high pressure applied by machinery traffic.

Considering the short-term benefit and cost of operation, the sugarcane sector is starting a transition from CT to localized or reduced tillage (i.e., subsoiling only the planting row) and even to a NT system in areas without chemical constraints (acidity in depth) and weed and pest infestation. Reduced tillage or NT promotes better conditions for preserving soil aggregation and C stocks and reducing soil erosion risks. Whereas CT disturbs 100% of the soil volume (0–40 cm), a NT system adopted in sugarcane fields disturbs around 13% of the soil volume to open-planting furrows (Tenelli et al., 2019) every five years (Figure 5.4).

A recent study conducted by Luz et al. (2022b) revealed that the adoption of NT did not negatively affect water availability, air permeability, and air-filled porosity in relation to CT in a clayey soil and a sandy loam soil in central southern Brazil. However, the soil physical advantages provided by NT required the adoption

Figure 5.4 Sugarcane planting under a no-tillage system consists of a single operation of planting furrow opened to a 30-cm depth over the cover crop biomass (e.g., *Crotalaria spectabilis*) using a two-row planter with spacing of 150 cm. Photo: João L. N. Carvalho.

of traffic-free seedbed zones. In the same experimental area, Luz et al. (2022a) assessed soil structure changes by a diversity of assessment scales. These authors found that CT and reduced tillage resulted in similar soil structure conditions and functioning after the fourth cycle of sugarcane ratoon. Therefore, CT did not bring additional benefits in alleviating soil compaction compared with NT. In addition, there is a trend toward a gradual restoration of SH due to the recovery of soil physical and chemical attributes under NT (Farhate et al., 2022), as well as a reduction of soil erosion rates (Thomaz et al., 2022) and an increase in SOC in the soil surface (Oliveira et al., 2022b). According to Tenelli et al. (2019), the adoption of NT resulted in an average rate of SOC accumulation of 1.64 Mg C ha^{-1} yr^{-1}, whereas CT resulted in an average rate of 0.05 Mg C ha^{-1} yr^{-1} in a clayey soil. In the same way, NT has the potential to improve soil N due to increases in soil organic matter (SOM) (Awe et al., 2020). Therefore, NT is an advantageous strategy to reduce soil C losses with the potential for additional soil C sequestration and SH restoration in sugarcane fields. In terms of sugarcane production, recent studies have shown that CT does not increase sugarcane yield (Barbosa et al., 2019; Farhate et al., 2022; Lovera et al., 2021) and in some cases decreases yield (e.g., Oliveira et al., 2022b) compared with NT or reduced tillage. Finally, a recent literature review confirmed these results, revealing similar sugarcane yields between CT and NT in 36 experiments across the country (Sá et al., 2022).

Despite that, CT is still widely adopted in sugarcane fields because there are some challenges for farmers adopting NT. In addition to soil compaction (a physical constraint), some soil chemical and biological constraints are also managed using CT. Highly weathered soils in Brazil, in general, are acidic and have low nutrient concentrations in surface and subsurface layers. Therefore, CT operations are also used to incorporate lime and fertilizers to correct soil acidity, neutralize toxic aluminum, and replenish nutrients. Also, soil mobilization is used as a mechanical practice to control some soil pests (e.g., *Sphenophorus levis*) and weeds (e.g., *Cynodon* spp.).

Based on this scenario, the transition of CT to a reduced tillage or NT system should be associated with other management practices and a judicious decision-making process. The NT system is only a feasible option when soil acidity is already corrected at depth and when integrated practices of machinery traffic control and management of pests and weeds are adopted in parallel. The rate of conservation management systems adoption is expected to increase in the coming years, driven by greater awareness of the importance of SH to produce more and better sugarcane, economic incentives to adopt management practices that reduce GHG emissions, and the necessity to optimize the operation and reduce production costs.

Crop Rotation and Cover Crops in Sugarcane Fields

Sugarcane is a semi-perennial crop cultivated as monocropping over a period of around five to six years, based on rainfed conditions in Brazil. The adoption of crop rotation or cover crops occurs during the crop renovation period, mainly cultivating legume crops. Crop rotation is adopted in the traditional system (i.e., cover crops are cultivated in the full area), but a significant number of farmers have used the Simultaneously Occurring Intercropping and Rotation (MEIOSI) system (Figure 5.5). The MEIOSI system, created in the 1980s (Barcelos, 1984), combines the planting rows (strips) of sugarcane for seedling production with other crops, such as cover crops or cash crops (soybean [*Glycine max* (L.) Merr.], corn [*Zea mays* L.], and peanuts [*Arachis hypogaea* L.]) between sugarcane rows.

The planting of legume crops is a promising strategy due to biological N fixation and other benefits to the overall SH. The most typical legume crops used as cover or cash crops in sugarcane fields are sunn hemp species (*Crotalaria juncea* L., *Crotalaria ochroleuca* G. Don, *Crotalaria spectabilis* Roth), soybean, and peanuts (Figure 5.5). However, in some areas legume crops are also cultivated with corn and other grass as cover crops [*Pennisetum glaucum* (L.) R. Br., *Sorghum bicolor* (L.) Moench, *Urochloa ruziziensis* (Hochst. ex A. Rich.) R. Webster] or even a mix of several species (Figure 5.5). There are two main benefits promoted by legume crops in sugarcane fields: the N effects (i.e., the direct aspect of N supply from the symbiotic association with rhizobia atmospheric N_2 fixation) and the non-N effects [i.e., indirect aspects associated with the promotion of a favorable environment for the plants growth, such as an increase in soil physico-chemical (Barbosa et al., 2018; Fernandes et al., 2012) and biological attributes (Stirling et al., 2001); nutrient recycling (Caceres & Alcarde, 1995; Perin et al., 2004); weed suppression (Ambrosano et al., 2005); and control of pests, diseases, and soil erosion (Garside & Bell, 2001)].

Overall, cover crops exhibit fast growth and high biomass production that protect the soil surface against erosion [especially during the heavy rains that occur in summer (November–March) in the Central-South region] and normally accumulate high amounts of N, K, and other macronutrients—(P, Ca, Mg, S) and micronutrients (B, Cu, Fe, Mn, Zn), which can improve soil fertility by releasing nutrients after crop residue decomposition. Ambrosano et al. (2005) found that combining *C. juncea* rotation with mineral N fertilizer resulted in positive changes on soil chemical attributes. More recently, some studies have also revealed that sunn hemp cover crop is able to reduce sugarcane responsiveness to inorganic N fertilizer, being a potential alternative for the total and partial supply of N for following crop cycles (Ambrosano et al., 2013; Otto et al., 2020; Tenelli et al., 2021). In the meantime, this practice has been crucial to promote yield gains of subsequent sugarcane ratoons by 15%–25% in Australia (Garside & Bell, 2001) and 30% in

Figure 5.5 Simultaneously Occurring Intercropping and Rotation ("MEIOSI") system in sugarcane fields in Central-Southern Brazil. (a) Corn in Ribeirão Preto–SP, From Mohamad Riad Perrone Sammour, (b) cover crops–*Crotalaria ochroleuca*, and (c) mix of cover crops, From Carlos Eduardo Avelar (*C. ochroleuca*, *C. spectabilis*, *Cajanus cajan*, *Raphanus sativus*, *Pennisetum glaucum*, and *Fagopyrum esculentum*) in Pradópolis, SP, Brazil, intercropped with sugarcane for production of cane seedling.

Brazil (Ambrosano et al., 2011), whereas sugarcane was able to produce more tons of stalks for a given amount of N fertilizer applied (Tenelli et al., 2021). By isolating the N and non-N effects of sunn hemp on sugarcane yields over a four-year period, a recent study showed that the accumulated yield gains were more related to non-N effects in comparison to N effects (Tenelli et al., 2021).

Regarding other effects related to SH, crop rotation, including cover crops before sugarcane planting, is known to improve aspects of soil physical quality, such as high porosity, low bulk density, low resistance to penetration, and high aggregation (Farhate et al., 2022; Fernandes et al., 2012). Soil biological activity is also recognized to be affected by a diverse crop system due to the increase in nutrients and water supply for the microorganisms and the capacity of some species in controlling nematode infestations in sugarcane fields. The effects of sunn hemp were reported to increase microbial biomass C compared with fallow during the first sugarcane ratoon cycle (Tenelli et al., 2019), whereas some sunn hemp species were able to reduce nematode populations (*Meloidogyne* spp. and *Pratylenchus* spp.) in sugarcane crop cycles (Macedo & Botelho, 1995; Moura, 1995; Stirling et al., 2001). Using a life-cycle assessment approach, Chagas et al. (2016) showed that the implementation of crop rotation with sunn hemp and soybean in sugarcane fields can reduce GHG emissions, resulting in several economic and environmental benefits for the sugarcane sector. Therefore, the inclusion of cover crops and cash crops integrated with sugarcane is a useful tool to sustain SH and mitigate the environmental impacts (e.g., erosion, GHG emissions from N fertilizer, biodiversity loss) in the sugarcane production system.

In recent years, a new movement has started in sugarcane fields: spreading crop mixes (a mix of two or more species) (Figure 5.6). Like that occurring for annual crops, seed-producing companies of cover crops have tested personalized mixes of species combining different botanical families, such as Fabaceae, Poaceae, Polygonaceae, Asteraceae, Brassicaceae, and so on. The goal is to explore the benefits brought by each individual species and the synergies among the different species growing together. Beside the novelty, several sugarcane farmers have already adopted cover crop mixes based on the benefits previously proved in annual crop fields (see Chapter 4). However, there is an avenue of research opportunities to evaluate the impacts of these different configurations of mixes (e.g., number of species, sowing time, seed proportion, interspecies competition, etc.) on SH and sugarcane yield.

Recycling the Sugarcane By-products into the Field

In a typical Brazilian sugarcane ethanol plant, different agroindustry residues, such as bagasse, filter cake, and vinasse, are generated. Except for bagasse, which

Figure 5.6 Crop rotation using a mix of cover crops (*Crotalaria spectabilis*, *Raphanus sativus*, *Urochloa ruziziensis*, *Helianthus annuus*) in Uberaba, MG, southeastern Brazil. Photo: Tedson L. F. Azevedo.

is used exclusively as biomass to produce steam and electricity, supplying energy to the whole sugarcane mill energy, all of the by-products are recycled back into the field in order to improve the agronomic aspects of crop production, reduce environmental impacts, and contribute to alleviating the unwanted effects on SH and global climate change (Cherubin et al., 2021b).

The filter cake, also known as press mud, is a residue of sugar and ethanol production generated during the clarification process of sugarcane juice (Rossetto et al., 2018). Composed basically of bagasse and sludge, filter cake is usually applied to the field furrows (in natura or composted) when planting sugarcane as a partial replacement of mineral P fertilizer. Its use, when combined with mineral fertilizer, showed positive results in SOM improvement as well as soil moisture, consequently increasing sugarcane productivity and seedling growth (Prado et al., 2013). The main effects of filter cake on soil chemical attributes are to increase N, P, and Ca concentrations; increase cation exchange capacity; and reduce concentrations of exchangeable Al^{3+}, which is toxic to plants. Beneficial effects on physical and biological soil properties are also observed. Thus, filter cake plays a fundamental role in sugarcane production, in the maintenance of soil fertility, and in maintaining SH (Rossetto et al., 2018).

Another important by-product of ethanol production is vinasse. Also called stillage, vinasse is the high-volume residue of the sugarcane industry, being produced at a rate of 10–15 l per liter of ethanol. Vinasse is applied to sugarcane fields (ratoon stage) mainly as source of K fertilizer, but it also contains small amounts of other nutrients, such as N, P, Ca, S, and organic matter (Christofoletti et al., 2013; Mutton et al., 2014). Historically, the fertigation of vinasse in sugarcane fields started after the prohibition of its discharge into rivers and "sacrifice" areas (i.e., areas adjacent to the sugarcane industry that received large amounts of vinasse for several years). These areas under intense application of vinasse became useless for agriculture, mostly due to the high soil salinity, turning it barren and difficult to remediate (Mutton et al., 2014). For that reason, the appllication of vinasse in "sacrifice" areas was prohibited in the late 1960s. The attractiveness of fertigation to decision-makers in the sugarcane industry is based on the minimal investments required to transportation of vinasse; it is easily transported via pipes or trucks (Fuess et al., 2021), in the same way that water and some nutrients are provided to the crop.

Many studies in Brazil have revealed the crucial impacts of vinasse application on SH indicators, associating the advantages to the soil fertility. Among the benefits on SH indicators, increases in SOC, water retention, cation exchange capacity, and microbiota activity are frequently observed (Jiang et al., 2012; Resende et al., 2006; Yang et al., 2013). A recent study showed that the K, Ca, and Mg in vinasse improve soil macro-aggregation (Fuess et al., 2021) because the higher electrical charges result in well-structured soils. In terms of soil pH, despite its acidic pH (< 5.0), vinasse fertigation causes a fast increase in soil pH, promoted by organic matter oxidation, followed by a decrease, although of lower magnitude (Oliveira et al., 2022a). The application of vinasse usually increases SOC (Fuess et al., 2021); this increase is related to the high decomposition rate of the organic residues, associated with the inputs of organic C and N presented in the vinasse, shaping, and promoting an increase in soil microbial activity (Oosterkamp et al., 2016), particularly those related to N metabolism (Lourenço et al., 2018; Suleiman et al., 2018). However, this input could trigger drawbacks, such as a decrease in the abundance of arbuscular mycorrhizal fungi (Sanchez-Lizarraga et al., 2018) and an increase in the abundance of bacteria of N cycle (Lourenço et al., 2018) and subgroups of acidobacteria (Chaves et al., 2019), further enhancing GHG emissions, particularly of N_2O (Oliveira et al., 2022a).

The negative effects of vinasse fertigation on SH (e.g., salinity, clay dispersion, ion leaching, GHG emissions) are dependent on the amount applied to the field; when following the environmental standards and agronomic indications, these applications usually do not cause damage to the soil, crop, and environment. More recently, alternative strategies of vinasse use and management, such as the concentration by evaporation (reducing the applied volume from 150–300 to

7–40 $m^3\ ha^{-1}$), use as raw material for organo-mineral fertilizer production, or anaerobic digestion, have been adopted in the sugarcane sector. We hypothesize that these new methods of vinasse management could be more economical and effective to improve SH indicators in sugarcane areas. However, there is no relevant information, and more specific studies are needed. Therefore, understanding the mechanisms of vinasse use as biofertilizer and its effect on soil chemical, physical, and biological indicators is an essential step to stimulate farmers and biofuel producers to adopt practices that improve SH, lower production costs, and reduce negative impacts.

In summary, recycling sugarcane by-products into the field is a sustainable management practice that reduces the dependance on synthetic fertilizers and, when combined with the new technologies toward SH improvement (e.g., the addition of nitrification and urease inhibitors, organo-mineral formulation use, and application of biostimulants that naturally promote plant growth and SH) should be encouraged as part of the circular economy.

Harvesting Systems and Straw Management

In the modern era of agriculture mechanization (after the colonial period, i.e., the twentieth century), sugarcane harvesting was characterized by burning straw, manual stalk cutting, and mechanized collection and transportation. Since the early 2000s, this system has been replaced by a green harvesting system, which is completely mechanized (cutting, collection, and transportation), and a large amount of crop residue (10–20 $Mg\ ha^{-1}$ straw) is maintained on the soil surface (Figure 5.7). This harvesting transition has occurred due to social-economic pressure (e.g., to improve work and health conditions), environmental laws (e.g., reduce GHG emissions and other negative impacts of burning straw), and the advancement of mechanization. Currently, the green harvesting system is adopted in approximately 90% of Brazilian sugarcane-producing areas (Conab, 2023). Whereas in the Central-South region (i.e., 90% of production) this system has been adopted in 97% of sugarcane fields, in the Northeast region (i.e., 10% of production), due to availability of labor and unfavorable topographical conditions, the semi-mechanized system is adopted in only 88% of sugarcane fields.

In the semi-mechanized system, adopted predominantly in the Northeast region, the use of uncontrolled traffic and fire are the main concerns regarding SH. The harvesting machinery includes loaders and transport trucks; the mechanized operations occurring randomly over the soil induces soil compaction (Cavalcanti et al., 2020), and complete straw removal by burning considerably reduces soil water content (Santos et al., 2022). Cavalcanti et al. (2020) examined the impact of the semi-mechanized harvesting system on the soil physical quality index (SPQI), which considers soil physical function linked to plant growth, water fluxes, air diffusion, and resistance to degradation. Based on the SPQI, sugarcane

Figure 5.7 Green harvesting system adopted in 90% of sugarcane fields in Brazil. Photo: Maurício Roberto Cherubin.

soil was performing its physical functions at 67%–72% of its full capacity, whereas native vegetation soil (Atlantic Forest) was functioning at 88%. In other words, sugarcane cultivation has reduced soil physical quality due to compaction in a semi-mechanized system, negatively affecting soil air and water flow functionality, in addition to increasing penetration resistance. Therefore, in that region, traffic control and management of crop residues are still a challenge to enhance SH.

In the Central-South region of Brazil, the adoption of the green harvesting system was a complete transformation for the whole sector. The use of straw to cover the soil brought unprecedented positive impacts to SH and the environment, such as increased soil C stocks, nutrient cycling, water retention, and soil biodiversity as well as reduced GHG emissions and erosion risk (Bordonal et al., 2018), but it also raised concerns about soil compaction and the proliferation of pests (Bordonal et al., 2018; Rossetto et al., 2022).

Despite the benefits delivered by leaving straw on the soil surface, in the last decade, the sugarcane sector has begun to use available biomass as a feedstock for bioelectricity cogeneration or cellulosic ethanol production. Although there are multiple benefits with the use of sugarcane straw for bioenergy production, there is consensus that if straw removal for any purpose is not properly managed, there may be a degradation in SH, thus threatening the sustainability of the entire bioenergy production system (Cherubin et al., 2021a). Keeping the sugarcane straw on the soil surface has an important role in chemical, physical, and biological processes in the soil (Carvalho et al., 2017a; Cherubin et al., 2018). Such processes strongly influence the suitable functioning of soil, affecting sugarcane production and other environmental processes. The impacts of sugarcane straw removal management on multiple chemical, physical, and biological indicators of SH are synthesized in Figure 5.8.

Decomposition of crop residues is considered a basic component of the nutrient cycle, and thus intensive removal of sugarcane straw can significantly affect the dynamics and availability of nutrients in the soil (Fortes et al., 2013). Cherubin et al. (2019) reported that the potential nutrient recycled by straw (tops and bottom leaves) could reach 69, 7, 92, 45, 16, and 14 kg ha^{-1} of N, P, K, Ca, Mg, and S, respectively, and thus can significantly affect short- and long-term fertilizer demands. Studies assessing the short-term impacts of straw removal in Brazilian soils reported that nutrient changes took longer to be detected and that only subtle alterations were found after four years of straw removal (Lisboa et al., 2019; Satiro et al., 2017). It is likely that the straw effects on chemical attributes are reduced by soil buffering capacity and the frequent application of lime and fertilizer (N–P–K) to sustain sugarcane production along the crop cycle (Cherubin et al., 2019). The underlying reasons for the lack of significant straw removal effects on soil nutrient changes are likely associated with a bidirectional relationship. Although excessive straw removal can deplete soil nutrient levels and reduce nutrient cycling (Cherubin et al., 2019), it may also impair plant growth and reduce crop yield (Carvalho et al., 2019). This reduces plant nutrient uptake and leaves more available nutrients in the soil. However, understanding these hypothesized bidirectional causal relationships requires further research. Also, soil fertility must be monitored for adjustment of fertilizer rates with straw removal.

Crop residue removal for bioenergy production also affects soil physical properties, such as soil moisture content, aggregate formation, bulk density, and soil porosity (Castioni et al., 2018, 2019, 2021; Satiro et al., 2017). Most studies have shown that sugarcane straw removal is associated with soil structural degradation, mainly because of lower C inputs into the soil (Castioni et al., 2019) and the absence of mechanical protection that disperses the pressure caused by machinery traffic (Cherubin et al., 2021c). Structural degradation of soils induced by straw removal is related to decreased aggregate stability, which leads to soil compaction (Castioni et al., 2018). Additionally, sugarcane straw covers the soil and dissipates the energy of raindrops and prevents soil disaggregation and surface sealing, thereby improving the water infiltration capacity (Carvalho et al., 2022). Peres et al. (2010) observed that sugarcane straw preservation on the field reduced the water evaporation rate in the plow layer by almost half of that observed under conditions without straw. In this context, the removal of inadvisably high rates of straw may intensify runoff and water evaporation in addition to compromising soil health in sugarcane fields.

Indicators	Impact level	References[a)]
Nutrient removal	(−−−)	Cherubin et al. (2019)
pH	(−)	Satiro et al. (2017); Lisboa et al. (2019); Cherubin et al. (2021a)
P	(−−−)	Satiro et al. (2017); Lisboa et al. (2019); Cherubin et al. (2021a); Soltangheisi et al. (2021)
K^+	(−−−)	Satiro et al. (2017); Lisboa et al. (2019); Cherubin et al. (2021a)
Ca^{2+}	(−−)	Satiro et al. (2017)
Mg^{2+}	(−−)	Satiro et al. (2017)
Aggregate stability	(−−−)	Castioni et al. (2019)
Bulk density	(−−−)	Satiro et al. (2017); Castioni et al. (2019); Castioni et al. (2018); Castioni et al. (2021); Barbosa et al. (2019); Carvalho et al. (2022)
Penetration resistance	(++)	Satiro et al. (2017); Castioni et al. (2019); Castioni et al. (2018); Castioni et al. (2021); Barbosa et al. (2019); Carvalho et al. (2022)
Porosity	(−−)	Castioni et al. (2019); Castioni et al. (2018); Castioni et al. (2021); Barbosa et al. (2019); Carvalho et al. (2022)
Water retention	(−−)	Carvalho et al. (2022); Correa et al. (2019); Gmach et al. (2019);Santos et al. (2021)
Temperature regulation	(−−−)	Correa et al. (2019); Santos et al. (2021)
Soil loss	(−−−)	Silva et al. (2012)
C stock	(−−−)	Satiro et al. (2017); 15: Sousa Júnior. et al. (2018); Tenelli et al. (2019); Tenelli et al. (2021); Popin et al. (2020)
Macrofauna abundance/diversity	(−−)	Menandro et al. (2019)
Microbial abundance	(−)	Morais et al. (2019)
Earthworms	(−−)	Castioni et al. (2018); Menandro et al. (2019)
Microbial biomass C	(−)	Popin et al. (2020); Morais et al. (2019); Tenelli et al. (2019); Morais et al. (2020)
β-glucosidase activity	(−)	Tenelli et al. (2019)

a) Numbers represent the following references Note. Negative (−) or positive (+) signs indicate a tendency of reduced or increased impacts found in the literature.

Figure 5.8 Straw removal management implications on soil health indicators (chemical, physical, and biological) and its related functions. Source: Cherubin et al. (2019); 2: Satiro et al. (2017); 3: Lisboa et al. (2019); 4: Cherubin et al. (2021a); 5: Soltangheisi et al. (2021); 6: Castioni et al. (2019); 7: Castioni et al. (2018); 8: Castioni et al. (2021); 9: Barbosa et al. (2019); 10: Carvalho et al. (2022); 11: Correa et al. (2019); 12: Gmach et al. (2019); 13: Santos et al. (2022); 14: Silva et al. (2012); 15: Souza Júnior. et al. (2018); 16: Tenelli et al. (2019); 17: Tenelli et al. (2021); 18: Popin et al. (2020); 19: Menandro et al. (2019); 20: Morais et al. (2019); 21: Tenelli et al. (2019), 22: Morais et al. (2020)./John Wiley & Sons.

Retention of crop residues also affects soil biological attributes, which are more sensitive to soil management changes when compared to chemical and physical attributes (Cherubin et al., 2018). In the last decade, there have been efforts to characterize the microbial groups responsible for the soil activities under various management practices in sugarcane cropping areas (Navarrete et al., 2015; Pitombo et al., 2016; Rachid et al., 2016). Rachid et al. (2016) observed that straw removal modified the fungal communities present in soil but did not affect bacterial communities. Soil macrofauna are closely linked to other soil attributes and ecosystem functions. Menandro et al. (2019) reported that no straw removal was positively related to higher soil moisture, macropores, C concentration, and soil fertility, which were associated with a higher abundance of most macrofauna organisms. In contrast, total removal of straw was associated with higher soil compaction and lower macrofauna abundance. Straw is also the main C input in sugarcane soils (Carvalho et al., 2017b), and indiscriminate removal of straw causes the reduction of SOM cycling, leading to gradual depletion of soil C stocks (Tenelli et al., 2021). Several studies have shown that sugarcane straw promotes increases in microbial biomass (Souza et al., 2012; Paredes Jr et al., 2015). Souza et al. (2012) showed that the C and N microbial biomass of areas covered with sugarcane straw were 102% and 34% higher than those with no straw coverage, respectively. Straw cover improves soil microclimate conditions (moisture and temperature) and provides shelter and an energy source to support a higher abundance and diversity of soil biota (Menandro et al., 2019; Morais et al., 2019; Pimentel et al., 2019).

Using the Soil Management Assessment Framework (SMAF) developed by Andrews et al. (2004), Cherubin et al. (2021a) evaluated the SH response to sugarcane straw removal in 12 areas in the Central-Southern region of Brazil. This study showed that SH responses to straw removal were accurately detected by SMAF scoring curves. Based on the SHI scores, sandy and clayey soils were functioning at 41%–56% and 67%–86% of their full potential, respectively, and SH degradation was mostly induced by intensive straw removal rates. Conclusions from this study suggest that comprehensive SH assessments by integrating chemical, physical, and biological indicators are fundamental to designing more sustainable bioenergy production in Brazil.

Machinery Traffic Control

Mechanization is highly intensive in the sugarcane production system and involves large and heavy machinery for planting, fertilizing, spraying, and harvesting the crop along the cycle. However, the most important impacts occur in the harvesting operation. The excessive weight of harvesters and transloaders trafficking over the soil causes compaction in the traffic lane, inducing rut formation, a first signal of compaction (Figure 5.9).

Figure 5.9 Rut formation due to machinery traffic in mechanized harvesting of sugarcane fields. Photo: Maurício Roberto Cherubin.

At the tire soil interface, there are reports that vehicles used for sugarcane production can apply approximately 800 kPa to the soil (Jimenez et al., 2021; Silva et al., 2022). Beneath the tire, where stress is induced, soil compaction is visually diagnosed. The applied stress is distributed over the tire–soil contact area and then propagated into the soil (Figure 5.10). Once the stress reaches the profile, the soil may or may not resist deformation. This depends on its resistance state, which is mainly induced by water content. When the applied stress is greater than the load-bearing capacity, soil can permanently deform, causing compaction (i.e., altering the mass/volume ratio and thereby altering bulk density). Soil physical quality is dependent on soil compaction. Figure 5.10 shows the impact of a simulated 5000-kg wheel load sugarcane harvesting transloader on bulk density. First, the stress is propagated and deforms the soil, increasing bulk density. In this scenario, initial soil bulk density is 1.50 Mg m^{-3} and reaches 1.60 Mg m^{-3} near the soil surface.

There are management strategies and technologies that can prevent soil compaction or mitigate its effects to the crop. The term "prevention" indicates that compaction can be avoided altogether, whereas "mitigation" includes a reduction of the deleterious effects of compaction, making it softer, smoother, and less intense on soil functionality. The prevention and mitigation of soil compaction involves field-scale strategies, such as soil compaction risk analysis and controlled traffic (Lamandé et al., 2018).

Risk analysis can be understood as an alert of the risk of compaction and includes knowledge of the risk agent (machine) and the target system (soil).

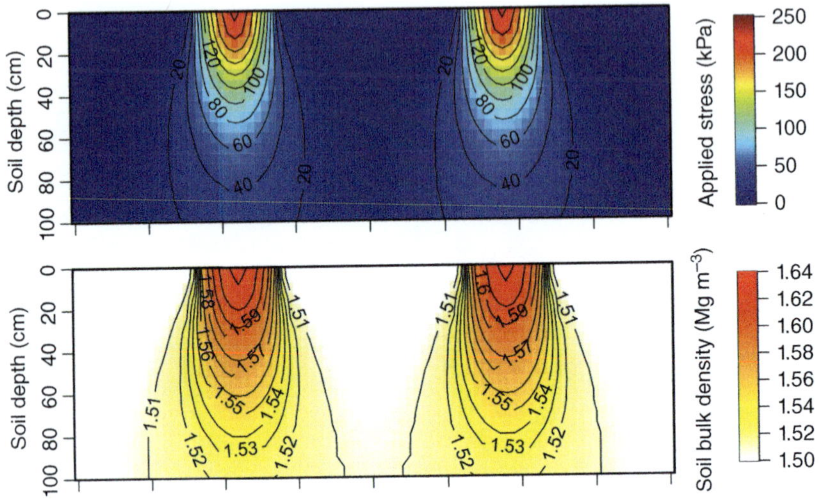

Figure 5.10 Stress propagation and impact on sulk density induced by harvesting vehicles in sugarcane fields.

Traffic control is a concept that consists of confining all machine loads to specific and known traffic lanes, such that the risk of compaction is restricted to traffic lanes. In sugarcane production, with single-row harvesters, all inter-row lanes are wheel traffic lanes. Therefore, traffic control in sugarcane production basically consists of Global Navigation Satellite System–guided autosteering for parallel traffic, avoiding accidental wheel traffic over the crop rows (causing row gaps). It creates a seedbed region (i.e., traffic-free zones) and zones fully dedicated to wheel traffic (Figure 5.11). Prevention, mitigation, and traffic control are part of the set of tools to avoid compaction, which can be included in the traffic control strategy in sugarcane fields.

Recent studies have shown the positive impacts of traffic-free seedbed regions on soil physical quality under CT (e.g., Esteban et al., 2019; Sousa et al., 2019) and reduced tillage (e.g., Barbosa et al., 2021; Luz et al., 2022a; Luz et al., 2022b). According to Esteban et al. (2019), the adoption of controlled traffic has the potential to preserve soil physical quality and increase sugarcane yield. These authors found that controlled traffic increased root dry biomass and sugarcane yield by 18% and 8.2 Mg ha^{-1}, respectively, and reduced soil penetration resistance and bulk density. In the same way, the adoption of traffic-free seedbed zones increased water availability, air permeability, and pore continuity in sandy loam soil as well as air permeability and pore continuity in clayey soil (Luz et al., 2022b). In clayey soil, Luz et al. (2022a) found that the preservation and recovery of soil structure under reduced tillage are conditioned by the additional

Figure 5.11 Controlled traffic in sugarcane fields creates a traffic-free seedbed region (at the sugarcane row position) and permanent traffic zones (at the sugarcane inter-row position). Photo: Guilherme A. Castioni.

adoption of traffic-free seedbed zones. This management practice has the potential to improve soil functions related to habitat for a myriad of organisms, C accumulation, soil structure stability, plant growth, and production of biomass (Luz et al., 2022a).

The impact of agriculture traffic control on soil health was measured in a commercial sugarcane field under contrasting soil textures located near the municipality of Lençóis Paulista, São Paulo state. In this study, the adoption of controlled traffic farming (CTF) and random traffic farming (RTF) in two soils (sandy clay soil and clayey soil) was compared. The CTF was characterized by the adoption of Global Navigation Satellite System–guided autosteering in planting and harvesting. Harvesters and wagons worked using an automatic steering system with permanent traffic lanes at the sugarcane inter-row position. The SH was measured using the SMAF (Andrews et al., 2004). The minimum dataset included five soil indicators: bulk density to represent soil physical quality; SOC to represent soil biological quality; and pH, P, and K to represent soil chemical quality and related soil functions. The interpretation of these indicators was performed using the SMAF algorithm that considers soil type, inherent organic matter content, texture, climate, slope, region, mineralogy, weathering class, crop, sampling time, and the analytical method used for P assessments (Andrews et al., 2004; Cherubin et al., 2021a).

The overall SHI and scores related to chemical, biological, and physical components are presented in Figure 5.12. In the fourth sugarcane ratoon, SHI was 12% higher under CTF in sandy clay soil. The lowest SHI associated with random traffic is specifically explained by the degradation of soil physical components (scores of 0.45 and 0.22 in CTF and RTF, respectively; i.e., a reduction of 23% in soil physical quality due to random traffic in the row position). In clayey soil, although random

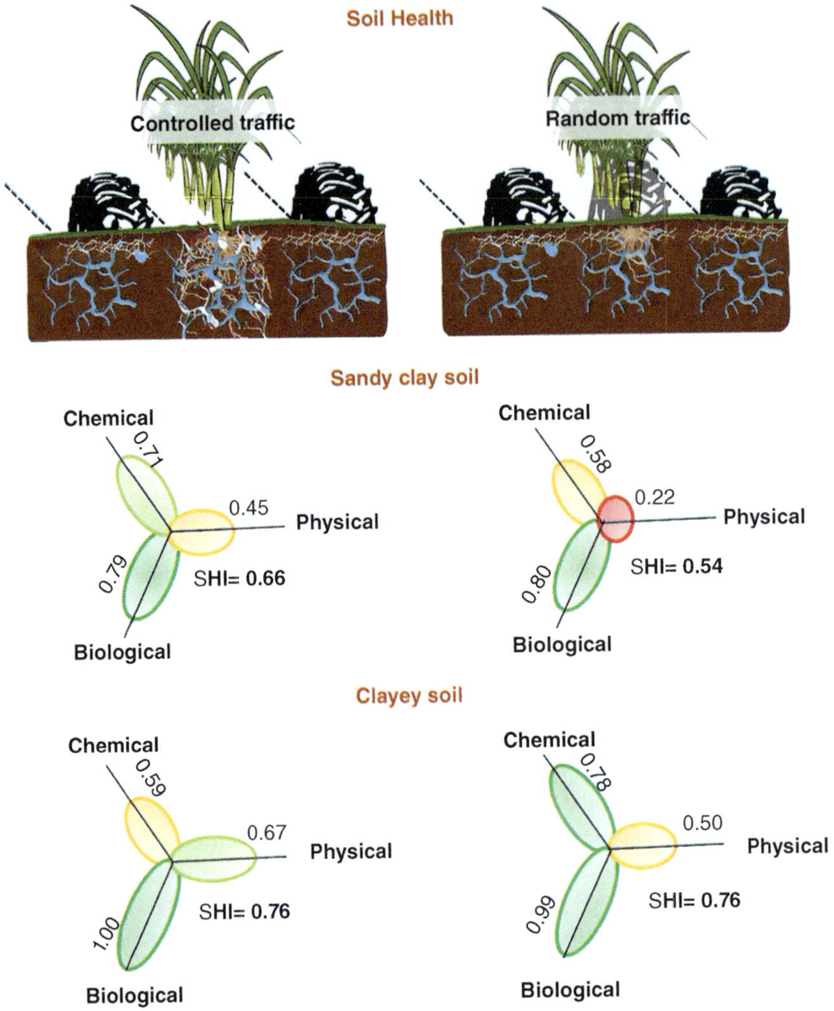

Figure 5.12 Soil health index (SHI) and scores of the chemical, physical, and biological components at 0–10 cm soil depth in sandy clay soil and clayey soil managed under controlled traffic farming and random traffic farming.

traffic did not reduce SHI, this farming system reduced soil physical quality by 17%. In that case, the higher score of the chemical component under random traffic explains the similar SHI values in both farming systems. High soil chemical quality may be related to the yield reduction in the random traffic area. However, it was only a case study, and more information is needed to compare these two types of management in other areas of Brazil.

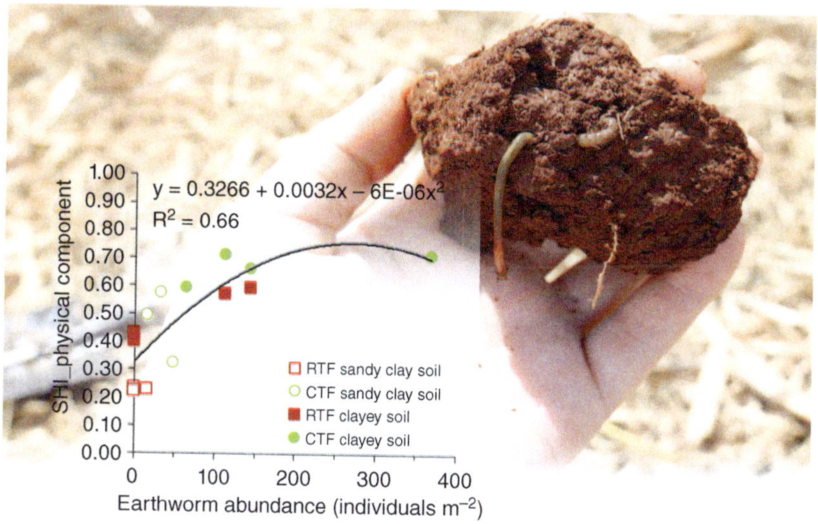

Figure 5.13 Relationship between SHI physical component and abundance of earthworms in clayey soil and sandy clay soil managed under controlled traffic farming (CTF) and random traffic farming (RTF) in southeast Brazil. Photo: Beatriz da Silva Vanolli.

In the same experimental area, a relationship between the physical component of SHI and the abundance of earthworms was found (Figure 5.13), indicating that CTF benefited biological activity and consequently soil physical health. Therefore, the adoption of traffic-free seedbed zones not only maintains soil physical health but also can improve soil functions related to biological activity, nutrient cycling, and soil C storage.

Final Remarks and Future Perspectives

Sugarcane is a semi-perennial crop for which good SH is essential for sustainable production. The production of sugarcane has substantially increased in the last decades in Brazil; this increase is associated with an expansion of the area and increased crop yields. The expansion of sugarcane areas occurred mainly in low-productivity pastures, which in general resulted in improvements in SH indicators and land productivity. Associated with this increase in the production area, the sugarcane sector also experienced relevant management changes in the last decades. Historically, the sugarcane sector was managed under CT, monoculture, burned, and manual harvesting systems. More recently, management strategies that have been adopted have contributed to improving SH indicators. The transition from a burned to an unburned mechanized harvesting system is the most relevant management change in the sugarcane sector. In this new system, a large amount of crop residues was maintained on the soil surface, with multiple benefits on SH indicators. The main drawback of this system is the increased soil compaction caused by intense and disorderly machinery traffic. In this sense, the adoption of reduced tillage and traffic-free seedbed zones is essential for reducing soil compaction in sugarcane fields. In addition, these management practices can provide other benefits to SH, such as soil erosion control, soil C storage, nutrient cycling, primary productivity, and habitat for a variety of organisms. In the same way, the integration of legumes (with cover crops or cash crops) or cover crop mixes in the renovation period has been more frequently used either in MEIOSE or in total area. It results in several positive effects on SH and, consequently, increases crop yields and crop resilience to adverse environmental conditions.

The correct use of agroindustry residues, like straw filter cake and vinasse, is essential to improve SH and support the growing demand for sugarcane-based products. In this sense, the indiscriminate removal of sugarcane straw for bioenergy production should be analyzed with caution. The recycling of vinasse and filter cake, as well as sugarcane straw management, must be done to maximize nutrient use efficiency and to reduce the dependency on chemical fertilizers. We advocate that the adoption of new technologies of residues recycling aiming to improve nutrient recycling and SH should be a priority. The adoption of such technologies should improve sugarcane sustainability, reducing the costs and environmental impacts.

In terms of advances on SH assessments, by 2023/2024 interpretative algorithms may be available for the soil enzymes arylsulfatase and β-glucosidase (which are the pillars of the SoilBio technology; see Chapter 10) specific for areas under sugarcane to enhance SH monitoring at farm scale.

Finally, it is fundamental to promote and strengthen sectoral programs and public policies, such as the Environmental Protocol (Greener Ethanol Protocol) and the Brazilian National Biofuel Policy (RenovaBio), to support sustainable sugarcane expansion in the near future and the adoption of sustainable management practices that enhance SH and the sustainability of large-scale sugarcane production in Brazil.

Acknowledgments

This work was supported by Fundação Agrisus (Process no. 2563/18), the National Council for Scientific and Technological Development—CNPq (Process no. 427170/2018-4; 406922/2013-6; 311787/2021-5), Sugarcane Renewable Electricity project—SUCRE/PNUD (Grant BRA/10/G31); the Brazilian Development Bank (BNDES) (Project no. 14.2.0773.1), and the São Paulo Research Foundation (Process no. 2017/23978-7; 2018/09845-7).

References

Ambrosano, E. J., Cantarella, H., Ambrosano, G. M. B., Schammas, E. A., Dias, F. L. F., Muraoka, T., Trivelin, P. C. O., Rossi, F., Schammass, E. A., & Sachs, R. C. C. (2011). Productivity of sugarcane after previous legumes crop. (In Portuguese.). *Bragantia*, 70(4), 810–818. https://doi.org/10.1590/S0006-87052011000400012

Ambrosano, E. J., Cantarella, H., Bovi Ambrosano, G. M., Ferreira Dias, F. L., Rossi, F., Trivelin, P. C. O., & Muraoka, T. (2013). The role of green manure nitrogen use by corn and sugarcane crops in Brazil. *Agricultural Science*, 04(12), 89–108. https://doi.org/10.4236/as.2013.412a008

Ambrosano, E. J., Trivelin, P. C. O., Cantarella, H., Ambrosano, G. M. B., Schammass, E. A., Guirado, N., Rossi, F., Mendes, P. C. D., & Muraoka, T. (2005). Utilization of nitrogen from green manure and mineral fertilizer by sugarcane. *Scientia Agricola*, 62(6), 534–542. https://doi.org/10.1590/s0103-90162005000600004

Andrews, S. S., Karlen, D. L., & Cambardella, C. A. (2004). The soil management assessment framework. *Soil Science Society of America Journal*, 68(6), 1945–1962. https://doi.org/10.2136/sssaj2004.1945

Awe, G. O., Reichert, J. M., & Fontanela, E. (2020). Sugarcane production in the subtropics: Seasonal changes in soil properties and crop yield in no-tillage, inverting and minimum tillage. *Soil and Tillage Research*, 196, 104447. https://doi.org/10.1016/j.still.2019.104447

Barbosa, L. C., de Souza, Z. M., Franco, H. C. J., Otto, R., Rossi Neto, J., Garside, A. L., & Carvalho, J. L. N. (2018). Soil texture affects root penetration in Oxisols under sugarcane in Brazil. *Geoderma Regional*, 13(March), 15–25. https://doi.org/10.1016/j.geodrs.2018.03.002

Barbosa, L. C., Magalhães, P. S. G., Bordonal, R. O., Cherubin, M. R., Castioni, G. A. F., Rossi Neto, J., Franco, H. C. J., & Carvalho, J. L. N. (2021). Untrafficked furrowed seedbed sustains soil physical quality in sugarcane mechanized fields. *European Journal of Soil Science*, 72(5), 2150–2164. https://doi.org/10.1111/ejss.13107

Barbosa, L. C., Magalhães, P. S. G., Bordonal, R. O., Cherubin, M. R., Castioni, G. A. F., Tenelli, S., Franco, E. C. J., & Carvalho, J. L. N. (2019). Soil physical quality associated with tillage practices during sugarcane planting in south-Central Brazil. *Soil and Tillage Research*, 195, 104383. https://doi.org/10.1016/j.still.2019.104383

Barcelos, J. E. (1984). "Meiosi—Cana e Alimentos"—Método interrotacional ocorrendo simultaneamente. Saccharum. (In Portuguese.). *Rev. Tecnológica da Indústria Açucareira Alcooleira*, 31, 10–18.

Caceres, N. T., & Alcarde, J. C. (1995). Adubação verde com leguminosas em rotação com cana-de-açúcar (Saccharums sp). *STAB*, 13, 16–20.

Carvalho, J. L. N., Hudiburg, T. W., Franco, H. C. J., & DeLucia, E. H. (2017a). Contribution of above—and belowground bioenergy crop residues to soil carbon. *GCB Bioenergy*, 9(8), 1333–1343. https://doi.org/10.1111/gcbb.12411

Carvalho, J. L. N., Menandro, L. M. S., Castro, S. G. Q., Cherubin, M. R., Bordonal, R. O., Barbosa, L. C., Gonzaga, L. C., Tenelli, S., Franco, H. C. J., Kolln, O. T., & Castioni, G. A. F. (2019). Multilocation straw removal effects on sugarcane yield in south-Central Brazil. *Bioenergy Research*, 12(4), 813–829. https://doi.org/10.1007/s12155-019-10007-8

Carvalho, J. L. N., Nogueirol, R. C., Menandro, L. M. S., Bordonal, R. O., Borges, C. D., Cantarella, H., & Franco, H. C. J. (2017b). Agronomic and environmental implications of sugarcane straw removal: A major review. *GCB Bioenergy*, 9(7), 1181–1195. https://doi.org/10.1111/gcbb.12410

Carvalho, M. L., Luz, F. B., De Lima, R. P., Maria, K., Cavalieri-Polizeli, K. M. V., Carvalho, J. N. L., & Cherubin, M. R. (2022). Assessment of soil physical quality and water flow regulation under straw removal Management in Sugarcane Production Fields. *Sustainability*, 14(2), 841.

Castioni, G. A., Cherubin, M. R., Menandro, L. M. S., Sanches, G. M., Bordonal, R. O., Barbosa, L. C., Franco, H. C. J., & Carvalho, J. L. N. (2018). Soil physical quality response to sugarcane straw removal in Brazil: A multi-approach assessment. *Soil and Tillage Research*, 184, 301–309. https://doi.org/10.1016/j.still.2018.08.007

Castioni, G. A. F., Cherubin, M. R., Bordonal, R. O., Barbosa, L. C., Menandro, L. M. S., & Carvalho, J. L. N. (2019). Straw removal affects soil physical quality and sugarcane yield in Brazil. *Bioenergy Research*, 12(4), 789–800. https://doi.org/10.1007/s12155-019-10000-1

Castioni, G. A. F., Lima, R. P., Cherubin, M. R., Bordonal, R. O., Rolim, M. M., & Carvalho, J. L. N. (2021). Machinery traffic in sugarcane straw removal operation: Stress transmitted and soil compaction. *Soil and Tillage Research*, 213, 105122. https://doi.org/10.1016/j.still.2021.105122

Cavalcanti, R. Q., Rolim, M. M., Lima, R. P., Tavares, U. E., Pedrosa, E. M. R., & Cherubin, M. R. (2020). Soil physical changes induced by sugarcane cultivation in the Atlantic Forest biome, northeastern Brazil. *Geoderma*, 370, 114353. https://doi.org/10.1016/j.geoderma.2020.114353

Chagas, M. F., Bordonal, R. O., Cavalett, O., Carvalho, J. L. N., Bonomi, A., & La Scala Jr, N. (2016). Environmental and economic impacts of different sugarcane production systems in the ethanol biorefinery. *Biofuels, Bioproducts and Biorefining*, 10, 89–106. https://doi.org/10.1002/bbb.1623

Chaves, M. G., Silva, G. G., Rossetto, R., Edwards, R. A., Tsai, S. M., & Navarrete, A. A. (2019). Acidobacteria subgroups and their metabolic potential for carbon degradation in sugarcane soil amended with vinasse and nitrogen fertilizers. *Frontiers in Microbiology*, 10, 1680. https://doi.org/10.3389/fmicb.2019.01680

Cherubin, M., da Silva Oliveira, D. M., Feigl, B. J., Pimentel, L. G., Lisboa, I. P., Gmach, M. R., Varanda, L. L., Morais, M. C., Satiro, L. S., Popin, G. V., de Paiva, S. R., Belarmino dos Santos, A. K., Soares de Vasconcelos, A. L., de Melo, P. L. A., Cerri, C. E. P., & Cerri, C. C. (2018). Crop residue harvest for bioenergy production and its environmental implications: A review. *Scientia Agricola*, 75, 255–272. https://doi.org/10.1590/1678-992X-2016-0459

Cherubin, M. R., Bordonal, R. O., Castioni, G. A., Guimarães, E. M., Lisboa, I. P., Moraes, L. A. A., Menandro, L. M. S., Tenelli, S., Cerri, C. E. P., Karlen, D. L., & Carvalho, J. L. N. (2021a). Soil health response to sugarcane straw removal in Brazil. *Industrial Crops and Products*, 163, 113315. https://doi.org/10.1016/j.indcrop.2021.113315

Cherubin, M. R., Carvalho, J. L. N., Cerri, C. E. P., Nogueira, L. A. H., Souza, G. M., & Cantarella, H. (2021b). Land use and management effects on sustainable sugarcane-derived bioenergy. *Land*, 10(1), 72. https://doi.org/10.3390/land10010072

Cherubin, M. R., Franchi, M. R. A., Lima, R. P., Moraes, M. T., & Luz, F. B (2021c). Sugarcane straw effects on soil compaction susceptibility. *Soil and Tillage Research*, 212(105066). https://doi.org/10.1016/j.still.2021.105066

Cherubin, M. R., Franco, A. L. C., Cerri, C. E. P., Oliveira, D. M. S., Davies, C. A., & Cerri, C. E. P. (2015). Sugarcane expansion in Brazilian tropical soils-effects of land use change on soil chemical attributes. *Agriculture, Ecosystems and Environment*, 211, 173–184. https://doi.org/10.1016/j.agee.2015.06.006

Cherubin, M. R., Karlen, D. L., Franco, A. L. C., Cerri, C. E. P., Tormena, C. A., & Cerri, C. C. (2016a). A soil management assessment framework (SMAF) evaluation of Brazilian sugarcane expansion on soil quality. *Soil Science Society of America Journal*, 80(1), 215–226. https://doi.org/10.2136/sssaj2015.09.0328

Cherubin, M. R., Karlen, D. L., Franco, A. L. C., Tormena, C. A., Cerri, C. E. P., Davies, C. A., & Cerri, C. C. (2016b). Soil physical quality response to sugarcane expansion in Brazil. *Geoderma*, 267, 156–168. https://doi.org/10.1016/j.geoderma.2016.01.004

Cherubin, M. R., Lisboa, I. P., Silva, A. G. B., Varanda, L. L., Bordonal, R. O., Carvalho, J. L. N., Otto, R., Pavinato, P. S., Soltangheisi, A., & Cerri, C. E. P. (2019). Sugarcane straw removal: Implications to soil fertility and fertilizer demand in Brazil. *Bioenergy Research*, 12(4), 888–900. https://doi.org/10.1007/s12155-019-10021-w

Chessman, D., Moebius-Clune, B. N., Smith, B. R., & Fisher, B. (2019). *The basics of addressing resource concerns with conservation practices within integrated soil health management systems on cropland* (Soil Health Technical Note No. 450-04). USDA-NRCS. https://directives.sc.egov.usda.gov/OpenNonWebContent.aspx?content=44293.wba

Christofoletti, C. A., Escher, J. P., Correia, J. E., Marinho, J. F. U., & Fontanetti, C. S. (2013). Sugarcane vinasse: Environmental implications of its use. *Waste Management*, 33(12), 2752–2761. https://doi.org/10.1016/J.WASMAN.2013.09.005

Companhia Nacional de Abastecimento—(CONAB). (2023). *Acompanhamento de safra brasileira de cana-de-açúcar, v. 9 – Safra 2022/23, n° 3 – Terceiro Levantamento*. (In Portuguese.) www.conab.gov.br/info-agro/safras/cana

Corrêa, S. T. R., Barbosa, L. C., Menandro, L. M., Scarpare, F. V., Hernandes, T. A. D., Franco, H. C. J., & Carvalho, J. L. N. (2019). Straw removal effects on soil water dynamics, soil temperature, and sugarcane yield in south-Central Brazil. *Bioenergy Research*, 12, 749–763.

de Bordonal, R. O., Carvalho, J. L. N., Lal, R., Figueiredo, E. B., Oliveira, B. G., & La Scala, N., Jr., (2018). Sustainability of sugarcane production in Brazil: A review. *Agronomy for Sustainable Development*, 38(2), 13. https://doi.org/10.1007/s13593-018-0490-x

Esteban, D. A. A., Souza, Z. M., Tormena, C. A., Lovera, L. H., Souza Lima, E. S., Oliveira, I. N., & Ribeiro, N. P. (2019). Soil compaction, root system and productivity of sugarcane under different row spacing and controlled traffic at harvest. *Soil and Tillage Research*, 187, 60–71. https://doi.org/10.1016/j.still.2018.11.015

Farhate, C. V. V., Souza, Z. M., Cherubin, M. R., Lovera, L. H., Oliveira, I. N., Junnyor, W. S. G., & La Scala Jr, N. (2022). Soil physical change and sugarcane stalk yield induced by cover crop and soil tillage. *Revista Brasileira de Ciencia do Solo*, 46, 1–24. https://doi.org/10.36783/18069657rbcs20210123

Fernandes, C., Corá, J., & Marcelo, A. (2012). Soil uses in the sugarcane fallow period to improve chemical and physical properties of two latosols (oxisols). *Revista Brasileira de Ciencia do Solo*, 36, 283–294.

Food and Agriculture Organization of the United Nations. (2023). *FAOSTAT – Production – crops: Sugarcane*. FAO. http://wwwfaoorg/faostat/en

Fortes, C., Vitti, A. C., Otto, R., Ferreira, D. A., Franco, H. C. J., & Trivelin, P. C. O. (2013). Contribution of nitrogen from sugarcane harvest residues and urea for crop nutrition. *Scientia Agricola*, 70(5), 313–320. https://doi.org/10.1590/S0103-90162013000500005

Franco, A. L., Cherubin, M. R., Pavinato, P. S., Cerri, C. E., Six, J., Davies, C. A., & Cerri, C. C. (2015). Soil carbon, nitrogen and phosphorus changes under sugarcane expansion in Brazil. *Science of the Total Environment*, 515–516, 30–38. https://doi.org/10.1016/j.scitotenv.2015.02.025

Franco, A. L. C., Bartz, M. L. C., Cherubin, M. R., Baretta, D., Cerri, C. E. P., Feigl, B. J., Wall, D. H., Davies, C. A., & Cerri, C. C. (2016). Loss of soil (macro) fauna due to the expansion of Brazilian sugarcane acreage. *Science of the Total Environment*, 563–564, 160–168. https://doi.org/10.1016/j.scitotenv.2016.04.116

Fuess, L. T., Altoé, M. E., Felipe, M. C., & Garcia, M. L. (2021). Pros and cons of fertirrigation with in natura sugarcane vinasse: Do improvements in soil fertility offset environmental and bioenergy losses? *Journal of Cleaner Production*, 319, 128684. https://doi.org/10.1016/J.JCLEPRO.2021.128684

Garside, A. L., & Bell, M. J. (2001). Fallow legumes in the Australian sugar industry: review of recent research findings and implications for the sugarcane cropping system. In *Proceedings of the 2001 Conference of the Australian Society of Sugar*

Cane Technologists held at Mackay, Queenslad, Australia (pp. 230–235). P.K. Editorial Services Pty Ltd.

Gmach, M. R., Scarpare, F. V., Cherubin, M. R., Lisboa, I. P., dos Santos, A. B., Cerri, C. E. P., & Cerri, C. C. (2019). Sugarcane straw removal effects on soil water storage and drainage in southeastern Brazil. *Journal of Soil and Water Conservation*, 74(5), 466–476. https://doi.org/10.2489/jswc.74.5.466

Gunkel, G., Kosmol, J., Sobral, M., Rohn, H., Montenegro, S., & Aureliano, J. (2007). Sugar cane industry as a source of water pollution—case study on the situation in Ipojuca river, Pernambuco, Brazil. *Water, Air, & Soil Pollution*, 180(1–4), 261–269. https://doi.org/10.1007/s11270-006-9268-x

Hernandes, T. A. D., Duft, D. G., dos Santos Luciano, A. C., Leal, M. R. L. V., & Cavalett, O. (2021). Identifying suitable areas for expanding sugarcane ethanol production in Brazil under conservation of environmentally relevant habitats. *Journal of Cleaner Production*, 292. https://doi.org/10.1016/j.jclepro.2020.125318

IBGE. (2022). *Levantamento sistemático da produção agrícola*. (In Portuguese.) Instituto Brasileiro de Geografia e Estatística. https://sidra.ibge.gov.br/tabela/1618

Jiang, Z. P., Li, Y. R., Wei, G. P., Liao, Q., Su, T. M., Meng, Y. C., Zhang, H. T., & Lu, C. Y. (2012). Effect of long-term vinasse application on Physico-chemical properties of sugarcane field soils. *Sugar Tech*, 14(4), 412–417. https://doi.org/10.1007/s12355-012-0174-9

Jimenez, K. J., Rolim, M. M., Gomes, I. F., de Lima, R. P., Berrío, L. L. A., & Ortiz, P. F. (2021). Numerical analysis applied to the study of soil stress and compaction due to mechanised sugarcane harvest. *Soil and Tillage Research*, 206, 104847. https://doi.org/10.1016/j.still.2020.104847

Lamandé, M., Greve, M. H., & Schjønning, P. (2018). Risk assessment of soil compaction in Europe: Rubber tracks or wheels on machinery. *Catena*, 167, 353–362. https://doi.org/10.1016/j.catena.2018.05.015

LAPIG. (2022). *Atlas das pastagens*. (In Portuguese.) Federal University of Goiás. https://atlasdaspastagens.ufg.br

Lisboa, I. P., Cherubin, M. R., Satiro, L. S., Siqueira-Neto, M., Lima, R. P., Gmach, M. R., Wienhold, B. J., Schmer, M. R., Jin, V. L., Cerri, C. C., & Cerri, C. E. P. (2019). Applying soil management assessment framework (SMAF) on short-term sugarcane straw removal in Brazil. *Industrial Crops and Products*, 129, 175–184. https://doi.org/10.1016/j.indcrop.2018.12.004

Lourenço, K. S., Cassman, N. A., Pijl, A. S., Van Veen, J. A., Cantarella, H., & Kuramae, E. E. (2018). Nitrosospira sp. govern nitrous oxide emissions in a tropical soil amended with residues of bioenergy crop. *Frontiers in Microbiology*, 9, 674. https://doi.org/10.3389/fmicb.2018.00674

Lovera, L. H., de Souza, Z. M., Esteban, D. A. A., de Oliveira, I. N., Farhate, C. V. V., de Souza Lima, E., & Panosso, A. R. (2021). Sugarcane root system: Variation over three cycles under different soil tillage systems and cover crops. *Soil and Tillage Research*, 208, 104866. https://doi.org/10.1016/j.still.2020.104866

Luz, F. B., Carvalho, M. L., Castioni, G. A. F., Bordonal, R. O., Cooper, M., Carvalho, J. L. N., & Cherubin, M. R. (2022a). Soil structure changes induced by tillage and reduction of machinery traffic on sugarcane: A diversity of assessment scales. *Soil and Tillage Research*, 223, 105469. https://doi.org/10.1016/j.still.2022.105469

Luz, F. B., Castioni, G. A. F., Tormena, C. A., Freitas, R. S., Carvalho, J. L. N., & Cherubin, M. R. (2022b). Soil tillage and machinery traffic influence soil water availability and air fluxes in sugarcane fields. *Soil and Tillage Research*, 223, 105459. https://doi.org/10.1016/j.still.2022.105459

Macedo, N., & Botelho, P. S. M. (1995). Leguminosas em rotação com cana-de-açúcar e o controle de térmitas. (In Portuguese.). *STAB*, 13(6), 16–20.

MapBiomas. (2022). *Uso e cobertura do solo*. (In Portuguese.) MapBiomas plataform. https://plataforma.mapbiomas.org.

Mello, F. F., Cerri, C. E., Davies, C. A., Holbrook, N. M., Paustian, K., Maia, S. M., Galdos, M. V., Bernoux, M., & Cerri, C. C. (2014). Payback time for soil carbon and sugar-cane ethanol. *Nature Climate Change*, 4(7), 605–609. https://doi.org/10.1038/nclimate2239

Menandro, L. M. S., de Moraes, L. O., Borges, C. D., Cherubin, M. R., Castioni, G. A., & Carvalho, J. L. N. (2019). Soil macrofauna responses to sugarcane straw removal for bioenergy production. *Bioenergy Research*, 12(4), 944–957. https://doi.org/10.1007/s12155-019-10053-2

Morais, M. C., Ferrari, B. M., Borges, C. D., Cherubin, M. R., Tsai, S. M., Cerri, C. C., Cerri, C. E. P., & Feigl, B. J. (2019). Does sugarcane straw removal change the abundance of soil microbes? *Bioenergy Research*, 12(4), 901–908. https://doi.org/10.1007/s12155-019-10018-5

Morais, M. C., Siqueira-Neto, M., Guerra, H. P., Satiro, L. S., Soltangheisi, A., Cerri, C. E., Feigl, B. J., & Cherubin, M. R. (2020). Trade-offs between sugarcane straw removal and soil organic matter in Brazil. *Sustainability*, 12(22), 9363. https://doi.org/10.3390/su12229363

Moura, R. M. (1995). Dois anos de rotação de cultura em campos de cana-de-açúcar para controle de nemotóides. (In Portuguese.). *Fitopatologia Brasileira*, 20(4), 597–600.

Mutton, M. A., Rossetto, R., & Mutton, M. J. R. (2014). Agricultural use of stillage. Sugarcane bioethanol — R&D product. *Sustainability*, 1, 423–440. https://doi.org/10.5151/blucheroa-sugarcane-sugarcanebioethanol_40

Navarrete, A. A., Diniz, T. R., Braga, L. P. P., Silva, G. G. Z., Franchini, J. C., Rossetto, R., Edwards, R. A., & Tsai, S. M. (2015). Multi-analytical approach reveals potential microbial indicators in soil for sugarcane model systems. *PLoS One*, 10(6), e0129765. https://doi.org/10.1371/journal.pone.0129765

Oliveira, B. G., Lourenço, K. S., Carvalho, J. L. N., Gonzaga, L. C., Teixeira, M. C., Tamara, A. F., & Cantarella, H. (2022a). Soil pH does not interfere with nitrification inhibitor efficiency for reducing N2O emissions from soils treated with concentrated vinasse and urea. *Geoderma*, 426, 116087. https://doi.org/10.1016/J.GEODERMA.2022.116087

Oliveira, D. M., Cherubin, M. R., Franco, A. L., Santos, A. S., Gelain, J. G., Dias, N. M., … Cerri, C. E. (2019). Is the expansion of sugarcane over pasturelands a sustainable strategy for Brazil's bioenergy industry? *Renewable and Sustainable Energy Reviews*, 102, 346–355. https://doi.org/10.1016/j.rser.2018.12.012

Oliveira, D. M., Paustian, K., Davies, C. A., Cherubin, M. R., Franco, A. L. C., Cerri, C. C., & Cerri, C. E. P. (2016). Soil carbon changes in areas undergoing expansion of sugarcane into pastures in south-central Brazil. *Agriculture, Ecosystems and Environment*, 228, 38–48. https://doi.org/10.1016/j.agee.2016.05.005

Oliveira, I. N., de Souza, Z. M., Bolonhezi, D., Totti, M. C. V., de Moraes, M. T., Lovera, L. H., Lima, E. S., Esteban, D. A. A., & Oliveira, C. F. (2022b). Tillage systems impact on soil physical attributes, sugarcane yield and root system propagated by pre-sprouted seedlings. *Soil and Tillage Research*, 223, 105460. https://doi.org/10.1016/j.still.2022.105460

Oosterkamp, M. J., Méndez-García, C., Kim, C. H., Bauer, S., Ibáñez, A. B., Zimmerman, S., Hong, P. Y., Can, I. K., & Mackie, R. I. (2016). Lignocellulose-derived thin stillage composition and efficient biological treatment with a high-rate hybrid anaerobic bioreactor system. *Biotechnology for Biofuels*, 9, 1–15. https://doi.org/10.1186/s13068-016-0532-z

Ortiz, P. F., Rolim, M. M., de Lima, R. P., Tormena, C. A., Cavalcanti, R. Q., & Pedrosa, E. M. (2022). A soil physical assessment over three successive burned and unburned sugarcane annual harvests. *Sugar Tech*, 25, 518–530. https://doi.org/10.1007/s12355-022-01136-0

Otto, R., Pereira, G. L., Tenelli, S., Carvalho, J. L. N., Lavres, J., Castro, S. A. Q., Lisboa, I. P., & Sermarini, R. A. (2020). Planting legume cover crop as a strategy to replace synthetic N fertilizer applied for sugarcane production. *Industrial Crops and Products*, 156, 112853. https://doi.org/10.1016/J.INDCROP.2020.112853

Paredes, F. P. P., Portilho, I. I. R., & Mercante, F. M. (2015). Atributos microbiológicos de um latossolo sob cultivo de cana-de-açúcar com e sem queima da palhada. (In Portuguese.). *Semina: Ciencias Agrárias*, 36(1), 151–163. https://doi.org/10.5433/1679-0359.2015v36n1p151

Peres, J. G., Souza, C. F., & Lavorenti, N. A. (2010). Avaliação dos efeitos da cobertura de palha de cana-de-açúcar na umidade e na perda de água do solo. (In Portuguese.). *Revista Engenharia Agrícola*, 30, 875–886.

Perin, A., Santos, R. H. S., Urquiaga, S., Guerra, J. G. M., & Cecon, P. R. (2004). Produção de fitomassa, acúmulo de nutrientes e fixação biológica de nitrogênio por adubos verdes em cultivo isolado e consorciado. (In Portuguese.). *Pesquisa Agropecuária Brasileira*, 39(1), 35–40. https://doi.org/10.1590/s0100-204x2004000100005

Pimentel, L. G., Gumiere, T., Oliveira, D. M., Cherubin, M. R., Andreote, F. D., Cerri, C. E. P., & Cerri, C. C. (2019). Soil bacterial community changes in sugarcane fields under straw removal in Brazil. *Bioenergy Research*, 12(4), 830–842. https://doi.org/10.1007/s12155-019-10010-z

Pitombo, L. M., do Carmo, J. B., De Hollander, M., Rossetto, R., López, M. V., Cantarella, H., & Kuramae, E. E. (2016). Exploring soil microbial 16S rRNA sequence data to increase carbon yield and nitrogen efficiency of a bioenergy crop. *GCB Bioenergy*, 8(5), 867–879. https://doi.org/10.1111/gcbb.12284

Popin, G. V., Santos, A. K., Melo, P. L., Cherubin, M. R., Cerri, C. E., & Siqueira-Neto, M. (2020). Importance of sugarcane straw maintenance to prevent soil organic matter depletion in a nitisol in the central-southern region of Brazil. *Soil Research*, 59, 119–129.

Prado, R. M., Caione, G., & Campos, C. N. S. (2013). Filter cake and vinasse as fertilizers contributing to conservation agriculture. *Applied and Environmental Soil Science*, 2013, 581984. https://doi.org/10.1155/2013/581984

Qiao, L., Wang, X., Smith, P., Fan, J., Lu, Y., Emmett, B., … Fan, M. (2022). Soil quality both increases crop production and improves resilience to climate change. *Nature Climate Change*, 12(6), 574–580. https://doi.org/10.1038/s41558-022-01376-8

Rachid, C. T. C. C., Pires, C. A., Leite, D. C. A., Coutinho, H. L. C., Peixoto, R. S., Rosado, A. S., … Balieiro, F. D. C. (2016). Sugarcane trash levels in soil affects the fungi but not bacteria in a short-term field experiment. *Brazilian Journal of Microbiology*, 47(2), 322–326. https://doi.org/10.1016/j.bjm.2016.01.010

Resende, A. S., Xavier, R. P., de Oliveira, O. C., Urquiaga, S., Alves, B. J., & Boddey, R. M. (2006). Long-term effects of pre-harvest burning and nitrogen and vinasse applications on yield of sugar cane and soil carbon and nitrogen stocks on a plantation in Pernambuco, N.E Brazil. *Plant and Soil*, 281(1–2), 339–351. https://doi.org/10.1007/s11104-005-4640-y

Rossetto, R., Crusciol, C. A. C., Cantarella, H., Carmo, J. B., & Nascimento, C. A. C. (2018). Residues uses and environmental sustainability. In A. K. Tiwari & P. Singh (Eds.), *Sustainable sugarcane production* (pp. 161–187). Apple Academic Press.

Rossetto, R., Ramos, N. P., Pires, R. C. M., Xavier, M. A., Cantarella, H., & Landell, M. G. A. (2022). Sustainability in sugarcane supply chain in Brazil: Issues and way forward. *Sugar Tech*, 24(3), 941–966. https://doi.org/10.1007/s12355-022-01170-y

Sá, M. A. C., Rein, T. A., Santos Jr, J. D. G., Souza, K. W., & Franz, C. A. B. (2022). Produtividade da cana-de-açúcar e atributos de um Latossolo Acriférrico do Cerrado sob plantio direto e convencional. (In Portuguese.). *STAB*, 40(1), 23–30. http://stabregsul.ddns.net:8080/site/pdf/40.1.1.pdf

Sanchez-Lizarraga, A. L., Arenas-Montaño, V., Marino-Marmolejo, E. N., Dendooven, L., Velazquez-Fernandez, J. B., Davila-Vazquez, G., Rodriguez-Campos, J., Hernández-Cuevas, L., & Contreras-Ramos, S. M. (2018). Vinasse irrigation: Effects on soil fertility and arbuscular mycorrhizal fungi population. *Journal of Soils and Sediments*, 18(11), 3256–3270. https://doi.org/10.1007/s11368-018-1996-1

Santos, A. K. B. D., Popin, G. V., Gmach, M. R., Cherubin, M. R., Siqueira Neto, M., & Cerri, C. E. P. (2022). Changes in soil temperature and moisture due to sugarcane straw removal in central-southern Brazil. *Scientia Agricola*, 79(6). https://doi.org/10.1590/1678-992X-2020-0309

Satiro, L. S., Cherubin, M. R., Safanelli, J. L., Lisboa, I. P., da Rocha Jr, P. R., Cerri, C. E. P., & Cerri, C. C. (2017). Sugarcane straw removal effects on Ultisols and Oxisols in south-central Brazil. *Geoderma Regional*, 11, 86–95. https://doi.org/10.1016/J.GEODRS.2017.10.005

Segnini, A., Carvalho, J. L. N., Bolonhezi, D., Milori, D. M. B. P., Silva, W. T. L. D., Simões, M. L., Cantarella, H., Maria, I. C., & Martin-Neto, L. (2013). Carbon stock and humification index of organic matter affected by sugarcane straw and soil management. *Scientia Agricola*, 70(5), 321–326. https://doi.org/10.1590/S0103-90162013000500006

Shen, R. F., & Teng, Y. (2022). The frontier of soil science: Soil health. *Pedosphere*, 33, 6–7. https://doi.org/10.1016/J.PEDSPH.2022.06.007

Silva, G. R. V. D., Souza, Z. M. D., Martins Filho, M. V., Barbosa, R. S., & Souza, G. S. D. (2012). Soil, water and nutrient losses by interrill erosion from green cane cultivation. *Revista Brasileira de Ciência do Solo*, 36(3), 963–970. https://doi.org/10.1590/s0100-06832012000300026

Silva, R. P., Rolim, M. M., Gomes, I. F., Maciel, B. M., Ortiz, P. F., Pedrosa, E. M., & de Lima, R. P. (2022). Comparing pseudo-analytical solutions with 2D and 3D numerical simulations of soil stress induced by agricultural field traffic. *Biosystems Engineering*, 221, 195–207. https://doi.org/10.1016/j.biosystemseng.2022.07.005

Soltangheisi, A., Haygarth, P. M., Pavinato, P. S., Cherubin, M. R., Teles, A. P. B., de Bordonal, R. O., Carvalho, J. L. N., Withers, P. J. A., & Martinelli, L. A. (2021). Long term sugarcane straw removal affects soil phosphorus dynamics. *Soil and Tillage Research*, 208, 104898. https://doi.org/10.1016/J.STILL.2020.104898

Sousa, A. C. M., Farhate, C. V. V., de Souza, Z. M., Torres, J. L. R., & da Silva, R. B. (2019). Soil soil load-bearing capacity and development of root system in area under sugarcane with traffic control in Brazil. *Sugar Tech*, 21(1), 153–161. https://doi.org/10.1007/s12355-018-0636-9

Sousa Jr, J. G. D. A., Cherubin, M. R., Oliveira, B. G., Cerri, C. E., Cerri, C. C., & Feigl, B. J. (2018). Three-year soil carbon and nitrogen responses to sugarcane straw management. *Bioenergy Research*, 11(2), 249–261. https://doi.org/10.1007/s12155-017-9892-x

Souza, R. A., Telles, T. S., Machado, W., Hungria, M., Tavares Filho, J., & de Fátima Guimarães, M. (2012). Effects of sugarcane harvesting with burning on the chemical and microbiological properties of the soil. *Agriculture, Ecosystems and Environment*, 155, 1–6. https://doi.org/10.1016/J.AGEE.2012.03.012

Stirling, G. R., Blair, B. L., Pattemore, J. A., Garside, A. L., & Bell, M. J. (2001). Changes in nematode populations on sugarcane following fallow, fumigation and crop rotation, and implications for the role of nematodes in yield decline. *Australasian Plant Pathology*, 30(4), 323–335. http://dx.doi.org/10.1071/AP01044

Suleiman, A. K. A., Lourenco, K. S., Pitombo, L. M., Mendes, L. W., Roesch, L. F. W., Pijl, A., Carmo, J. B., Cantarella, H., & Kuramae, E. E. (2018). Recycling organic residues in agriculture impacts soil-borne microbial community structure, function and N2O emissions. *Science of the Total Environment*, 631, 1089–1099. https://doi.org/10.1016/J.SCITOTENV.2018.03.116

Taniwaki, R. H., Cassiano, C. C., Filoso, S., de Barros Ferraz, S. F., de Camargo, P. B., & Martinelli, L. A. (2017). Impacts of converting low-intensity pastureland to high-intensity bioenergy cropland on the water quality of tropical streams in Brazil. *Science of the Total Environment*, 584, 339–347. https://doi.org/10.1016/J.SCITOTENV.2016.12.150

Tenelli, S., Bordonal, R. O., Barbosa, L. C., & Carvalho, J. L. N. (2019). Can reduced tillage sustain sugarcane yield and soil carbon if straw is removed? *Bioenergy Research*, 12(4), 764–777. https://doi.org/10.1007/s12155-019-09996-3

Tenelli, S., Bordonal, R. O., Cherubin, M. R., Cerri, C. E. P., & Carvalho, J. L. N. (2021). Multilocation changes in soil carbon stocks from sugarcane straw removal for bioenergy production in Brazil. *GCB Bioenergy*, 13(7), 1099–1111. https://doi.org/10.1111/gcbb.12832

Thomaz, E. L., Marcatto, F. S., & Antoneli, V. (2022). Soil erosion on the Brazilian sugarcane cropping system: An overview. *Geography and Sustainability*, 3(2), 129–138. https://doi.org/10.1016/j.geosus.2022.05.001

UNICA. (2022). *Observatório da Cana, 2022.* (In Portuguese.) The Brazilian Sugarcane Industry Association. https://observatoriodacana.com.br

Vanolli, B. S., Canisares, L. P., Franco, A. L., Delabie, J. H., Cerri, C. E., & Cherubin, M. R. (2021). Epigeic fauna (with emphasis on ant community) response to land-use change for sugarcane expansion in Brazil. *Acta Oecologica*, 110, 103702. https://doi.org/10.1016/J.ACTAO.2021.103702

Yang, S. D., Liu, J. X., Wu, J., Tan, H. W., & Li, Y. R. (2013). Effects of vinasse and press mud application on the biological properties of soils and productivity of sugarcane. *Sugar Tech*, 15(2), 152–158. https://doi.org/10.1007/s12355-012-0200-y

Youlton, C., Wendland, E., Anache, J. A. A., Poblete-Echeverría, C., & Dabney, S. (2016). Changes in erosion and runoff due to replacement of pasture land with sugarcane crops. *Sustainability*, 8(7), 685. https://doi.org/10.3390/su8070685

6

Soil Health under Arabica Coffee Plantations in the Cerrado Biome

Carla Eloize Carducci, Geraldo César de Oliveira, Samara Martins Barbosa, Yuri Lopes Zinn, Daiane Pereira de Souza, Clandio Favarini Ruviaro, Joyce Cristina Costa, and Eduardo Costa Severiano

Chapter Overview

Typical of tropical climates, coffee plantations were adapted to different Brazilian regions, especially the southeast and the State of Minas Gerais. Among the existing coffee species, Arabica (*Coffea arabica* L.) is the most cultivated in Brazil, and different soil and crop management systems are used. However, only systems using a combination of soil conservationist practices, whether cultural, edaphic, or mechanical, present high productivity and beverage quality; high soil C stocks; lower greenhouse gas (GHG) emissions; and better biological, physical, and chemical soil quality, leading to increased soil health (SH). In this chapter, we emphasize the production of Arabica coffee in the Cerrado biome, which presents stronger soil and climatic restrictions than other areas in Brazil, and the use of management practices that improve SH.

A Brief History of Coffee in Brazil

Arabica coffee is the species most widely cultivated and commercialized worldwide. It originated in eastern Africa, in the Ethiopian mountains, where mean annual temperature is approximately 20°C, with well-distributed annual rainfall in excess of 1600 mm and a dry season not exceeding four months. Originally, this species evolved under dense tropical forests, sheltered from high temperatures (Camargo & Telles Jr., 1953; Rena & Maestri, 1986). Another species of economic expression in the world is *Coffea canephora* L., but it is less widespread due to its

Soil Health Series: Volume 3 Soil Health and Sustainable Agriculture in Brazil, First Edition.
Edited by Ieda Carvalho Mendes and Maurício Roberto Cherubin.

lower beverage quality, with Vietnam responsible for approximately 40% of the world's production. It originated in the Congo River basin, in altitudes varying from sea level to 1200 m as, where rainfall normally exceeds 2000 mm, with a dry period of two to three months and a mean temperature of around 26°C, with maxima around 30°C. Arabica coffee was brought to Yemen in the early seventeenth century but did not adapt there due to the arid climate. However, the popularity of the coffee drink quickly spread to the Middle East and to the main European cities.

Arabica coffee was introduced to the American continent in the early eighteenth century, with the first coffee trees being planted in Suriname, brought from Java (Camargo & Teles Junior, 1953). Subsequently, the species was introduced to Cayenne (a French colony), and the coffee trees grown there are considered the ancestors of the primitive Brazilian fields because it was from there that in 1727 some seeds were brought and planted in Belém, in the Eastern Brazilian Amazon. In 1770, coffee trees were planted in the northeastern province of Bahia, and in 1774 some seedlings were planted in the foothills of the Serra do Mar in Rio de Janeiro. From there, coffee plantations occupied the Parnaiba valley in the provinces of Rio de Janeiro and São Paulo in 1825 and later reached the province of Minas Gerais, which became Brazil's main coffee production area (Brazilian Coffee Institute, 1981; CONAB, 2022).

Minas Gerais has stood out as the largest coffee producer in the country since the 1970s, largely due to the severe frosts in 1975 in the state of Paraná, which accounted for 32% of the entire Brazilian production, when more than 200 million coffee plants were decimated and another 700 million damaged. In addition, fiscal incentives subsidized by the State of Minas Gerais, coupled with low land, production, and marketing costs, contributed to the displacement of the bulk of coffee production from the south of Brazil to its southeast (Andrade, 1994).

Brazilian Coffee-Producing Regions

In the 2021/2022 harvest, Brazil sold US$8.11 billion of processed coffee, with the state of Minas Gerais as the largest producer (ca. 1 M ha), with more than 45% of the national bean acreage of 1.8 M ha of the Arabica and Conilon varieties combined (Comex, 2022; Conab, 2022), mostly within small properties with family labor. The coffee culture can be found in 15 Brazilian States and the Federal District, occupying mainly the Atlantic and Amazon rainforests, as well as the Cerrado biome. Therefore, it was successfully adapted to contrasting climatic and soil conditions (Oliveira et al., 2019). Such adaptation to the different Brazilian regions was only possible due to the introduction of different species and cultivars with their intrinsic characteristics, especially Arabica coffee (1.45 M ha) and Conilon or Robusta (*C. canephora*) (ca. 388,800 ha). Along with Minas Gerais, the

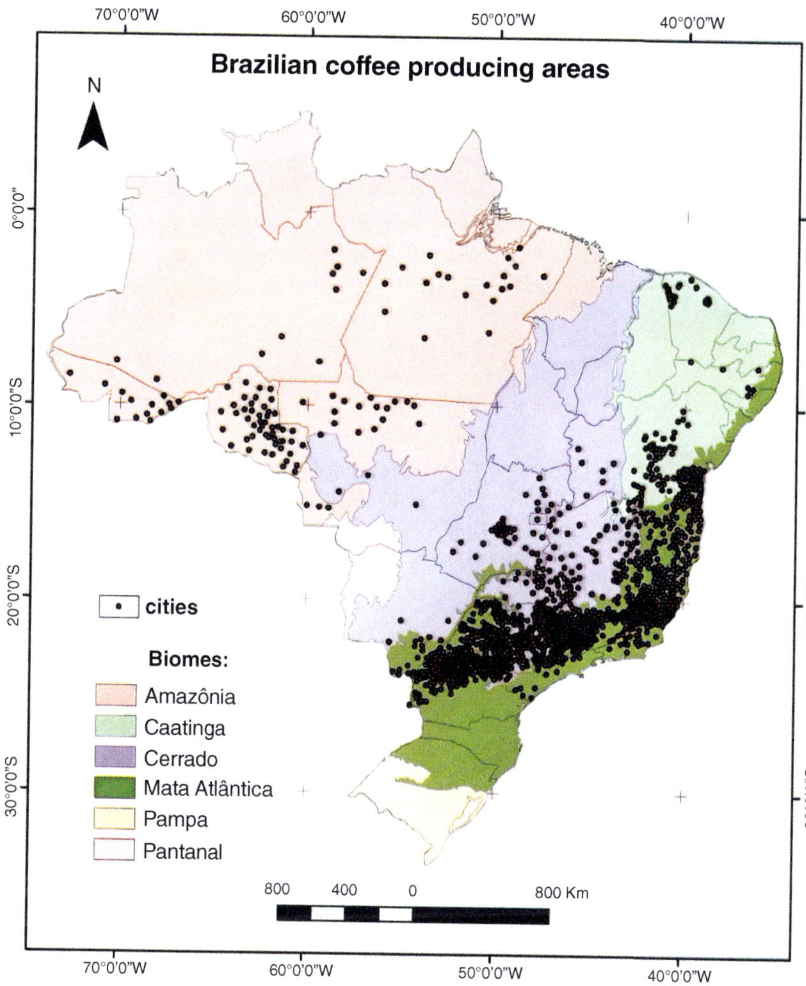

Figure 6.1 Brazilian coffee-producing areas. Adapted from Oliveira et al. (2019).

states of São Paulo, Paraná, Rio de Janeiro, Goiás, and Bahia (Atlantic rainforest and Cerrado region) are Arabica producers, whereas Conilon coffee is mostly grown in Espírito Santo, Bahia, Rondônia, Mato Grosso, and Amazonas (Atlantic and Amazon rainforest regions) (Conab, 2022) (Figure 6.1).

Considering the highly variable climates (Figure 6.2) and soils where coffee trees are grown in Brazil, the management of this crop faces many challenges when it comes to maintaining SH. In the survey by Oliveira et al. (2019) on the geographical distribution of coffee crops (Figure 6.1), it was evident that the Cerrado biome, which has a large territorial expression, comprises mostly the Arabica species.

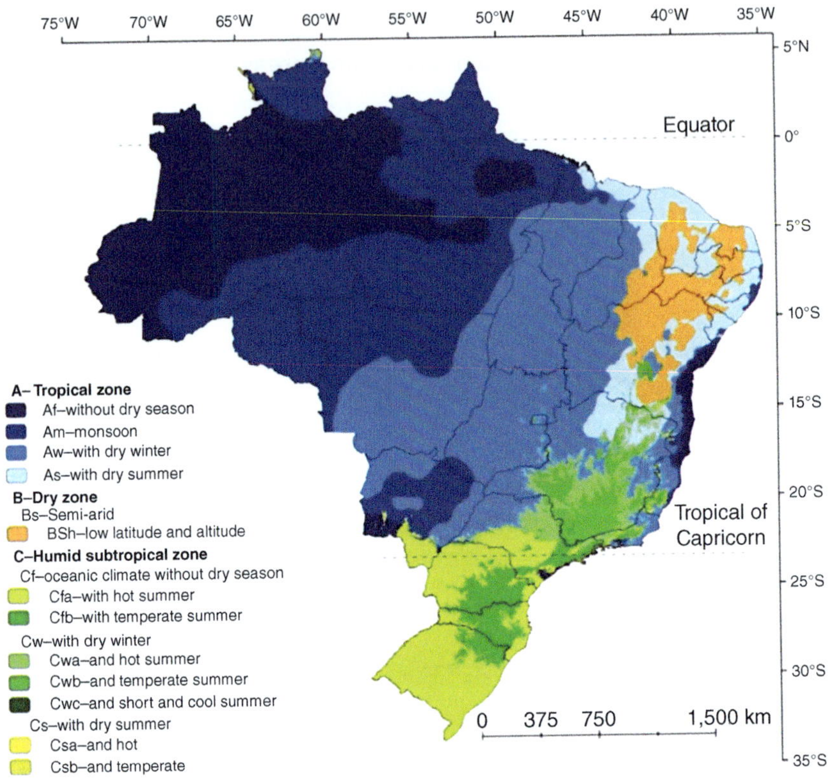

Figure 6.2 Köppen climate classification from Brazil. Adapted from Alvares et al. (2014).

In this biome, most soils located on gentle slopes are very deep Oxisols as well as Quartzipsamments, whereas oxidic Ultisols and Inceptisols typically occur on steeper slopes or near rock outcrops. Cerrado Oxisols generally have high macroporosity due to the fine granular structure responsible for their typical physical properties favoring cultivation, namely low bulk densities and easy cultivation, which has been often termed "physical fertility."

In the Atlantic Rainforest Biome, where mountainous areas are common, Arabica cultivation occupies various soils, from deep and porous Oxisols to shallower and denser Ultisols and Inceptisols. In areas of lower altitudes and higher temperatures, only Conilon is grown, typically on dense, cohesive Oxisols and Ultisols that present blocky structures and consistency that varies from very hard when dry to solid or friable when wet, which makes soil management very complex. In the Amazon rainforest Biome, due to the year-round high temperatures, Conilon coffee cultivation prevails.

Soil Properties and SH on Coffee Production Systems

Considering that soils are living systems vital to humanity, their use and management must take into account their sustainable use for both agricultural and environmental needs (Baulcombe et al., 2009). One of the greatest challenges related to SH in coffee plantations is the efficient use of soil water because irrigation is seldom used. To overcome this challenge, management practices resulting in soil organic matter (SOM) increases are crucial, such as the use of cover crops, surface straw residues (protecting the soil against erosion and helping to maintain soil moisture), and improved soil structure and soil water storage capacity. In addition, agronomic techniques that increase the efficiency of production systems in the face of climatic changes are increasingly necessary for competitive and sustainable agriculture.

Coffee crops have been planted mainly without irrigation, thus presenting greater vulnerability to climate variations. Between 2013 and 2017, rainfed agricultural crops in Brazil underwent an average water deficit of 37%, with 30% occurring in the critical period of vegetative development and 7% near harvest (ANA, 2021). In such cases, inadequate soil management associated with irregular rainfall can decrease production, particularly during flowering, and fruit formation, because the Arabica coffee tree has an average annual water requirement of 490 mm, and annual deficits of more than 150 mm affect its longevity, particularly in the early stages, when root growth is incipient and thus more sensitive to water stress (ANA, 2021). This effect can be observed especially in conventional coffee farming systems that generally present wider row spacing (ca. 3,000 to 5,000 plants ha^{-1}), soil preparation in the total area at 0.35-m depth, shallow planting furrows (0.40 m), mineral N–P–K fertilizer, and bare soil at the inter-row space by manual weeding or herbicides (Figure 6.3).

Increasing the number of coffee trees per hectare has shown many benefits, including mitigation of elevated soil and air temperatures. According to Morais et al. (2011), who compared coffee stands with 8000, 6000, and 4000 plants ha^{-1}, denser spacings provided milder average and maximum temperatures, in addition to higher relative humidity and lower wind speed, with significant gains in productivity. Another important aspect of soil temperature is coffee tree morphology. Rosa (2015) reported that the architecture of densely spaced coffee stands significantly altered the temperature to a soil depth of 5 cm, due to the canopy size and area.

A very successful strategy in coffee stand management in the Cerrado savannas is the formation of a deep and well-distributed root system using agricultural gypsum, limestone, and mechanical practices. Coffee trees follow no specific root system patterns, and thus management is critical to their development. In areas under irrigation, for example, the plant roots are concentrated near the surface.

Figure 6.3 Conventional and conservationist soil management systems for Arabica coffee production at Vargem Bonita, MG; Londrina, PR; Luís Eduardo Magalhães, BA, Brazil. Oliveira, A., Carducci, C.E., Araujo-Junior, C. F., Mendes, I. C.

In densely planted coffee stands, there is generally greater horizontal growth of the root system, and when different soil conservation practices are used in association, they also promote adequate root growth in depth (Barbosa et al., 2020; Carducci et al., 2015).

In this regard, one challenge is the mechanical impediment arising from naturally dense or compacted subsurface layers in soils, such as those common in many Inceptisols of southern Minas Gerais (Barbosa et al., 2020; Silva et al., 2021) and the cohesive, very dense Xanthic Oxisols of Espírito Santo State (Silva et a., 2021; Oliveira et al., 2019). In addition, the fine granular structure of the Brazilian Cerrado Oxisols, when subject to inadequate management, is highly susceptible to compaction (Severiano et al., 2013). Coffee farming in the Cerrado region has been notable due to the advanced technology used, with most cultural practices performed using mechanization (Oliveira et al., 2019), which can potentially create soil compaction as a result of heavy machine traffic, thus restricting root development and water availability in deeper soil layers.

Therefore, with the need to improve SH and reduce soil compaction, different types of management are used in coffee plantations, using at least one conservationist practice. Thus, stands of coffee trees intercropped with forage grasses and fruit and forest species, as well as the use of mulch made from forage plant residues or polyethylene plastic along the planting line (in surface), are becoming increasingly popular.

A novel tillage and management system has proven to efficiently reduce mechanical restrictions to root growth in the Cerrados and involves deep soil preparation and application of lime and fertilizers, enhancing both physical and chemical soil properties (Figure 6.4). Soil fertility management for coffee plantations in Brazil must start with liming to correct the typically strong soil acidity (Barbosa et al., 2020; Carducci et al., 2015; Ramos et al., 2019; Silva et al., 2021), followed by fertilization. The combined use of organo-mineral fertilizers, gypsum, limestone, and plastic mulching can considerably increase water uptake by plants. Such operations were shown to increase the effective depth of roots to >1 m, especially those with diameter <1 mm, responsible for water and nutrient uptake, as observed by Barbosa (2018) and Carducci et al. (2015) in young coffee trees (less than three years old). A large part of this effect was due to the Ca^{2+}, present in liming and/or gypsum, that also promoted a new stable aggregate hierarchy after 1.5 years of preparation, even in the dense, shallow Inceptisols.

To assess how such management practices affect an acidic Inceptisol under coffee crops, Barbosa et al. (2020) used physical quality indicators in an area

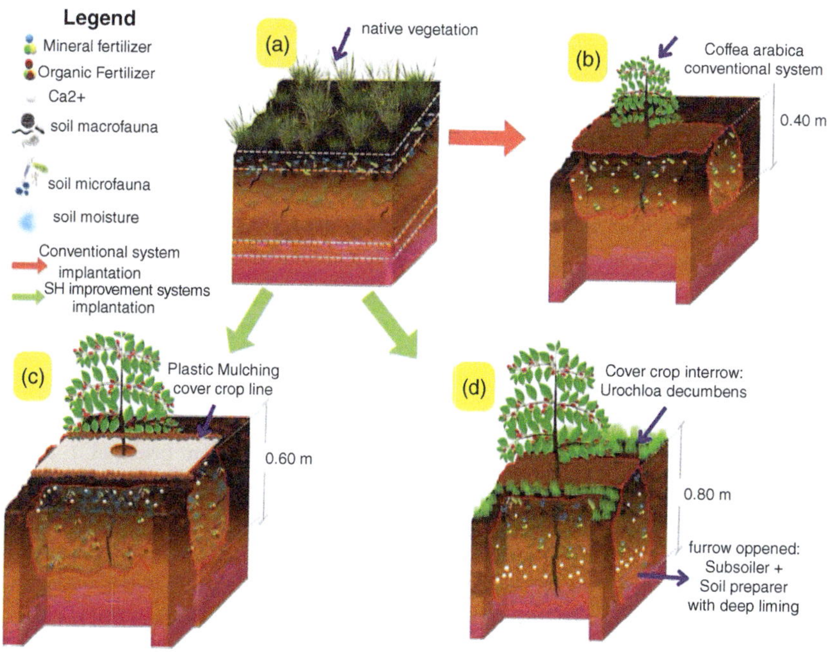

Figure 6.4 Soil management practices used to improve soil health in coffee crops. (a) Low-fertility soil under native vegetation; (b) mechanical tillage to a 0.40-m depth and mineral fertilization in conventional coffee farming systems (SC); (c) novel system with subsoiler use to furrow to a 0.60-m depth, organo-mineral fertilization and soil cover with plastic mulch in the row; (d) furrowing to 0.80 m, with soil inversion (subsoiler) after Ca^{2+} application to 0.60 m and cover crops on inter-rows with deep-rooted grasses such as *Urochloa decumbens*. Adapted from Silva et al. (2021).

subsoiled to a depth of 0.80 m, followed by soil cultivation and liming to a depth of 0.60 m on the planting row (Figure 6.4a–d). The results showed significant improvements, compared with conventional preparation, in soil aeration (0.15–0.27 $m^3\ m^{-3}$) and water contents (0.18–0.25 $m^3\ m^{-3}$) along the soil profile, due to a significant increase in pores with a mean diameter of 300 μm, responsible for drainage and aeration, with other agronomic and environmental gains. The same experiment (Figure 6.4) also showed improvements in the S-index, defined as the absolute value of the slope of the water retention curve at its inflection point (Barbosa et al., 2020), and basic infiltration rate, in addition to increased soil water storage capacity (Silva et al., 2021).

In addition, aiming to reduce soil and water losses and seeking improvements in the crop microclimate, coffee growers associate crop stand densification with different types of soil cover, aiming to decrease evaporation, especially in the dry season. A plastic soil cover along the coffee planting row is used during the deployment of the new coffee plantation (setting it out by hand or machinery), and, when associated with organo-mineral fertilization, increases water storage in an Ultisol, even in the driest months, when compared to conventional fertilization with soil cover (Figure 6.4b,c). In addition, physiological benefits were noted in the crops, such as greater tree height, number of leaves, and stem thickening. Qualitative and quantitative improvements, such as soil water conservation, better aggregation, lower soil compaction, and lower nutrient losses due to N leaching or volatilization, have also been observed (Barbosa, 2018; Ramos et al., 2019; Schminthals & Kuhn, 2017). Jointly, these factors effectively contribute to SH.

The intercropping of coffee trees with deep-rooted *Urochloa* sp. grasses (syn. *Brachiaria*) between the rows promotes strong soil cover, mainly in high-density coffee plantations (Figure 6.4d). This live mulching (cover crop) reduces soil temperature and water loss by runoff and increases nutrient cycling for further use by the coffee tree, as well as other soil improvements, because it can affect the germination of photoblastic weeds (Cerri et al., 2017; Flávio Neto et al., 2015; Serafim et al., 2011; Silva et al., 2013). However, the main benefits of this operation are associated with the root system effects on soil structure because tropical forages have a high potential to create biological pores from both root growth and increased biological activity (Baptistella et al., 2020; Carducci & Oliveira, 2021; Carducci et al., 2022), notably with *Urochloa brizantha* and *Urochloa decumbens* (Flávio Neto et al., 2015). The high plant density, the periodic renewal of the roots, and the tolerance to soil physical constraints are the keys to these benefits.

The physical effect of the root activity in binding and stabilizing soil aggregates must also be highlighted because, in these coffee plantations, the tropical forage (*Urochloa* spp.) and coffee trees interact. In these mixed stands, a pore network with a significant presence of tubular and irregular (prolate and oblate) biopores is formed, as noted by Carducci et al. (2022) using X-ray tomography on an Oxisol subjected to multi-practice conservationist soil management system (MPCS). Those pores are considered structural and diversified in diameter and thus are

Figure 6.5 Spatial variability of root length (mm) of *Coffea arabica* (less than three years) and the porous system of an Oxisol cultivated with a multi-practice conservationist soil management system (MPCS). Adapted from Carducci et al. (2015) and Carducci et al. (2022).

preferential pathways for water infiltration, gas exchange, organism habitat, and root growth for the coffee trees while simultaneously favoring lower temperatures in deeper layers (Figure 6.5).

Cultivation of spreading liverseed grass [*Urochloa decumbens* (Stapf) R. Webster] on the inter-rows of recently planted coffee trees (less than three years) on a naturally dense Inceptisol (Figure 6.4d) not only prevented surface crusting and soil erosion but also contributed greatly to the permeability of the soil profile and the final infiltration rate, which was higher than under the native Cerrado vegetation and conventional coffee stands, to a depth of 0.40 m (Figure 6.6). The improvement in soil structure promoted by *Urochloa* grasses, which occurred uniformly throughout the layer explored by its root system, eliminated the subsurface compaction commonly caused by tillage implements (Silva et al., 2021).

Besides the chemical (C sequestration and nutrient cycling, among others) and physical (biological decompaction, higher water availability, and groundwater recharge) benefits, intercropping with *Urochloa* grasses also has a direct impact on soil biological functioning. Rodrigues et al. (2022) observed that intercropped *Urochloa* positively affected the activities of soil enzymes β-glucosidase and arylsulfatase (associated with the C and S cycles, respectively) either in the presence or absence of liming. The increased input of plant residues and root

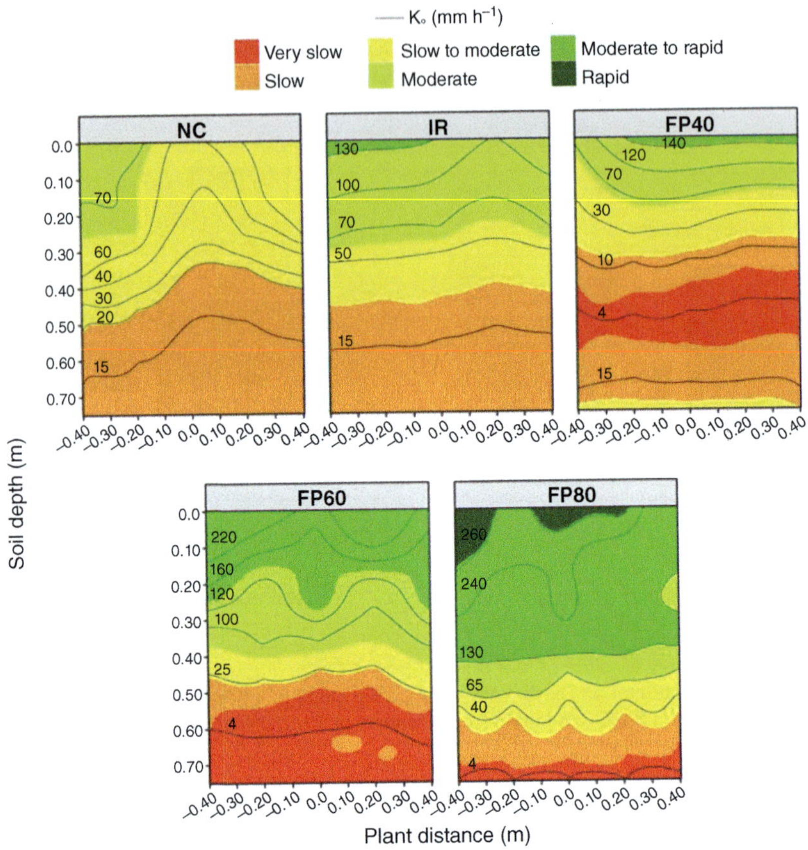

Figure 6.6 Soil profile permeability maps (K_0, mm h^{-1}) in an Inceptisol in Minas Gerais state, Brazil. Treatments: FP40, furrowing to a depth of 0.40 m; FP60, furrowing and mixing to a depth of 0.60 m; FP80, furrowing to a depth of 0.80 m and mixing to a depth of 0.60 m; IR, inter-rows of coffee plants intercropped with *Urochloa* grass; NC, native Cerrado vegetation. Adapted from Silva et al. (2021).

exudates, in addition to soil protection during the dry season, favors a more biologically active edaphic environment in the inter-rows of coffee production areas (Table 6.1). These soil enzymes are the pillars of the SoilBio technology (see Chapter 10), and since 2020 they have been used in large-scale on-farm SH assessments in Brazil, representing an opportunity to engage producers in soil testing beyond the standard chemical analyses.

In sandy and clayey Cerrado soils, increased activities of β-glucosidase and arylsulfatase in coffee production areas in response to intercropped *Urochloa* also were reported by Mendes et al. (2018) (Figure 6.7). As expected, on average, the activities of β-glucosidase and arylsulfatase were, respectively, two and four times higher in

Table 6.1 β-Glucosidase and arylsulfatase activities in Typic Haplustox, cultivated with coffee either with and without intercropping with *Urochloa* (*Brachiaria* spp.), in the presence and absence of liming.

	With lime	Without lime
	–μg *p*-nitrophenol g^{-1} soil h^{-1}–	
	β-glucosidase	
With *Urochloa*	176aA	151bA
Without *Urochloa*	127aB	120aB
	Arysulfatase	
	–μg *p*-nitrophenol g^{-1} soil h^{-1}–	
With *Urochloa*	145aA	115bA
Without *Urochloa*	94aB	84aB

Note: Soil samples were collected at the rainy season to a depth of 0–0.10 m. Means followed by different letters (lowercase in rows and uppercase in columns) are significantly different at $P < 0.05$ (Tukey's test).
Adapted from Rodrigues et al. (2022).

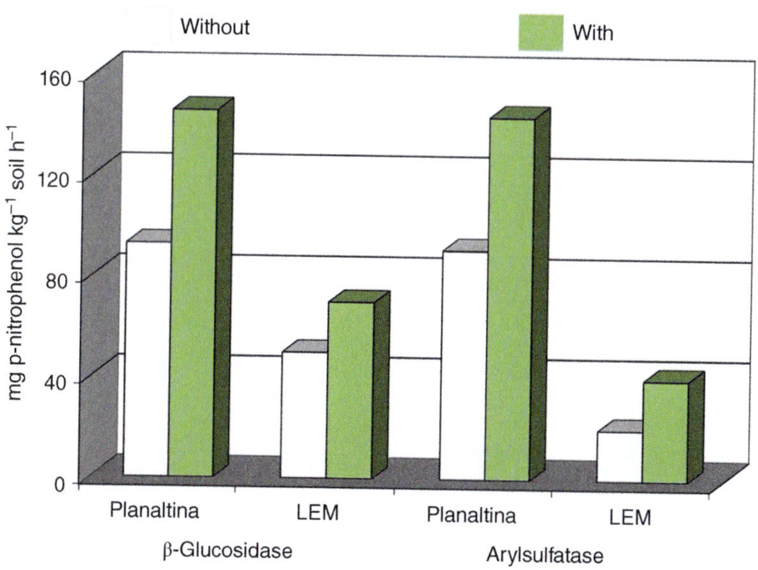

Figure 6.7 β-glucosidase and arylsulfatase activities in coffee plantations with and without *Urochola* (0-0.10 m depth) in two different sites: Planaltina, GO state where the soil is a Haplutox (60% clay content, soil organic matter content, 25 g kg^{-1}) and, Luis Eduardo Magalhães (LEM), BA state, where the soil is a Typic Quartzipsamment (14% clay content, soil organic matter 8 g kg^{-1}). Adapted from: Rodrigues et al. (2022).

Figure 6.8 Main Brazilian soils used for coffee cultivation with their common physical constraints, showing the roots in the conventional management (top) and with conservationist management (bottom). The authors and Icograns website (https://icograms.com).

the clayey Haplustox (Planaltina, DF; 60% clay; SOM, 25 g kg^{-1}) than in the Typic Quartzipsamment (Luís Eduardo Magalhães, BA; 14% clay; SOM, 8 g kg^{-1}).

All the mechanical and cultural practices described above have multiple implications that, in addition to SH, also benefit the whole coffee production system. Chiefly, the combination of independent practices, beyond the benefits for soil conservation (Serafim et al., 2011) and increased crop productivity, also promotes the improvement of site microclimate. This process was aptly called “soil revival” by Camargo and Telles Jr. (1953) (Figure 6.8), and the use of those practices is gradually becoming more popular among coffee farmers.

Healthy soils support sustainable agriculture and positively affect productivity but also improve the environmental services provided by farmlands. By improving the physical, chemical, and biological soil attributes, the management practices presented here, in a broader sense, also contribute to the achievement of the Sustainable Development Goals (FAO, 2022). Soil physical quality directly influences responsible and sustainable food production and contributes to the mitigation of deleterious effects associated with climate change, as discussed below.

Soil Organic C and SH in Coffee Plantations

Soil organic C (SOC) is a minor (generally <3%) component of most soils, but it can arguably be considered the most important single factor involved in SH because it affects practically all indicators involved and all ecological services provided by soils, including agricultural production. SOC comprises about half of SOM, which in tropical humid soils is the most important “pool” of nutrients such as N and S, often is the most available source of P, and also responds for most of the cation exchange capacity for other critical nutrients because such soils are typically dominated by low-activity clays, such as kaolinite and oxides of Fe and Al. However, SOC is also the main source of C and energy to the decomposer soil biota, which ranges from soil fauna to fungi and bacteria (i.e., the soil biological diversity itself, which mediates innumerous environmental services). In terms of soil physical properties, the maintenance of adequate SOC concentrations promotes lower bulk and particle densities, resulting in good porosity and thus soil tilth (i.e., the desired friable consistence of topsoils that favors cultivation and the retention of water at low suction). Finally, SOC is a major global C pool on Earth, containing more C than the atmosphere and biomass combined, and with a longer residence time. Thus, increasing SOC concentrations improves SH and quality but also help mitigate anthropogenic global warming, which has been called a win-win situation (Lal, 2004).

The concentration of SOC in the soil is mostly the result of a steady-state equilibrium between C inputs to soils (typically litter, crop residues, root growth and exudation, and in some cases organic amendments) vis à vis its depletion by biological decomposition (soil respiration), leaching, and erosion of SOC-rich topsoil. However, the processes involved in the alteration of fresh, easily decomposable organic residues into more stable, highly altered SOC (humus) are mediated by many organisms and require a time lapse from weeks to months, and typically only a small fraction (e.g., 2–20%) of total inputs becomes incorporated as SOC (Lal, 2004). In addition, the intensity and efficiency of SOC sequestration into soils varies widely with climate and residue type as well as with soil properties (Zinn, 2016). As a general rule of thumb, SOC concentrations and retention in soils are favored by moister, cooler climates; by incorporation of grass and forest residues; and by well-structured soils of loamy and finer textures, including those where Fe and Al oxides abound in the clay fraction. Although there is no standard definition of adequate SOC concentrations for good SH, a 2% value (mass/mass, or 20 g kg^{-1}) has been referred to as the minimal recommended for adequate quality of loamy to clayey topsoils (Zinn & Silva, 2021).

However, a critical factor affecting SOC retention and dynamics is the land use system. Soils under climax native vegetation have been considered the standard of steady-state equilibrium SOC concentrations, and most often are considered the maximum SOC that can be stored in an area, and thus the starting point from which conversion to agriculture almost always starts lowering SOC concentrations. This eventual SOC depletion is mostly due to lower inputs and accelerated decomposition and erosion when compared to the native vegetation. Nevertheless, agricultural land use systems are highly variable in their ability to deplete, maintain, and even increase SOC concentrations. In comparison to annual crops, perennial crops offer a considerably different soil environment that is often more favorable to SOC retention because (a) mechanical soil preparation is generally much less intensive and frequent because annual replanting is not done, (b) perennial crops occupy the site for more than one year and thus provide more continuous soil cover through time and space, and (c) total harvest is not done annually, and thus biomass production and accretion to soils are more abundant and frequent (Ledo et al., 2020). Therefore, as noted by these authors in their global meta-analysis, perennial crops offer an average 20% SOC increase after conversion from annual crops and a nonsignificant change or just a slight decrease after conversion from forests and natural pastures, respectively, especially when the perennial crop involves woody plants.

Coffee is among the most important commodities exported by Brazil and many other tropical countries and thus is a major perennial tree crop occupying large areas of prime agricultural land. However, coffee plantations are also very different from other perennial woody crops because (a) they require fertilization, pesticides, and management practices in intensity and frequency comparable to or even higher than those of the most demanding horticultural crops; (b) coffee stands typically occupy the same tracts of land for cycles up to 30 years, with intensive soil preparation and disturbance only previously to planting; and (c) even the short-sized coffee varieties promote considerable year-round litter inputs, at least on the planting row, which is often reinforced by cover crops or weeds on inter-rows, and sometimes by consortia with shade or wind-break trees. Thus, perhaps more than any other perennial crop, coffee stands are unique in which they merge intensive nutrient inputs (which favor biomass and SOC accretion) typical of grain crops, with features typical of forest plantations, such as limited soil disturbance (thus not accelerating SOC depletion) and high, constant residue inputs (also favoring biomass and SOC accretion). The most visible consequence of such features of coffee plantations is that SOC is often conserved by this land use system, and in some cases sequestered, even when compared to SOC concentrations and stocks under native or secondary forest vegetations (e.g., Cogo, 2012; Cogo et al., 2013; Cerri et al., 2017; França et al., 2023; Santos, 2012). This is a highly auspicious prospect for SH in coffee plantations, and further research and development can improve this potential for SOC sequestration in Brazil and other countries. The following is a collection of management practices that can further increase SOC conservation or sequestration, and thus enhance SH, in coffee crops:

Tall varieties: More recent coffee varieties reach short (1 m or lower) heights (although if irrigated, they can reach 2 m height), a trait that greatly favors manual harvesting but also results in lower biomass and residue accretion to soils. However, most of the coffee planted in Brazil and elsewhere comprise tall varieties, often >2 m in height.

Dense spacings: Coffee trees are typically planted along rows distant at least 2.75–3.2 m (to allow passage of mechanized traction), and the various distances between trees on a row result in densities varying from 1000 to 10,000 trees per ha (Zinn & Silva, 2021). For each environmental setting, there must be an optimal spacing that allows for maximal coffee production, which often will provide maximum leaf and litter production and thus optimal SOC inputs. In some countries of Central America and other areas, coffee trees are planted in very sparse spacings, which, combined with strict weed control on inter-rows, can intensely deplete SOC over time.

Intercalation with other tree species on agroforestry systems: Such practices have the potential to produce more organic residues and thus sequester SOC (e.g., Tumwebaze & Byakagaba, 2016) because most trees are taller and more branched than coffee trees, but this is not necessarily the rule. On a clayey Oxisol in Minas Gerais, Brazil, Cogo (2013) found similar SOC levels under conventional production systems (CS) and coffee + macadamia consortia, although soil pH under the latter was significantly lower, probably due to protonation of organic acids derived from macadamia tree residues or exudations.

Weed and cover crops on inter-rows: The wide (3.2-m) inter-rows occupy most of the coffee stand area and thus represent a critical potential for soil cover and SOC inputs or, if left bare, for soil erosion and SOC depletion. In Brazil, inter-rows are often covered with weeds or planted with Brachiaria (*Urochloa* sp.) grass. However, when grasses and weeds are not allowed to grow, using pre-emergency herbicides, SOC and SH are severely depleted; conversely, when weeds are allowed to grow and mechanically cut and left on the inter-row, SOC is conserved in levels similar to those of secondary forests (Cogo et al., 2013).

Fertilization with organic or organo-mineral amendments: These are becoming more common in the selected sector of organic foods but also on CS and have the potential to sequester SOC if consistent throughout the years. França et al. (2023) reported considerable SOC sequestration under short coffee trees with application of 3 Mg ha^{-1} yr^{-1} of peat-based organo-mineral fertilizer, with *Urochloa* sp. on the inter-row, on a gravelly Inceptisol on a mountain range in Minas Gerais.

SH in Life Cycle Assessments Applied to Coffee Production Systems in Brazil

The process of agricultural production, as well as the application of life cycle assessments (LCAs) for agricultural products, presents complexities and particularities when compared to the application of LCAs to industrial products. The use of LCAs in food production chains faces specific challenges related to the functional unit, the influences of geographical and climatic changes, and the major influence of consumer behavior, as well as the chain structure for a large number of small-scale production units. These difficulties imply a high variability in the environmental effects and problems associated with data collection. When applying LCAs for food production chains, there are several important reasons for generating and using the information on the origin and characteristics of different food products, including the environmental aspects that influence SH as well.

The LCA methodology is the only approach that allows for the identification of the transfer of environmental impacts from one place to the other (e.g., the elimination of air emissions can be performed at the expense of increased wastewater emissions) and/or from one life cycle stage to another (e.g., from the phase of raw material acquisition to the phase of product usage) (SETAC, 1991).

Studies relating LCAs to the dynamics of coffee and agricultural production in general, focusing on soil C and soil quality, are still scarce in the literature. Table 6.2 lists some scientific papers found that address Brazilian coffee production from the perspective of the LCA tool.

Table 6.2 Some results from studies of practical life cycle assessment work in the coffee chain.

Author	Manuscript	Goal	Impact categories	Results
Usva et al. (2020)	CF and water footprint of coffee consumed in Finland—life cycle assessment	Assess the CF and water scarcity impacts throughout the life cycle of the coffee chain and identify the most impactful sources	CF and water scarcity	There was considerable variability in the CF, with fertilization being the main contributor. Water scarcity was low for non-irrigated systems in Central America and on Brazilian farms irrigation was a major contributor.
Nab and Maslin (2020)	LCA synthesis of the CF of Arabica coffee: Case study of Brazil and Vietnam conventional and sustainable coffee production and export to the United Kingdom Carmen	Identify and quantify GHG emissions and the CF of coffee	(a) GHG, (b) CF	The CF of Arabica coffee from both countries was approximately 15.33 kg CO_2eq kg^{-1} green coffee for conventional coffee production and 3.51 kg CO_2eq kg^{-1} for sustainable coffee production. Conventional coffee from Vietnam showed a CF (16.04 kg CO_2eq kg^{-1} green coffee) followed by conventional coffee from Brazil (14.61 kg CO_2eq kg^{-1} green coffee).

(Continued)

Table 6.2 (Continued)

Author	Manuscript	Goal	Impact categories	Results
Tavares and Mourad (2020)	Coffee beverage preparation by different methods from an environmental perspective	Assess the environmental performance of the preparation step of the most representative methods of coffee preparation, in beverage form	Environmental efficiency of energy, water consumption, waste generation, environmental impacts related to global warming, eutrophication, acidification, abiotic depletion, and human toxicity	Preparing a single-serve soft capsule using an automatic machine resulted in the lowest emissions and generated non-biodegradable packaging waste to landfill about 11 times less than single-serve plastic capsules with aluminum top seal, which had the highest energy consumption, water, and waste generation in the single-serve category. In the out-of-home consumption category, expresso, produced under pressure, and at higher temperatures, had the highest impact.
Coltro et al. (2006)	Environmental Profile of Brazilian Green Coffee	Present the LCA of green coffee produced in Brazil for the reference crops 2001/2002 and 2002/2003, in order to generate detailed production inventory data	There were no impact categories because the article made the construction of an inventory	Pioneering Brazilian study on LCA in coffee cultivation. The use of fertilizers and pesticides specific to each agricultural unit shows an opportunity to reduce these inputs. LCA associated with climate data allows for sustainable management of natural resources. Different farming practices can produce different environmental performance, and the amount of pesticides is directly related to farming practices such as crop rotation, plant density, etc.

Abbreviations: CF, carbon footprint; GHG, greenhouse gas; LCA, life cycle assessment

Table 6.3 The principal management practices used in conventional (CS) and multi-practice conservationist coffee systems (MPCS).

Practices	CS	MPCS
Plant stand	3000 plants ha^{-1}	8000 plants ha^{-1}
Row preparation	0.35 cm depth	0.60 cm depth
Interrow	Bare soil	*Urochloa* sp.
	Hand weeding and control with herbicides	Kept cut
Gypsum	1–2 Mg ha^{-1}	14–28 Mg ha^{-1}
	Soil mixed	Applied on soil surface
N–P–K	20–10–20[a)]	8–44–0; 1% Zn, 0.5% B
Fertilization at planting	700 kg ha^{-1}	980 kg ha^{-1}
Irrigation system	–	–

a) Values for maintaining optimal soil nutrient levels, depending on plant spacing and fertilizer source.

We present a case study by Anis (2021) where a LCA was performed in a coffee crop chain, involving the comparison between two coffee production systems in the State of Minas Gerais, Brazil (Table 6.3). The system boundary is defined by the GHG emissions associated with coffee production from "cradle to farm-gate." The LCA of the production systems included cultivated coffee and the resources used to produce it (e.g., diesel, fertilizers, limestone, gypsum, crop protection products, and agricultural machinery) is important to know that the secondary inputs, such as transport and agrochemicals were imported from the ecoinvent® database v.3.8 (ECOINVENT, 2023) (Figure 6.9).

The complete coffee production system—from the inputs extraction to bagged coffee at the farm gate—was evaluated in order to estimate its global warming potential, known as the "carbon footprint" (CF). The CF was first performed by assessing the inputs and outputs of CO_2 equivalents (CO_2eq) per hectare and later normalized by kilogram of coffee harvested. The CF over the entire life cycle of a crop can provide information on its contribution to climate change, help identify possible GHG mitigation options (Cheng et al., 2019), and improve soil management practices aiming at C sequestration (Table 6.4).

Moving forward on this topic, Anis (2021) proposed as C balance the relationship between the emissions detected in the LCA, with the IPCC (2007) protocol, of the coffee crops with the C stock promoted by the MPCS used.

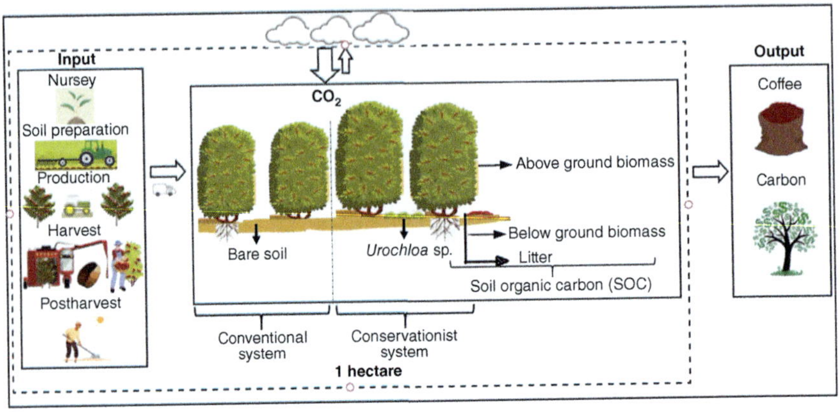

Figure 6.9 System boundary. Inputs (agricultural phases of production) and outputs of conventional (CS) and multi-practice conservationist coffee systems (MPCS). The analysis was equivalent to 1 ha. Adapted from Anis (2021).

Table 6.4 Evaluation of environmental impacts of conventional systems (CS) and multi-practices conservationist coffee system (MPCS) using a life cycle assessment.

Environmental Impact Category	Unit	Plant Seedling	Planting	Production	Total
		CS			
GWP	Mg CO_2eq ha^{-1}	4.10	33.26	17.41	54.77
		MPCS			
GWP	Mg CO_2eq ha^{-1}	5.16	14.44	14.52	34.12

Abbreviation: GWP, global warming potential.
Note: Adapted from Anis (2021) and Carducci et al. (2023).

The C stock and CO_2eq at the 0.30-m depth in the soil are presented in Table 6.5. The MPCS was superior to CS. Thus, the C increment was over 60% in the MPCS (Tables 6.4 and 6.5). In other words, the CS emitted 37.70% more CO_2eq ha^{-1} than the MPCS, considering 1 kg of coffee produced (Tables 6.4 and 6.5). The C balance proposed by Anis (2021) unified soil C stocks and cultivation emissions according to the LCA method. The MPCS resulted resulted in −144.88 Mg CO_2eq ha^{-1} yr^{-1}, whereas the CS resulted in −55.23 Mg CO_2eq ha^{-1} yr^{-1}. The negative C value means a net gain in C from the LCA standpoint. These results add more evidence of the multiple impacts of MPCS on SH.

The MPCS produced yield greater than 2340 kg ha^{-1} (Oliveira, 2019), high above the Brazilian average of 1476 kg ha^{-1} (Conab, 2022). In addition to the

Table 6.5 Soil C stocks and emission of conventional systems (CS) and the multi-practices conservationist system (MPCS).

Attributes	Unit	CS	MPCS
Soil C Stock,	Mg C ha^{-1} yr^{-1}	30	49
Soil CO_2eq Stock,	Mg CO_2 ha^{-1} yr^{-1}	110	179
LCA Emission,	Mg CO_2 ha^{-1} yr^{-1}	54.77	34.12
Coffee Yield,	kg ha^{-1}	1476	2340

Abbreviation: LCA, life cycle assessment.
Note: Carducci et al. (2023)

intrinsic characteristics of the coffee cultivars used in this study, the significant gain in production in the multi-practice system was associated with improvements in SH (biological, physical, and chemical properties), which resulted in lower CO_2eq emissions and, according to Silva et al. (2015), better water use efficiency. In summary, these results highlight the outstanding potential that the MPCS has regarding its capacity for C storage, with direct consequences to SH. In addition, LCA can add value to the coffee business by including the eventual commercialization of C in both voluntary and regulated markets. By investing in MPCS, farmers can increase their annual income and reduce environmental impacts associated with the external inputs (e.g., fossil fuel, electricity, fertilizers [chemical and organic], agricultural gypsum, limestone, and pesticides).

SH in Coffee Production Systems in the Brazilian Cerrados: A Case Study of the AP Romero System

Coffee production systems in Brazil have evolved rapidly in recent years in order to achieve more profitability and sustainability. Coffee growers need to adopt conservationist agricultural practices that will help to mitigate the physical and chemical limitations of tropical agricultural soils and promote SH, with positive impacts for the environment.

In this sense, the experience reported here began in 1999, with the acquisition of a 335-ha farm near the city of São Roque de Minas, in the Canastra Range region, Minas Gerais, at the headwaters of a major Brazilian river, the São Francisco, in the Cerrado Biome. Limited water availability for irrigation and rainfed production, in association with the naturally low chemical fertility of the clayey Oxisols, represented the main challenges for the establishment of coffee production systems in this region. On the other hand, when soil chemical problems were solved, there was a great potential for large-scale and mechanized coffee production due to the gently rolling topography of the farm (favoring mechanization) associated with good soil physical conditions (physical fertility), resulting in a high potential for water retention in the profile.

At the beginning of farming in the Canastra region, there were many climatic challenges due to the dry season, typically from four to five months. Thus, the farm owners searched for solutions to develop sustainable and profitable entrepreneurial, large-scale coffee farming and with the intention of expanding the knowledge about this system to other coffee farmers.

After consulting with mountain coffee experts, they developed an innovative system involving multiple soil management practices that allowed access to water at greater depths, even during the critical dry season (May–September), at the flowering stage. The development of deep roots would enhance the performance of coffee plants during the critical periods of water stress, typical in the Cerrado region.

This is how the soil management system for dryland coffee production (non-irrigated conditions), called the AP Romero system (for its creator's namesake), was initiated in 2000 (Figure 6.10). This system is based on conservationist agriculture, in which coffee trees are densely planted (8000 plants ha^{-1}) with an inter-row spacing of only 2.65 and 0.65 m (Pavan et al., 1997; Serafim et al., 2011). Planting takes place between mid-October and mid-November, unlike conventional farming, which takes place in December and January. Coffee seedlings are grown in perforated plastic bags, guaranteeing greater initial soil moisture. The area was systematically leveled to optimize mechanization, with minimum movements, reducing production costs and favoring maximum light exposure to the entire leaf area of the coffee trees.

An important factor to promote rapid root development for coffee trees is soil preparation to a 0.60-m depth with a subsoiler and the use of massive amounts of agricultural gypsum (14–28 Mg ha^{-1} to correct at a 2 m depth), which results in nutrients (mostly Ca and S) to move deeper in the soil profile, favoring the formation of a large root network (Figure 6.11). Another important characteristic of the AP Romero system is the addition of B via ulexite [$NaCaB_5O_6(OH)_6$], a slow-release B source for plants. The development of an abundant and deep root system was necessary to enhance fertilizer use efficiency and to make the coffee trees more drought tolerant (Carducci et al., 2015; Serafim et al., 2011; Silva et al., 2013). There is constant nutrient monitoring via leaf analysis. Mini-terraces were also built in each planting row (Figure 6.10). All the equipment used in coffee handling, whether mechanized or by animal traction (mules), was adapted or developed exclusively on the farm, so that it was possible to carry out all the management besides harvesting, including the "ground coffee" even

Figure 6.10 The AP Romero system stages. From left to right: seedlings, soil preparation, planting, fertilization, *Urochloa* mowing, and harvesting. Oliveira, A., Carducci, C. E.

Figure 6.11 Root distribution on soil profile under AP Romero system. Oliveira, A.

in densely planted areas (8000 plants ha^{-1}), which reduced operational costs by 50%.

To protect the inter-row soil from erosion and insolation, *U. decumbens* was planted as a cover crop and periodically cut to reduce soil water losses by 49% in the dry months (Pedrosa, 2013). In addition, it enhances soil biological functioning, protects the soil structure from the effects of intense machine traffic (thus maintaining physical fertility), and increases SOM and nutrients cycling (Pavan et al., 1997; Rodrigues et al., 2022; Silva et al., 2015). There is also the effect of lowering soil temperatures, always below 30°C, especially in the first years of coffee cultivation, when the young coffee plants do not cover the surrounding soil, with benefits to root performance and diversity of arbuscular mycorrhizal fungi and enzymatic activities (Cogo et al., 2020; Naves et al., 2020; Rodrigues et al., 2022). Coffee husks are also distributed along the inter-rows after drying (Figure 6.10).

Another highlight of the system is financial and social management. There are incentives for professional development with a view to better crop management and yield. The AP Romero system has promoted average yields of 2340 kg ha^{-1} (39 bags of 60 kg ha^{-1}), well above the national average for Arabica coffee of 1,476 ka ha^{-1} (24.6 bags ha^{-1}) (Conab, 2022; Oliveira, 2019), which is remarkable because it does not use prime agricultural land but rather uses marginal, low-fertility soils.

Thus, with this coffee farmers group and collaborators with innovative ideas, coffee farming started in the Canastra Range and has been concentrated in the Central-West area of Minas Gerais state (42,000 ha). However, the low technical knowledge of most farmers (because it is a very intensive system with high plant density and requires specialized machinery and, in some cases, animal traction) limits its adoption in a larger number of coffee-growing areas.

In this path of innovation, opportunities arise to publicize the system in conventional media, such as the Brazilian agribusiness TV show "Globo Rural," as well as in many scientific papers (Carducci et al., 2022; Cogo et al., 2020; Naves et al., 2020; Serafim et al., 2011; Ramos et al., 2019; Silva et al., 2013, 2015). All the experience of the region's farmers and the research done on this system (>20 years) demonstrate the Cerrado of Minas Gerais has indeed a vocation for dryland coffee production (non-irrigated condition), using good soil management practices that maintain SH, allowing the production of high-quality coffees with environmental, economic, and social sustainability in addition to protecting the region's headwaters (i.e., the São Francisco River).

Concluding Remarks

Brazilian coffee production is constantly innovating and adapting, especially in soil and crop management, which has mostly improved SH in terms of soil biological, chemical, and physical conditions. In terms of advances in SH assessments, by 2023/2024 it is expected the development of interpretative algorithms for soil enzymes arylsulfatase and β-glucosidase (which are the pillars of the SoilBio technology; see Chapter 10) specific for areas under coffee production to enhance on farm SH monitoring. The intercropping of coffee and *Urochloa* sp. has been increasingly adopted by various coffee farmers.

In addition to increased coffee yields, the adoption of conservationist practices contributes to mitigating adverse climatic effects on coffee production areas, reducing GHG emissions and increasing C stocks, which will enable farmers to participate effectively in the agricultural C market. As a consequence, coffee farmers are adopting conservationist coffee systems that will promote, in the long term, the maintenance of healthy soils, better yields of high-quality coffee, and resilience against drought and other climate constraints.

References

ANA – National Water Corporation (2021). Water use in Dryland farming. Available: <http://biblioteca.ana.gov.br/asp/download.asp?codigo=140872&tipo_midia=2&iIndexSrv=1&iUsuario=0&obra=82410&tipo=1&iBanner=0&iIdioma=0> (in Portuguese).

Andrade, R. G. R. (1994). *The expansion of coffee growing in Minas Gerais: From state intervention to market liberalization* (Master's thesis). (In Portuguese.). Federal University of Minas Gerais.

Alvares, C. A., Stape, J. L., Sentelhas, P. C., Gonçalvez, J. L. M., & Sparovek, G. (2014). Köppen's climate classification map for Brazil. *Meteorologische Zeitschrift*, 22(6), 711–728. https://doi.org/10.1127/0941-2948/2013/0507

Anis, C. F. (2021). *Conservation coffee production system and the generation of carbon credits* (Master's thesis). (In Portuguese.). Federal University of Grande Dourados.

Barbosa, S. M. (2018). *Management of Inceptisols and Ultisols in coffee plantations* (Doctoral dissertation). (In Portuguese.). Federal University of Lavras.

Barbosa, S. M., Silva, B. M., Oliveira, G. C., Benevenute, P. A. N., Silva, R. F., Curi, N., Moreti, B. S., Silva, S. H. G., Norton, L. D., & Pereira, V. M. (2020). Deep furrow and additional liming for coffee cultivation under first year in a naturally dense Inceptisol. *Geoderma*, 357, 113934. https://doi.org/10.1016/j.geoderma.2019.113934

Baptistella, J. L. C., de Andrade, S. A. L., Favarin, J. L., & Mazzafera, P. (2020). Urochloa in tropical agroecosystems. *Frontiers in Sustainable Food Systems*, 4, 119.

Baulcombe, D., Crute, I., Davies, B., Dunwell, J., Gale, M., Jones, J., Pretty, J., Sutherland, W., & Toulmin, C. (2009). *Reaping the benefits: Science and the sustainable intensification of global agriculture*. The Royal Society.

Brazilian Coffee Institute. (1981). *Cultura do café no Brasil*. (In Portuguese.). IBC/GERCA.

Camargo, R., & Telles, A. Q., Jr., (1953). *Coffee in Brazil.* (In Portuguese.). Ministério da Agricultura.

Carducci, C. E., Oliveira, G. C., Curi, N., Heck, R. J., Rossoni, D. F., De Carvalho, T. S., & Costa, A. L. (2015). Gypsum effects on the spatial distribution of coffee roots and the pores system in oxidic Brazilian latosol. *Soil & Tillage Research*, 145(1), 171–180. https://doi.org/10.1016/j.still.2014.09.015

Carducci, C. E., & Oliveira, G. E. (2021). *Manejo do solo na cafeicultura: produtividade e sustentabilidade.* UFLA. (in Portuguese). http://repositorio.ufla.br/handle/1/48424

Carducci, C. E., Anis, C.F., de Souza, D. P., Costa, J. C., & Ruviaro, C. F. (2023). Life cycle assessment and carbon balance in coffee production systems in Cerrado Mineiro, Brazil. *Recima21 – Revista Científica Multidisciplinar*, 12, e4124690, ISSN 2675-6218. https://doi.org/10.47820/recima21.v4i12.4690. In Portuguese.

Carducci, C. E., Pinto, L. C., Barbosa, S. M., Costa, J. C., Zinn, Y. L., & Heck, R. J. (2022). Micromophology and x-ray CTscan: Gibbsitic oxisol porosity under coffee multi-practice management system. *Recima*, 3(6), e361520. https://doi.org/10.47820/recima21.v3i6.1520

Cerri, C. E., Moreira, C. S., Alves, P. A., Toledo, F. H. R. B., Castigioni, B. A., Rodrigues, G. A. A., Cerri, D. G. P., Cerri, C. E. P., Teixeira, A. A., Candiano, C. A. C., Reis, M. R., D'Alessandro, S. C., & Turello, L. (2017). Estoques de carbono e nitrogêncio no solo devido a mudança do uso da terra em áreas de cultivo de café em Minas Gerais. *Coffee Science*, 12(1), 30–41.

Cheng, X., Van Dammes, S., Li, L., & Uyttenhove, P. (2019). Evaluation of cultural ecosystem services: A review of methods. *Ecosystem Services*, 37, 100925. https://doi.org/10.1016/j.ecoser.2019.100925

Cogo, F. D., Araujo-Junior, C. F., Zinn, Y. L., Dias Junior, M. S., Alcântara, E. N., & Guimarães, P. T. G. (2013). Estoques de carbono orgânico do solo em cafezais sob diferentes sistemas de controle de plantas invasoras (In Portuguese.). *Semina Ciências Agrárias*, 34, 1089–1097. https://doi.org10.5433/1679-0359.2013v34n3p1089

Cogo, F. D. (2012). *Carbono orgânico em Latossolo sob lavoura cafeeira e outros usos da terra no sul de Minas Gerais* (Master's thesis). (In Portuguese.). Universidade Federal de Lavras.

Cogo, F. D., Saggin Júnior, O. J., Guimarães, P. T. G., Siqueira, J. O., & Carneiro, M. A. C. (2020). High rates of agricultural gypsum affect the arbuscular mycorrhiza fungal community and coffee yield. *Bragantia*, 79(4), 612–622. https://doi.org10.1590/1678-4499.20200014

Coltro, L., Mourad, A. L., Oliveira, P. P. L. V., Baddini, J. P. O. A., & Kletecke, R. M. (2006). Environmental profile of Brazilian green coffee. *International Journal of Life Cycle Assessment*, 11(1), 16–21. https://doi.org/10.1065/lca2006.01.230

Comex. (2022). *Foreign trade chamber*. https://www.comexdobrasil.com/brasil-exporta-396-milhoes-de-sacas-de-cafe-na-safra-2021-22-e-obtem-receita-recorde-de-us-81-bilhoes-diz-cecafe

Companhia Nacional De Abastecimento. (Conab). (2022). *Brazilian coffee crop*. Boletim Café 2022–3° Levantamento. www.conab.gov.br/info-agro/safras/cafe

ECOINVENT. (2023). On-line. http://www.ecoinvent.org

Food and Agriculture Organization of United Nations (FAO). (2022). *Sustainable development goals*. https://www.fao.org/sustainable-development-goals/en

Flávio Neto, J., Severiano, E. C., Costa, K. A. P., Guimarães Júnnyor, W. S., Goncalves, W. G., & Andrade, R. (2015). Biological soil loosening by grasses from genus Brachiaria in crop-livestock integration. *Acta Scientiarum: Agronomy*, 37(3), 375–383. https://doi.org/10.4025/actasciagron.v37i3.19392

França, E. M., Silva, C. A., & Zinn, Y. L. (2023). Coffee plantations can strongly sequester soil organic carbon at high altitudes in Brazil. *Soil Research*, 61, 198–207.

IPCC. (2007). *Climate change 2007: Synthesis report. Contribution of working groups I, II and III to the fourth assessment report of the intergovernmental panel on climate change*. IPCC.

Lal, R. (2004). Soil carbon sequestration impacts on global climate change and food security. *Science*, 304(5677), 1623–1627. https://doi.org/10.1126/science.1097396

Ledo, A., Smith, P., Zerihun, A., Whitaker, J., Vicente-Vicente, J. L., Qin, Z., McNamara, N. P., Zinn, Y. L., Llorente, M., Liebig, M., Kuhnert, M., Dondini, M., Don, A., Diaz-Pines, E., Datta, A., Bakka, H., Aguilera, E., & Hillier, J. (2020). Changes in soil organic carbon under perennial crops. *Global Change Biology*, 26, 4158–4168. https://doi.org/10.1111/gcb.15120

Morais, H., Sera, T., Nagashima, G. T., Caramori, P. H., Andre, J., Marur, C. J., Soares, A. G. G., & Nagai, F. F. (2011). Microclima, fisiologia e produção de café em dois espaçamentos de plantio. In *VII Simpósio de Pesquisas dos Cafés do Brasil, Araxá, MG*. (In Portuguese.)

Mendes, I. C., Sousa, D. M. G., Reis Junior, F. B., & Lopes, A. A. C. (2018). Indicadores de qualidade biológica para manejo sustentável de solos arenosos. (In Portuguese.). *Boletim Informativo SBCS*, 44(1), 20–25.

Nab, C., & Maslin, M. (2020). Life cycle assessment synthesis of the carbon footprint of arabica coffee: Case study of Brazil and Vietnam conventional and sustainable coffee production and export to the United Kingdom. *Geo: Geography and Environment*, 11, 1–19. https://doi.org/10.1002/geo2.96

Naves, A. P., Silva, A. O., Barbosa, M. V., Pinto, F. A., Santos, J. V., Saggin Junior, O. J., Guimarães, P. T. G., & Carneiro, M. A. C. (2020). Biochemical parameters of an Oxisol submitted to high doses of gypsum in the coffee culture. *Coffee Science*, 15, e151756. https://doi.org/10.25186/.v15i.1756

Rena, A. B., & Maestri, M. (1986). *Fisiologia do cafeeiro*. In *Cultura do cafeeiro: Fatores que afetam a produtividade* (pp. 13–85). (In Portuguese.)). Associação Brasileira para Pesquisa da Potassa e do Fosfato.

Oliveira, G. C., Serafim, M. E., Carducci, C. E., Silva, B. M., SIlva, E. A., Barbosa, S. M., Silva, S. H. G., Melo, L. B. B., & Benevenute, P. A. N. (2019). Melhoria físico-hídrica do ambiente radicular de cafeeiro em condições de sequeiro e implicações no uso da água. (In Portuguese.). In E. C. Severiano, M. F. Moraes, & A. M. Paula (Eds.), *Tópicos em ciência do solo* (Vol. 10). Sociedade Brasileira de Ciência do Solo.

Oliveira, L. (2019). *A relação entre altas dosagens de gesso agrícola e a produtividade de cafeeiros adensados*. (In Portuguese.) Universidade de São Paulo Esalq/USP.

Pavan, M. A., Chaves, J. C. D., Siqueira, R., & Androcioli Filho, A. (1997). *Cultura do cafeeiro: O Sistema de plantio adensado e a melhoria da fertilidade do solo*. (In Portuguese.) International Plant Nutrition Institute.

Pedrosa, A. W. (2013). *Eficiência da adubação nitrogenada no consórcio entre cafeeiro e Brachiaria brizantha* (Doctoral dissertation, Universidade de São Paulo). (In Portuguese.) https://doi.org/10.11606/T.11.2013.tde-15032013-101246

Ramos, B. Z., Lima, J. M., Serafim, M. E., Coscione, A. R., Ferraz, R. M., Amorim, L. M., & Lopes, G. (2019). Ionic speciation in a dystrophic red latosol under coffee crop and high doses of gypsum. *Coffee Science*, 14(3), 281–290.

Rodrigues, R. N., Reis Junior, F. B., Lopes, A. A. C., Rocha, O. C., Guerra, A. F., Veiga, A. D., & Mendes, I. C. (2022). Soil enzymatic activity under coffee cultivation with different water regimes associated to liming and intercropped brachiaria. *Ciência Rural*, 52(3), e20200532. http://doi.org/10.1590/0103-8478cr20200532

Rosa, F. R. T. R. (2015). *Physical properties of a Oxisol under a densely coffee plantation system* (Master's thesis). (In Portuguese.). Agronomic Institute of Paraná.

Santos, L. L. (2012). *Soil organic carbon: Effect of landscape position and reforestation in succession to coffee in Western Minas Gerais* (Master's thesis). (In Portuguese.). Universidade Federal de Lavras.

Schmithals, A., & Kühn, N. (2017). To mulch or not to mulch? Effects of gravel mulch toppings on plant establishment and development in ornamental prairie plantings. *PLoS ONE*, 12(2), e0171533. https://doi.org/10.1371/journal.pone.0171533

Serafim, M. E., Oliveria, G. C., Oliveria, A. S., Lima, J. M., Guimarães, P. T. G., & Costa, J. C. (2011). Sistema conservacionista e de manejo intensivo do solo no cultivo de cafeeiros na região do alto São Francisco, MG: Um estudo de caso. *Bioscience Journal*, 27(6), 964–977.

SETAC. (1991). *A technical framework for life cycle assessment*. Society of Environmental Toxicology and Chemistry.

Severiano, E. C., Oliveira, G. C., Dias Junior, M. S., Curi, N., Costa, K. A. P., & Carducci, C. E. (2013). Preconsolidation pressure, soil water retention characteristics, and texture of latosols in the Brazilian Cerrado. *Soil Research*, 51, 193–202. https://doi.org/10.1071/SR12366

Silva, B. M., Oliveira, G. C., Serafim, M. E., Silva, E. A., Ferreira, M. M., Norton, L. D., & Curi, N. (2015). Critical soil moisture range for a coffee crop in an oxidic latosol as affected by soil management. *Soil & Tillage Research*, 154, 103–113. https://doi.org/10.1016/j.still.2015.06.013

Silva, E. A., Oliveira, G. C., Carducci, C. E., Silva, B. M., Oliveira, L. M., & Costa, J. C. (2013). Increasing doses of agricultural gypsum, aggregate stability and organic carbon in Cerrado Oxisol under coffee crop. *Revista Ciências Agrárias*, 56(1), 25–32. https://doi.org/10.4322/rca.2013.012

Silva, R. F., Severiano, E. C., Oliveira, G. C., Barbosa, S. M., Peixoto, D. S., Tassinari, D., Silva, B. M., Silva, S. H. G., Dias Júnior, M. S., & Figueiredo, T. D. A. F. R. (2021). Changes in soil profile hydraulic properties and porosity as affected by deep tillage soil preparation and *Brachiaria* grass intercropping in a recent coffee plantation on a naturally dense Inceptisol. *Soil & Tillage Research*, 213, 105127. https://doi.org/10.1016/j.still.2021.105127

Tavares, M. P. F., & Mourad, A. L. (2020). Coffee beverage preparation by different methods from an environmental perspective. *International Journal of Life Cycle Assessment*, 25(7), 1356–1367. https://doi.org/10.1007/s11367-019-01719-2

Tumwebaze, S. B., & Byakagaba, P. (2016). Soil organic carbon stocks under coffee agroforestry systems and coffee monoculture in Uganda. *Agriculture, Ecosystems & Environment*, 216, 188–193. https://doi.org/10.1016/j.agee.2015.09.037

Usva, K., Sinkko, T., Silvenius, F., Riipi, I., & Heusala, H. (2020). Carbon and water footprint of coffee consumed in Finland—Life cycle assessment. *International Journal of Life Cycle Assessment*, 25(10), 1976–1990. https://doi.org/10.1007/s11367-020-01799-5

Zinn, Y. L. (2016). *Soil profile characteristics and carbon sink capacity*. In L. Rattan (Ed.), *Encyclopedia of soil science* (3rd ed.). Taylor and Francis.

Zinn, Y. L., & Silva, C. A. (2021). Preservação e seqüestro de carbono orgânico em solos de cafezais no Brasil. In C. E. Carducci & G. C. Oliveira (Eds.), *Manejo do solo na cafeicultura: produtividade e sustentabilidade* (pp. 54–65). (In Portuguese.) UFLA

7

Soil Health in No-tillage Vegetable Production Systems—SPDH

Jucinei José Comin, Fabiane Vezzani, Monique Souza, Claudinei Kurtz, Álvaro Luiz Mafra, Paulo Emilio Lovato, Cledimar Rogério Lourenzi, and Arcângelo Loss

Chapter Overview

No-tillage (NT) vegetable production systems (referred to herein as SPDH, an acronym in Portuguese for "Sistema de Plantio Direto de Hortaliças") are an alternative developed in Brazil for transitioning from conventional to agroecological agriculture that combines productive efficiency and environmental quality. The SPDH movement was born in the late 1990s from the union of farmers and popular education agents, technical assistance personnel, and researchers from the vegetable-producing region of the state of Santa Catarina (South Brazil) as a response to severe land degradation, high production costs, and low prices of agricultural outputs. This chapter presents the fundamentals of SPDH and its relationship with soil health (SH). The mainstay of SPDH is the high production of diversified biomass in a system that uses technical and scientific principles to attain the health of cash crops. Soil management under SPDH meets the foundations of SH based on the addition of energy and matter through plant production in a process that is dependent on the system components. The process promotes multifunctional flows and the emergence of soil properties at higher levels of function. Long-term data from vegetable-producing areas with a history of advanced land degradation and those converted to SPDH point to an improvement in soil properties and cash crop yields, supporting the relevance of SPDH for SH in Brazilian horticultural production systems.

Soil Health Series: Volume 3 Soil Health and Sustainable Agriculture in Brazil, First Edition.
Edited by Ieda Carvalho Mendes and Maurício Roberto Cherubin.

Introduction

Brazil is the world's sixth largest producer of vegetables, with 23 million tons per year in an area of 900,000 ha. The most widely grown vegetables are potatoes, tomatoes, onions, watermelons, carrots, sweet potatoes, lettuce, and cabbage (Camargo Filho & Camargo, 2017). In the state of Santa Catarina, which is a pioneer in the development of NT vegetable production systems, the main vegetables grown are onions, squash, Andean potatoes, tomatoes, garlic, sweet potatoes, and cabbage, totaling around 32,000 ha in 2021 (Toresan et al., 2022).

Brazilian horticultural stands out for its intensive management, usually characterized by high production costs, intense labor demands, high yields, and profitability subject to market variations. It is highly susceptible to adverse weather conditions, which cause sharp fluctuations in the income of farming families. The conventional production system is used in most vegetable-producing regions of Brazil. It is characterized by intense soil tillage, a high need for irrigation water, and increased addition of inputs such as fertilizers and pesticides, usually in short cycles of crop growth (Andriolo, 2020).

SPDH is a Brazilian alternative for horticultural production systems to achieve soil health (SH) in the transition from conventional to agroecological agriculture. SPDH is an efficient form of production given adverse environmental conditions in horticultural farming, evidenced by the rugged terrain in many mountain-farming areas; intensified soil degradation, especially the loss of the topsoil due to the intensive conventional tillage (CT); high incidence of diseases, pests, and weeds, which are difficult to manage under monocropping; and losses in production associated with the occurrence of extreme weather events, especially periods of drought and concentrated rainfall. In Santa Catarina, SPDH is used in ~5000 ha, corresponding to about 10% of the horticultural production in the state (Epagri, 2022). The main crops are tomatoes, chayote, (hybrid) Tetsukabuto squash, cauliflower, broccoli, cabbage, onion, parsley, and passion fruit (Fayad et al., 2019).

The precepts in soil management of SPDH are conservationist and seek high production of diversified plant biomass and no soil tillage. Thus, based on high photosynthetic activity, it is possible to ensure permanent soil coverage, which reduces soil degradation by erosion. Reduced tillage combined with the action of the roots of cover crops (CCs) and cash crops as well as soil biota maintain soil structure and increase organic matter (OM) stability and soil fertility (Lima et al., 2017). The adoption of other management practices in the cultivation of vegetables, such as the use of organic residues, crop rotation, and CCs, makes it possible to increase biomass production and reduce dependence on inputs and energy from non-renewable sources due to the improvement of the ecological processes that increase resource use efficiency. These practices result in SH construction, which is closely related to its productive capacity and resilience (Norris & Congreves, 2018).

Soil health refers to a set of functional aspects provided by the soil that maintain the productive capacity in agriculture and provide ecosystem services involved in the dynamics of C, energy, water, and nutrients capable of sustaining plants, animals, and humans (Rodriguez et al., 2022). Although commonly used as a synonym for soil quality, SH focuses more on the biological dimension acting on soil functioning and its influence on the health and sustainability of the biosphere (Bünemann et al., 2018; Lehmann et al., 2020).

Soil functions involve supporting plant growth; biomass production, also in agriculture and forestry; storing, filtering, and transforming nutrients, substances, and water; detoxifying pollutants; regulating water movement; promoting the biodiversity pool (i.e., habitats, species, and genes); providing raw materials; acting as a C pool; generating a physical and cultural environment for humans and human activities; and acting as an archive of geological and archeological (cultural) heritage (European Commission, 2006; Guo, 2021; USDA-NRCS, 2022). Soil functions enable the soil to perform ecological processes such as nutrient cycling, decomposing residues, maintaining soil structure stability, regulating soil biological population, gas exchange, and water and solute movement. Soil functions are the integrated expression of its biological, physical, and chemical properties.

The construction of SH occurs through vital processes caused by the addition of plant biomass that influences the development of organisms (Chaparro et al., 2012; Finney et al., 2017). Consequently, plants and organisms act in nutrient cycling and in forming and stabilizing the soil's physical structure, providing environmental conditions for biochemical reactions. Thus, soil properties are constructed in a self-organizing process that is co-dependent on system components and triggered by plants. Plant diversity enhances this process by producing multifunctional flows and the emergence of soil properties at higher functioning levels (Anghinoni & Vezzani, 2021).

In short, the management of horticultural production systems under SPDH meets the conditions for achieving SH. Next, we briefly present the historical background and characterization of SPDH, followed by the results related to SH thus far in two experimental areas.

SPDH History and Characterization

SPDH emerged in the late 1990s in the Contestado region of Santa Catarina in Southern Brazil (Figure 7.1) from the union of farmers, popular education agents, and rural extension and research personnel. The system's first experimental results were obtained in 1998 at the Caçador Experimental Station, owned by the Santa Catarina State Agricultural Research and Rural Extension Agency (Empresa de Pesquisa Agropecuária e Extensão Rural de Santa Catarina, Epagri). The system sought to oppose the cultivation of vegetables based on the conventional model

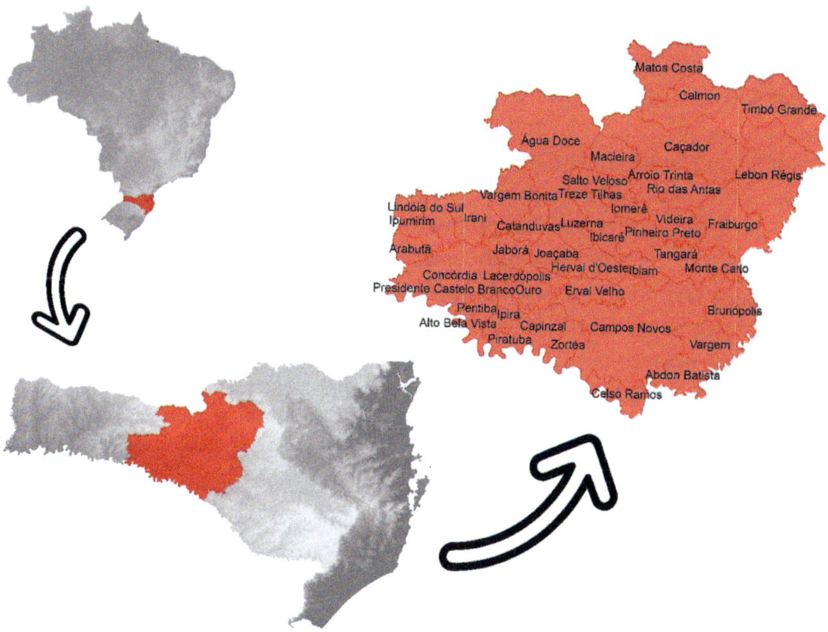

Figure 7.1 Map of Brazil highlighting the location of the state of Santa Catarina (in red). Source: https://pt.wikipedia.org/wiki/Santa_Catarina#/media/Ficheiro:Santa_Catarina_in_Brazil.svg. Daniel Heberle.

that is dominant in the region and to encourage minimum cultivation (CM) and a NT system on straw, focused on soil conservation. That coincided with the beginning of movements by farming families organized around the fight for the release from debts with a Brazilian bank, caused by the high cost of production due to the soaring demand for inputs to produce in degraded areas and the low prices of suppress commercialized tomatoes (Masson et al., 2019).

After successfully renegotiating the debts, the need to change the production system was evident. Action coordinators met and planned several activities that characterized the SPDH movement. It was defined by the installation of five "study fields" with tomatoes, which differed from the design of demonstrative units for being implemented on rural properties and conducted by the farmers themselves based on the technical and scientific principles of SPDH, frequency of technical visits, study trips to share experiences, and training courses and meetings in the "study fields" for the five farmers to share their findings with other tomato-producing farmers. It has a robust political-pedagogical nature to serve as a transition method from agriculture that is extremely dependent on inputs (soluble fertilizers and pesticides) to a new way of production based on agroecology (Masson et al., 2019), focusing on human and environmental well-being.

As of 2004, SPDH activities were concentrated in the municipalities of Ituporanga, Rio do Sul, Criciúma, and the metropolitan region of Florianópolis, in the state of Santa Catarina, with study crops of tomato, watermelon, hybrid squash, chayote, cauliflower, cabbage, broccoli, onion, parsley, lettuce, and passion fruit.

The SPDH is based on the high production of diversified biomass in the system, which uses the technical-scientific principles to achieve the health of cash crops, characterized by stress minimization, adequate nutrition, temperature, luminosity, water availability, pH, and salinity. The technical-scientific principles of SPDH have been constructed over time as follows: crop rotation with cultivated or spontaneous CCs and green manure; rotations with animal production; biomass addition (>dry matter) by intercropping or rotation/succession; soil tillage restricted to sowing or plant rows or planted seedbeds; management of spontaneous cover plants using knife-rollers and mowers to improve the system with plants adapted to local conditions and without harming yield of the cash crop, evolving to NT with live cover (an efficient strategy to eliminate herbicide use); and balanced plant nutrition management based on daily nutrient uptake rates (Mafra et al., 2019).

SPDH differs from the traditional vegetable and crop cultivation models due to the academic and practical qualifications of the stakeholders involved in the production system. It is based on substantially reducing or eliminating agrochemicals from production, aiming to promote soil, plant, and water health (Masson et al., 2019). Its fundamental aspect is the concept of plant health, based on the principle of integrated and co-dependent functioning of soil, plant, water, and atmosphere, where the soil is an open, self-organizing system and the basis for the generation of properties that reflect on the multifunctionality of relationships, promoting SH.

Plants producing abundant and diversified biomass through photosynthetic activity are the active protagonists in the structural quality, nutrient cycling, and deposition of organic residues that are the food source for the soil biota. Thus, plants cultivated in a planned manner in efficient management systems (Furey & Tilman, 2021; Liu et al., 2022) combined with soil organisms are the foundations of SH, promoting the ability to receive, store, and release energy, represented by organic compounds (Janzen, 2015).

SPDH aims to reduce soil, water, and nutrient losses by erosion, with increased soil surface protection (Santos et al., 2022). Live cover by plants and dry biomass on the soil reduce heating and thermal amplitude, improving the growth microclimate. Abundant biomass production can maintain and even increase soil OM content while promoting nutrient cycling (Roters et al., 2021). Roots and soil biota are active agents of structural improvement, providing aggregate stability, associated with continuity, abundance, and balanced pore distribution, improving water and aeration functions. Water use efficiency and irrigation savings in this system can be related to increased infiltration and soil water storage as well as reduced evaporation and runoff (Lima et al., 2017). The soil conservationist practices that makeup SPDH have the potential to increase fertilizer use efficiency, improve weed management, prevent the occurrence of diseases and pests, and reduce the emission of greenhouse gasses, in addition to reducing the need for labor, water, and fuel, which result in increased yields, reduced production costs, and a productive system that is less dependent on external inputs (Rodriguez et al., 2022).

Thus, improvement in SH leads to increased yields and production stability in horticultural crops, combined with enhanced ecosystem services, including nutrient supply, water production, C storage in the soil, and the maintenance of biodiversity, which is increasingly crucial in the face of climate change (Karlen et al., 2021).

In the following section, we present the results of two long-term experiments with onion (*Allium cepa* L.) under SPDH, aiming to show that this management system promotes SH as it positively affects soil biological, physical, and chemical properties, in addition to increasing onion yields. One experiment was conducted according to agroecological principles with a CC succession, and the other was based on agroecological transition with crop succession and rotation as well as the use of agrochemicals.

Long-term Adoption of SPDH Principles for Onion Production: A Case Study in Southern Brazil

The two experiments were carried out at the experimental station (belonging to Santa Catarina Agricultural Research and Rural Extension Agency, Empresa de Pesquisa Agropecuária e Extensão Rural de Santa Catarina—Epagri), located in Ituporanga, state of Santa Catarina (latitude, 27° 24′ 52″; longitude, 49° 36′ 9″; 475 m asl). The climate at the site is humid subtropical, oceanic with no dry season, with hot summers (Alvares et al., 2013). The soil in both experiments is a "Cambissolo Húmico" in the Brazilian Soil Classification System (Santos et al., 2018), Cambic Umbrisol in IUSS Working Group (2015), and Typic Humudept in Soil Taxonomy (Soil Survey Staff, 2014).

The experiment, managed according to agroecological principles, was installed in 2009 with the following treatments: a control in fallow under NT and CC in SPDH of black oat (*Avena strigosa* Schreb), rye (*Secale cereale* L.), oilseed radish (*Raphanus sativus* L.), oilseed radish + black oats, and oilseed radish + rye preceding onion cultivation (Figure 7.2). After the onion harvest (Figure 7.3), black mucuna [*Mucuna aterrima* (Piper & Tracy) Holland NA] is sown in every treatment, remaining until the sowing of the winter species each year. The

Figure 7.2 Winter cover crops grown in an agroecological no-tillage system (40 days after sowing).

Figure 7.3 Onion (*Allium cepa* L.) growing in an agroecological no-tillage system (60 days after seedling transplanting).

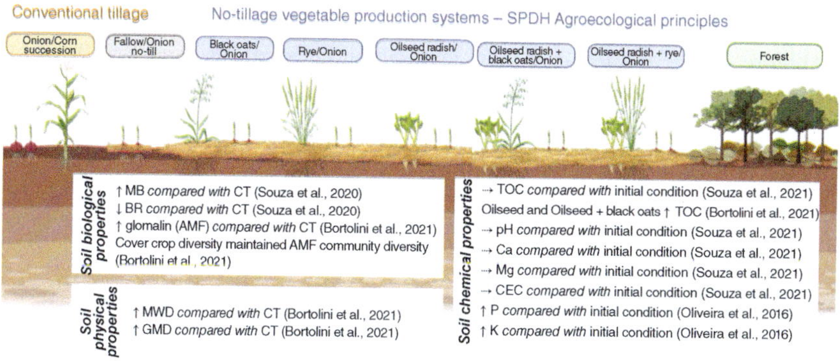

Figure 7.4 Schematic representation of the experiment based on agroecological principles with the main effects on soil health. ↑, increase; ↓, decrease; → constant. AMF, arbuscular mycorrhizal fungi; BR, basal respiration; CEC, cation exchange capacity; CT, conventional tillage; GMD, aggregate geometric mean diameter; MB, microbial biomass; MWD, aggregate mean weight diameter; TOC, total organic carbon. Illustration by Jéssica Pereira de Souza, using BioRender.com.

experimental design has randomized blocks with four replicates. Each experimental unit was 5 m by 5 m, totaling 25 m^2. Adjacent areas of crop succession onion and corn under CT and forest are analyzed for treatment comparison. Figure 7.4 summarizes the results from the experiment, with a schematic design of the treatments.

The winter species were broadcast on the soil surface every April, and then a grain seed drill went over the area twice to lightly incorporate the seeds into the soil. No fertilization, irrigation, or phytosanitary treatments were performed during the CC cycle. Every July, the winter species were shredded with a knife roller. The fertilizers were added before onion seedlings were transplanted by hand with Gafsa rock phosphate and poultry litter. As of the 2011 harvest, no natural phosphate was applied because P contents were interpreted as "very high" (CQFS-RS/SC, 2016). Hoeing was done yearly after seedling planting to reduce the number of weeds. Further details regarding this experiment can be found in Souza et al. (2021).

The experiment under agroecological transition was installed in April 2007 and has the following treatments under SPDH: onion–corn succession; commercial crop rotation (corn [*Zea mays* L.] and beans [*Phaseolus vulgaris* L.]) with winter CCs and biennial onion, crop rotation of corn–winter grasses and annual onion, summer legume succession and annual onion, rotation of summer and winter grasses and annual onion, summer–winter grass rotation and annual onion, summer legume–winter grass succession and annual onion, intercropped sunflower (*Helianthus annuus* L.) + black mucuna + millet [*Pennisetum glaucum* (L.) R. Brown] and annual onion, and onion–corn succession in SPDH until

2010 and in a CT system with plowing (once) and harrowing (twice) from 2011 on (Figure 7.5). A randomized block experimental design was used, with eight treatments and four replications. The area of each plot was 8.7 m^2, with seven rows of 30 onion plants each. Figure 7.6 shows a summary of the results of this experiment, with a schematic design of the treatments.

Fertilization occurred only in onion and corn cultivation periods and followed regional recommendations (CQFS-RS/SC, 2016). Irrigation was done via a

Figure 7.5 Cover crops and commercial crops grown in SPDH in agroecological transition.

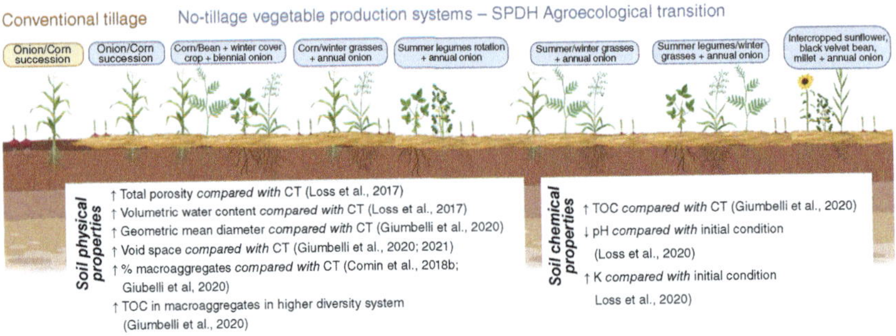

Figure 7.6 Schematic representation of the experiment based on agroecological transition with the main effects on soil health. ↑, increase; ↓, decrease; TOC, total organic carbon. Illustration by Jéssica Pereira de Souza, using BioRender.com.

Figure 7.7 Onion (Allium cepa L.) grown in SPDH in agroecological transition with sprinkler irrigation.

sprinkler system (Figure 7.7). Further details on this experiment can be found in Loss et al. (2020).

Long-term SPDH Adoption and Its Effects on Soil Health

Next, we present the main results of soil properties developed throughout SPDH that characterize the SH of both experiments.

Effects on Soil Biological Properties

After 8 years of SPDH under agroecological principles, oilseed radish, grown single or intercropped with black oat, increased total organic C (TOC) in soil compared with the other CCs (Bortolini et al., 2021). This partially affected microbial biomass, which was higher in most seasons in areas with CCs than in the adjacent area under CT (Souza et al., 2020). Basal soil respiration was higher in the area under CT than in areas under SPDH, which resulted in a higher metabolic quotient and indicated that soil biota could be under stress (Souza et al., 2020). The use of CCs in SPDH increased total glomalin by 84% and easily extractable glomalin by 60% in the surface soil layers compared with

CT, indicating better development of arbuscular mycorrhizal fungi (AMF) in the system with higher biomass addition (Bortolini et al., 2021).

During the onion cycle, the highest mycorrhizal colonization rates occurred in the intermediate stages and decreased at the end of the crop cycle. Similar behavior was observed in spontaneous vegetation (fallow under NT). The yearly presence of a non-mycotrophic crop reduced glomalin fractions in the soil. Genetic sequencing showed a predominance of the *Glomus* genus of AMF (Giovanetti, 2022). The diversity of AMF was affected by CCs and sampling times, which indicates that plants affect this fungal community. Oilseed radish, a non-mycorrhizal plant, reduced AMF richness, which was recovered by subsequent growth of mycotrophic crops (onion and black mucuna). In short, non-mycorrhizal crops did not reduce AMF diversity in systems that include mycorrhizae-forming species. That shows that using different CCs promotes the rapid recovery of soil microbial diversity.

Effects on Soil Physical Properties

The experiment conducted with agroecological principles showed similar values of mean weight diameter, and the largest aggregate geometric mean diameter (GMD) occurred in treatments with CCs, either single or intercropped, and even in fallow, compared with the native forest (data in Bortolini et al., 2021; Loss et al., 2015). That shows that SPDH with CCs efficiently improves and increases soil aggregation in an area previously managed under CT. The increase is due to physical processes associated with soil exploration by roots, which exert pressure on mineral particles and promote approximation during growth (Gracia & Rosolem, 2010; Guedes Filho et al., 2013), and chemical processes related to the exudation of compounds that have a binding effect on soil particles (Liu et al., 2005).

The similar effect between treatments with CCs and fallow for aggregation indexes indicates that SPDH with CCs promotes direct and indirect action of plant biomass on soil ecological processes. Soil cover by vegetation prevents raindrop impact and keeps soil moisture and temperature more uniform (Panachuki et al., 2011). Consequently, this provides ideal conditions for biological activity that will act directly and indirectly on the formation of macroaggregates.

Loss et al. (2017) evaluated total soil porosity (TSP) and volumetric water content in the agroecological transition experiment. They found that TSP was generally higher in the native forest ($0.67\,cm^3\,cm^{-3}$) and that CT had lower values ($0.47\,cm^3\,cm^{-3}$) in comparison to SPDH ($0.52\,cm^3\,cm^{-3}$) at the 0- to 5-cm soil layer just 6 months after installation. In SPDH, the black oat treatment showed the highest TSP values ($0.57\,cm^3\,cm^{-3}$), which supports the highest volumetric water content value ($0.48\,m^3\,m^{-3}$ at a tension of 0.6 kPa). The higher TSP values in SPDH compared with CT show that, due to the constant presence of roots, CCs acted on aggregate formation and stabilization as well as on the construction of biopores. Furthermore, vegetation cover helped soil biota, promoting higher TSP in the soil surface layer.

Soil volumetric water contents in treatments under SPDH were 0.33 $m^3\ m^{-3}$, whereas CT had 0.27 $m^3\ m^{-3}$ at the 10 cm layer at a tension of 60 kPa. The lower value in CT results from lower aggregation indexes, which reduce TSP and water content (Loss et al., 2017). On the other hand, higher aggregation indexes, TSP, and water content at tensions of 60 and 600 kPa in treatments under SPDH compared with those under CT indicate an improvement in soil physical condition with SPDH and the use of CCs. This combination favors soil via the root system and reduced soil tillage, preserving aggregates from breaking down and preventing decreases in porosity and water storage, as observed in CT.

In the same experiment conducted with agroecological transition, after 9 years of installation, the lowest GMD values were found in treatments with corn–onion succession, both in SPDH and CT, and the lowest value (2.551 mm) was observed in CT (Giumbelli et al., 2020). The other treatments with various CCs presented GMD values between 3.718 and 4.193 mm. The corn–onion succession treatments also had the smallest pores, determined by soil density (1.24 $Mg\ m^{-3}$ in SPDH and 1.27 $Mg\ m^{-3}$ in CT).

The data show that the absence of CCs in summer and winter results in lower biological activity in the root system and negatively affects soil physical properties. The low plant diversity of corn–onion succession leads to decreased root activity, which results in reduced release of root exudates (Lange et al., 2015). Thus, it is possible to infer low mechanical force exerted by root pressure and decreased cementing agents in the soil, which do not favor the stability of large aggregates. That was observed in the treatment under SPDH and especially in CT (Giumbelli et al., 2020), where plowing and harrowing accelerated OM oxidation and decreased aggregate stability in water.

Lower levels of TOC in the corn–onion succession treatments are also associated with lower plant diversity in these two treatments (Giumbelli et al., 2020). With lower TOC contents, aggregates are less stable in water, which favors the breakdown of large aggregates (macroaggregates) into smaller aggregates (microaggregates). This is especially evident on the soil surface layer, where succession treatments in both SPDH and CT had the lowest and the highest amounts of microaggregates, respectively (Comin et al., 2018a; Giumbelli et al., 2020).

The intercropping of sunflower, black mucuna, and millet associated with onion cultivation under SPDH provided higher GMD at 10–20 cm due to the higher addition of TOC in the soil via rhizodeposition, resulting in higher TOC contents in macroaggregates (Giumbelli et al., 2020). The diversity of root systems provided by intercropping can promote better use of water and nutrients. Intercropping promoted the best aggregation and the highest aggregate stability in depth compared with the other treatments (Giumbelli et al., 2020). The plant diversity of treatments with grasses and legumes in winter and summer associated with onion cultivation under SPDH also resulted in the lowest soil density values and, consequently, higher void spaces (Giumbelli et al., 2020, 2021).

Effects on Soil Chemical Properties

Soil pH, TOC, exchangeable Ca and Mg contents, and potential cation exchange capacity were evaluated at the 0- to 20-cm layer for 10 years in the experiment with agroecological principles. These chemical properties did not differ among CCs or evaluation years (Souza et al., 2021), indicating that SPDH maintained these characteristics stable over the years. The effective nutrient cycling during the experiment and the maintenance of TOC levels with the constant supply of plant residues helped to stabilize and maintain good conditions (CQFS-RS/SC, 2016) for crop development for 10 years. That emphasizes that plant development, which promotes biological processes in the system, is vital to maintain soil functioning regardless of the type of CC. However, even in soils under SPDH, when chemical fertilizers are used, there may be variations in chemical properties because there is increased interference in the soil, as reported by Loss et al. (2020) in the experiment with the agroecological transition. These authors evaluated soil chemical properties and found reductions in pH values after 7 years in every treatment compared with the initial soil conditions. They emphasized that soil pH decreased due to using fast-release N fertilizers, such as urea and ammonium nitrate, which release H^+ during nitrification processes.

In the experiment with agroecological principles, the main differences were found over time with the evaluations, with minor differences among treatments caused by fertilization with Gafsa rock phosphate until 2011 and then with poultry litter. Even with the addition of only poultry litter, soil P and K contents remained at levels considered "very high" (CQFS-RS/SC, 2016) due to slower mineralization of added nutrient sources and to nutrient cycling promoted by CCs (Oliveira et al., 2016). In the experiment with agroecological transition, Loss et al. (2020) did not find any differences in P contents at 0–5 cm among treatments. However, those with greater plant diversity in rotation/succession had the highest K contents.

Total organic C contents were constant over 10 years in the experiment with the agroecological principles (Souza et al., 2021) due to the use of CCs and the constant contribution of plant residues on the soil surface with no tillage. On the other hand, Loss et al. (2020) found the highest TOC contents in the experiment with agroecological transition with the succession of summer legumes/winter grasses and annual onion under SPDH and found the lowest TOC with the corn–onion succession under CT. Therefore, using CCs combined with reduced soil disturbance is a vital strategy to activate biological processes that result in adequate soil functioning and promote SH.

Effects on Onion Yield

From the start of the experiment with agroecological principles in 2009 until 2018, single and intercropped CCs (black oat, rye, and oilseed radish) had higher dry matter yields than those of the control treatment (Bortolini et al., 2021; Souza et al., 2021; Ventura et al., 2021). Dry matter yields of the winter CCs in single and intercropped cultivation ranged from 3 to 5 t ha^{-1}, with an average of 4.5 t ha^{-1}, whereas yields in fallow areas were on average 1.8 t ha^{-1} (Figure 7.8) (Bortolini et al., 2021; Souza et al., 2021; Ventura et al., 2021).

Residues of single and intercropped CC species in SPDH contributed to increasing onion yields over time, starting at 9.8 t ha^{-1} and reaching 15–24 t ha^{-1} 8 years later (Figure 7.9). These data are evidence of the action of plant biomass in promoting ecological processes of energy and C capture, residue decomposition, nutrient storage and cycling, soil structural stability, and biomass production that improve SH. Poaceae, such as oats and rye, absorb K and P from deeper soil layers and accumulate them in shoots. They decompose more slowly and gradually mineralize nutrients during the onion cycle. On the other hand, oilseed radish has a higher amount of N in tissues, which is quickly released into the soil, increasing the supply to succession crops (Oliveira et al., 2016).

However, one should highlight the importance of cultivating black mucuna after the onion harvest because it helps to increase dry matter production of summer

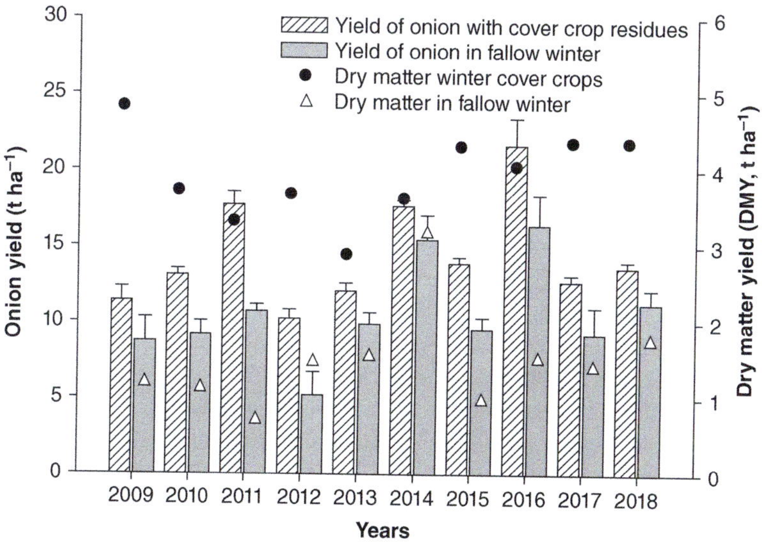

Figure 7.8 Spontaneous plant dry matter yield in winter fallow or in single or intercropped cover crops (*Avena strigosa* Schreb, Secale cereale L., and *Raphanus sativus* L.), and onion (*Allium cepa* L.) yield in areas in SPDH for 10 years.

Figure 7.9 Onion growing in an agroecological no-tillage system. Control plot with spontaneous vegetation and onion (bottom); plot with black oats (*Avena strigosa* Schreb), oilseed radish (*Raphanus sativus* L.), and onion (*Allium cepa* L.) (top).

CCs. As a result, an average of 4 t ha^{-1} of dry matter (Comin et al., 2018b) is added, serving as a strategy in agroecological management so that the SPDH keeps the soil covered throughout the entire cycle with different species, exerting more control pressure on the spontaneous vegetation in addition to raising and maintaining soil nutrient levels (Souza et al., 2021).

The lower onion yields in areas with spontaneous vegetation can be attributed to lower production and deposition of residues, less time on the soil surface; releasing lower amounts of nutrients; and increasing the losses of soil, water, and nutrients, reducing onion bulb yields by 2–6 t ha^{-1} (Oliveira et al., 2016).

In general, these results show that, even with the low use of inputs and no agrochemicals, the agroecological NT system for vegetables increased onion yields and reduced the emergence of spontaneous vegetation. That is due to the continuous use of CC species, which promoted development of the crop of interest.

The soil in the onion areas was managed according to the principles of agroecology, with an integrated approach to the system's biological, physical, and chemical aspects, involving the CCs to sustain constant onion yields in the long term instead of aiming for maximum yields in the short term. Thus, constant yields were obtained with growth potential for years to come, weed control without herbicides, weeding the area twice during the onion cycle, as well as maintaining or improving soil fertility.

The Practical Guide for Participative Evaluation of Soil Quality

SPDH is based on farmers' participatory action in building its practical bases, which aligns with the co-creative agroecology principle (Committee on World Food Security, 2019). The Practical Guide for Participative Evaluation of Soil Quality (PGPE-SH) methodology was developed by SPDH researchers to assess SH in horticultural systems, taking into account the active participation of farmers (Comin et al., 2016). PGPE-SH is an expeditious method using indicators that can be evaluated in the field. The process participants choose indicators from a suggested list, presented in Table 7.1. The participatory evaluation of SH presupposes the choice of indicators based on the participants' knowledge. Such indicators must be adapted to the local reality and suitable for each evaluation type.

After the participants have chosen the indicators and defined how to evaluate them, it is suggested to assign grades to each indicator on a scale that can vary from 1 to 10 or from 1 to 5. In the case of the range from 1 to 10, grade 1 corresponds to the worst condition for that indicator (i.e., an undesirable scenario); a score of 5 is assigned to a minimally acceptable condition, and a score of 10 indicates a desirable state (i.e., the ideal situation, similar to an area of native vegetation). Intermediate grades can be attributed to the indicators.

The PGPE-SH methodology recommends that assessments be carried out on the soil profile. However, the evaluation of each indicator has its peculiarities. For example, the OM indicator will commonly be assessed in a layer of 0–20 cm depth, whereas the evaluation of the rooting indicator will be in a layer of 0–40 cm. This layer may be variable according to the different scenarios that will be evaluated and the different soil classes.

The compaction and water infiltration indicators were grouped but can be evaluated separately. In water infiltration assessment, it is recommended to introduce a PVC pipe 20–30 cm long and 10–15 cm in diameter into the ground, pour a known volume of water, and mark the time that the water takes to infiltrate completely.

Assessing OM content associated with microbiological activity requires adding a given volume of H_2O_2 (10%) to a transparent container with a few grams of soil. Intensity and duration of effervescence are observed. When there is little or almost no effervescence, during a short time, soil biological activity and OM content are low. On the other hand, abundant and lasting effervescence will indicate soil with high biological activity and OM content.

Table 7.1 A suggestion of soil health indicators with their minimum, medium, and maximum values, and evaluation traits.

Indicators	Score	Traits
Color, odor, and organic matter content	1	Light colors, foul odor, very low organic matter content
	5	Darker color, no marked odor, low organic matter content
	10	Dark color, forest soil odor, high organic matter content
Depth of soil explored by roots (rooting)	1	Layer of explored soil does not exceed 10-cm depth
	5	Layer of explored soil between 10- and 20-cm depth
	10	Layer of explored soil beyond 40-cm depth
Soil structure	1	Pulverized soil, without visible aggregates, and/or with an abundance of compacted clods
	5	Soil with few visible aggregates, which break with light pressure, and/or with medium presence of compacted clods
	10	Soil with many aggregates, that maintain their shape after light pressure, and contain no compacted clods
Soil compaction and infiltration	1	Strongly compacted soil layer, with high resistance to knife penetration and little or no water infiltration
	5	Compacted layer, with medium resistance to knife penetration and slow water infiltration
	10	No compacted layer, knife tip penetrates soil easily, fast water infiltration
Erosion	1	Severe erosion, with grooves and channels
	5	Barely visible (laminar), erosion, runoff does not create furrows
	10	No visible signs of erosion
Soil moisture retention	1	No apparent moisture, soil dries fast
	5	Low moisture retention during prolonged drought
	10	High moisture retention even with prolonged drought
Soil biological activity (macrofauna)	1	No signs of earthworms and/or arthropods
	5	Presence of some earthworms and/or arthropods
	10	Abundance of earthworms and/or arthopods
State of plant remains and soil cover (litter)	1	Little or no plant residues, no signs of decomposition
	5	Thin litter layer, <50% soil cover
	10	Thick litter layer, plant residues in different stages of decomposition, over 90% soil cover

(Continued)

Table 7.1 (Continued)

Indicators	Score	Traits
Spontaneous plant diversity	1	Presence of a single species
	5	Presence of two to four species, with predominance of one species
	10	Presence of five or more species, with no predominance of one species

The abbreviation of each indicator is shown in parentheses.
Note. Adapted from Comin et al. (2020).

Many spontaneous plants are indicators of a specific soil condition, and farmers have extensive knowledge about this aspect that contributes to evaluating SH.

To facilitate assigning grades to the indicators, we suggest using a reference scenario, such as native vegetation located on the same type of soil and topography as the area of the system under evaluation. After attributing scores to the indicators, diagrams in the form of network, pie, radar, or umbrella are drawn up, as shown in the example in Figure 7.10. This type of representation allows easy visualization of the state of SH because the values closest to the outer side of the graph indicate that the system is nearer to the ideal condition.

The PGPE-SH contemplates the criteria for constituting a SH index (Doran & Parkin, 1994). It integrates biological, physical, and chemical properties, which

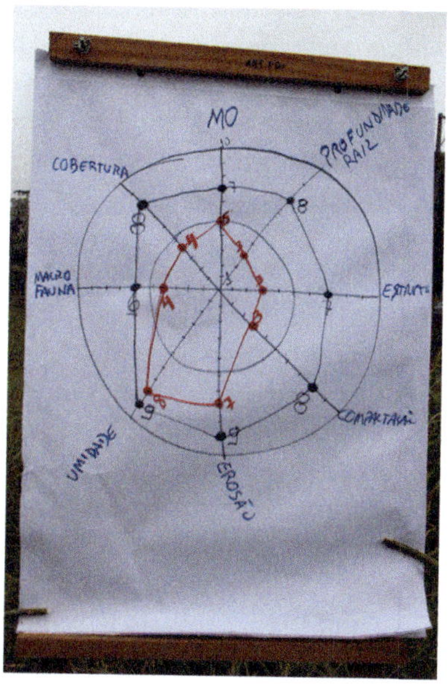

Figure 7.10 Soil quality evaluation diagram of plots in succession under minimum cultivation of black oat (*Avena strigosa* Schreb) and no-tillage of soybean [*Glycine max* (L.) Merr.] (red line) and plot under fallow and vegetated with plantain signalgrass [*Brachiaria plantaginea* (Link) R. Webster] (black line). Comin et al. (2020).

represent relevant processes for the soil functions, and has the sensitivity to detect changes resulting from management practices. Participatory assessments also help choose easy-to-use indicators, providing adequate analysis and interpretation, which is the most appropriate way to arrive at comprehensive soil assessments (Vezzani & Mielniczuk, 2009; Vezzani et al., 2019).

When evaluating SH, farmers consider the form of production, the use of the soil, and the consequences of their actions for the soil to adequately fulfill its functions and the implications of this on SH through the use of indicators that can be evaluated in the field (e.g., OM [color, odor, and content], rooting [soil depth explored by the roots], soil structure [dry strength of aggregates], compaction and infiltration, erosion, water retention [humidity, biological activity [macrofauna], straw [state of plant remains and soil cover], and spontaneous plants [diversity]). It is possible to distinguish, in agricultural properties, in different production systems, the adverse effects or positive effects of management practices on SH. In addition, assessments can be carried out repeatedly, allowing soils to be monitored during the cultivation system or even during the transition from conventional to agroecological agriculture, in which different soil management practices are introduced over time (Vezzani et al., 2019). While assessing the SH of their production systems, the PGPE-SH methodology encourages farmers to question their realities and needs.

The following example appears in the PGPE-SH manual (Comin et al., 2020). It refers to a plot in Oxisol (Latossolo Vermelho distroférrico, in Brazilian Soil Classification System) managed under CM with the incorporation of black oat seed with light harrowing and NT drilling of soybean [*Glycine max* (L.) Merr.], with use of herbicides to desiccate black oat and weeds throughout the soybean cycle, in addition to other pesticides for crop management. The contrasting plot of land evaluated, beside the evaluated plot, was in fallow, with a predominance of plantain signalgrass [*Brachiaria plantaginea* (Link) R. Webster]. It has a condition similar to the management adopted in the SPDH because this grass has been used as a summer spontaneous cover plant in the SPDH for production of plants from the Brassicaceae family, (hybrid) Tetsukabuto squash, dragon fruit [*Hylocereus undatus* (Haw.) Britton & Rose], and passion fruit (*Passiflora edulis* Sims). The diagram resulting from the evaluation is shown in Figure 7.10. In the plot managed with black oats and soybean succession, only the erosion and moisture retention indicators had values above 5, a condition higher than the minimum acceptable value. In the plot with plantain signalgrass, the values were between 7 and 9 for all indicators, being closer to the outer side of the diagram, demonstrating better SH than the plot with black oats and soybean succession.

Figure 7.11 shows the presence of a non-compacted surface layer, reaching a maximum 10 cm depth, in which fine earth and few aggregates are present, with the original rounded shape typical in such soil. In the superficial layer, revolved

Figure 7.11 View of the soil profile from a plot managed in succession under minimum cultivation of black oat (*Avena strigosa* Schreb) and no-tillage of soybean [*Glycine max* (L.) Merr.]. Comin et al. (2020).

Figure 7.12 Detailed view of the soil profile from a plot managed in succession under minimum cultivation of black oats (*Avena strigosa* Schreb) and no-tillage of soybean [*Glycine max* (L.) Merr.]. Note the effect of agricultural machine tires in the center of the figure. Comin et al. (2020).

Figure 7.13 Soil profile view of land under fallow and vegetated with plantain signalgrass [*Brachiaria plantaginea* (Link) R. Webster]. Comin et al. (2020).

Figure 7.14 Detailed view of the soil profile from a plot under fallow and vegetated with plantain signalgrass [*Brachiaria plantaginea* (Link) R. Webster], with the distribution of roots reaching a depth over 50 cm. Comin et al. (2020).

with a light harrow to incorporate black oat seeds, oat roots occur at the beginning of the cycle. Between depths of 5 and 35 cm, there is a compacted layer with a massive structure and very little or no macropores and roots. The details of this compacted zone are presented in Figure 7.12. In the image of the cultural profile of the land under fallow and vegetated with plantain signalgrass (Figure 7.13), a well-formed and aggregated structure is visible, associated with an adequate distribution of roots, which reach ~55 cm in depth (Figure 7.14).

The second edition of the PGPE-SH (Comin et al., 2020) includes further examples of SH evaluation and an item for planning new actions based on the evaluations. Currently, the methodology is widely used due to its easy application and the relevance of the information provided (Valani et al., 2020). The PGPE-SH is meant to be a tool to help farmers to make decisions on soil management, aiming to promote SH.

Final Remarks

High yields of diversified biomass in production systems that follow the SPDH's technical-scientific principles were evidenced in our experiments. That was clear in the SPDH system with CCs in both experiments. Plant species richness activates processes that increase microbial biomass and activity and promote mycorrhizal fungal diversity. Different plant species, which affected the associated soil biota, promoted the formation and stabilization of soil aggregates and pores, increasing soil C stock. Chemical properties improved over time as a result of this self-organizing process. Developing integrated properties and multifunctional flows generated conditions that increased plant biomass production and onion yields. Increased plant production and diversity are linked to changes in biological diversity in bulk soil and around roots. Knowing how soil biota, modulated by different CCs, affect cash crop growth, development, and health is a promising, even necessary, avenue of research.

The agronomic practices that makeup SPDH are under the management principles to achieve SH presented by the USDA-NRCS (2022): (a) maximize the presence of living roots (i.e., the constant presence of growing plants), (b) minimize disturbance (i.e., no-tillage), (c) maximize soil cover (i.e., use of CCs), and (d) maximize biodiversity (i.e., increase crop species richness).

We emphasize the agroecological context in which SPDH is inserted. Principles of agroecology support practices that improve biological processes through plant diversity and the absence of soil disturbance (Wezel et al., 2014). Considering the human component as a co-creator of systems (FAO, 2018) that enhance the natural processes occurring in productive ecosystems is vital. Therefore, a call for diversity within an agroecological context is fully aligned with the assumptions of SH.

The results presented for onions in the state of Santa Catarina show the potential for other vegetable crops and use in other Brazilian regions. Based on the expertise of SPDH pioneers, joint actions of farmers and rural extension agents are increasing in Brazil, as exemplified by several research areas under SPDH on farms in Paraná state. Some specific measures are needed, such as developing farm equipment adapted for this management system and ensuring a supply of healthy, well-developed cash crop plantlets or seedlings. We emphasize that advances in this system result, and will always result, from the collaborative work of several agricultural research, extension, and education institutions, along with communities and social organizations.

Acknowledgments

The study was funded by the National Council for Scientific and Technological Development (CNPq), the Coordination for the Improvement of Higher Education Personnel (CAPES), the Research and Innovation Support Foundation of the State of Santa Catarina (FAPESC), the Ministry of Science, Technology and Innovation (MCTI), the Ministry of Agriculture, Livestock and Food Supply (MAPA), the Ministry of Agrarian Development (MDA), the Ministry of Fisheries and Aquaculture (MPA), the Ministry of Education (MEC) and the Agrisus Foundation. We thank the Agricultural Research and Rural Extension Company of Santa Catarina (EPAGRI) for the lasting and fruitful partnership.

References

Alvares, C. A., Stape, J. L., Sentelhas, P. C., Gonçalves, J. D. M., & Sparovek, G. (2013). Köppen's climate classification map for Brazil. *Meteorologische Zeitschrift*, 22(6), 711–728. https://doi.org/10.1127/0941-2948/2013/0507

Andriolo, J. L. (2020). *Olericultura geral*. Universidade Federal de Santa Maria.

Anghinoni, I., & Vezzani, F. M. (2021). Systemic soil fertility as product of system self-organization resulting from management. *Revista Brasileira de Ciência do Solo*, 45, e0210090. https://doi.org/10.36783/18069657rbcs20210090

Bortolini, J. G., Soares, C. R. F. S., Muller, M. J., Ferreira, G. W., Meyer, E., Vieira, C. V., Souza, M., Kurtz, C., Lourenzi, C. R., Lovato, P. E., Loss, A., & Comin, J. J. (2021). Soil carbon, glomalin, and aggregation in onion crop under no-tillage with cover crops or conventional tillage systems for eight years. *Journal of Agricultural Studies*, 9, 130–150. https://doi.org/10.5296/jas.v9i2.18196

Bünemann, E. K., Bongiorno, G., Bai, Z., Creamer, R. E., Deyn, G. D., de Goede, R., Fleskens, L., Geissen, V., Kuyper, T. W., Mäder, P., Pulleman, M., Sukkel, W., van

Groenigen, J. W., & Brussaard, L. (2018). Soil quality: A critical review. *Soil Biology and Biochemistry*, 120, 105–125. https://doi.org/10.1016/j.soilbio.2018.01.030

Camargo Filho, W. P., & Camargo, F. P. (2017). A quick review of the production and commercialization of the main vegetables in Brazil and the world from 1970 to 2015. *Horticultura Brasileira*, 35, 160–166. https://doi.org/10.1590/S0102-053620170202

Chaparro, J. M., Sheflin, A. M., Manter, D. K., & Vivanco, J. M. (2012). Manipulating the soil microbiome to increase soil health and plant fertility. *Biology and Fertility of Soils*, 48, 489–499. https://doi.org/10.1007/s00374-012-0691-4

Comin, J. J., Fayad, J. A., Kurtz, C., Mafra, Á. L., & Curmi, P. (2016). *Guia prático de avaliação participativa da qualidade do solo em Sistema de Plantio Direto de Hortaliças (SPDH)*. UFSC. https://bit.ly/359ESNE

Comin, J. J., Fayad, J. A., Kurtz, C., Mafra, A. L., & Curmi, P. (2020). *Guia prático de avaliação participativa da qualidade do solo em Sistema de Plantio Direto de Hortaliças (SPDH)* (2nd ed.). UFSC. https://www.researchgate.net/publication/372500061

Comin, J. J., Ferreira, L. B., Santos, L. H., Koucher, L. P., Machado, L. N., dos Santos Junior, E., Mafra, A. L., Claudinei Kurtz, C., Souza, M., Brunetto, G., & Loss, A. (2018a). Carbon and nitrogen contents and aggregation index of soil cultivated with onion for seven years using crop successions and rotations. *Soil and Tillage Research*, 184, 195–202. https://doi.org/10.1016/j.still.2018.08.002

Comin, J. J., Vilanova, C. C., Kurtz, C., Muller Junior, V., Ventura, B. S., Dos Reis, M. S., Brunetto, G., Lovato, P. E., & Souza, M. (2018b). Avaliação fitossociológica de plantas invasoras em cultivo de cebola sob sistema plantio direto sem uso de agrotóxicos. *Revista de la Facultad de Agronomía*, 117, 197–206. https://revistas.unlp.edu.ar/revagro/article/view/7335

Comissão de química e fertilidade do solo RS/SC (CQFS-RS/SC). (2016). *Manual de adubação e calagem para os Estados do Rio Grande do Sul e de Santa Catarina* ((10th ed.). ed.). SBCS - Núcleo Regional Sul/UFRGS.

Committee on World Food Security. (2019). *Agroecological and other innovative approaches for sustainable agriculture and food systems that enhance food security and nutrition*. FAO.

Doran, J. W., & Parkin, T. B. (1994). Defining and assessing soil quality. In J. W. Doran, D. C. Coleman, D. F. Bezdicek, & B. A. Stewart (Eds.), *Defining soil quality for a sustainable environment* (pp. 1–20). SSSA.

Epagri. (2022). *Sistema sustentável de produção de hortaliças desenvolvido pela Epagri prova valor nas chuvas de novembro*. Empresa de Pesquisa Agropecuária e Extensão Rural de Santa Catarina. https://blog.epagri.sc.gov.br/index.php/sistema-sustentavel-de-producao-de-hortalicas-desenvolvido-pela-epagri-prova-valor-nas-chuvas-de-novembro

European Comission. (2006). *Thematic strategy for soil protection (COM 2006, 231 final).* https://eur-lex.europa.eu/EN/legal-content/summary/thematic-strategy-for-soil-protection.html

FAO. (2018). *The 10 elements of agroecology: Guiding the transition to sustainable food and agricultural systems.* https://www.fao.org/3/i9037en/i9037en.pdf

Fayad, J. A., Arl, V., Comin, J. J., Mafra, A. L., & Marchesi, D. R. (2019). *Sistema de Plantio Direto de Hortaliças: método de transição para um novo modo de produção.* Expressão Popular.

Finney, D. M., Buyer, J. S., & Kaye, J. P. (2017). Living cover crops have immediate impacts on soil microbial community structure and function. *Journal of Soil and Water Conservation*, 72, 361–373. https://doi.org/10.2489/jswc.72.4.361

Furey, G. N., & Tilman, D. (2021). Plant biodiversity and the regeneration of soil fertility. *Proceedings of the National Academy of Sciences*, 118(49), e2111321118. https://doi.org/10.1073/pnas.2111321118

Giovanetti, L. K. (2022). *Diversidade de plantas espontâneas e de fungos micorrízicos arbusculares em sistema de plantio direto com cebola e culturas de cobertura micotróficas e não micotróficas.* [Doctoral dissertation]. Universidade Federal de Santa Catarina.

Giumbelli, L. D., Loss, A., Kurtz, C., Mafra, Á. L., Piccolo, M. C., Torres, J. L. R., Lourenzi, C. L., Brunetto, G., & Comin, J. J. (2021). Combinations of plant species for rotation with onion crops: Effects on the light fraction, carbon, and nitrogen contents in granulometric fractions of the soil organic matter. *Journal of Agricultural Studies*, 9, 202–230. https://doi.org/10.5296/jas.v9i1.17930

Giumbelli, L. D., Loss, A., Ventura, B. S., Santos Junior, E., Almeida, J., de Cássia Piccolo, M., Mafra, A. L., Kurtz, C., Brunetto, G., & Comin, J. J. (2020). Aggregation index, carbon, nitrogen, and natural abundance of 13C and 15N in soil aggregates and bulk soil cultivated with onion under crop successions and rotations. *Soil Research*, 58, 622–635. https://doi.org/10.1071/SR19346

Gracia, R. A., & Rosolem, C. A. (2010). Agregados em um Latossolo sob sistema plantio direto e rotação de culturas. *Pesquisa Agropecuária Brasileira*, 45, 1489–1498. https://doi.org/10.1590/S0100-204X2010001200021

Guedes Filho, O., Silva, A. P., Giarola, N. F. B., & Tormena, C. A. (2013). Structural properties of the soil seedbed submitted to mechanical and biological chiseling under no-tillage. *Geoderma*, 204/205, 94–101. https://doi.org/10.1016/j.geoderma.2013.04.017

Guo, M. (2021). Soil health assessment and management: Recent development in science and practices. *Soil Systems*, 5, 61. https://doi.org/10.3390/soilsystems5040061

IUSS Working Group. (2015). *World Reference Base for Soil Resources (WRB), sistema universal reconhecido pela.* International Union of Soil Science and FAO. https://www.fao.org/3/i3794en/I3794en.pdf

Janzen, H. H. (2015). Beyond carbon sequestration: Soil as conduit of solar energy. *European Journal of Soil Science*, 66, 19–32. https://doi.org/10.1111/ejss.12194

Karlen, D. L., De, M., McDaniel, M. D., & Stott, D. E. (2021). Evolution of the soil health movement. In L. K. Douglas, D. E. Stott, & M. M. Mikha (Eds.), *Approaches to soil health analysis* (pp. 21–48). SSSA.

Lange, M., Eisenhauer, N., Sierra, C. A., Bessler, H., Engels, C., Griffiths, R. I., Mellado-Vazquez, P. G., Malik, A. A., Roy, J., Scheu, S., Steinbeiss, S., Thomson, B. C., Trumbore, S. E., & Gleixner, G. (2015). Plant diversity increases soil microbial activity and soil carbon storage. *Nature Communications*, 6(6707). https://doi.org/10.1038/ncomms7707

Lehmann, J., Bossio, D. A., Kögel-Knaber, I., & Rillig, M. C. (2020). The concept and future prospects of soil health. *Nature Reviews Earth & Environment*, 1(10), 544–553. https://doi.org10.1038/s43017-020-0080-8

Lima, C. E. P., Madeira, N. R., Silva, J., Fontenelle, M. R., Castro e Melo, R. A., & Guedes, I. M. R. (2017). *Benefícios da adoção do sistema de plantio direto de hortaliças*. Embrapa Hortaliças.

Liu, A., Ma, B. L., & Bomke, A. A. (2005). Effects of cover crops on soil aggregate stability, total organic carbon, and polysaccharides. *Soil Science Society of America Journal*, 69, 2041–2048. http://doi.org/10.2136/sssaj2005.0032

Liu, C., Bonilla, D. P., Coulter, J. A., Kutcher, H. R., Beckie, H. J., Wang, L., Floch, J. B., Hamel, C., Siddique, K. H. M., Li, L., & Gan, Y. (2022). Diversifying crop rotations enhances agroecosystem services and resilience. *Advances in Agronomy*, 173, 299–335. http://doi.org/10.1016/bs.agron.2022.02.007

Loss, A., Basso, A., Oliveira, B. S., Koucher, L. P., Oliveira, R. A., Kurtz, C., Lovato, P. E., Curmi, P., Brunetto, G., & Comin, J. J. (2015). Carbono orgânico total e agregação do solo em sistema de plantio direto agroecológico e convencional de cebola. *Revista Brasileira de Ciência do Solo*, 39, 1212–1224. http://doi.org/10.1590/01000683rbcs20140718

Loss, A., Ferreira, L. B., Gonzatto, R., Giumbelli, L. D., Mafra, A. L., & Goedel, A. (2020). Efeito da sucessão ou rotação de culturas sobre a fertilidade do solo após sete anos de cultivo com cebola. *Brazilian Journal of Development*, 6, 16587–16606. http://doi.org/10.34117/bjdv6n3-507

Loss, A., Schmitz, D., Santos Junior, E., Kurtz, C., da Veiga, M., Kurtz, C., & Comin, J. J. (2017). Atributos físicos do solo sob sistemas de plantio direto e preparo convencional em cultivo de cebola. *Revista Colombiana de Ciências Horticolas*, 11, 105–113. https://doi.org/10.17584/rcch.2017v11i1.6144

Mafra, A. L., Comin, J. J., Lana, M. A., Bittencourt, H. v. H., Lovato, P. E., & do Prado Wildner, L. (2019). Iniciando o sistema de plantio direto de hortaliças: adequações do solo e práticas de cultivo. In J. A. Fayad, V. Arl, J. J. Comin, A. L. Mafra, & D. R. Marchesi (Eds.), *Sistema de Plantio Direto de Hortaliças: Método de transição para um novo modo de produção* (1st ed., pp. 215–226). Expressão Popular.

Masson, I., Arl, V., & Wuerges, E. W. (2019). Trajetória, concepção metodológica e desafios estratégicos junto ao sistema de plantio direto de hortaliças (SPDH). In A. Fayad, V. Arl, J. J. Comin, A. L. Mafra, & D. R. Marchesi (Eds.), *Sistema de Plantio Direto de Hortaliças: Método de transição para um novo modo de produção* (1st ed., pp. 25–38). Expressão Popular.

Norris, C. E., & Congreves, K. A. (2018). Alternative management practices improve soil health indices in intensive vegetable cropping systems: A review. *Frontiers in Environmental Science*, 6, 50. https://doi.org/10.3389/fenvs.2018.00050

Oliveira, R. A., Brunetto, G., Loss, A., Gatiboni, L. C., Kurtz, C., Müller Júnior, V., Lovato, P. E., Oliveira, B. S., Souza, M., & Comin, J. J. (2016). Cover crops effects on soil chemical properties and onion yield. *Revista Brasileira de Ciência do Solo*, 40, 1–17. https://doi.org/10.1590/18069657rbcs20150099

Panachuki, E., Bertol, I., Alves Sobrinho, T., Oliveira, P. T. S., & Rodrigues, D. B. B. (2011). Perdas de solo e de água e infiltração de água em Latossolo Vermelho sob sistemas de manejo. *Revista Brasileira de Ciência do Solo*, 35, 1777–1786. https://doi.org/10.1590/S0100-06832011000500032

Rodriguez, B. C., Zuazo, V. H. D., Rodriguez, M. S., Tejero, I. F. G., Ruiz, B. G., & Tavira, S. Q. (2022). Conservation agriculture as a sustainable system for soil health: A review. *Soil Systems*, (6), 87. https://doi.org/10.3390/soilsystems6040087

Roters, D. F., Mafra, A. L., Cunha, F. F., Macedo, R. P., Silveira, C. P. L., & Kruker, G. (2021). Carbon increase and soil physical improvement in an onion no-tillage system. *Revista Brasileira de Ciências Agrárias*, 16, 1–8. https://doi.org/10.5039/agraria.v16i2a9017

Santos, H. G., Jacomine, P. K. T., Anjos, L. H. C., Oliveira, V. A., Lumbreras, J. F., Coelho, M. R., Almeida, J. A., Araujo Filho, J. C., Oliveira, J. B., & Cunha, T. B. (2018). *Sistema Brasileiro de Classificação de Solos* (pp. 5). Embrapa.

Santos, V. P., Bertol, I., Santos, A. P., Kurtz, C., Wolschick, N. H., Bagio, B., Wroblescki, F. A., & Prazeres, M. S. (2022). Erosão hídrica no cultivo da cebola influenciada pelo manejo do solo e cobertura por resíduo cultural. *Ciencia del Suelo*, 40, 185–195.

Soil Survey Staff. (2014). *Keys to soil taxonomy* (12th ed.). USDA-Natural Resources Conservation Service.

Souza, M., Müller Júnior, V., Kurtz, C., Ventura, B. S., Lourenzi, C. L., Lazzari, C. J. R., Ferreira, G. W., Brunetto, G., Loss, A., & Comin, J. J. (2021). Soil chemical properties and yield of onion crops grown for eight years under no-tillage system with cover crops. *Soil and Tillage Research*, 208, 104897. https://doi.org/10.1016/j.still.2020.104897

Souza, M., Vargas, M. M. V., Ventura, B. S., Muller Junior, V., Soares, C. R. F. S., Kurtz, C., Comin, J. J., & Lovato, P. E. (2020). Microbial activity in soil with onion grown in a no-tillage system with single or intercropped cover crops. *Ciência Rural*, 50, 12. https://doi.org/10.1590/0103-8478cr20190849

Toresan, L., Padrão, G. A., Goulart Junior, R., Alves, J. R., & Mondardo, M. (2022). *Indicadores de desempenho da agropecuária e do agronegócio de Santa Catarina: 2020 e 2021*. Epagri.

USDA Natural Resources Conservation Service (USDA-NRCS). (2022). *Soil health*. https://www.nrcs.usda.gov/wps/portal/nrcs/main/soils/health

Valani, G. P., Vezzani, F. M., & Cavalieri-Polizeli, K. M. V. (2020). Soil quality: Evaluation of on-farm assessments in relation to analytical index. *Soil and Tillage Research*, 198, 104565. https://doi.org/10.1016/j.still.2019.104565

Ventura, B. S., Meyer, E., Souza, A. M., Vieira, A. S., Scarsanella, J. A., Comin, J. J., & Lovato, P. E. (2021). Soil phosphorus availability and uptake by mycorrhizal and non-mycorrhizal plants in an onion no-tillage system. *Ciência Rural*, 51, 1. https://doi.org/10.1590/0103-8478cr20200740

Vezzani, F. M., Ferreira, G. W., Souza, M., & Comin, J. J. (2019). Conceitos, métodos de avaliação participative e o SPDH como promotor de qualidade do solo H. In A. Fayad, V. Arl, J. J. Comin, A. L. Mafra, & D. R. Marchesi (Eds.), *Sistema de Plantio Direto de Hortaliças: Método de transição para um novo modo de produção* (1st ed., pp. 105–123). Expressão Popular.

Vezzani, F. M., & Mielniczuk, J. (2009). Uma visão sobre qualidade do solo. *Revista Brasileira de Ciência do Solo*, 33, 743–755. https://doi.org/10.1007/s13593-013-0180-7

Wezel, A., Casagrande, M., Celette, F., Vian, J.-F., Ferrer, A., & Peigné, J. (2014). Agroecological practices for sustainable agriculture: A review. *Agronomy for Sustainable Development*, 34, 1–20. https://doi.org/10.1007/s13593-013-0180-7

8

Family Farming and Conservation Agriculture: Lessons from a Long-term Experiment on a Sandy Soil in Southwestern Amazonia, Brazil

Falberni de Souza Costa, Claudenor Pinho de Sá, Deborah Pinheiro Dick, and Ieda Carvalho Mendes

Chapter Overview

Understanding the historical development of tropical agricultural activities by family farmers in the Amazon region is a challenge. The human, cultural, economic, social, and environmental aspects represent a challenge that surpasses and transcends the mere adoption of technical innovations. This chapter presents a case study on family farming and conservation agriculture (CA) in a long-term experimental and participatory study in the Southwestern Brazilian Amazon. Since 2006, agricultural systems of conventional tillage (CT), no-tillage (NT), permanent soil cover crops (legume species [L], grass [G], or intercropped legume and grass [LG]), diversification of cash and cover crops through rotation and/or intercropping, and soil liming (Lm) and fertilization (phosphorus fertilizer [P]) with or without the use of fire were established on a sandy Ultisol. Conventional and NT systems in various combinations with LG, P, and Lm were compared with the traditional regional slash-and-burn (SB) agriculture. Thirteen years after the land-use conversion, the mean accumulated yield of cassava (*Manihot esculenta* Crantz) together with maize (*Zea mays* L.) increased from 6.5 Mg ha^{-1} (CT-SB) to 21 Mg ha^{-1} (NT-LGPLm) under CA (i.e., an increase of 223%), along with additional income from cassava, maize, and dwarf pigeon pea (*Cajanus cajan* L. Mill sp.) due to the greater diversification. In spite of the significant increases in crop productivity and family income as a result of the adoption of CA, the positive impact on the health of the sandy soils of the region of Juruá during the first 16 years was limited. In the chemical analysis of soil organic matter (SOM), the molecular proxies and isotopic indices showed a relative enrichment in microbial products under NT, whereas under CT a greater mineralization of

Soil Health Series: Volume 3 Soil Health and Sustainable Agriculture in Brazil, First Edition.
Edited by Ieda Carvalho Mendes and Maurício Roberto Cherubin.

endogenous SOM was observed after CA introduction. After 16 years, the different cropping systems had a limited impact on β-glucosidase (GLU) and arylsulfatase (ARYL) activity levels. The absence of significant effects on SOM and soil enzyme activity levels in the long-term field experiment of Mâncio Lima highlights the difficulties of changing the soil biological functioning in sandy soils in a region with mean precipitation levels of >2000 mm and limited dry matter input from cover crops.

Introduction

Efforts to improve the conditions of traditional agriculture for family farmers in Amazonia by means of politics, scientific research, and technological solutions have been limited. Therefore, it is important to reassess the economic growth of the Amazonia region over time (Gomes & Vergolino, 1997; Kato et al., 2020; Rocha Neto, 2021). One aspect that should be taken into consideration is agricultural intensification in already deforested areas (Marin et al., 2022).

Slash-and-burn agriculture is still the most common form of land use, regardless of the economic resources available (Figure 8.1). Family farming, according to

Figure 8.1 Slashing and burning of native forest in the Amazon region of Brazil. Photograph: F. S. Costa (2011).

Garner and O Campos (2014), is "a means of organizing agricultural, forestry, fisheries, pastoral and aquaculture production which is managed and operated by a family and predominantly reliant on family labor: both women's and men's. Also, the family and the farm are linked, coevolve and combine economic, environmental, reproductive, social and cultural functions."

Despite agricultural innovation, there are families who live in extreme poverty and are vulnerable in aspects such as basic health; education; and social, economic, and environmental conditions. Any innovation implemented on family farms in Amazonia should take all these aspects into account. Over generations, the smallholder family farmers in the Amazon region of Brazil and other regions of Latin America have learned how to exploit the resources of nature and to create their own innovations. Among other knowledge areas, they empirically learned about soil fertility and short-term amelioration by fire, medicinal use of forest plants, and weather forecasting.

The Amazonia biome consists of vigorous trees, rivers, wild animals, and, of course, humans (Toledo et al., 2017), which, from a holistic point of view, are connected with each other by processes, phenomena, dynamics, and interrelationships. Irrespective of the scale of an area and its use in agricultural activities, a low-fertility soil cannot sustain intensive agriculture, whereas management alternatives (e.g., SOM input, mulching, and NT) improve soil fertility (Moura et al., 2016) and may thus indirectly contribute to reducing deforestation of native forests.

In view of the increasing world demand for food, feed, and fuel, how can agricultural activities be performed successfully mitigating climate change or at least not aggravating it? Answers to the questions of how to interrupt or break the cycle of crop production causing intensification of climate change with extreme events that affect crop production must be found. One possibility is to reduce the effect of crop production on climate change by reducing its relation with SOM depletion while enhancing crop diversity and reducing greenhouse gas (GHG) emissions without affecting food production and supply (Brown, 1981; Tamburini et al., 2020). Conservation agriculture (CA) may be an alternative to achieve these goals, ensuring the maintenance of soil health (SH) and crop productivity and benefitting social and environmental aspects (Kassam et al., 2022; Landers et al., 2021; Wittwer et al., 2021). In addition, the payment for ecosystem services, such as that of C sequestration in agricultural soils, may be another approach to climate change mitigation (Oldfield et al., 2022; Prado et al., 2016). Tipping points related to forest, climate, socio-economic, and environmental aspects are in the opposite direction (Boulton et al., 2022; Gatti et al., 2021; Nepstad et al., 2008), and, depending on the location, parts of the Amazon may be converted from net C sinks to sources (Gatti et al., 2021).

Conservation agriculture is a farming system with three principles (FAO, 2014) that can promote SH. However, the application of a single principle would not warrant the maintenance or improvement of SH (Fuentes-Llanillo et al., 2021). In areas with improved physical and chemical properties under CA, biological activity and diversity, which are bioindicators of soil management quality, are stimulated (Passinato et al., 2021).

In Acre, around 15% of the state territory has been deforested (INPE, 2022), providing an opportunity to study how to promote human well-being with the objective of maintaining family farming in the field. Our study, initiated in 2006, investigated the effects of CA on SH, crop production, and GHG emissions. The objectives were to find solutions to reduce farm costs, raise agricultural productivity, and mitigate GHG emissions by eliminating burning practices, resulting in an improved quality of life of the farmers by showing ways to diversify the sources of and increase their financial income.

This chapter presents a case study on family farming in Brazilian Amazonia, with a particular focus on the sandy Ultisols of the western part of Acre. Data were obtained from an innovative participatory research and an ongoing long-term field experiment, initiated in 2006. The main hypothesis tested over the years was that CA is an alternative to the contemporary use of sandy soils in western Acre, to maintain and increase agricultural productivity and the profitability of family farming.

To support our hypothesis, the following objectives were established: (a) evaluate CA effects on profitability of family farming, (b) check if switching to CA can guarantee maintenance and increase of cassava and maize yields in response to improvement of soil quality (biological, physical, and chemical properties), and (c) assess SOM and enzyme activity levels in response to shifting from conventional farming to CA.

Our research follows Sustainable Development Goals 2 (zero hunger and sustainable agriculture), 12 (responsible consumption and production), 13 (action against global climate change), and 17 (partnerships and means of implementation), established by the United Nations General Assembly and supported by Embrapa.

Geopolitics and Soil Use in the State of Acre and Region of Juruá

The state of Acre has five regions of development: Baixo Acre, Alto Acre, Purus, Tarauacá-Envira, and Juruá (Acre, 2021). The Juruá region comprises five districts: Cruzeiro do Sul, Rodrigues Alves, Marechal Thaumaturgo, Porto Walter, and Mâncio Lima. The on-farm long-term experiment of this study is installed in Mâncio Lima, in western Acre (Figure 8.2), where the population density is still low, but occupation of the area is expanding. The regional geology and geomorphology

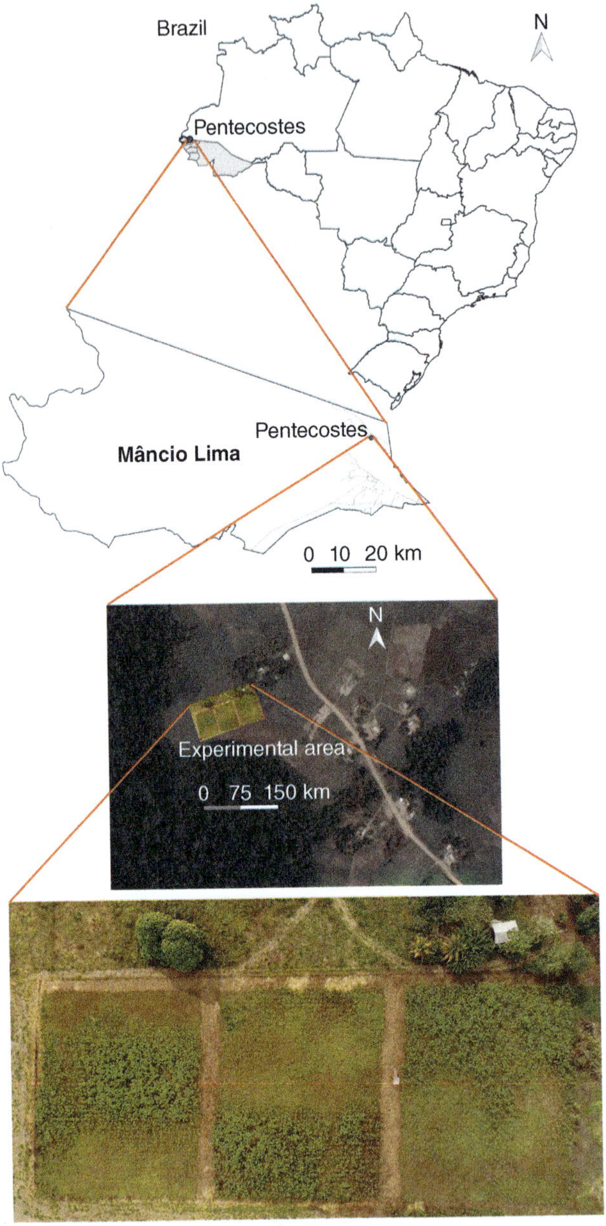

Figure 8.2 Study area in the community of Alto Pentecostes, district of Mâncio Lima, Acre, Brazil.

are highly diverse, owing mainly to the Solimões Formation, resulting in distinct environments (Bigarella, 1973; Latrubesse et al., 2007; Quesada et al., 2011).

In Juruá, soils derived from sandy and medium-textured sediments predominate. This is especially so on the west bank of the Juruá River and around the towns of Cruzeiro do Sul, Mâncio Lima, and Rodrigues Alves, which favors the development of Ultisols, or, in smaller areas, of Oxisols and Alfisols, which can be associated or not with Entisols or Spodosols (Bardales et al., 2020; Silva et al., 2021a). The soils in the region are deeper and better drained than other soils in Acre, although natural fertility is lower (Acre, 2006; Anjos et al., 2013).

Regarding soil use and management in Juruá, the parent material, landscape, and rainfall regime and annual distribution influence farmers' decision to usually prefer CT. Fine sand with the textures sandy to medium is predominant. The relief is gently to strongly undulating (Acre, 2006; Bardales et al., 2020). From 2006 to 2019, mean annual precipitation was 2160 mm (standard deviation, 357 mm; coefficient of variation, 17%), with over 60% of the total rainfall occurring between November and March (INMET, 2020). These associated variables determine the natural susceptibility of these soils to erosion.

Conventional tillage in Juruá consisted of heavy disking after forest burning and successive clearing of the area (Figure 8.3). These practices are not applied every

Figure 8.3 Conventional tillage in Juruá with heavy disking after forest burning and successive clearing of the area. Photograph: F. S. Costa (2006).

year because a tractor and tools are not always available when needed. Plowing is not a common practice in family farming in the Juruá region.

In general, soil liming and fertilization are also uncommon in Juruá, as was the use of herbicides when the experiment began. However, herbicide application has become a common practice to reduce labor in the management of areas prior to cultivation, as have desiccation or pre-harvest weed control or operations to combat weeds during cultivation, especially in cassava.

According to regional habits and following the regulatory norm 4/2013 of the Environment Institute of the state of Acre, which standardizes the use of fire for subsistence farming, the crop residues of the control treatment were burned in a 200-m^2 area for scientific purposes, with preventive safety measures. Similarly to CT, even by the traditional cultivation mode, burning is not used every year and has been used less and less since 2006.

The intrinsic nature of the soils of the Juruá region and their current management further accelerate the deterioration of the physical and chemical quality, which requires a technical solution to reverse the situation. One alternative is the management of already deforested and burned areas with technologies of soil quality recovery and maintenance, with agronomic, economic, and environmental gains that contribute to the reduction of deforestation and to abolish the use of fire in family production in the region.

Conservation agriculture is based on NT farming or minimum soil disturbance, permanent soil cover by living plants for its protection, and the addition of nutrients, combined with crop rotation and/or intercropping that complement each other in the function of recycling nutrients through the decomposition of cash crop and cover crop residues (Abrol et al., 2005; FAO, 2014; Hobbs et al., 2007).

Land use in Juruá is similar to that of other regions of Brazilian Amazonia. The land use change from forest to agriculture involves slashing and burning both native and secondary forest (Aragão et al., 2018; Libonati et al., 2021; Silva et al., 2021b) for monoculture crops for up to 5 years after the conversion. In the first 2 years, crops are chosen that require more nutrients (e.g., rice, common bean, and maize). Afterward, cassava becomes the predominant crop because of its low nutrient requirement and adaptation to acidic soils. Cultural aspects, such as family traditions and eating habits, are involved as well. However, over time, even cassava is negatively affected by poor natural resources and incorrect soil management.

Fire is used for clearing the areas to be cultivated, and the efficiency of the practice depends on factors such as burning season, the quantity and diversity of forest species, and the weather conditions in the pre-burning period (Carvalho Junior et al., 2016; Leal et al., 2018).

Practices of liming and fertilization to increase the content, replacement, and maintenance of the exported nutrients are not common.

Nutrient loss from the SB system is slow between felling and burning, fast during and immediately after burning, and slow if the area is abandoned and not used for farming after the fire. In the latter case, in Amazonia, losses are accelerated by soil tillage during the rainy season.

Mechanized soil tillage is recent and a result of government public policy programs but has not achieved the results expected by farmers. In Juruá, where the SOM levels in the sandy surface layers are naturally low (~1% in the 0.0- to 0.1-m layer) (Anjos et al., 2013), rainfall is well distributed throughout the year, and the annual mean air temperature is 25°C, in general, SOM decomposition is accelerated. The scenario of no organic or mineral replenishment further degrades the regional soils. Mechanical tilling of the soil (or mechanized soil tillage) may reduce the weed seed bank if the relief is not a limiting factor to the implementation of soil conservation practices. However, the degradation stage and size of the area are key factors to consider for recovery. The risk of runoff in the regional dry season is negligible because there is practically no rainfall.

The timing of soil tillage is another aspect that needs adjustment. In this sense, conventional land use in Amazonia and Acre follows the meteorological calendar of the region, as in any other region. In the state of Acre, the rainy season lasts from October to April, with December to March being the wettest months; the rain-to-dry transition is in May, the dry season lasts from June to August, and the dry-to-rain transition is in September (Duarte, 2006). Thus, the mechanical operations that precede cultivation are performed in September. If the soil is conventionally tilled (plowed and harrowed) when rainfalls begin in October, it is not covered by the developing crops, intensifying surface erosion processes. This is most critical if sowing or planting occurs relatively late, close to the rainiest period. For example, if cassava, commonly planted by family farmers, is planted in October, the soil will not be fully covered in December, when the most erosive rains set in. This scenario is common in the region, where the cultivation of cover crops, whatever their purpose, is not common. Weeds that were not burned are then incorporated into the soil by tillage. As soil loses quality under this mismanagement, weed infestation increases, requiring repeated operations of manual weeding and CT. Weeds are still controlled by hand due to the high costs of herbicides on the regional markets and the low income of family farmers.

After 5 years of soil use for family farming, the area is left fallow for a 5-year period. Thereafter, a new crop cycle begins with slashing and burning. A signal to initiate the fallow period is low cassava yield. In the third year of cultivation, the reduction of crop productivity varies from 30% to 50%, compared with the yield of the first year. The cultivation period and productivity reduction are related to the soil nutritional status in the first year of cultivation.

Regardless of the wide adoption of the SB system in terms of area, small rural producers in Amazonia still do not have a technique that would allow the management of the residues that remain after the fire or the clearing of areas for cropping. The alternatives available cannot be applied to large areas (Kato et al., 2020). These are the reasons for the need for burning, which also serves to increase the availability of nutrients that benefit crops in the early development stages, even under conditions that will favor continuous nutrient losses from the soil (Mackensen et al., 1996).

After clearing the area by burning, the nutrients exported by monoculture and the absence of liming and fertilization during cultivation gradually generate critical conditions for new crops. The successive use of fire for clearing areas for subsequent cultivation accelerates this dynamic (Davidson & Martinelli, 2009).

Weed desiccation with herbicide before each crop cycle can reduce the seed bank of spontaneous plants if herbicide cost is not a limiting factor. It is important to emphasize that the decision for one practice or the other should take into consideration that residues of desiccants remain in the soil, with environmental consequences, such as the possible transport of residual amounts to rivers in soil erosion (Flores et al., 2020; Labrière et al., 2015; Lense et al., 2021; Santos & Augustin, 2015). These risks are reduced by the use of soil conservation practices, such as NT management, the continuous cultivation of cover crops, and diversification of crop species.

Sixteen years after the beginning of this research, the scenario described above remains practically the same, which highlights the challenges related to technology transfer in this region and the lack of attention from public policy-makers and decision-makers for the family farming sector in the region of Juruá and others in Acre.

The Long-term Field Experiment of Mâncio Lima, Region of Juruá

The research project in Mâncio Lima from 2006 to 2019 was described in detail by Costa et al. (2014, 2020). The data presented here cover the period from 2019 to the present. The research has been carried out on a participatory and interinstitutional basis (Federal University of Acre, Campus of Cruzeiro do Sul; districts of Mâncio Lima, Cruzeiro do Sul and Rodrigues Alves; Federal University of Rio Grande do Sul; University of São Paulo/School of Agriculture Luiz de Queiroz).

The traditional SB cultivation technique was the reference for comparison with CA. For this reason, the alternative treatments do not use fire. Using the SB system as reference allows the determination of the impacts on soil quality and crop productivity and their consequences for economic, social, and environmental aspects in the region.

The criteria for selecting the experimental site and considerations regarding the empirical understanding of farmers about their choice of cultivation sites are described by Costa et al. (2014). However, it is worth emphasizing that the set of criteria for site selection allows applying the results of the experiment in Mâncio Lima to other districts in the Juruá region, Acre and Amazonia, where there are similar sandy soils.

The experiment was conducted in an area of rural family farmer partners located in the community of Alto Pentecostes (7°28′S, 72°56′W, 190 m asl) on a dystrophic yellow Argissol (Santos et al., 2018), corresponding to a Ultisol (Soil Survey Staff, 1999) and an Acrisol (WRB/FAO, 2002), with 134 g kg^{-1} clay, 77 g kg^{-1} silt, and 789 g kg^{-1} total sand (0–20 cm soil layer). In 2006, the soil had a pH (H_2O) of 4.0; 2.1 $cmol_c\ dm^{-3}$ Al; 0.8 $cmol_c\ dm^{-3}$ sum of bases; and 72% Al saturation. According to the Köppen classification, the regional climate is Af, with no dry season (Alvares et al., 2014). Total annual rainfall and mean annual air temperature of the region from 2005 to 2019 are shown in Figure 8.4. In this period, the air temperature was relatively stable, varying between 25.4 and 26.1°C. The years with rainfall above 2150 mm (mean of the 2005–2019 series) were 2008, 2009, 2013, 2014, 2015, and 2019. In the other years of the series, rainfall was below this value.

The experiment was set up in a split-plot arrangement on a randomized block design with three blocks. The two tillage systems (CT and NT) were compared in the main plots (20 m by 50 m) (Figure 8.5).

Five cropping systems were evaluated in the subplots (10 m by 20 m): (a) SB, which is the traditional mode of soil cultivation in the region, without liming (Lm)

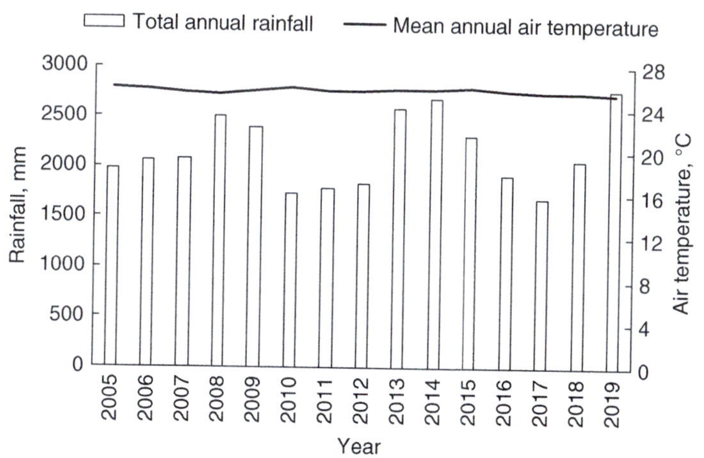

Figure 8.4 Total annual rainfall and mean annual air temperature in the region of Mâncio Lima, Juruá, Acre, southwestern Brazilian Amazonia from 2005 to 2019. Source: INMET (2020).

Figure 8.5 View of conventional tillage (left) and no-tillage (right) of on-farm long-term experiment in the community of Pentecostes, Mâncio Lima, region of Juruá, Acre, southwestern Brazilian Amazonia. Photograph: F. S. Costa (2018).

and fertilization; (b) legume species and/or grass (LG) cover crops without Lm and P; (c) LG with P (LGP) and without Lm; (d) LG without P and with Lm using dolomitic limestone (LGLm); and (e) LG with P and Lm (LGPLm) (Figure 8.6). These treatments represent an increasing gradient of technological inputs: SB < LG < LGP < LGLm < LGPLm. The sources of P and lime were super-triple phosphate and dolomitic limestone, respectively, and the rate applied varied according to soil analyses and the experimental year. The P and lime rates in 2006 were, respectively, 80 kg P_2O_5 ha^{-1} and 4.5 Mg ha^{-1}. The last application occurred in 2018, when P and lime were re-applied at rates equivalent to 80 kg ha^{-1} and 1.65 Mg ha^{-1}. In the treatments without liming, lime was re-applied at a rate of 3.2 Mg ha^{-1}.

Eleven crop cycles were grown in the period from 2006 to 2019, five of which were cassava (2007–2008, 2009–2010, 2014–2015, 2016–2017, and 2018–2019) and six maize, of which three were planted in the main growing season (2011–2012, 2012–2013, and 2013–2014) and three in second crop season (2014, 2016, and 2018). In Juruá, the main maize growing season lasts from September to January; sowing and harvesting are performed during the regional rainy season, when the risk of soil erosion is high. At sowing, the aboveground biomass is insufficient to

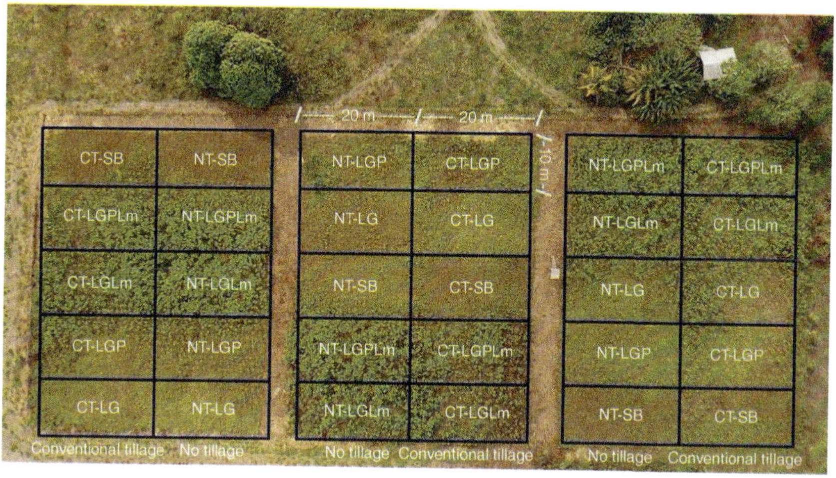

Figure 8.6 Aerial view of the long-term field experiment in the community of Alto Pentecostes, Mâncio Lima, region of Juruá, Acre, Brazil. CT, conventional tillage; LG, subplot with legume species and/or grass; LGLm, subplot with LG and liming; LGP, subplot with LG and phosphorus (P); LGPLm, subplot with LG with phosphorus (P) and liming (L); NT, no tillage; SB, subplot with a slash and burn system.

protect the soil against early torrential and frequent erosive rains. Rains at the end of the cropping season can cause harvest losses. For this reason, maize was also sown in a second planting season (from March to April) and harvested in July–August, according to the ordinance for Agricultural Zoning of Climate Risk (Zoneamento Agrícola de Risco Climático - Zarc in Portuguese) of Maize in Acre of the Ministry of Agriculture, Livestock and Food Supply.

No maize was produced in 2011–2012 due to the end of the residual effect of the lime applied in 2006.

The cover crops in the experiment were velvet bean (*Mucuna aterrima* Piper & Tracy), sown in the first year and naturally reseeded until the cultivation of forage sorghum [*Sorghum bicolor* (L.) Moench] in September 2014, jack bean [*Canavalia ensiformis* (L.) DC], intercropped with cassava in the 2014/2015 growing season and dwarf pigeon pea in the winter growing season of 2017–2018 after cassava harvest in the 2016/2017 growing season. It was assumed that the cover crops would supply at least some (if not all) of the N required for cassava crops by biological fixation. The N source for maize was always urea (80 kg ha^{-1}).

After cassava harvest in 2019, a cover crop mixture of jack bean and pearl millet (*Pennisetum glaucum* L.) were grown in 2019–2020, followed by cowpea [*Vigna unguiculata* (L.) Walp.] in 2020. Cassava was also grown in 2020–2021, followed by sunn hemp (*Crotalaria juncea* L.) until March 2022 (data not shown). Thereafter,

Table 8.1 Experimental soil management in Mâncio Lima from 2006 to 2018.

Year	Management/semester	Observation
2006	Soil tillage, liming, and fertilization[a]/second	Beginning of the experiment
2012	Soil tillage/second	No liming or fertilization
2013	Soil tillage, liming, and fertilization/second	End of residual effect of liming in 2006; expected yield of cash crop not reached, despite NT
2014	Soil tillage/second	According to the regionally increasing use of plow harrow.
2018	Soil tillage, liming, and fertilization/first	According to liming residues, before the second maize crop of 2018

Abbreviation: NT, no-tillage.

a) Fertilization required to reach adequate levels of soil nutrients (N–P–K), according to soil fertility analysis.

the soil was left fallow until September 2022, when cassava was planted again with harvest estimated for July/August 2023.

A description of the main soil management practices from 2006 to 2018 is given in Table 8.1. Five soil tillage and three liming and fertilization procedures were carried out from 2006 to 2018.

The same cassava cultivar was planted from 2006 to 2019, regionally known as “mansibraba,” a line selected by family growers. The maize cultivars used during 2006–2018 are described in Table 8.2.

Table 8.2 Maize cultivars in Mâncio Lima from 2006 to 2018 in the main and second crop.

Year	Maize cultivar
Conventional crop	
2011–2012	BRS 1040, Embrapa
2012–2013	AL Bandeirantes, Embrapa
2013–2014	BRS Sol da Manhã, Embrapa
Second crop	
2014	CD 308, Coodetec
2016	BM 2B655HX, Dow Agroscience
2018	BRS Caimbé, Embrapa

Economic Analyses

The economic analysis included cassava and maize cultivation over the total experimental period, based on the premise that family-scale agricultural production should be diverse in space and time, making use of opportunities to ensure a regular income according to the market supply and demand for the raw products. Costs and economic indicators were calculated (Guiducci et al., 2012).

The total production cost took all measurable expenses into account, such as costing, family labor remuneration, depreciation, and return on capital. Costing represents the calculation of the expenses incurred during crop cultivation; family labor remuneration corresponds to the alternative or opportunity cost, represented by the local daily rate of R$50.00 (US$9.50) (mean cost across January/August 2019). The cost for soil tillage for the first planting and the expenses for the cassava crop constitute the initial investment. Depreciation comprises the indirect costs incurred in loss due to physical decay of assets with a limited useful life (tractors and farming implements).

The remuneration of land capital (land) calculation considered a rate of 4% per year over the market value. The capital cost used in activities of costing and investment takes into account a discount rate of 6% per year, representing how much the producer invested in the establishment and production of the crop instead of in the financial market.

Economic analyses were performed for the production models with CT and NT for cassava and maize production, with cultivation of cover crops between these crops, in the period 2006–2019, covering the period from soil preparation of the area in 2006 to cassava harvest in 2019.

The total experimental period was divided into two phases for economic analyses. The first phase, between 2006 and 2007, was considered an investment in recovery of soil chemical properties, without commercial crops. In 2006, the experimental area was cleared of weeds and soil sampled for laboratory analyses; soil tillage, liming, and fertilization were carried out, and velvet bean was sown prior to cassava in 2007. This phase lasted 1 year. The second phase was between 2011 and 2012 and was considered the fallow year of the experiment. Thus, the years considered for the economic analysis of the commercial crops were 2007–2008 (Year 1), 2009 (Year 2), 2010 (Year 3), 2013 (Year 4), and so on until 2019 (Year 10). The divisions were necessary because the short-term timing effects of the applied agricultural practices in 2006 or the residual time of liming in 2006 were unknown.

The prices of inputs, services, and products corresponded to the market prices in the districts of Cruzeiro do Sul, Mâncio Lima, and Rodrigues Alves, in March 2019. The prices of maize and cassava were those paid to farmers on their property.

The total revenue was the sum of the sale of cassava and maize production per hectare in the analyzed period. Net income was computed as the total income minus all expenses with the agribusiness and minus depreciation, capital, and land costs. As a theoretical result, agribusiness is not sustainable in the long term if the net income is negative. If it is positive, the agribusiness can remunerate all production factors. In this case, the activity is stable and can expand.

The family income is equal to the sum of the net income, with income associated with the family labor used in the activity and the remuneration of the invested capital (i.e., the interest on the producer's own resources invested in the crop). The total productivity factor is the ratio between total revenue and total cost and should be at least equal to 1. Thus, the higher the total productivity factor, the better the profitability of the investment and the more efficient is the studied production system.

The family labor remuneration per day worked on the activity was obtained by dividing the family income by the number of days worked. This indicator represents the maximum daily rate that can pay for family labor, presenting as an advantage for the analysis a direct comparison between the remuneration that the farmer can obtain from the sale of his labor (his opportunity cost) and remuneration for having worked on his property.

All economic results are presented in Brazilian Real (R$1.00), equivalent to US$0.19 (February, 2023).

Agronomic Analyses

At harvest, a 72-m^2 (6 m by 12 m) section was removed from the central rows of each subplot (Figure 8.6). The cash crops and cover crop residues were always chopped with a knife roller pulled by a microtractor, preceded or followed by desiccation (glyphosate 2–4 L ha^{-1}) according to the cover crop and the agricultural calendar for each crop sequence. The control treatment was cut and burned, as usually done by the family farmers in Juruá, where burning and soil tillage are not repeated annually. In the next step, crops were sown based on an association of empirical observations of the farmers and recommendations of the Agricultural Zoning of Climate Risk (Zarc) (MAPA, 2022).

The mean (2007–2019) accumulated (cassava plus maize) productivity was calculated (Mg ha^{-1}) based on the crop seasons 2007–2008 (cassava) and 2012–2013 (maize), 2009–2010 (cassava) and 2013–2014 (maize), 2014 (maize) and 2014–2015 (cassava), 2016 (maize) and 2016–2017 (cassava), 2018 (maize), and 2018–2019 (cassava). The ratio of NT LGPLm to CT LGPLm was used to demonstrate whether the mean cumulative yield of cassava and maize crops tended to stabilize in the period from 2006 to 2019.

Analysis of Soil Organic Matter Quality

The SOM quality was studied in soil samples collected at the beginning of the rainy season (November) in 2016 (10 years) and 2019 (13 years) from three 100-cm-deep pits in each block of the treatments ($n = 3$) from the layers 0–5, 5–10, 10–15, 30–40, 50–75, and 75–100 cm. Reference samples were also collected from an adjacent native forest area, also from three pits. The chemical composition of SOM was analyzed by solid-state ^{13}C nuclear magnetic resonance (^{13}C NMR) cross-polarization magic angle spinning (model DD2, Agilent Technologies) in composite samples previously treated with hydrofluoric acid 10% solution (Gonçalves et al., 2003). The chemical shift assignments of the ^{13}C NMR spectra were performed according to Knicker (2011). The same composite samples were analyzed for free lipid distribution, where n-alkane extraction was carried out with a dichloromethane/methanol solution (3:1 [v/v]) ratio for 24 hours in a Soxhlet extractor (Wiesenberg et al., 2004). From the resulting n-alkane pattern distribution, molecular proxies were calculated.

Soil Enzyme Analyses

Soil samples were prepared for GLU and ARYL determinations using the FERTBIO soil sample concept, described by Mendes et al. (2019). Briefly, soil samples collected after harvest were air-dried at room temperature for at least 2 weeks, sieved (<2 mm), and stored at room temperature until the analyses were performed within 1 week after air-drying. GLU and ARYL activity levels were assayed in duplicate including one control as described by Tabatabai (1994). The method is based on the colorimetric determination of the p-nitrophenol (PNP) formed after the addition of specific colorless substrates to each enzyme. Due to the short incubation period (1 hour), toluene was omitted from the assays. Enzyme activity levels were expressed in mg PNP kg^{-1} h^{-1}.

Since 2020, GLU and ARYL have been used in large-scale on-farm SH assessments in Brazil (see Mendes et al. [2021] and Chapter 10), representing an opportunity to engage producers in soil testing beyond the standard chemical analyses.

The tillage systems selected for the enzyme analyses were NT and CT. The cropping systems included only SB and LGPLm. Soil samples (0- to 10-cm layer) were collected in October 2021 (local dry season) after cassava harvest and in April 2022 (local rainy season) under sunn hemp residues. Before the analyses, the soil was air-dried and sieved (<2 mm).

Lessons Learned in the First 16 Years

For this study, only the results for the control SB the traditional production system of the Juruá region and LGPLm, both in CT and NT, are presented and discussed from a technical and economic point of view because they represent the extremes for decision-making.

Economic Perspectives

The total costs increased from SB to LGPLm by 52% under NT and by 51% in CT management systems in the mean of 16 years. Service costs, materials, and inputs (consumables) represented the highest portion of the total cost for SB and LGPLm under NT and CT, followed in descending order by crop formation (cost of planting an agricultural crop) and capital opportunity (Table 8.3).

Conventional tillage had higher service costs than NT for SB (9%) and LGPLm (11%) due to the greater use of family labor and the renting of machines used for tillage and indirectly to GHG emission by burning fossil fuel at tilling, which is avoided under NT. In addition, tillage accelerates the decomposition of the remaining or native SOM, intensifying emissions. Inverse effects are expected under NT, especially in sandy soils, at a mean air temperature of 25°C and with regular rainfall distribution throughout the year in the region of Juruá.

The total cost of components, except services, increased from SB to LGPLm in CT and NT, which is related to the technological intensification in LGPLm (Table 8.3).

Table 8.3 Comparison of total costs and components for the slash-and-burn (SB) system and the conservation agriculture model (intercropped legume and grass with phosphorus fertilization and soil liming [LGPLm]) under no-tillage (NT) and conventional tillage (CT) in Mâncio Lima, Region of Juruá, Acre, southwestern Brazilian Amazonia (mean from 2006 to 2019).

	NT		CT	
Cost component	**SB**	**LGPLm**	**SB**	**LGPLm**
	–R$1.00 ha^{-1} yr^{-1}–			
Services	1364.70	1279.70	1490.90	1426.00
Materials and consumables	445.80	1292.20	433.40	1294.50
Capital opportunity	185.70	229.90	192.50	239.10
Crop formation	245.90	610.90	260.90	624.80
Total cost	2242.00	3412.70	2377.80	3584.40

Table 8.4 Profitability of traditional (slash-and-burn [SB]) and conservation (intercropped legume and grass with phosphorus fertilization and soil liming [LGPLm]) agriculture in family farming systems under no-tillage (NT) and conventional tillage (CT) in Mâncio Lima, Region of Juruá, Acre, Southwestern Brazilian Amazonia (mean from 2006 to 2019).

	NT		CT	
Cost component	**SB**	**LGPLm**	**SB**	**LGPLm**
		–R$1.00 ha^{-1} yr^{-1}–		
Total income	807	3444	1084	3297
Labor remuneration	392	1754	531	1603
Net income	−1435	31	−1294	−288
Daily rate	14.68	77.75	21.01	66.60
Total productivity of production factors	0.36	1.01	0.46	0.92

On the other hand, total income increased significantly in response to technical input from SB to LGPLm under NT (327%) and CT (204%). The total income obtained with NT was 60% greater than that for CT (Table 8.4).

Family income or labor remuneration was positive from SB to LGPLm (Table 8.4), with a favorable percentage difference (LGPLm – SB) of 72% for NT over CT. Under NT, LGPLm increased labor remuneration by 347% in relation to SB. Under CT, this increase was 202%. These results show that, from the economical point of view, it is worthwhile for the producer to work in agriculture on his own property in both cases and that income is higher from NT systems because the value per workday is higher than the daily rate paid in the region (R$50.00).

Although the mean (2007–2019) accumulated cassava and maize productivity in the technologically advanced LGPLm was similar for NT and CT, the net income was positive only in the LGPLm-NT (Table 8.4), which indicated that this treatment was the best alternative to provide economic stability in the long term (Figure 8.7).

The daily rate increased with technology inputs from SB to LGPLm under NT and CT. The NT-LGPLm and CT-LGPLm were 430% and 217%, respectively, higher than the SB (Table 8.4). This result indicates (a) that the value of the daily rate in 13 years of the study was higher than the opportunity cost of labor for the region (R$50.00 d^{-1}) and (b) that it is better for the family farmer to work on his own property than to sell his labor. In situations where there is no demand for labor, as is the case in isolated locations in Amazonia, even under SB (CT greater than NT), the daily rate was positive but below R$50.00.

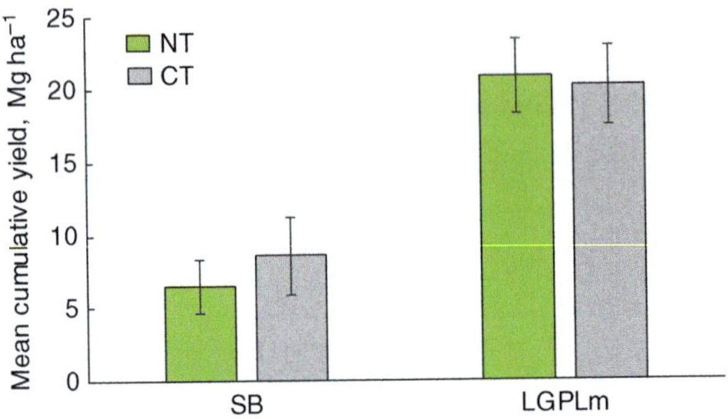

Figure 8.7 Mean cumulative yield (2007–2019) of cassava and maize under the traditional soil cultivation mode of the region based on slashing and burning (SB) and the alternative conservation production system (LGPLm) under no-till (NT) and conventional tillage (CT) in Mâncio Lima, region of Juruá, Acre, southwestern Brazilian Amazonia, 2019. Vertical bars represent the mean standard error.

The total productivity of production factors (TPF) was above 1.0 due to the technology contribution from SB to LGPLm under NT and CT, with percentage increases of 181% and 100%, respectively (Table 8.4). The TPF is the ratio of the total revenue to the total cost, with the expected result of at least 1.0. The higher the TPF, the better the profitability of the investment and the more efficient the studied production system.

In summary, the results indicate that during these first 13 years, the most conservation-oriented system NT-LGPLm provided the best economic viability and advantages over CT. To expand the spatial physical scale of the economically validated CA, the peculiarities of each site (e.g., climate, soil, and composition of alternative production systems) must be considered. Moreover, the abilities of the family farmers to change their technological production pattern should be taken into consideration. These considerations must also take into account the agricultural, social, and environmental regional public policies at the different locations.

Agronomic Perspectives

The mean cumulative yield of cassava and maize of the period from 2007 to 2019 raised the following questions: (a) Was there a yield increase of cassava and maize in the first 13 years of the experiment? (b) If so, was there a tendency toward stability of productivity, and at which levels? and (c) At what point did the agricultural production under CA exceed the production obtained in traditional ways for the regional climate and soil conditions?

The mean cumulative yield (2007–2019) of cassava and maize in SB was 6.5 Mg ha^{-1} under NT and 8.6 Mg ha^{-1} under CT, whereas in LGPLm it increased by 221% to 21 Mg ha^{-1} under NT and by 136% to 20 Mg ha^{-1} under CT (Figure 8.7).

Based on the cumulative yield until 2019, the above questions can be answered as follows. Technological contributions raised the mean cumulative yield of cassava and maize in the experiment in 13 years, although it was not possible to confirm a tendency of stability of this result. In addition, with respect to the question posed since the beginning of the study in 2006, it is possible to say that, already at the beginning of its adoption, under local climate and soil conditions, crop yields were higher under CA than under traditional SB systems for both NT and CT. However, a comparison of NT with CT shows a reduction of production costs and higher profitability of NT, which suggests that this system should be recommended and preferred.

The Soil Health Perspective: Organic Matter Quality and Enzymes

After 10 years of agriculture, the C content did not differ among treatments and ranged from 13.9 to 17 g kg^{-1} (0–5 cm), from 11.2 to 15.6 g kg^{-1} (5–10 cm), from 11.1 to 12.3 g kg^{-1} (10–15 cm), from 7.9 to 8.7 g kg^{-1} (30–40 cm), from 4.6 to 5.5 g kg^{-1} (50–75 cm), and from 3.9 to 5.1 g kg^{-1} (75–100 cm).

However, SOM quality was affected by treatments in the 0- to 10-cm surface layer, so that SOM under NT-LGPLm had more carbohydrate-like structures (O-alkyl C) and fewer lipid structures (alkyl C) than under CT-LGPLm (Table 8.5). This result may suggest different microbial activity under NT that promotes the stabilization of biochemically labile structures, such as O-alkyl C groups and inhibits their further degradation into alkyl C chains. The molecular proxies calculated from the n-alkane distribution supported this assumption: the short/long chain ratio of SOM in the 0- to 5-cm layer was greater under NT-LGPLm than under CT-LGPLm (Table 8.5), indicating a greater contribution of microbial-derived lipids (comparatively shorter n-alkane chain) to the SOM composition in the former. Additionally, in the surface layer (0–5 cm), SOM from NT-LGPLm had a lower C preference index for long chain (which indicates the preference of odd alkane homologs over even counterparts), confirming that the remaining vegetal inherited lipids (comparatively longer n-alkane chain) were more degraded than under CT-LGPLm.

Overall, SOM accumulation was less influenced by CA, probably because of the weak organo-mineral interaction to hold C of this coarse-textured Acrisol and the hot and wet climate that stimulates mineralization. However, SOM stabilization and quality were affected by conservation practices. The establishment of

Table 8.5 Distribution of C functional groups alkyl/O-alkyl ratio determined by ^{13}C RMN CP/MAS and molecular proxies (n-alkanes) of the SOM_{HF} of the 0- to 5-cm and 5- to 10-cm layers of an Acrisol under native forest, and maize and cassava cultivation under conventional tillage (CT) or no-tillage (NT), evaluated 10 years after experiment implementation (Aquino, 2019).

	C groups (%)/chemical shift (ppm)					Molecular proxies	
Treatment	Alkyl C, 0–45 ppm	N/O- alkyl C, 45–110 ppm	Aryl C, 110–160 ppm	Carboxyl/carbonyl, 160–220 ppm	Alkyl/O-alkyl	$R_{S/L}$ [a)]	CPI_L [b)]
				0- to 5-cm depth			
Forest	34.4	36.1	12.7	16.6	1.0	2.6	11.2
CT-SB	26.4	41.1	21.1	11.4	0.6	3.1	3.7
CT-LGPLm	30.1	37.5	18.9	13.5	0.8	2.8	5.1
NT-SB	25.7	38.7	24.3	11.3	0.7	5.2	4.3
NT-LGPLm	22.9	41.3	24.4	11.4	0.6	4.3	3.6
				5- to 10-cm depth			
Forest	34.5	37.6	12.1	15.9	0.9	3.5	10.4
CT-SB	30.1	34.8	22.5	12.6	0.9	3.9	3.2
CT-LGPLm	33.9	34.8	19.7	11.6	1.0	5.7	3.0
NT-SB	31.4	36.1	21.1	11.4	0.9	3.3	3.9
NT-LGPLm	27.1	38.0	24.0	11.0	0.7	4.8	3.5

a) $R_{S/L}$, short/long chain ratio.
b) CPI_L, carbon preference index for long chain.

Table 8.6 Soil activity levels of β-glucosidase and arylsulfatase under different management systems (no-tillage [NT] and conventional tillage [CT] and control [slash-and burn, SB], legume species and/or grass [LG], and LG plus fertilization with phosphorus [LGPLm]) in soil samples (0- to 10-cm layer) in October 2021 (local dry season) and April 2022 (local rainy season).

	NT		CT	
Cropping system	**Dry**	**Rainy**	**Dry**	**Rainy**
		–mg PNP kg^{-1} h^{-1}–		
		Arylsulfatase		
SB	122ns	141ns	98ns	94ns
LG	108ns	111ns	74ns	92ns
LGPLm	122ns	118ns	93ns	93ns
		β-glucosidase		
SB	18ns	41ns	21ns	38ns
LG	19ns	38ns	18ns	41ns
LGPLm	20ns	34ns	17ns	35ns
	Dry		**Rainy**	
		Arylsulfatase		
Soil tillage				
NT	117ns		123a	
CT	88ns		93b	
		β-glucosidase		
NT	19ns		38ns	
CT	19ns		38ns	

Note. In each season, means followed by different letters in the columns differ significantly at $P < 0.05$ (Tukey's test). ns, not significant by F test ($P > 0.05$).

an adequate conservation soil management system for this particular region is paramount in order to maintain and improve C stocks and SH.

In terms of soil enzyme activity levels, regardless of the sampling season (rainy or dry), GLU and ARYL activity levels did not differ between the cropping systems SB, LG, and LGPLm after 15 years (Table 8.6). Compared with CT, increased ARYL activity levels under NT were only observed in the rainy season. Soil enzymes act as early-warning indicators of soil changes in response to management practices. The absence of significant effects on GLU and ARYL activity levels in the long-term field experiment of Mâncio Lima demonstrate the difficulties of changing the soil biological functioning of sandy soils in a region with mean precipitation levels of >2000 mm and limited dry matter input from cover crops.

It should be highlighted that GLU activity levels were twice as high in the rainy season as in the dry season. This might be associated with the fact that sampling in the dry season occurred immediately after cassava harvest, which involves soil disturbance, resulting in reduced GLU activity levels. In the rainy season, the presence of sun hemp residues probably causes positive effects on GLU activity levels (Table 8.6).

Concluding Remarks

During the first two decades of the twenty-first century, the goal of Brazilian agriculture to contribute to the sustainable production of food, with economic, environmental, and social benefits, is met by the unique opportunity to improve the production systems of a huge contingent of small rural farms, making them more competitive. Despite the development of knowledge and appropriate technologies available for immediate adoption and incorporation in different production arrangements, small rural farms in the Amazon region still have little access to the techniques required for sustainable production. As observed in the long-term field experiment of Mâncio Lima, simple technologies such as liming, P fertilization, and cover crops can be incorporated into production systems at relatively low investment costs and with proportionally great economic benefits.

The results of this 16-year study indicate that the evaluated CA model is adequate for agriculture without burning practices and minimal soil disturbance on small family farms of western Amazonia. The use of cover crops, P fertilization, and liming on the sandy soil of Mâncio Lima proved to be economically and technically viable. To implement this production system, some adjustments to the history of soil use, texture, and relief are necessary. Aspects associated with the capacity of investment of each farmer family should also be taken into account.

The application of this economically validated model of CA to the conditions described in other regions of the state of Acre, and even in the wider Brazilian Amazonia, should consider the regional peculiarities of climate, soil, composition of production system alternatives, and, equally important, the conditions a farmer has to initiate the transition according to his technological level of production. These conditions should be aligned with the agricultural, social, and environmental public policies for each region.

The proposed solutions to the problems caused by SB agriculture on the sandy soils of the Juruá region require adjustments in the field, such as (a) the inclusion of more plant species in the rotation or intercropping of cash and cover crops and (b) adapting the level of mechanization to family farming for all stages of crop development. Plant diversity improves soil quality and cash crop yields, and mechanization reduces production costs. In addition, by reducing or even eliminating burning practices and the use of fossil fuels, this agricultural production model can reduce the emission potential of GHG and, in the long term, increase soil C stocks, directing the conservation production system toward low-C agriculture.

The results of this study can be separated into lessons learned and those to be learned. During these first 16 years, although the best management practices had a limited impact on the health of the sandy soils in the Juruá region, crop productivity was clearly significantly increased and income was ensured, which determines the profitability of family farming. Future studies will evaluate the impacts of this CA model on soil properties (e.g., SOM fractionation and aggregate stability) and GHG emissions and the hidden C costs of each system. The continuity of this long-term field experiment is certainly a valuable and critical investment. It will allow, for instance, to determine when the balance of C stock and GHG emissions will invert the production system from C source to sink and when SH indicators will be able to differentiate agronomic management systems from each other.

Acknowledgments

The authors thank Sebastião Oliveira do Nascimento for allowing the research on his private property. They are also indebted to the research funding agencies Brazilian Agricultural Research Corporation - Embrapa (03. 09.06.022.00.00, 23.16.05.027.00.00 and 24.17.01.014.03.00), Technology Foundation of the State of Acre - Funtac (T.O 001/2009 and 03/2012), National Council for Scientific and Technological Development - CNPq (575795/2008-5), Amazon Fund (Embrapa 24.17.01.014.03.00), and Agrisus Foundation (process 3115/21).

References

Abrol, I. P., Gupta, R. K., & Malik, R. K. (2005). *Conservation agriculture: Status and prospects*. Centre for Advancement of Sustainable Agriculture.

Acre. (2006) Acre, Governo do Estado Zoneamento ecológico-econômico do Acre: fase III: escala 1:250.000: documento-síntese/Secretária de Estado do Meio Ambiente. – Rio Branco: Semapi, 2021. 162 p.: il. col.

Alvares, C. A., Stape, J. L., Sentelhas, P. C., de Moraes Gonçalves, J. L., & Sparovek, G. (2014). Koppen's climate classification map for Brazil. *Meteorologische Zeitschrift*, 22, 711–728. https://doi.org/10.1127/0941‐2948/2013/0507

Anjos, L. H. C., Silva, L. M., Wadt, P. G. S., Lumbreras, J. F., & Pereira, M. G. (2013). *Guia de Campo da IX Reunião Brasileira de Classificação e Correlação de Solos* [Field Guide to the IX Brazilian Soil Classification and Correlation Meeting]. Embrapa.

Aquino, I. O. (2019). *Teor e composição química da matéria orgânica em Argissolo do Acre em diferentes sistemas de manejo* [Doctoral dissertation, Universidade Federal do Rio Grande do Sul]. https://www.ufrgs.br/agronomia/materiais/itauaneoliveiradeaquino_mestrado.pdf

Aragão, L. E. O. C., Anderson, L. O., Fonseca, M. G., Rosan, T. M., Vedovato, L. B., Wagner, F. H., Silva, C. V. J., Silva Junior, C. H. L., Arai, E., Aguiar, A. P., Barlow, J., Berenguer, E., Deeter, M. N., Domingues, L. G., Gatti, L., Gloor, M., Malhi, Y., Marengo, J. A., Miller, J. B., … Saatchi, S. (2018). 21st century drought-related fires counteract the decline of Amazon deforestation carbon emissions. *Nature Communications*, 9, 536. https://doi.org/10.1038/s41467-017-02771-y

Bardales, N. G., Araújo, E. A., Amaral, E. F., Ker, J. C., Maia, G. R., Araújo, D. R., Amaral, E. F., Oliveira, T. K., Franke, I. L., Lani, J. L., Martorano, L. G., Melo, A. W. F., & Negreiros, J. S. (2020). *Solos e geopaisagens do município de Cruzeiro do Sul, estado do Acre: Potencialidades e fragilidades* [Soils and geolandscapes in the county of Cruzeiro do Sul, state of Acre: Potentialities and weaknesses]. Ipam Amazônia.

Bigarella, J. J. (1973). Geology of the Amazon and Parnaiba basins. In A. E. M. Nairn & F. G. Stehli (Eds.), *The ocean basins and margins* (pp. 25–86). Springer.

Boulton, C. A., Lenton, T. M., & Boers, N. (2022). Pronounced loss of Amazon rainforest resilience since the early 2000s. *Nature Climate Change*, 12, 271–278. https://doi.org/10.1038/s41558-022-01287-8

Brown, L. R. (1981). World population growth, soil erosion, and food security. *Science*, 214, 995–1002. https://doi.org/10.1126/science.7302578

Carvalho Junior, J. A., Amaral, S. S., Costa, M. A. M., Soares Neto, T. G., Veras, C. A. G., Costa, F. S., van Leeuwen, T. T., Krieger Filho, G. C., Tourigny, E., Forti, M. C., Fostier, A. H., Siqueira, M. B., Santos, J. C., Lima, B. A., Cascão, P., Ortega, G., & Frade, E. F., Jr., (2016). CO_2 and CO emission rates from three forest fire controlled experiments in Western Amazonia. *Atmospheric Environment*, 135, 73–83. https://doi.org/10.1016/j.atmosenv.2016.03.043

Costa, F. S., Albuquerque, J. A., Bayer, C., Fontoura, S. M. V., & Wobeto, C. (2003). Propriedades físicas de um Latossolo bruno afetadas pelos sistemas plantio direto e preparo convencional. (In Portuguese, with English abstract). *Revista Brasileira de Ciência do Solo*, 27, 527–535. https://doi.org/10.1590/S0100-06832003000300014

Costa, F. S., Campos Filho, M. D., Sá, C. P., Lambertucci, D. M., Tavella, L. B., Brito, E. S., Klein, M. A., & Dick, D. P. (2020). *Agricultura conservacionista: Solução de inovação tecnológica e econômica para a produção diversa em solos arenosos do Juruá, Acre, Sudoeste da Amazônia – resultados integrados de 13 anos.* [Conservation agriculture: Technological and economic innovation solution for diverse production in sandy soils of Juruá, Acre, Southwest Amazon – integrated results of 13 years]. Embrapa Acre.

Costa, F. S., Campos Filho, M. D., Santiago, A. C. C., Magalhães, I. B., Cordeiro, L. S., Lima, A. P., Maia, G. R., Silva, E. P., Klein, M. A., Silva, F. A. C., Bardales, N. G., Queiroz, L. R., & Brito, E. S. (2014). *Agricultura conservacionista na produção familiar de mandioca e milho no Juruá, Estado do Acre: Efeitos da adoção nos resultados de safras de 2006 a 2014* [Conservation agriculture in family production of cassava and maize in Juruá, State of Acre: Effects of adoption on crop results from 2006 to 2014]. Embrapa Acre.

Davidson, E. A., & Martinelli, L. A. (2009). Nutrient limitations to secondary forest regrowth. In M. Keller, M. Bustamante, J. Gash, & P. S. Dias (Eds.), *Amazonia and global change* (pp. 229–309). American Geophysical Union.

Duarte, A. F. (2006). Aspectos da climatologia do Acre, Brasil, com base no intervalo 1971-2000. (In Portuguese, with English abstract). *Revista Brasileira de Meteorologia*, 21, 308–317.

FAO. (2014). *What is conservation agriculture?* FAO. https://www.fao.org/conservation-agriculture/overview/what-is-conservation-agriculture/en

FAO. (2002). *World reference base for soil resources*. FAO. https://www.fao.org/soils-portal/data-hub/soil-classification/world-reference-base/en

Flores, B. M., Staal, A., Jakovac, C. C., Hirota, M., Holmgren, M., & Oliveira, R. S. (2020). Soil erosion as a resilience drain in disturbed tropical forests. *Plant and Soil*, 450, 11–25. https://doi.org/10.1007/s11104-019-04097-8

Fuentes-Llanillo, R., Telles, T. S., Soares Junior, D., Melo, T. R., Friedrich, T., & Kassam, A. (2021). Expansion of no-tillage practice in conservation agriculture in Brazil. *Soil & Tillage Research*, 208, 104877. https://doi.org/10.1016/j.still.2020.104877

Garner, E., & O Campos, A. P. de la (2014). *Identifying the "family farm": An informal discussion of the concepts and definitions*. FAO.

Gatti, L. V., Basso, L. S., Miller, J. B., Gloor, M., Domingues, L. G., Cassol, H. L. G., Tejada, G., Aragão, L. E. O. C., Nobre, C., Peters, W., Marani, L., Arai, E., Sanches, A. H., Corrêa, S. M., Anderson, L., Von Randow, C., Correia, C. S. C., Crispim, S. P., & Neves, R. A. L. (2021). Amazonia as a carbon source linked to deforestation and climate change. *Nature*, 595, 388–393. https://doi.org/10.1038/s41586-021-03629-6

Gomes, G. M., & Vergolino, J. R. (1997). *Trinta e cinco anos de crescimento econômico na Amazônia (1960/1995)* [Thirty-five years of economic growth in the Amazon (1960/1995)]. Ipea.

Gonçalves, C. N., Dalmolin, R. S. D., Dick, D. P., Knicher, H., Klamt, E., & Koogel-Knabner, I. (2003). The effect of 10% HF treatment on the resolution of CPMAS 13C NMR spectra and on the quality of organic matter in Ferralsols. *Geoderma*, 116, 373–392. http://dx.doi.org/10.1016/S0016-7061(03)00119-8

Guiducci, R. C. N., Alves, E. R. A., Lima Filho, J. R., & Mota, M. M. (2012). Aspectos metodológicos da análise de viabilidade econômica de sistemas de produção [Methodological aspects of the economic feasibility analysis of production systems]. In R. C. N. Guiducci, J. R. de Lima Filho, & M. M. Mota (Eds.), *Viabilidade econômica de sistemas de produção agropecuários: metodologia e estudos de caso* (pp. 17–78). Embrapa.

Hobbs, P. R., Sayre, K., & Gupta, R. (2007). The role of conservation agriculture in sustainable agriculture. *Philosophical Transactions of the Royal Society B*, 363, 543–555. https://doi.org/10.1098/rstb.2007.2169

Instituto Nacional de Meteorologia. (2020). *Estações automáticas e convencionais.* INMET. https://bdmep.inmet.gov.br#

Instituto Nacional de Pesquisas Espaciais. (2022). *Prodes.* INPE. http://www.dpi.inpe.br/prodesdigital/prodesmunicipal.php

Kassam, A., Friedrich, T., & Derpsch, R. (2022). Successful experiences and lessons from conservation agriculture worldwide. *Agronomy*, 12, 769. https://doi.org/10.3390/agronomy12040769

Kato, O. R., Borges, A. C. M. R., Azevedo, C. M. B. C., Aragão, D. V., Matos, G. B., Matos, L. M. S., Shimizu, M. K., Vasconcelos, S. S., & Sá, T. D. A. (2020). *Projeto Tipitamba – Transformando paisagens e compartilhando conhecimento na Amazônia* [Tipitamba Project – Transforming landscapes and sharing knowledge in the Amazon]. CEPAL.

Knicker, H. (2011). Solid state CPMAS 13C and 15N NMR spectroscopy in organic geochemistry and how spin dynamics can either aggravate or improve spectra interpretation. *Organic Geochemistry*, 42, 867–890. http://dx.doi.org/10.1016/j.orggeochem.2011.06.019

Labrière, N., Locatelli, B., Laumonier, Y., Freycon, V., & Bernoux, M. (2015). Soil erosion in the humid tropics: A systematic quantitative review. *Agriculture, Ecosystems & Environment*, 203, 127–139. https://doi.org/10.1016/j.agee.2015.01.027

Landers, J. N., Freitas, P. L., Oliveira, M. C., da Silva Neto, S. P., Ralisch, R., & Kueneman, E. A. (2021). Next steps for conservation agriculture. *Agronomy*, 11, 2496. https://doi.org/10.3390/agronomy11122496

Latrubesse, E. M., Silva, S. A. F., Cozzuol, M., & Absy, M. L. (2007). Late Miocene continental sedimentation in southwestern Amazonia and its regional significance: Biotic and geological evidence. *Journal of South American Earth Sciences*, 23, 61–80. https://doi.org/10.1016/j.jsames.2006.09.021

Leal, O. A., Dick, D. P., Costa, F. S., Knicker, H., Carvalho Júnior, J. A., & Santos, J. C. (2018). Carbon in physical fractions and organic matter chemical composition of an Acrisol after Amazon Forest burning and conversion into pasture. *Journal of Brazilian Chemical Society*, 30, 413–424. http://dx.doi.org/10.21577/0103-5053.20180191

Lense, G. H. E., Parreiras, T. C., Spalevic, V., Avanzi, J. C., & Mincato, R. L. (2021). Soil losses in the state of Rondônia, Brazil. *Ciência Rural*, 51, e20200460. http://doi.org/10.1590/0103-8478cr20200460

Libonati, R., Pereira, J. M. C., da Camara, C. C., Peres, L. F., Oom, D., Rodrigues, J. A., Santos, F. L. M., Trigo, R. M., Gouveia, C. M. P., Machado-Silva, F., Enrich-Prast, A., & Silva, J. M. N. (2021). Twenty-first century droughts have not increasingly exacerbated fire season severity in the Brazilian Amazon. *Scientific Reports*, 11, 4400. https://doi.org/10.1038/s41598-021-82158-8

Mackensen, J., Hölscher, D., Klinge, R., & Fölster, F. (1996). Nutrient transfer to the atmosphere by burning of debris in eastern Amazonia. *Forest Ecology and Management*, 86, 121–128. https://doi.org/10.1016/S0378-1127(96)03790-5

Marin, F. R., Zanon, A. J., Monzon, J. P., Andrade, J. F., Silva, E. H. F. M., Richter, G. L., Antolin, L. A. S., Ribeiro, B. S. M. R., Ribas, G. G., Battisti, R., Heinemann, A. B., & Grassini, P. (2022). Protecting the Amazon forest and reducing global warming via agricultural intensification. *Nature Sustainability*, 5, 1018–1026. https://doi.org/10.1038/s41893-022-00968-8

Mendes, I. C., Sousa, D. M. G., Chaer, G. M., Reis Junior, F. B., Lopes, A. A. C., Dantas, O. D., & Souza, L. M. (2021). Soil bioanalysis: a simple and effective tool to access and interpret soil health. *Proceedings of the Global Symposium on Soil Biodiversity 2021* (pp. 256–261). FAO. https://doi.org/10.4060/cb7374en

Mendes, I. C., Souza, L. M., Sousa, D. M. G., Lopes, A. A. C., Reis Junior, F. B., Lacerda, M. P. C., & Malaquias, J. V. (2019). Critical limits for microbial indicators in tropical Oxisols at postharvest: The FERTBIO soil sample concept. *Applied Soil Ecology*, 139, 85–93. https://doi.org/10.1016/j.apsoil.2019.02.025

Ministério da Agricultura, Pecuária e Abastecimento. (2022). *ZARC - Zoneamento agrícola de risco climático* [ZARC - Agricultural zoning of climate risk]. MAPA. https://indicadores.agricultura.gov.br/zarc/index.htm

Moura, E. G., Gehring, C., Braun, H., Ferraz Junior, A. S. L., Reis, F. O., & Aguiar, A. C. F. (2016). Improving farming practices for sustainable soil use in the humid tropics and rainforest ecosystem health. *Sustainability*, 8, 841. https://doi.org/10.3390/su8090841

Nepstad, D. C., Stickler, C. M., Soares-Filho, B., & Merry, F. (2008). Interactions among Amazon land use, forests and climate: Prospect for a near-term forest tipping point. *Philosophical Transactions of the Royal Society B*, 363, 1737–1746. https://doi.org/10.1098/rstb.2007.0036

Oldfield, E. E., Eagle, A. J., Rubin, R. L., Rudek, J., Sanderman, J., & Gordon, D. R. (2022). Crediting agricultural soil carbon sequestration: Regional consistency is necessary for carbon credit integrity. *Science*, 375, 1222–1225. https://doi.org/10.1126/science.abl7991

Passinato, J. H., Amado, T. J. C., Kassam, A., Acosta, J. A. A., & Amaral, L. d. P. (2021). Soil health check-up of conservation agriculture farming systems in Brazil. *Agronomy*, 11, 2410. https://doi.org/10.3390/agronomy11122410

Prado, R. B., Fidalgo, E. C. C., Monteiro, J. M. G., Schuler, A. E., Vezzani, F. M., Garcia, J. R., Oliveira, A. P., Viana, J. H. M., Pedreira, B. C. C. G., Mendes, I. C., Reatto, A., Parron, L. M., Clemente, E. P., Donagemma, G. K., Turetta, A. P. D., & Simões, M. (2016). Current overview and potential applications of the soil ecosystem services approach in Brazil. *Pequisa Agropecuária Brasileira*, 51, 1021–1038.

Quesada, C. A., Lloyd, J., Anderson, L. O., Fyllas, N. M., Schwarz, M., & Czimczik, C. I. (2011). Soils of Amazonia with particular reference to the RAINFOR sites. *Biogeosciences*, 8, 1415–1440. https://doi.org/10.5194/bg-8-1415‐2011

Rocha Neto, J. M. (2021). Sobre expectativas e frustrações: Uma interpretação da governança do plano regional de desenvolvimento da Amazônia. (In Portuguese, with English abstract.). *Revista Tempo do Mundo*, 27, 273–303. https://doi.org/10.38116/rtm27art10

Santos, H. G., Jacomine, P. K. T., Anjos, L. H. C., Oliveira, V. A., Lumbreras, J. F., Coelho, M. R., Almeida, J. A., Araujo Filho, J. C., Oliveira, J. B., & Cunha, T. J. F. (2018). *Sistema Brasileiro de classificação de solos* [Brazilian system of soil classification] (5th ed.). Embrapa. https://www.embrapa.br/solos/sibcs

Santos, W. L., & Augustin, C. H. R. R. (2015). Water and sediment loss through runoff in areas of forest and pasture cover in southwestern Amazonia – Acre– Brazil. *Zeitschrift für Geomorphologie*, 59, 23–39. https://doi.org/10.1127/zfg_suppl/2015/S-59203

Secretaria Estadual de Meio Ambiente. (2006). *Programa Estadual de Zoneamento Ecológico-Econômico do Acre. Zoneamento ecológico econômico do Acre: fase II: documento síntese: escala 1: 250.000* [State Program for Ecological-Economic Zoning of Acre. Economic ecological zoning of Acre: phase II: synthesis document: scale 1: 250,000]. Governo do Estado do Acre. SEMA.

Silva, C. P., Araújo, E. A., Silva, J. F., Moreira, W. C. L., Negri, F. R., & Oliveira, W. M. S. (2021a). Caracterização de perfis geológicos desenvolvidos em geoambientes da Formação Cruzeiro do Sul, Amazônia sul-ocidental. (In Portuguese, with English abstract). *Boletim do Museu Paraense Emílio Goeldi. Ciências Naturais*, 16, 115–127. http://doi.org/10.46357/bcnaturais.v16i1.304

Silva, S. S., Oliveira, I., Morello, T. F., Anderson, L. O., Karlokoski, A., Brando, P. M., Melo, A. W. F., Costa, J. G., Souza, F. S. C., Silva, I. S., Nascimento, E. S., Pereira, M. P., Almeida, M. R. N., Alencar, A., Aragão, L. E. O. C., Brown, I. F., Graça, P. M. L. A., & Fearnside, P. M. (2021b). Burning in southwestern Brazilian Amazonia, 2016–2019. *Journal of Environmental Management*, 286, 112189. https://doi.org/10.1016/j.jenvman.2021.112189

Soil Survey Staff. (1999) https://www.nrcs.usda.gov/sites/default/files/2022-06/Soil%20Taxonomy.pdf

Tabatabai, M. A. (1994). Soil enzymes. In R. W. Weaver, et al. (Eds.), *Methods of soil analysis. Part 2. Microbiological and biochemical properties* (pp. 775–833). SSSA.

Tamburini, G., Bommarco, R., Wanger, T. C., Kremen, C., van der Heijden, M. G. A., Liebman, M., & Hallin, S. (2020). Agricultural diversification promotes multiple ecosystem services without compromising yield. *Science Advances*, 6, eaba1715. https://doi.org/10.1126/sciadv.aba1715

Toledo, P. M., Dalla-Nora, E., Vieira, I. C. G., Aguiar, A. P. D., & Araújo, R. (2017). Development paradigms contributing to the transformation of the Brazilian Amazon: Do people matter? *Current Opinion in Environmental Sustainability*, 26–27, 77–83. https://doi.org/10.1016/j.cosust.2017.01.009

Wiesenberg, G. L. B., Schawark, L., & Schimidt, M. W. I. (2004). Sources and turnover of organic matter in agricultural soils derived from n-alkane/n-carboxylic acid compositions and C-isotope signatures. *Organic Geochemistry*, 35, 1371–1393. https://doi.org/10.1016/j.orggeochem.2004.03.009

Wittwer, R. A., Bender, S. F., Hartman, K., Hydbom, S., Lima, R. A. A., Loaiza, V., Nemecek, T., Oehl, F., Olsson, P. A., Petchey, O., Prechsl, U. E., Schlaeppi, K., Scholten, T., Seitz, S., Six, J., & van der Heijden, M. G. A. (2021). Organic and conservation agriculture promote ecosystem multifunctionality. *Science Advances*, 7, eabg6995. https://doi.org/10.1126/sciadv.abg6995

WRB/FAO. (2002) https://www.fao.org/soils-portal/data-hub/soil-classification/world-reference-base/en/

9

Soil Health in Brazilian Forestry Systems

Elke Jurandy Bran Nogueira Cardoso, José Leonardo de Moraes Gonçalves, Victor Lucas Vieira Prudêncio de Araújo, Antonio Marcos Miranda Silva, Ademir Sérgio Ferreira de Araujo, and Arthur Prudêncio de Araújo Pereira

Chapter Overview

The forestry sector plays an important economic and ecological role in Brazil, providing wood-based products for industrial purposes and reducing pressure on native forest systems. Traditional and innovative strategies have been applied to promote soil health (SH) in planted forests in Brazil, with a particular focus on eucalypt plantations. For example, the minimum tillage system (MTS) was introduced in 1990 and is now widely used in various production sectors. Several beneficial effects of the MTS on soil have been shown (e.g., maintaining the soil physical structure, increasing soil C, reducing erosion, among others). Mixed plantations including both eucalypt and N_2-fixing plants (e.g., acacia [*Acacia* spp.]) have been tested in several eco-regions across Brazil over the last decade. Initial results demonstrated that mixed plantations of acacia with eucalypt can increase soil C pools and fertility status while improving the soil physical quality. Mixed eucalypt plantations have also been shown to significantly improve soil biological health by increasing enzyme activity and microbial diversity, including bacteria, fungi, and archaea. However, these benefits were only observed when forest residues were left on the soil. Residue removal has been shown to negatively affect the chemical composition of soil organic fractions, consequently affecting SH. Although these advances are significant, little research has been done on SH assessment of planted forests in Brazil by integrated approaches that include chemical, physical, and biological indicators. In some local studies, SH has been addressed by multivariate analysis and standard algorithms for land use change, by which mixed plantations can be compared with native forests, eucalypt monoculture, and pasture. However, because Brazil is the world's

Soil Health Series: Volume 3 Soil Health and Sustainable Agriculture in Brazil, First Edition.
Edited by Ieda Carvalho Mendes and Maurício Roberto Cherubin.

largest eucalypt producer, national initiatives are urgently needed to access and compare SH across regions. Monitoring SH in the forestry sector would be an important tool for evaluation of silvicultural practices, with a view to improving fertilizer use efficiency and most importantly, increasing soil biodiversity and sustainability.

Introduction

Worldwide, a total of 294 million ha are covered with planted forests (~7% of the Earth's surface). Planted forests are used for wood and wood products, energy generation, and cellulose for different purposes (FAO, 2021). Currently, industrial forests cover ~9.59 million ha in Brazil. Of this total area, around 7.45 million ha are used for *Eucalyptus* sp. (i.e., ~30% of the global eucalypt plantations). In addition, pine (*Pinus* sp.) is planted on 1.7 million ha and rubber [*Hevea brasiliensis* (Willd. ex A. Juss.) Müll. Arg.], acacia, teak (*Tectona grandis* L. f.), Paricá (*Schizolobium amazonicum* Huber ex Ducke), and other species on ~0.38 million ha (Figure 9.1). In 2020, these forest plantations produced 216.6 million m^3 of wood, of which 74.5% was eucalypt and 23.9% pine wood.

Eucalypts have been successful in adapting to Brazilian soils, but this success was no coincidence. This Australian species has been the subject of research studies in Brazil for more than 60 years, mostly focused on pulp and paper production (Castro et al., 2016). Some reasons for the broad use of eucalypts include its fast growth, market demand, low cost, seedling availability, and tree architecture, which facilitates harvesting and transportation.

In general, eucalypt forests in Brazil are planted on soils with low natural fertility, requiring the application of adequate soil management practices to

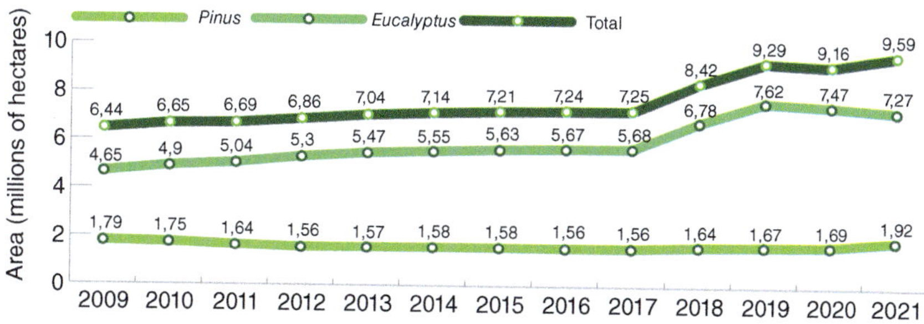

Figure 9.1 Evolution of the planted area per tree species in Brazil. APRE (2022). Figure extracted from the public report of APRE (2022) (https://apreflorestas.com.br/publicacoes/estudo-setorial-apre-2022).

achieve high yields (da Silva et al., 2020) and to ensure the long-term sustainability of forest production (McMahon et al., 2019). Despite the short rotation period of most plantations (6–8 years), a diverse range of species with distinct climatic and soil adaptation capacities, associated with easy propagation by seeds or cloning, allows for a successful adaptation to different Brazilian regions.

On average, eucalypt and pine have annual wood growth rates of 37 and 30 $m^3\ ha^{-1}\ yr^{-1}$, respectively (IBÁ, 2021). However, the key challenge to maintaining or increasing productivity while minimizing environmental impacts and ensuring SH lies in enhancing the efficiency of natural resource use by breeding, site-specific allocation of genotypes, and appropriate silvicultural practices (Gonçalves et al., 2013).

The forest sector in Brazil is a global leader in efficiency and advanced technology for the development of forest plantations (Pena-Vergara et al., 2022). This has resulted in the world's lowest cost of pulp and wood production, based on high productivity (Pena-Vergara et al., 2022). Despite covering ~10 million ha, which is a relatively small area compared with the total of natural forests (~485 million ha), forest plantations can supply ~90% of the wood produced for industrial purposes in Brazil (IBÁ, 2017).

However, the ecosystems where most of the forest plantations are established in Brazil are particularly sensitive to anthropogenic disturbance due to various ecological and historical reasons. These regions are characterized by very intensive climatic processes (e.g., high rainfall and thermal indices), which result in frequent highly erosive rains in the rainy season. Additionally, a significant part of the plantations stand on uneven relief, and the natural fertility of soils is generally low, with poor contents of primary minerals (low nutrient reserves) (Gonçalves et al., 2013).

Soil health evaluation has become a focus of several areas of soil science, including the forestry sector, due to the need to assess the impact of land use on SH and forest productivity and to maintain the environmental and economic sustainability of the production chain (Page-Dumroese et al., 2021). However, despite the importance of this issue, few studies have comprehensively evaluated SH in planted forests in Brazil (Simon et al., 2022). Thus, the main objective of this chapter is to provide an overview of Brazilian research studies that examine the impact of planted forests with different soil management strategies on physical, chemical, and biological indicators of SH.

Minimum Soil Tillage System

Soil plays an essential role in the maintenance and functioning of biodiversity. Heavy disturbance can negatively affect soil structure and health. For example, conventional cultivation systems that involve tilling the soil surface layer with total or partial residue incorporation can affect the soil physical structure, C stocks, and functioning (dos Santos & Reichert, 2022). To mitigate these effects, the minimum soil tillage (MST) system has been developed and implemented in forest plantations across Brazil. This system was introduced in the 1990s and is currently being used in over 90% of the forest stands of Brazil (Gonçalves et al., 2013).

The MST system minimizes soil disturbance by tilling the soil only during the required planting operations and leaving the residual plant biomass on the soil surface as mulch. This approach limits soil tillage to only the planting row or pit (Figures 9.2 and 9.3). Due to the wide space between rows (usually 2.5 or 3.0 m), the volume of disturbed soil is much lower than in the case of annual crops. By this practice, more than 70% of the soil surface remains covered with residual plant biomass, making it resistant against soil and water losses through erosive processes (Gonçalves et al., 2013).

Figure 9.2 Windrowing of forest residues and subsoiling of the planting line in a typical area of eucalypt replanting in Brazil. Plants benefit from the greater effective soil depth promoted by subsoiling, improving their tolerance to water stress and competitiveness against weeds. Photos by José Leonardo de Moraes Gonçalves.

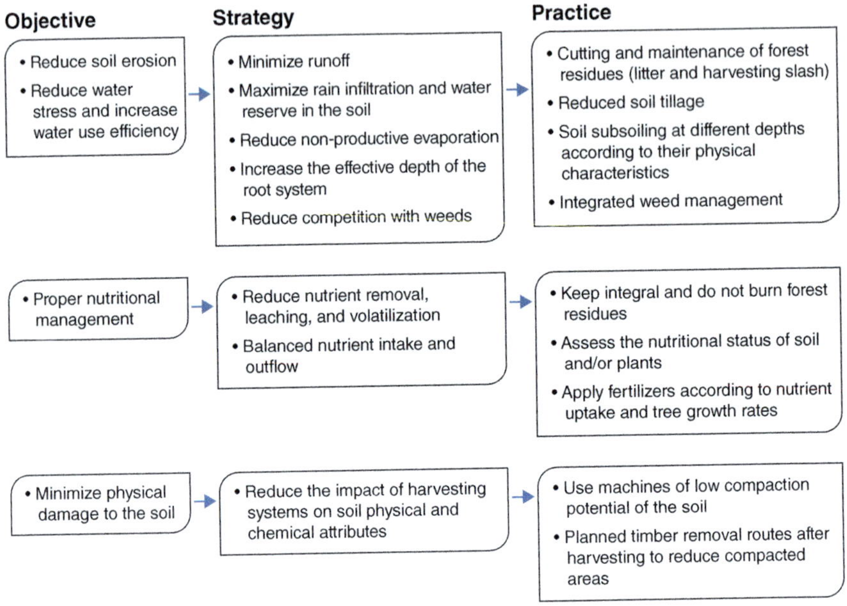

Figure 9.3 Objectives, strategies, and forestry practices in plantations under the Minimum Soil Tillage (MST) system. José Leonardo de Moraes Gonçalves.

The MST system has several benefits, including (a) interruption of the intense erosive processes that occurred when plantations were established with intensive practices of soil tillage; (b) reduction of nutrients and C losses in the ecosystem, which promotes a reduction in the fertilization costs; (c) efficient weed control; and (d) reduced surface runoff due to less compaction and higher water infiltration capacity in the soil, promoting water recharge in the subsurface soil layers and regularization of watercourses (Gonçalves et al., 2013; Santos & Reichert, 2022).

Brazilian Forestry and its Influence on Soil Health Indicators

Eucalypt plantations can increase soil organic C (SOC) pools. Soil organic C is an important indicator of SH because it provides microbial communities with energy, which boosts nutrient cycling and soil functioning (Liptzin et al., 2022). However, SOC storage in forest plantations depends on factors such as previous land use, soil-climatic conditions, and residue management (Rocha et al., 2016). In two eucalypt plantations on Latosols in the state of Minas Gerais, Brazil, Lima et al. (2008) evaluated the effects of four crop rotations (over 34 years) on SOC stocks in comparison with those of nearby native forest and degraded pasture. In both locations, SOC stocks at the 0- to 20-cm depth were highest in the native forest, followed by eucalypt and pasture. In the most clayey soil (~63–70% clay), SOC stocks were 82, 53, and 67 Mg ha^{-1} in the native forest, pasture, and eucalypt, respectively, but were 53, 32, and 42 Mg ha^{-1}, respectively, in the least clayey soil (~0–60% clay). These results suggest that more C is sequestered in eucalypt plantations than in degraded pastures, independent of the soil texture.

Over two decades, in 306 continuous-rotation eucalypt plantations (6- to 8-year harvest cycle), Cook et al. (2016) characterized soil C stocks and changes under a minimum soil cultivation system across a 1200-km gradient in Brazil. Across all sites, in 2010, soil C stocks were 16 ± 0.37 Mg ha^{-1} (mean ± SE) in the 0- to 15-cm layer and 13 ± 0.38 Mg ha^{-1} in the 15- to 30-cm layer, for a total of 29 ± 0.70 Mg ha^{-1}. From the original sampling until 2010 (ranging from 18 to 26 years or approximately three to four crop rotations), average soil C decreased slightly (−0.22 ± 0.05 Mg ha^{-1} yr^{-1}, $p = 0.0001$). At locations with Atlantic Forest vegetation, in Region 2, Espírito Santo, there was a loss of soil C stocks (−0.87 Mg ha^{-1} yr^{-1}, $p = 0.0001$), and at locations with savannah vegetation (Brazilian Cerrado), in Region 3, São Paulo state, the C stocks remained the same (0.06 Mg ha^{-1} yr^{-1}, $p = 0.3969$). Soil C change tended to increase with rainfall during the dry season and with high soil clay content and was less associated with soil order and mean annual temperature.

Thus, according to the results found by Lima et al. (2008) and Cook et al. (2016), in the medium or long term, the conversion of Atlantic Forest to eucalypt plantations reduces the C stock in soil horizon A and, from Cerrado to eucalypt, maintains or increases it a little. On the other hand, the conversion of degraded pastures into eucalypt plantations results in an increase in C stocks in the soil surface layer. These effects are directly related to the change in the net primary productivity of the ecosystem and the decomposition rate of soil organic matter (SOM) (Zhang et al. 2017).

The removal of forest residues can have negative impacts on SOC content, mainly in the topsoil layers (Figure 9.4), and, consequently, on the chemical composition of SOC. This reduction occurs initially in the most labile fractions, whereas the more recalcitrant fractions remain unaffected. Thus, the success of MTS in increasing SOC stocks depends on the maintenance of forest residues in the soil and on soil microbial activity (Maillard et al., 2019).

The impact of different forest plantations on the physical, chemical, and microbiological properties of a clayey Cerrado Oxisol in Brasília, Brazil, was evaluated by Silva et al. (2009). Soil samples from the 0- to 10-cm layer were collected in 20-year-old pine (*Pinus tecunumanii*), eucalypt (*Eucalyptus grandis* W. Hill ex Maid.), and Carvoeiro (*Sclerolobium paniculatum*) stands. An adjacent native Cerrado area was used as reference area of the original soil conditions (Figure 9.5). Soil density increased under pine and eucalypt compared with the native Cerrado area (Table 9.1), which affected the root system architecture and

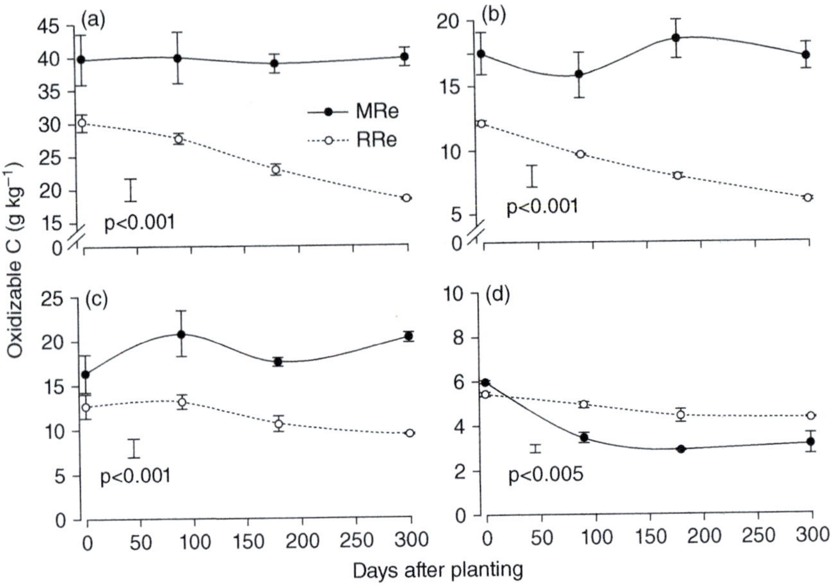

Figure 9.4 Total oxidizable organic C (a) and easily oxidizable (b), medium-oxidizable (c), and hardly oxidizable (d) organic C content in the 0- to 5-cm soil layer, in treatments where all residues were maintained on the soil (MRe) and in which all forest residues were removed (RRe). The bars next to the midpoints indicate the standard error, and the bar without the point indicates the minimum significant difference by the LSD test. Values under the bars indicate the probability of the F test. Rocha (2014). Figure extracted from the report previously published by Rocha (2014) (https://teses.usp.br/teses/disponiveis/11/11150/tde-21032014-135230/publico/Jose_Henrique_Tertulino_Rocha_versao_revisada.pdf).

resulted in less soil cover during the early growth stage of the planted forest. An increase in surface soil density may be attributed to the long-term use of the area; root system architecture; insufficient soil cover during the initial plant growth period; agricultural tractor and machinery traffic; soil moisture during seedling preparation and planting; and, more importantly, reduction in SOC content. A decrease in soil pH and base saturation was observed under pine and Carvoeiro (Table 9.2). The eucalypt plantation promoted an increase in K^+ content compared with native Cerrado (Table 9.2), whereas the Ca^{2+} and Mg^{2+} content of eucalypt soils did not differ significantly from the native Cerrado, although they were higher than in the Carvoeiro and pine plantations (Table 9.2). Because all forest stands were only fertilized at planting and there was no replanting, these results were attributed to nutrient uptake of eucalypt from the deeper soil layers and via litter decomposition.

Figure 9.5 Stands (>20 years) of pine (a), eucalypt (b), and Carvoeiro (c). An adjacent native Cerrado area was used as a reference of the original soil conditions (d). Photos by Ieda Mendes.

Table 9.1 Soil physical properties (0–10 cm) under different forest systems in the Cerrado region (Silva et al., 2009)

Forest system	Soil density	Microporosity	Macroporosity	Total porosity
	$g\ cm^{-3}$		–%–	
Eucalyptus	0.91a	32.2b	31.8a	64.7b
Carvoeiro	0.77b	41.0a	30.0ab	71.0a
Pine	0.87a	40.8a	26.2b	67.0ab
Cerrado	0.78b	35.8ab	34.8a	70.7a

Note: Values followed by the same letters within a column do not differ by Wilcoxon test ($p < 0.10$).

The changes associated with the quantity and quality of the plant residues in the forest systems evaluated in the study of Silva et al. (2009) were reflected in soil biological functioning (Table 9.3). Soil microbial biomass C (MBC) amounts under pine, eucalypt, and Carvoeiro was found to be only half of that under native

Table 9.2 Soil chemical properties (0–10 cm) under different forest systems in the Cerrado region (Silva et al., 2009).

Forest system	Organic matter	pH (H_2O)	m	P	K^+	Ca^{2+}	Mg^{2+}	CEC	V
	$g\,kg^{-1}$		%	–mg dm^{-3}–		–cmol$_c$ dm^{-3}–			%
Eucalyptus	29b	4.86a	60b	0.7	157a	0.47a	0.55a	11.2a	13a
Carvoeiro	42a	4.57b	72a	0.4	34c	0.16b	0.21b	9.7ab	4b
Pine	30b	4.48b	70a	0.7	14d	0.16b	0.18b	8.6b	4b
Cerrado	39a	4.91a	49b	0.5	49b	0.34a	0.32ab	9.6ab	8ab

Note: Values followed by the same letters in columns do not differ by the Wilcoxon test ($p < 0.10$). *Abbreviations*: CEC, cation exchange capacity: CEC = $Ca^{2+} + Mg^{2+} + K^+ + H^+ + Al^{3+}$; m = Al saturation: m = $100[Al^{3+}/(Ca^{2+} + Mg^{2+} + K^+ + Al^{3+})]$; V, base saturation: V, base saturation: V = $100[(Ca^{2+} + Mg^{2+} + K)/CEC]$.

Table 9.3 Soil microbial biomass C (MBC) and soil enzyme activities (0–10 cm) under different forest systems (Silva et al., 2009).

Forest system	MBC[a]	Acid phosphatase	Arylsulfatase	β-glucosidase
	$g\,kg^{-1}$	–μg *p*-nitrophenol g^{-1} h^{-1}–		
Eucalyptus	183b	1118b	79b	85a
Carvoeiro	223b	1452a	113a	71ab
Pine	169b	541c	33c	69b
Cerrado	453a	1171b	83b	76ab

Note: Values followed by the same letters in columns do not differ by Wilcoxon test ($p < 0.10$).
a) Microbial biomass C content determined by the chloroform fumigation-extraction method.

Cerrado vegetation. The activities of acid phosphatase and arylsulfatase were inhibited by pine and stimulated by Carvoeiro, whereas under eucalypt no significant difference to native Cerrado was observed. A principal component analysis (PCA) for the chemical, physical, and biological properties showed an increasing gradient of soil quality along the sequence: pine < eucalypt < Carvoeiro = Cerrado. Of the evaluated forest covers, pine was found to have the strongest negative impact on soil quality.

da Gama-Rodrigues et al. (2008) evaluated changes in the forest soil–litter interface resulting from replacement of native forest with commercial eucalypt plantations in four areas in southeastern Brazil, with soil clay contents between 6% and 59%. The study found that forest litter was higher in eucalypt than native forest stands due to the higher C/N ratio. However, the impact of the conversion of native forest to eucalypt stands on soil and forest floor properties varied depending on the site-specific characteristics. Differences between the contents of MBC and N in eucalypt and native vegetation were more frequently observed in the soil than in forest litter. Forest litter microbial biomass represented a larger reservoir of C and N than soil microbial biomass, indicating the relevance of the litter as a component for C and N cycling in these ecosystems. Apart from one site, soil microbial biomass at the 0- to 10-cm layer decreased significantly under eucalypt compared with native vegetation at the three other sites. Although no differences were observed in the soil chemical properties, significant reductions in SOC were recorded under eucalypt at all four evaluated sites (Table 9.4).

The litter layer plays an important role to maintain nutrient cycling in forest plantations, yet it is generally neglected in studies. According to da Gama-Rodrigues et al. (1997), the qCO_2 (metabolic quotient, i.e., the basal respiration rate per unit of microbial biomass) of litter decreased exponentially with the age of the eucalypt stands. At the beginning, qCO_2 was five times higher than at the end of the rotation (~7 years). According to the authors, this occurred because leaf deposition in the litter is high in the early growth phase and shortly after canopy closure, making this substrate less recalcitrant. In this phase, the bacterial community is more efficient in incorporating C into the biomass. However, in the final phase of rotation, the decay process induces a relative increase in branch,

Table 9.4 Litter production, litter microbial biomass C (LMBC), soil microbial biomass C (SMBC), soil organic C (SOC), and total N in forest systems at four different sites in southeastern Brazil (da Gama-Rodrigues et al., 2008).

Site	Forest system	Litter production	LMBC	SMBC	SOC	Total N
		Mg ha^{-1}	–mg kg dry soil^{-1} –		–g kg^{-1}–	
Aracruz, ES	*Eucalyptus*	37.6a	4491a	81a	7.7b	0.7b
	Native	22.5b	7293a	135a	11.1a	1.0a
Guanhães, MG	*Eucalyptus*	6.4a	1828a	310b	12.3b	1.4b
	Native	3.8b	4313b	597a	18.4a	2.4 a
Luiz Antonio, SP	*Eucalyptus*	9.5a	1876a	95b	4.9a	0.4a
	Native	8.7b	2400a	137a	3.8b	0.4a
Lençois Paulista, SP	*Eucalyptus*	4.2a	4875a	62b	6.8b	0.4b
	Native	2.4b	8417a	270a	10.9b	1.5a

Note: Values followed by the same letters in columns do not differ by the Duncan test ($p < 0.05$).

a) Clay contents at Aracruz, ES; Guanhães, MG; Luiz Antonio, SP; and Lençois Paulista, SP, were 330, 590, 60, and 130 g kg^{-1}, respectively.
b) Microbial biomass C content determined using the chloroform fumigation-extraction method.

bark, and ligneous residues, making the substrate more recalcitrant. In this phase, the fungal community predominates in the litter (Pereira et al., 2021) because fungi are the primary decomposers of lignin-rich substrates. This represents a great challenge for SH assessment in forest plantations because the dynamics of soil litter properties vary over time during the rotation (6–7 years).

Residue management practices, particularly total removal, can affect several SH indicators, such as soil fertility (Menegale et al., 2016), microbial community activity (Maillard et al., 2019), soil organisms (Rousseau et al., 2018), erodibility (Wichert et al., 2018), and compaction (Cherubin et al., 2021). In a commercial eucalypt plantation, Chaer and Tótola (2007) established the physical, nutritional, and biochemical/microbiological quality indicators of a medium-textured soil (20% clay content), evaluated 5.5 years after the site had undergone different management practices during stand replanting. Soil samples were collected from the 0- to 5-cm layer and the 5- to 20-cm layer on a commercial farm in Botucatu, São Paulo state. The microbiological and biochemical properties were found to be more adequate to detect soil quality changes caused by management than chemical or physical indicators (Table 9.5). A PCA showed that soil quality under secondary forest, which was taken as a reference, was significantly different from that of eucalypt plantations, demonstrating that the introduction of eucalypt monoculture causes notable changes in soil quality. The soil quality of eucalypt stands left unharvested after the first normal 7-year cycle (11-year-old stand) was the closest to that of the native forest, followed by the soils under eucalypt subjected to management systems that prioritized organic residue conservation during stand replanting. Contrarily, areas from which the organic material was removed or burned from the soil surface appeared very distant from the reference area in the PCA plot. The microbiological indicators that were most sensitive to land use (natural vegetation × eucalypt) or soil management (pure eucalypt) were microbial biomass N and the activity of the enzymes dehydrogenase, β-glucosidase, acid phosphatase, and alkaline phosphatase (Table 9.5). These results demonstrate that the management system used during eucalypt stand replanting had a medium-term impact on the potential of soils to store and cycle nutrients via microbial biomass and associated biochemical activities.

In another study, de São José et al. (2023) evaluated the impact of different residue management practices on SH in eucalypt forests. The experiment was conducted in Barra do Ribeiro in the state of Rio Grande do Sul, southern Brazil, on a site with Arenosols ($33\,g\,kg^{-1}$ clay in the 0- to 20-cm layer). The authors applied the tool Soil Management Assessment Framework Karlen et al. (1997) to calculate the SH index. In brief, the SH index was best where all residues from the previous rotation, including bark, branches, leaves, and litter, were left on the soil (FRM treatment) or where only bark was removed (FRMB treatment)

Table 9.5 Activity of dehydrogenase, β-glucosidase, acid, and alkaline phosphomonoesterases and urease enzymes in soil samples under eucalypt plantation taken from plots subjected to different cultural managements and from an area with adjacent natural vegetation.

Area	Dehydrogenase		β-Glucosidase		Acid phosphatase		Alkaline phosphatase		Urease	
	0–5 cm	0–20 cm	0–5 cm	0–20 cm	0–5 cm	0–20 cm	0–5 cm	0–20 cm	0–5 cm	0–20 cm
	–μmol kg^{-1} h^{-1} TFF–		–mmol kg^{-1} h^{-1} *p*-nitrophenol–						–mmol kg^{-1} h^{-1} NH_4^+ –	
AVN	348.1a	54.0a	69.5a	41.9a	421.2a	165.1a	71.6a	16.2a	2.11a	2.05a
E11	54.5b	21.9b	45.9b	25.7b	223.7b	103.0b	29.9b	12.4ab	1.81a	1.86ab
RCM	108.0b	29.1b	50.1ab	18.2b	179.3b	77.1bc	24.7b	8.0b	1.54a	2.10a
RGC	76.1b	28.0b	46.5b	18.5b	175.5b	82.0bc	30.0b	9.6b	1.39a	1.59ab
RCS	70.8b	21.7b	42.8b	15.9b	146.6b	73.3c	24.3b	9.6b	1.55a	1.35ab
INC	74.5b	33.8b	44.4b	23.5b	174.7b	99.9bc	23.1b	8.4b	1.52a	1.69ab
RCQ	51.4b	24.6b	44.6b	14.9b	154.2b	74.4c	25.7b	7.9b	1.80a	1.06b

Means followed by the same letter within a column did not differ from each other by Tukey's test at 5%.
Abbreviations: AVN, natural vegetation area of secondary forest; INC, reform with incorporation of all residues with grid; E11, area where the first cycle trees were not cut (11-year-old eucalypt); RCM, renovation without removing residues (minimum cultivation); RCQ, reform with burning of all residues; RCS, reform with removal of cultural remains and litter; RGC, renovation with removal of cultural remains (branches and bark); TFF, triphenyl formazan.
Note. Chaer and Tótola (2007).

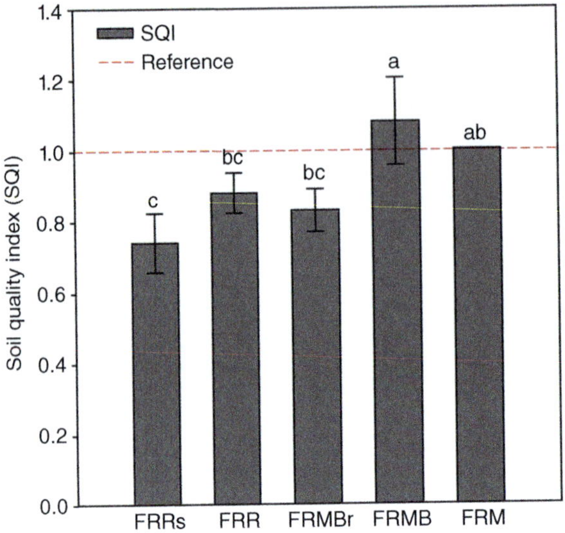

Figure 9.6 Soil health index under different eucalypt harvest residue management strategies. FRM: All residues (i.e., bark, branches, leaves, and litter layer of the previous rotation) left on the soil and removal of only trunk wood. FRMB: Same as FRM, but bark also removed. FRMBr: All residues except trunk wood and branches left on the soil. FRR: All residues removed. FRRs: Same as FRRs, but litter from the new plantation prevented from falling onto the soil surface by a shade net. Figure extracted from the paper previously published by de São José et al. (2023)/with permission of Springer Nature.

(Figure 9.6). The study highlights the importance of maintaining SH for eucalypt productivity, which is adversely affected by soil C depletion and the reduction of biological activity.

Mixed Forest Plantations as a Strategy to Improve Soil Health

The introduction of leguminous trees in mixed plantations represents an opportunity to reduce mineral fertilizer and pesticide applications, decrease production costs, and improve SH while reducing the risks of soil and water contamination (Bouillet et al., 2008a; Pereira et al., 2022). Several field experiments have demonstrated positive results of acacia trees intercropped in eucalypt plantations (Cardoso et al., 2020). Acacia is a leguminous plant efficient in biological N_2 fixation through root-associated bacteria groups, besides making associations with different types of endo/ectomycorrhizal fungi, leading to enhanced soil microbiome diversity and benefits for neighboring plants (Bini et al., 2018; Pereira et al., 2018a). The mixed system of intercropped *Acacia mangium* Willd. and *E. grandis* plantations has been recently studied, and most authors concluded that mixed systems are more productive than pure monoculture plantations (Brandani et al., 2020).

Eucalyptus spp. is economically and ecologically important in Brazil (the world's largest eucalypt producer). The species is well adapted to different climates, soil types, and water regimes (Gonçalves et al., 2013). However, when grown in monoculture, *E. grandis* can negatively affect SH indicators (Gonçalves et al., 2013; Pereira et al., 2018b). Previous studies have reported that *E. grandis* extracts large amounts of N from the soil during harvesting, depleting the soil N-pools and microbial diversity (Bouillet et al., 2008a,b; Pereira et al., 2017; Pulito et al., 2015). On the other hand, when intercropped with acacia, *E. grandis* can be less detrimental to the soil because N_2 fixed by acacia could mitigate the negative effects (Pereira et al., 2018b, 2019, 2021, 2022) (Figures 9.7 and 9.8). Particularly, *A. mangium* has been suggested as a suitable N_2-fixing plant species for intercropping with *E. grandis* to promote soil microbiome diversity and nutrient cycling (Koutika et al., 2014, 2020a,b, Paula et al., 2015, 2018; Santana et al., 2021). Thus, mixed-species plantations of *E. grandis* with *A. mangium* have the capacity to improve SH in tropical soils.

A pioneering study in Brazil, conducted by Rachid et al. (2013), analyzed soil fertility and bacterial communities on pure and mixed plantations of *Eucalyptus urograndis* and *A. mangium* in Seropédica, Rio de Janeiro state. The soil was classified as Haplic Planosol, with a sandy texture (~90% sand in the 0- to 20-cm layer). Sampling was done in December 2010 and 2011, during the summer, when the trees were 2 and 3 years old, respectively. Interestingly, the mixed system did not affect the soil fertility status, such as soil total soil C and N content, compared with the pure eucalypt plantation. However, there was a significant increase in soil available P and nitrate contents in both the *A. mangium* and mixed-species treatments.

In a similar experiment, Pereira et al. (2018b) analyzed pure and mixed plantations of *E. grandis* and *A. mangium* in Itatinga, São Paulo state. Briefly, the soil was classified as Ferralsol (FAO classification system) with a sand clay loam texture in the 0- to 20-cm layer (~204 g kg^{-1} clay). Differently from Rachid et al. (2013), Pereira et al. (2018b) showed that mixed plantations increase C and N concentrations in the labile fractions of SOM. Thus, the mixed plantation can be an effective strategy to improve SH in sandy soil areas with low levels of organic matter, enhancing nutrient availability in the soil and promoting eucalypt growth and development.

In subsequent studies, Rachid et al. (2015) and Pereira et al. (2021) investigated the impact of mixed eucalypt and acacia plantations on fungal communities in the same experimental fields described previously. Both studies demonstrated that mixed plantations significantly increased the number of fungal genera and diversity indices in comparison with the monoculture of the two species.

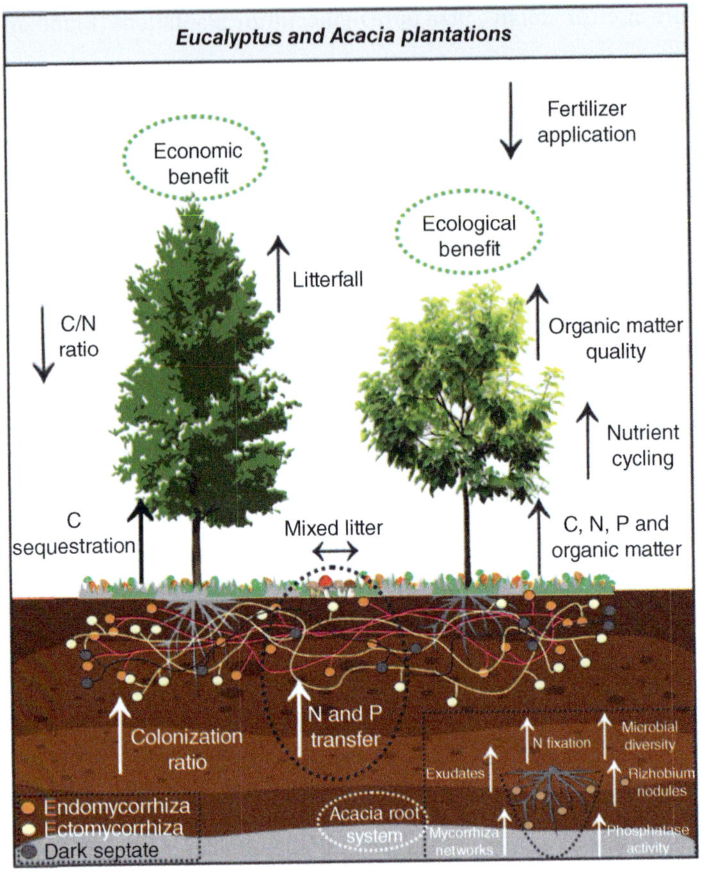

Figure 9.7 Benefits of intercropping eucalypt with acacia compared with pure monoculture. Upward arrows indicate higher values for the two-species intercropping; downward arrows indicate lower values under monoculture. Underground networks are mycorrhizal associations, including endo- and ectomycorrhiza and dark septate endophytes, as well as interactions between the two plants. Pereira et al. (2020). Photos by Arthur Prudêncio de Araújo Pereira. This figure was previously published (https://doi.org/10.1007/978-3-030-32365-3_9).

Interestingly, the richness and Shannon diversity of fungi in the litter were significantly higher in the intercropped and pure acacia plantations (Pereira et al., 2021). Also, a network co-occurrence analysis showed that the fungal taxa network was simpler under pure *E. grandis* plantations, whereas mixed systems were more connected and complex, mainly in the litter layer (Figure 9.9).

Another study by Bini et al. (2018) analyzed the dynamics of arbuscular mycorrhizal fungi (AMF) in these two tree species in both pure and mixed

Figure 9.8 Mixed forest of *Eucalyptus grandis* and *Acacia mangium* (3 m by 3 m; proportion 1:1) at 5 months after planting (a) and 30 months after planting (b) at the Experimental Station of Forest Sciences of Itatinga, São Paulo, Brazil. Photos by José Leonardo de Moraes Gonçalves.

plantations during the first 20 months after planting in Itatinga, São Paulo state. These authors reported an increase in AMF root colonization of *E. grandis* when intercropped with *A. mangium* as well as an increase in the activity of acid and alkaline phosphatase in the presence of the leguminous tree. However, it is important to consider the litter microbiome because microbial complexity and connections can be higher in this layer than in soil (Pereira et al., 2021).

As shown by Pereira et al. (2019), mixed eucalypt and acacia plantations increased not only bacterial diversity but also the number of *nifH* gene copies in the soil, which indicated a close correlation with total soil N. Moreover, the study showed that pure eucalypt plantations supplied with mineral N fertilizers reduced the abundance of functional genes, bacterial diversity, and microbial activities (Pereira et al., 2019). At the same site, Paula et al. (2015) injected ^{15}N directly into the xylem of acacia trees in a mixed plantation with eucalypt and demonstrated a short-term belowground transfer of N from acacia to eucalypt. This facilitation could lead to the release of N pools to non-fixing trees growing in unfertile and degraded tropical soils via common mycorrhizal networks (Paula et al., 2015).

In recent years, several different strategies have been applied to access soil quality and SH in pure and mixed eucalypt plantations of Brazil. Zagatto et al. (2019) collected soil samples from pure and mixed eucalypt plantations in Itatinga, São Paulo, during spring, in a second rotation (6 years). The soil was classified as a Geric Rhodic Ferralsol, with 84% sand and 12% clay. The authors analyzed chemical and microbiological soil properties as well as mesofauna

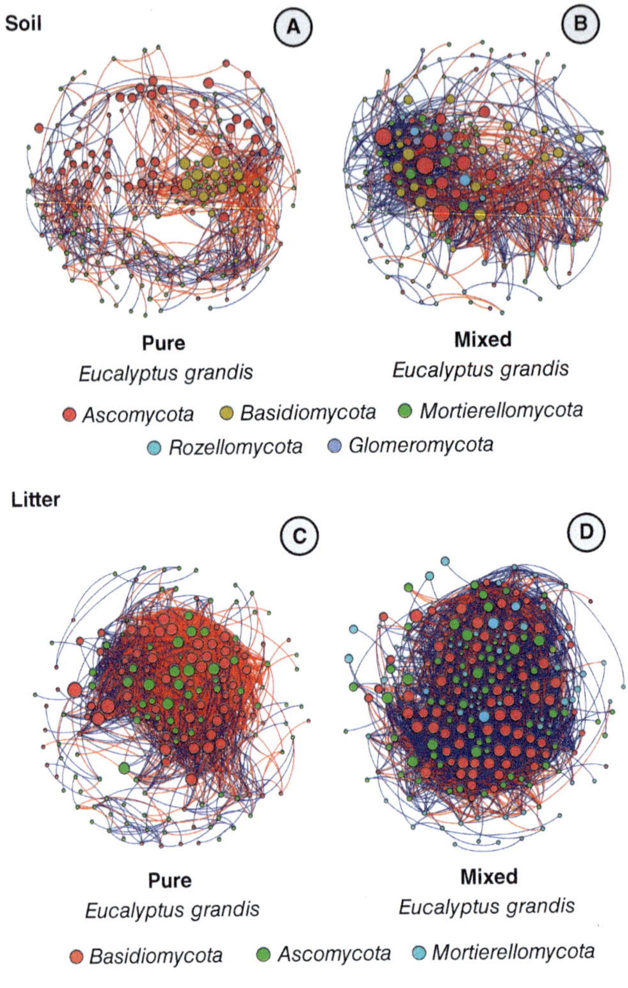

Figure 9.9 Network co-occurrence analysis of fungal communities associated with soil (a and b) and litter (c and d) of pure and mixed *Eucalyptus grandis* and *Acacia mangium* plantations. Blue and red edges are positive and negative correlation, respectively. Each node represents taxa affiliated at the phylum level (based on ITS region). The size of the node is proportional to the number of connections (degree), and each node was labeled at the phylum level. Pereira et al. (2021)/ With permission of Elsevier. Figure created by Arthur Prudêncio de Araújo Pereira, previously published (https://doi.org/10.1016/j.scitotenv.2020.143711).

abundance and diversity. Finally, the authors proposed a general indicator of soil quality (GISQ). The GISQ was initially based on PCA, applied individually to all soil variable datasets (mesofaunal, chemical, and microbiological soil properties), as described by Velasquez et al. (2007). In summary, the soil quality under pure *E. grandis* plantations was lower (GISQ = 0.32) than under pure *A. mangium* (GISQ = 0.48) and mixed *E. grandis* and *A. mangium* plantations (GISQ = 0.43). According to the PCA, soil CO_2-C emission, metabolic quotient (ratio between MBC and SOC), H + Al, Na^+, C/N ratio, and Al^{3+} were negatively associated with GISQ indicators, whereas total C, N, moisture, and faunal groups were positively associated with SH, mainly in mixed plantations (Zagatto et al., 2019).

In a recent study, in the same experimental area used by Pereira et al. (2019), Pereira et al. (2022) applied SH indices to compare monocropping and mixed plantations of *E. grandis* and *A. mangium* (Figure 9.10). The authors analyzed young stands 27 and 39 months after planting on a medium-texture, highly weathered Ferralsol (FAO/WRB). Soil samples were collected (0–20 cm) and analyzed for chemical (pH, available P, and K^+), physical (bulk density) and biological properties (SOC, MBC, and β-glucosidase activity). The SMAF (Andrews et al., 2004) was used to calculate SH indices with the values of the seven analyzed soil properties. Interestingly, SH scores were significantly higher in pure acacia and mixed eucalypt and acacia plantations in both periods (values >0.9) (Figure 9.10a). However, only chemical indicators differed significantly in SH scores (Figure 9.10b). In this case, the SH of acacia monoculture and the mixed plantation scored higher for chemical properties both 27 and 39 months after planting (Figure 9.10b). In contrast, no changes were detected between the physical (Figure 9.10c) and biological (Figure 9.10d) SH scores among the treatments, independent of the sampling period. Among the chemical indicators, pH and P were most affected by the treatments, indicating that these chemical properties were closely correlated with the SH indices. Phosphorus is essential for eucalypt growth and productivity and is part of several key macromolecules and plant functions (Margalef et al., 2017). The fact that P availability is typically low in tropical forest soils (Zavišić et al., 2018) reinforces the importance of intercropping *E. grandis* with *A. mangium*. In addition, a recent study reported that P cycling is stimulated in mixed forest plantations, mainly due to the presence of active microbial communities that affect the soil litter layer and organic fractions (Pereira et al., 2020).

Acacia mangium is known for its ability to fix atmospheric N when associated with symbiotic N-fixing bacterial communities in the rhizosphere, which play an important role in biological N fixation (Paula et al., 2015, 2018). Additionally, this species deposits nutrient-rich litter on the soil (Bini et al., 2013a,b, 2018), increasing SOM and microbial biomass (Koutika et al. 2020a,b, 2021). These factors can contribute to improve SH in mixed eucalypt and acacia plantations. Soil organic matter participates in several ecosystem services, and mixed plantations have been shown to be efficient in C sequestration (Cardoso et al., 2013). The increases

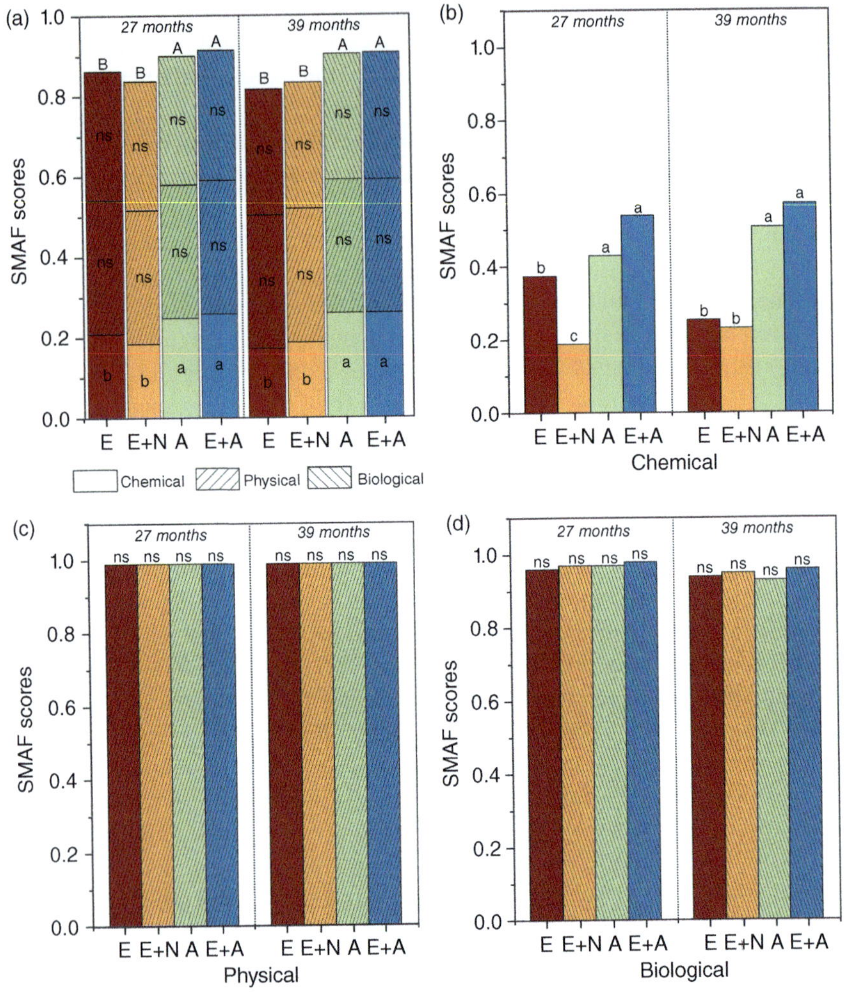

Figure 9.10 Soil Management Assessment Framework–based soil health (SH) index scores (a) and the contribution of chemical (b), physical (c), and biological (d) components to the overall SH (0- to 20-cm layer) under pure and mixed eucalypt–acacia plantations, 27 and 39 months after planting. Pure eucalypt plantation without (E) and with N fertilization (E + N), pure acacia (A) and mixed eucalypt–acacia plantation (E + A). *Bars with the same letter did not differ from one another by Tukey's test ($p < 0.05$); ns = nonsignificant effect. Pereira et al. (2022)/With permission of Elsevier. Figure created by Arthur Prudêncio de Araújo Pereira, previously published (https://dx.doi.org/10.2139/ssrn.4069626).

in SOC in mixed eucalypt–acacia plantations, mainly of the soil labile fractions (Pereira et al., 2018b), and their correlation with SH indices demonstrated the eco-functionality and contribution to soil ecosystem functions and biodiversity (Hoffland et al., 2020).

Final Considerations

Land use conversion from native forests, degraded pastures, and/or agricultural areas to planted forest systems affects several SH indicators, especially those related to C dynamics and biological activity. Also, mixed forest plantations of eucalypt and leguminous trees have been shown to significantly alter SH indicators. Companies in the forestry sector have a growing interest in the introduction of environmentally friendly management practices (e.g., MST system, residue management, mixed plantations, etc.), to warrant the sustainability of forest production on healthy soils. Tools for SH evaluation are therefore essential to monitor the quality and sustainability of the forestry sector in Brazil. In terms of advances in SH assessments, the development of interpretative algorithms for soil enzymes such as arylsulfatase and β-glucosidase specific for areas under eucalypt is expected by 2023/2024. These algorithms will enhance SH monitoring of planted forest stands in Brazil and support the use of the "SoilBio" technology presented in Chapter 10.

As shown throughout this chapter, Brazil has a thriving forestry sector. Planted forests, whether with exotic or native species, play a crucial role in the production of wood and other tree products, indirectly protecting the native tropical forests from deforestation. Prospects for the forest sector in Brazil look promising, with an increasing global demand for cellulose and certified wood, the possibility of expanding integrated crop–livestock–forestry systems, and the strategic use of forests for C sequestration within the scope of clean development mechanisms. All these factors highlight the importance of research studies and technologies toward the development of sustainable forestry production systems and understanding the impacts of this activity on the environment, particularly on SH and soil functioning.

References

Andrews, S. S., Karlen, D. L., & Cambardella, C. A. (2004). The soil management assessment framework: a quantitative soil quality evaluation method. Soil Science Society of America Journal, 68(6), 1945–1962. https://doi.org/10.2136/sssaj2004.1945

Associação Paranaense de Empresas de Base Florestal (APRE). (2022). *Estudo setorial*. https://apreflorestas.com.br/sobre-o-setor

Bini, D., dos Santos, C. A., Bouillet, J. P., de Gonçalves, J. L. M., & Cardoso, E. J. B. N. (2013a). *Eucalyptus grandis* and *Acacia mangium* in monoculture and intercropped plantations: evolution of soil and litter microbial and chemical properties during early stages of plant development. *Applied Soil Ecology*, 63, 57–66. https://doi.org/10.1016/j.apsoil.2012.09.012

Bini, D., dos Santos, C. A., da Silva, M. C. P., Bonfim, J. A., & Cardoso, E. J. B. N. (2018). Intercropping *Acacia mangium* stimulates AMF colonization and soil phosphatase activity in *Eucalyptus grandis*. *Scientia Agricola*, 75(2), 102–110. https://doi.org/10.1590/1678-992x-2016-0337

Bini, D., Figueiredo, A. F., da Silva, M. C. P., de Vasconcellos, R. L. F., & Cardoso, E. J. B. N. (2013b). Microbial biomass and activity in litter during the initial development of pure and mixed plantations of *Eucalyptus grandis* and *Acacia mangium*. *Revista Brasileira de Ciência do Solo*, 37(1). https://doi.org/10.1590/s0100-06832013000100008

Bouillet, J. P., Laclau, J. P., Gonçalves, J. L. M., Moreira, M. Z., Trivelin, P. C. O., Jourdan, C., Silva, E. V., Piccolo, M. C., Tsai, S. M., & Galiana, A. (2008a). Mixed-species plantations of *Acacia mangium* and *Eucalyptus grandis* in Brazil. *Forest Ecology and Management*, 255(12), 3918–3930. https://doi.org/10.1016/j.foreco.2007.10.050

Bouillet, J. P., Laclau, J. P., Gonçalves, J. L. M., Moreira, M. Z., Trivelin, P. C. O., Jourdan, C., Silva, E. V., Piccolo, M. C., Tsai, S. M., & Galiana, A. (2008b). Mixed-species plantations of *Acacia mangium* and *Eucalyptus grandis* in Brazil. 2: Nitrogen accumulation in the stands and biological N2 fixation. *Forest Ecology and Management*, 255(12), 3918–3930. https://doi.org/10.1016/j.foreco.2007.10.050

Brandani, C. B., Santos, F. M., de Oliveira, I. R., Bordon, B., Bheling, M., Silva, E. V., & de Gonçalves, J. L. M. (2020). Growth patterns at different sites and forest management systems. In E. J. B. N. Cardoso, J. L. de Moraes Gonçalves, F. de Carvalho Balieiro, & A. A. Franco (Eds.), *Mixed plantations of eucalyptus and leguminous trees* (pp. 15–43). Springer. https://doi.org/10.1007/978-3-030-32365-3_2

Cardoso, E. J. B. N., Gonçalves, J. L. M., Balieiro, F. C., Franco, A. A. (2020). Mixed Plantations of Eucalyptus and Leguminous Trees: Soil, Microbiology, and Ecosystem Services. Springer Cham. 280p. https://doi.org/10.1007/978-3-030-32365-3

Cardoso, E. J. B. N., Vasconcellos, R. L. F., Bini, D., Miyauchi, M. Y. H., dos Santos, C. A., Alves, P. R. L., de Paula, A. M., Nakatani, A. S., de Pereira, J. M., & Nogueira, M. A. (2013). Soil health: Looking for suitable indicators. What should be considered to assess the effects of use and management on soil health? *Scientia Agricola*, 70(4), 219–303. https://doi.org/10.1590/S0103-90162013000400009

Chaer, G. M., & Tótola, M. R. (2007). Impact of organic residue management on soil quality indicators during replanting of eucalypt stands. *Revista Brasileira de Ciência do Solo*, 31(6). https://doi.org/10.1590/s0100-06832007000600016

Cherubin, M. R., Franchi, M. R. A., de Lima, R. P., de Moraes, M. T., & da Luz, F. B. (2021). Sugarcane straw effects on soil compaction susceptibility. *Soil and Tillage Research*, 212, 105066. https://doi.org/10.1016/j.still.2021.105066

Cook, R. L., Binkley, D., & Stape, J. L. (2016). *Eucalyptus* plantation effects on soil carbon after 20 years and three rotations in Brazil. *Forest Ecology and Management*, 359, 92–98. https://doi.org/10.1016/j.foreco.2015.09.035

da Gama-Rodrigues, E. F., de Barros, N. F., Viana, A. P., & de Araújo Santos, G. (2008). Microbial biomass and activity in soil and forest litter of eucalyptus plantations and native vegetation in southeastern Brazil. *Revista Brasileira de Ciência do Solo*, 32(4). https://doi.org/10.1590/s0100-06832008000400013

da Silva, N. F., de Barros, N. F., Lima Neves, J. C., Schulthais, F., de Novais, R. F., & Mattiello, E. M. (2020). Yield and nutrient demand and efficiency of eucalyptus under coppicing regime. *Forests*, 11(8), 852. https://doi.org/10.3390/F11080852

da Simon, C. P., Gomes, T. F., Pessoa, T. N., Soltangheisi, A., Bieluczyk, W., de Camargo, P. B., Martinelli, L. A., & Cherubin, M. R. (2022). Soil quality literature in Brazil: A systematic review. *Revista Brasileira de Ciência do Solo*, 46, 1–21. https://doi.org/10.36783/18069657rbcs20210103

de Castro, C. A. O., Resende, R. T., Bhering, L. L., & Cruz, C. D. (2016). Brief history of *Eucalyptus* breeding in Brazil under perspective of biometric advances. *Ciência Rural*, 46(9). https://doi.org/10.1590/0103-8478cr20150645

de Gonçalves, J. L. M., Alvares, C. A., Higa, A. R., Silva, L. D., Alfenas, A. C., Stahl, J., de Ferraz, S. F. B., de Lima, W. P., Brancalion, P. H. S., Hubner, A., Bouillet, J. P. D., Laclau, J. P., Nouvellon, Y., & Epron, D. (2013). Integrating genetic and silvicultural strategies to minimize abiotic and biotic constraints in Brazilian eucalypt plantations. *Forest Ecology and Management*, 301, 6–27. https://doi.org/10.1016/j.foreco.2012.12.030

de São José, J. F. B., Cherubin, M. R., Vargas, L. K., Lisboa, B. B., Zanatta, J. A., Araújo, E. F., & Bayer, C. (2023). A soil quality index for subtropical sandy soils under different *Eucalyptus* harvest residue managements. *Journal of Forestry Research*, 34(1), 243–255. https://doi.org/10.1007/s11676-022-01507-z

dos Santos, K. F., & Reichert, J. M. (2022). Best tillage practices for eucalyptus growth and productivity: A review on the Brazilian experience. *In Revista Brasileira de Ciência do Solo*, 46. https://doi.org/10.36783/18069657rbcs20210091

FAO. (2021). *Forestry production and trade*. https://www.fao.org/forestry/energy/catalogue/search/detail/en/c/1305532

Gama-Rodrigues, E. F., Gama-Rodrigues, A. C., & Barros, N. F. (1997). Biomassa microbiana de carbono e de nitrogênio de solos sob diferentes coberturas florestais Revista Brasileira de Ciência do Solo, 21(3), 361–365. https://doi.org/10.1590/S0100-06831997000300002

Hoffland, E., Kuyper, T. W., Comans, R. N. J., & Creamer, R. E. (2020). Eco-functionality of organic matter in soils. *Plant and Soil*, 455, 1–22. https://doi.org/10.1007/s11104-020-04651-9

Indústria Brasileira de Árvores (IBÁ). (2017). *Annual report*. https://iba.org/images/shared/Biblioteca/IBA_RelatorioAnual2017.pdf

Indústria Brasileira de Árvores (IBÁ). (2021). Annual report. https://iba.org/datafiles/publicacoes/relatorios/relatorioiba2021-compactado.pdf

Karlen, D. L., Mausbach, M.J., Doran, J.W., Cline, R.G., Harris, R.F., & Schuman, G.E. (1997) Soil quality: a concept, definition, and framework for evaluation (A guest editorial). Soil Science Society of America Journal 61, 4–10. https://doi.org/10.2136/sssaj1997.03615995006100010001x

Koutika, L. S., Cafiero, L., Bevivino, A., & Merino, A. (2020a). Organic matter quality of forest floor as a driver of C and P dynamics in acacia and eucalypt plantations established on a Ferralic Arenosols, Congo. *Forest. Ecosystems*, 7(1). https://doi.org/10.1186/s40663-020-00249-w

Koutika, L. S., Epron, D., Bouillet, J. P., & Mareschal, L. (2014). Changes in N and C concentrations, soil acidity and P availability in tropical mixed acacia and eucalypt plantations on a nutrient-poor sandy soil. *Plant and Soil*, 379(1–2), 205–216. https://doi.org/10.1007/s11104-014-2047-3

Koutika, L. S., Fiore, A., Tabacchioni, S., Aprea, G., de Pereira, A. P. A., & Bevivino, A. (2020b). Influence of *Acacia mangium* on soil fertility and bacterial community in eucalyptus plantations in the Congolese coastal plains. *Sustainability*, 12(21), 8763. https://doi.org/10.3390/su12218763

Koutika, L. S., Zagatto, M. R. G., Pereira, A. P. D. A., Miyittah, M., Tabacchioni, S., Bevivino, A., & Rumpel, C. (2021). Does the introduction of N2-fixing trees in forest plantations on tropical soils ameliorate low fertility and enhance carbon sequestration via interactions between biota and nutrient availability? Case studies from Central Africa and South America. Frontiers in Soil Science, 1, 752747. https://doi.org/10.3389/fsoil.2021.752747

Lima, A. M. N., da Silva, I. R., Neves, J. C. L., de Novais, R. F., de Barros, N. F., de Sá Mendonça, E., de Sales Moreira Demolinari, M., & Leite, F. P. (2008). Soil organic matter fractions after three decades of eucalypt cultivation in the Rio Doce Valley. *Brazil. Revista Brasileira de Ciência do Solo*, 32(3). https://doi.org/10.1590/s0100-06832008000300014

Liptzin, D., Norris, C. E., Cappellazzi, S. B., Bean, G., mac Cope, M., Greub, K. L. H., Rieke, E. L., Tracy, P. W., Aberle, E., Ashworth, A., Bañuelos Tavarez, O., Bary, A. I., Baumhardt, R. L., Borbón Gracia, A., Brainard, D. C., Brennan, J. R., Briones Reyes, D., Bruhjell, D., Carlyle, C. N., … Honeycutt, C. W. (2022). An evaluation of carbon indicators of soil health in long-term agricultural experiments. *Soil Biology and Biochemistry*, 172, 108708. https://doi.org/10.1016/j.soilbio.2022.108708

Maillard, F., Leduc, V., Bach, C., de Moraes Gonçalves, J. L., Andreote, F. D., Saint-André, L., Laclau, J. P., Buée, M., & Robin, A. (2019). Microbial enzymatic activities and Community-Level Physiological Profiles (CLPP) in subsoil layers are altered by harvest residue management practices in a tropical *Eucalyptus grandis* plantation. *Microbial Ecology*, 78(2), 528–533. https://doi.org/10.1007/s00248-018-1298-6

Margalef, O., Sardans, J., Fernández-Martínez, M., Molowny-Horas, R., Janssens, I. A., Ciais, P., Goll, D., Richter, A., Obersteiner, M., Asensio, D., & Peñuelas, J. (2017). Global patterns of phosphatase activity in natural soils. *Scientific Reports*, 7(1), 1337. https://doi.org/10.1038/s41598-017-01418-8

McMahon, D. E., Vergütz, L., Valadares, S. V., da Silva, I. R., & Jackson, R. B. (2019). Soil nutrient stocks are maintained over multiple rotations in Brazilian *Eucalyptus* plantations. *Forest Ecology and Management*, 448, 364–375. https://doi.org/10.1016/j.foreco.2019.06.027

Menegale, M. L. C., Rocha, J. H. T., Harrison, R., de Goncalves, J. L. M., Almeida, R. F., de Piccolo, M. C., Hubner, A., Arthur Junior, J. C., de Ferraz, A. V., James, J. N., & Michelsen-Correa, S. (2016). Effect of timber harvest intensities and fertilizer application on stocks of soil C, N, P, and S. *Forests*, 7(12), 319. https://doi.org/10.3390/f7120319

Page-Dumroese, D. S., Sanchez, F. G., Udawatta, R. P., Hobie Perry, C., & González, G. (2021). Soil health assessment of forest soils. In D. L. Karlen, D. E. Stott, & M. M. Mikha (Eds.), *Approaches to soil health analysis* (pp. 100–138). https://doi.org/10.1002/9780891189817.ch6

Paula, R. R., Bouillet, J. P., José, J. L., Trivelin, P. C. O., de Balieiro, C. F., de Nouvellon, Y., de Oliveira, J. C., de Deus Júnior, J. C., Bordron, B., & Laclau, J. P. (2018). Nitrogen fixation rate of *Acacia mangium* wild at mid rotation in Brazil is higher in mixed plantations with *Eucalyptus grandis* hill ex maiden than in monocultures. *Annals of Forest Science*, 75(1), 14. https://doi.org/10.1007/s13595-018-0695-9

Paula, R. R., Bouillet, J. P., Ocheuze Trivelin, P. C., Zeller, B., de Moraes, L., Gonçalves, J., Nouvellon, Y., Bouvet, J. M., Plassard, C., & Laclau, J. P. (2015). Evidence of short-term belowground transfer of nitrogen from *Acacia mangium* to *Eucalyptus grandis* trees in a tropical planted forest. *Soil Biology and Biochemistry*, 91, 99–108. https://doi.org/10.1016/j.soilbio.2015.08.017

Pena-Vergara, G., Castro, L. R., Gasparetto, C. A., & Bizzo, W. A. (2022). Energy from planted forest and its residues characterization in Brazil. *Energy*, 239, 122243. https://doi.org/10.1016/j.energy.2021.122243

Pereira, A. P. A., Andrade, P. A. M., Bini, D., Durrer, A., Robin, A., Bouillet, J. P., Andreote, F. D., Cardoso, E. J. B. N. (2017). Shifts in the bacterial community composition along deep soil profiles in monospecific and mixed stands of Eucalyptus grandis and Acacia mangium. PLoS ONE, 12(7), e0180371. https://doi.org/10.1371/journal.pone.0180371

Pereira, A. P. A., Bini, D., Rodrigues, E. G., Santana, M. C., & Bran Nogueira Cardoso, E. J. (2020). Bioindicators of soil quality in mixed plantations of *Eucalyptus* and leguminous trees. In E. J. B. N. Cardoso, J. L. de Moraes Gonçalves, F. de Carvalho Balieiro, & A. A. Franco (Eds.), *Mixed plantations of eucalyptus and leguminous trees* (pp. 173–192). Springer. https://doi.org/10.1007/978-3-030-32365-3_9

Pereira, A. P. A., Cherubin, M. R., Araujo, A. S. F., Santana, M. C., Medeiros, É. V., Costa, D. P., Souza, A. J., Lima, A. Y. V., Estrada, P. A. C., & Cardoso, E. J. B. N. (2022). Soil health assessment in mixed eucalypt plantations with N2-fixing trees. *SSRN Electronic Journal*. https://doi.org/10.2139/ssrn.4069626

Pereira, A. P. A., Durrer, A., Gumiere, T., Gonçalves, J. L. M., Robin, A., Bouillet, J. P., Wang, J., Verma, J. P., Singh, B. K., & Cardoso, E. J. B. N. (2019). Mixed *Eucalyptus* plantations induce changes in microbial communities and increase biological functions in the soil and litter layers. *Forest Ecology and Management*, 433, 332–342. https://doi.org/10.1016/j.foreco.2018.11.018

Pereira, A. P. A., Santana, M. C., Bonfim, J. A., de Mescolotti, D. L., & Cardoso, E. J. B. N. (2018a). Digging deeper to study the distribution of mycorrhizal arbuscular fungi along the soil profile in pure and mixed *Eucalyptus grandis* and *Acacia mangium* plantations. *Applied Soil Ecology*, 128, 1–11. https://doi.org/10.1016/j.apsoil.2018.03.015

Pereira, A. P. A., Santana, M. C., Zagatto, M. R. G., Brandani, C. B., Wang, J. T., Verma, J. P., Singh, B. K., & Cardoso, E. J. B. N. (2021). Nitrogen-fixing trees in mixed forest systems regulate the ecology of fungal community and phosphorus cycling. *Science of the Total Environment*, 758, –143711. https://doi.org/10.1016/j.scitotenv.2020.143711

Pereira, A. P. A., Zagatto, M. R. G., Brandani, C. B., de Mescolotti, D. L., Cotta, S. R., Gonçalves, J. L. M., & Cardoso, E. J. B. N. (2018b). *Acacia* changes microbial indicators and increases C and N in soil organic fractions in intercropped *Eucalyptus* plantations. *Frontiers in Microbiology*, 9(655). https://doi.org/10.3389/fmicb.2018.00655

Pulito, A. P., de Gonçalves, J. L. M., Smethurst, P. J., Junior, J. C. A., Alvares, C. A., Rocha, J. H. T., Hübner, A., de Moraes, L. F., Miranda, A. C., Kamogawa, M. Y., Gava, J. L., Chaves, R., & Silva, C. R. (2015). Available nitrogen and responses to nitrogen fertilizer in Brazilian eucalypt plantations on soils of contrasting texture. *Forests*, 6(4), 973–991. https://doi.org/10.3390/f6040973

Rachid, C. T. C. C., Balieiro, F. C., Fonseca, E. S., Peixoto, R. S., Chaer, G. M., Tiedje, J. M., & Rosado, A. S. (2015). Intercropped silviculture systems, a key to achieving soil fungal community management in eucalyptus plantations. *PLoS One*, 10(2), e0118515. https://doi.org/10.1371/journal.pone.0118515

Rachid, C. T. C. C., Balieiro, F. C., Peixoto, R. S., Pinheiro, Y. A. S., Piccolo, M. C., Chaer, G. M., & Rosado, A. S. (2013). Mixed plantations can promote microbial integration and soil nitrate increases with changes in the N cycling genes. *Soil Biology and Biochemistry*, 66, 146–153. https://doi.org/10.1016/j.soilbio.2013.07.005

Rocha, J. H. T. (2014). *Reflexos do manejo de resíduos florestais na produtividade, nutrição e fertilidade do solo em plantações de* Eucalyptus grandis [Master's thesis, Universidade de São Paulo]. https://doi.org/10.11606/D.11.2014.tde-21032014-135230

Rocha, J. H. T., de Gonçalves, J. L. M., Gava, J. L., de Godinho, T. O., Melo, E. A. S. C., Bazani, J. H., Hübner, A., Arthur Junior, J. C., & Wichert, M. P. (2016). Forest residue maintenance increased the wood productivity of a *Eucalyptus* plantation over two short rotations. *Forest Ecology and Management*, 379, 1–10. https://doi.org/10.1016/j.foreco.2016.07.042

Rousseau, L., Venier, L., Hazlett, P., Fleming, R., Morris, D., & Handa, I. T. (2018). Forest floor mesofauna communities respond to a gradient of biomass removal and soil disturbance in a boreal jack pine (Pinus banksiana) stand of northeastern Ontario (Canada). *Forest Ecology and Management*, 407, 155–165. https://doi.org/10.1016/j.foreco.2017.08.054

Santana, M. C., de Pereira, A. P. A., de Souza, A. J., Zagatto, M. R. G., de Araújo, V. L. V. P., Wang, J. T., Verma, J. P., Singh, B. K., & Cardoso, E. J. B. N. (2021). Shifts on archaeal community structure in pure and mixed *Eucalyptus grandis* and *Acacia mangium* plantations. *Forest Ecology and Management*, 492, 119218. https://doi.org/10.1016/j.foreco.2021.119218

Silva, L., de Carvalho Mendes, I., Bueno Reis Junior, F., Ferreira Fernandes, M., de Teodoro Melo, J., & Kato, E. (2009). Atributos físicos, químicos e biológicos de um Latossolo de cerrado em plantio de espécies florestais. *Pesquisa Agropecuária Brasileira*, 44(6). https://doi.org/10.1590/S0100-204X2009000600010

Velasquez, E., Lavelle, P., & Andrade, M. (2007). GISQ, a multifunctional indicator of soil quality. *Soil Biology and Biochemistry*, 39, 3066–3080. https://doi.org/10.1016/j.soilbio.2007.06.013

Wichert, M. C. P., Alvares, C. A., Arthur, J. C., & Stape, J. L. (2018). Site preparation, initial growth and soil erosion in *Eucalyptus grandis* plantations on steep terrain. *Scientia Forestalis/Forest Sciences*, 46(117), 17–30. https://doi.org/10.18671/scifor.v46n117.02

Zagatto, M. R. G., de Pereira, A. P. A., de Souza, A. J., Pereira, R. F., Baldesin, L. F., Pereira, C. M., Lopes, R. V., & Cardoso, E. J. B. N. (2019). Interactions between mesofauna, microbiological and chemical soil properties in pure and intercropped *Eucalyptus grandis* and *Acacia mangium* plantations. *Forest Ecology and Management*, 433, 240–247. https://doi.org/10.1016/j.foreco.2018.11.008

Zavišić, A., Yang, N., Marhan, S., Kandeler, L., & Polle, A. (2018). Forest soil phosphorus resources and fertilization affect ectomycorrhizal community composition, beech p uptake efficiency, and photosynthesis. *Frontiers in Plant Science*, 9, 463. https://doi.org/10.3389/fpls.2018.00463

Zhang, X., Han, X., Yu, W., Wang, P., & Cheng, W. (2017). Priming effects on labile and stable soil organic carbon decomposition: Pulse dynamics over two years. *PLoS One*, 12(9), e0184978. https://doi.org/10.1371/journal.pone.0184978

10

Soil Bioanalysis (SoilBio): A Sensitive, Calibrated, and Simple Assessment of Soil Health for Brazil

Ieda Carvalho Mendes, Guilherme Montandon Chaer, Fábio Bueno dos Reis Junior, Ozanival Dario Dantas, Juaci Vitoria Malaquias, Maria Inês Lopes de Oliveira, Marco Antonio Nogueira, and Mariangela Hungria

Chapter Overview

In the past 20 years, the Embrapa research group on bioindicators of soil quality (SQ) addressed the selection of robust SQ/soil health (SH) bioindicators to improve commercially offered routine soil analyses in Brazil. The main objective of our studies was to develop a simple, effective, and practical monitoring tool of SH at the farm scale. Equipped in this way for SQ/SH evaluation, farmers would know exactly what, when, why, and how to proceed and, most importantly, how to interpret the results. In the course of these studies, two soil enzymes, arylsulfatase (ARYL) and β-glucosidase (GLU) (associated with the S and C cycles, respectively), were selected, and interpretative algorithms were developed. The proposed protocol of routine soil analysis completed with the evaluation of the two soil enzymes plus the calculation of soil quality indices (SQIs) was called "Soil Bioanalysis technology" (BioAS in Portuguese; SoilBio in English). By this approach, SQ/SH assessments are based on chemical (FERT) and biological (BIO) indicators, in a framework consisting of three soil functions (SFs): (F1) nutrient cycling (defined by the activities of GLU and ARYL), (F2) nutrient storage (given by soil organic C [SOC] and cation exchange capacity [CEC]), and (F3) nutrient supply (based on Ca^{2+}, Mg^{2+}, K^{+}, P, pH, H + Al; Al^{3+}, sum of bases, and base saturation). Since 2020, GLU and ARYL have been used in large-scale on-farm assessments of SH in Brazil, stimulating the interest of producers in enhanced soil testing over the traditional chemical analysis methods. To make SoilBio available to Brazilian producers, Embrapa capacitates commercial soil analysis laboratories (Rede Embrapa de BioAS in Portuguese; the Embrapa SoilBio Network in English).

Soil Health Series: Volume 3 Soil Health and Sustainable Agriculture in Brazil, First Edition.
Edited by Ieda Carvalho Mendes and Maurício Roberto Cherubin.

The reliability of the results across the country is ensured by the nationwide standardization of methods and protocols and appropriate proficiency testing.

Introduction

"Simplicity is the ultimate sophistication."

(Leonardo da Vinci)

In Brazil, until July 2020, soil samples sent in for laboratory analysis were only evaluated for chemical (acidity, organic matter, macro- and micronutrient contents) and some physical properties (usually clay, silt, and sand contents). Undoubtedly, these analyses, with no representation of soil biological components, which are the basis of SH, were incomplete and unsatisfactory (Lehman et al., 2015).

The expansion of conservation agriculture (CA) practices, such as no-tillage (NT) and integrated crop–livestock (ICL) systems, mainly with the planting of deep-rooted forage grasses, represented a milestone in the construction of biologically active and healthy agricultural soils in Brazil. Long-term CA adoption enabled verifying the limitations of the soil fertility reductionist approach to represent the real plant nutrition status in diversified agricultural and ICL production systems (Anghinoni & Vezzani, 2021). Different yield performances under adverse environmental conditions, observed in soils with similar chemical composition (Anghinoni et al., 2021; Mendes et al., 2017, 2020), and high crop yields under soil fertility levels not optimized according to standard interpretation tables for soil nutrients (Nicolodi et al., 2008) demonstrated the need to include parameters related to soil biological functioning (bioindicators) into routine analyses.

For all these reasons, an enhancement of the traditional soil analyses protocols became necessary to meet the demands of modern agriculture, to which soil bioanalysis (Portuguese: BioAS; English: SoilBio, used henceforth) is suggested as a response. Since 1999, the research group of Embrapa on Bioindicators of Soil Health has been selecting robust SH bioindicators to improve routine soil analyses in commercial laboratories. The main objective of this research was to provide a simple, effective, and practical tool that would allow SH monitoring at farm scale (Mendes et al., 2018a, 2019a). Based on this, farmers would know exactly what, why, how, and when to evaluate SQ/SH and, most importantly, how to interpret the results. Because of these studies, two soil enzymes, ARYL and GLU, associated to the S and C cycles, respectively, were selected, and interpretative algorithms were developed (Mendes et al., 2018a, 2019a,b, 2021a). The novelty of the SoilBio

Table 10.1 Summary of the SoilBio Protocol.

Step	Description
Soil sampling	0–10 cm depth; same sampling procedures as for soil chemical analyses
Sampling time	At the post-harvest stage of the second cash crop
Soil sample pre-treatment procedures	Air-drying and sieving ≤2 mm
Where to send soil samples	Member laboratories of the Embrapa SoilBio Network
Methodology	ARYL and GLU analyses according to Tabatabai (1994) protocol, omitting toluene
Interpretation	By algorithms based on the relations among GLU and ARYL with SOC and crop yield, considering clay content

Abbreviations: ARYL, arylsulfatase; GLU, β-glucosidase; SOC, soil organic C.

approach is the establishment of lower and upper critical limits (CLs) for ARYL, GLU, SF scores associated with plant nutrition (nutrient cycling, storage, and supply), and SQIs based on crop yields and soil organic matter (SOM).

Since July 2020, SoilBio has served extensionists, agronomists, and soil scientists as a supportive framework to assist Brazilian producers with SH assessments by clearly showing the relevance of SH/SQ for the economic performance and sustainability of a farm. As of December 2022, the scoring algorithms of the SoilBio technology were calibrated for areas under annual grain crops in the Cerrado biome and in the state of Paraná, southern Brazil. Based on a series of previous studies, the following important features were defined for the SoilBio protocol: (a) the choice of the 0- to 10-cm depth as the diagnostic soil layer, sampled by the same procedure as for soil chemical fertility analyses and using air-drying and sieving ≤2 mm for the pretreatment of the soil sample; (b) time of soil sampling after harvest of the second crop, together with that for chemical analysis, facilitating the procedure for farmers (in Brazil, two summer cash crops—soybean [*Glycine max* (L) Merr.] and maize [*Zea mays* L.]—are grown on most farms); and (c) use of the widely validated and accessible methodology developed by Tabatabai (1994), omitting toluene. A summary of the SoilBio protocol is presented in Table 10.1.

To make SoilBio practicable for Brazilian farmers, Embrapa has been training commercial soil analysis laboratories (Rede Embrapa de BioAS in Portuguese, Embrapa SoilBio Network in English). The standardization of methods and protocols along with appropriate proficiency testing makes nationwide comparisons of measurements within the network possible. Commercial laboratories interact with Embrapa through the Web platform "Soil Quality Interpretation

Module [SQIM]" (in Portuguese, Módulo de Interpretação da Qualidade do Solo [MIQS]), developed by Embrapa Cerrados and Embrapa Agrobiologia. The SQIM platform interprets the results of enzyme measurements and calculates scores of SQIs and SFs associated with nutrient cycling, storage, and supply, based on chemical and biological properties.

This chapter discusses the fundamentals of using soil enzymes to access soil memory and health status and describes the SoilBio technology in detail as well as some of the results since its release in July 2020.

Accessing Soil's "Memory" through Enzyme Determinations in Tropical Environments

The intensity of mechanical tillage operations, in association with the quality and quantity of plant residues added to soil, reflects in the biological functioning of agricultural systems. Thus, different management systems leave their "fingerprint" or "biological signature" in the soil environment (Figure 10.1). This "memory" of the cultivation system with which a soil had been managed is closely related with the soil biological component.

Selected enzymes, and in particular ARYL and GLU, have theoretical basis for reflecting microbial "memory" of recent soil management. Access to the soil "memory" by enzyme determinations is possible because these measurements represent the sum of activities associated with living organisms (microorganisms,

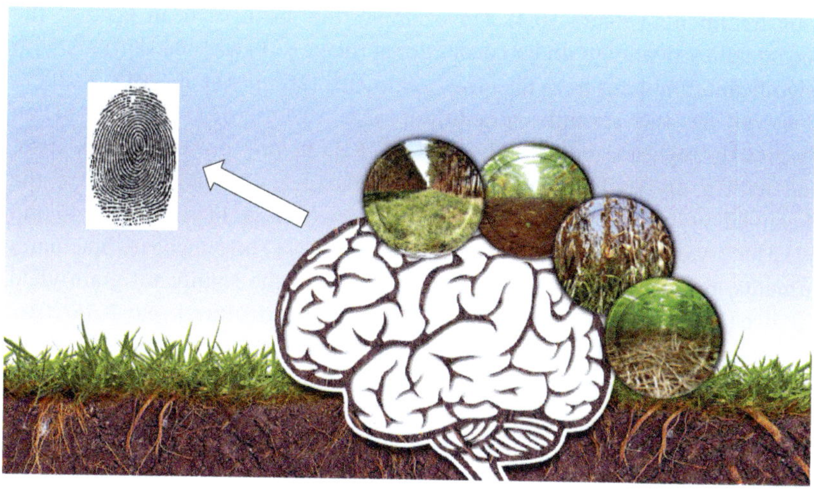

Figure 10.1 Management systems leave their fingerprint in the soil "memory." Art: Fabiano Bastos.

plants, and animals) and past generations of organisms that were present in the soil (abiontic component). The abiontic enzyme pool is associated with the non-living fraction, which accumulates in soil protected from proteases through their adsorption onto clay particles and organic matter (Skujiņš, 1976; Wallenstein & Burns, 2011). In the case of ARYL and GLU, a significant part of the enzyme activity is known to be associated with the abiontic fraction (Knight & Dick, 2004).

The soil capacity to stabilize and protect enzymes is an early-warning indication of conditions favoring SOM stabilization and storage. However, it may take years until changes in SOM are detected, in contrast to alterations in the enzymatic activity (Bandick & Dick, 1999; Dick & Burns, 2011; Ndaye et al., 2000). For this reason, systematic increases in soil enzymes activities (reflecting the increase in BIO activity) over time strongly suggest that the soil environment favors SOM accumulation.

An example of how GLU and ARYL can be used to access the soil "memory" and current SH status in tropical environments is the long-term Soybean-Crop Rotation experiment, initiated in 2008 by the Mato Grosso Foundation (Fundação MT), at the Experimental Station Cachoeira (Itiquira, MT), Central-Western Brazil. This experiment evaluated eight crop production systems, including monoculture soybean or soybean–fallow, soybean followed by pearl millet [*Pennisetum glaucum* (L.) R. Br.] and brachiaria (*Urochloa ruziziensis* syn. *Brachiaria ruziziensis*), soybean in crop rotation, and soybean followed by maize (*Zea mays* L.) (Figure 10.2).

Until the agricultural year of 2013/2014, soybean yields did not differ among treatments (Figure 10.3). For example, the mean soybean yield in the soybean–fallow treatment under conventional tillage (CT) was 3660 kg ha^{-1}, with no significant difference from the other treatments (Mendes et al., 2017, 2020).

However, in the seventh agricultural year (2014/2015), the use of an early-growth soybean cultivar (TMG 7262 RR, 98-day growth cycle), associated with a dry spell in January 2015, initiated a decline in yields of the soybean–fallow treatments under NT and CT. Soybean grain yields of 1740 and 3540 kg ha^{-1} were obtained in the soybean–fallow (NT) and soybean–brachiaria treatments, respectively (Figure 10.4). Regardless of the significant grain yield difference (1800 kg ha^{-1}), a chemical analysis in the agricultural year 2015/2016, 1 year after the dry spell, showed that, except for SOM with a 50% higher content in the soybean/brachiaria than in the soybean–fallow treatment, the soil chemical properties were similar in both (Table 10.2). This reinforced that results based exclusively on chemical properties—the parameters underlying the traditional soil fertility chemist concept—were not sufficient to explain the differences in yield performance, pointing out the inadequacy of the chemical soil-testing methods currently used (Anghinoni et al., 2021; Mendes et al., 2017, 2020).

Figure 10.2 Soil surface under eight treatments in a Soybean-Crop Rotation (SCR) experiment, in Itiquira, MT, Brazil. (NT -no-tillage; CT – conventional tillage; "+" indicates intercropping). Soybean (*Glycine max* (L.) Merr); Pearl millet (*Pennisetum glaucum* (L.) R. Br.); maize (*Zea mays* L.); crotalaria (*Crotalaria ochroleuca*) and brachiaria (*Urochloa ruziziensis*). Photos by Fabio Ono/MT Foundation (agricultural year 2015/2016).

The activity levels of ARYL and GLU determined in the agricultural year 2015/2016 for these two treatments are also shown in Table 10.2. In the soybean/brachiaria treatment, ARYL and GLU activities were, respectively, eight and four times higher than in the soybean–fallow treatment. Therefore, in spite of the similar chemical properties, enzyme activities indicated marked differences in the two treatments regarding the biological components. The biologically most active soil, with increased ARYL and GLU activities and SOM contents resulting from the inclusion of brachiaria in the production system, was also the most productive. The highest yield in this treatment can be explained by the presence of brachiaria straw residues on the soil surface (as protection against soil erosion and drying) and the deep-root system (that improved the soil structure and water storage capacity), which resulted in a more stable and resilient soil environment than under soybean–fallow.

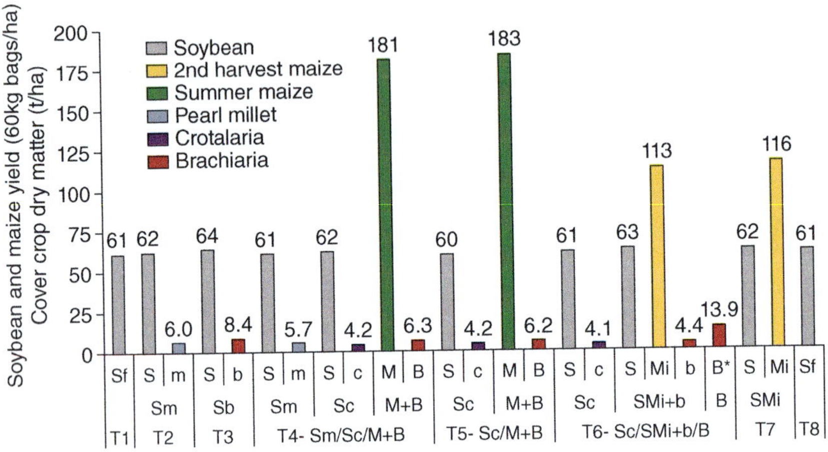

Figure 10.3 Mean grain yield of soybean (S) and maize (M) (60 kg bags ha^{-1}) and shoot dry matter production of pearl millet (m), *Chrotalaria ochroleuca* (c), and brachiaria (B) in the first six harvests under different production systems in Itiquira, MT, Brazil. T1: Sf, soybean–fallow under no-tillage (NT); T2: Sm, soybean–pearl millet (NT); T3: Sb, soybean–brachiaria (NT); T4: Sm, soybean–pearl millet; Sc, soybean–crotalaria; M+B, summer maize+brachiaria (NT); T5: Sc, soybean–crotalaria; M+b, summer maize intercropped with brachiaria (NT); T6: Sc, soybean–crotalaria; SMi+b, soybean second harvest maize+brachiaria; B*, 18-month-old brachiaria established after intercropping with maize (NT); T7: soybean/second harvest maize (NT); T8: Sf, soybean–fallow under conventional tillage (CT). MT Foundation (Mendes et al., 2017; 2020).

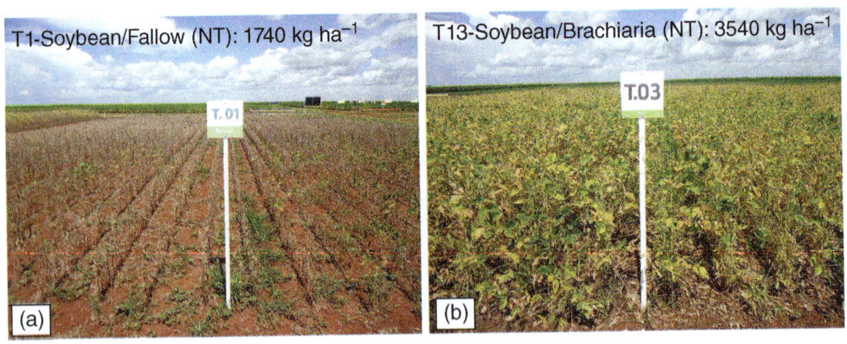

Figure 10.4 Soybean development ('TMG 7262 RR') after the dry spell in January 2015 (seventh agricultural year) in the soybean–fallow (a) and soybean–brachiaria (b) treatments in Itiquira, MT, Brazil. NT, no-tillage. Photos: MT Foundation.

Table 10.2 Soil chemical and biological properties (0–10 cm depth) in the soybean–fallow and soybean–brachiaria no-tillage treatments in a long-term soybean–crop rotation experiment (Fundação MT, Itiquira, MT, Brazil), agricultural year 2015/2016.

Treatment	SOM	pH (H_2O)	Al	H + Al	Ca	Mg	P	K	ARYL	GLU
	g kg^{-1}			–$cmol_c$ dm^{-3}–			–mg dm^{-3}–		–mg PNP kg^{-1} h^{-1}–	
Soybean–fallow	28.2	6.3	0	3.0	3.8	2.2	15	131	28	64
Soybean–brachiaria	42.3	6.4	0	2.5	4.4	3.1	16	151	223	233

Note. H + Al determined by calcium acetate at pH 7.0; Ca, Mg, and Al determined by KCl (1 mol L^{-1}); P and K determined by Mehlich-1.

Abbreviations: ARYL, arylsulfatase; GLU, β-glucosidase; PNP, *p*-nitrophenol; SOM, soil organic matter (Walkley–Black).

In summary, these results show that healthy soil environments in the tropics, with increased ARYL and GLU activities, are more tolerant to adverse environmental conditions (e.g., dry spells). Moreover, the soil "memory" could be assessed by ARYL and GLU determinations, which proved more sensitive to detect symptoms of SH degradation than the traditional soil chemical measurements.

Arylsulfatase and β-Glucosidase: Less Is More

β-Glucosidase (EC 3.2.1.21) is an important enzyme in C cycling. It is a carbohydrate hidrolase involved in the conversion of cellobiose to glucose, a major energy source for soil microbial communities (Deng & Popova, 2011). Thus, it plays a key role in the final step of cellulose degradation. From the perspective of SH indicators, this enzyme is influenced by organic amendments that contain cellulose, an important constituent of plant residues. Numerous studies reported elevated activity levels of GLU in areas under cover crops and crop rotations and with high biomass incorporation (as described above).

Arylsulfatases (sulfohydrolases EC 3.1.6) are enzymes of the esterase class. They are involved in the hydrolysis of ester sulfates to sulfate, a major S source for plant uptake (Castellano & Dick, 1991; Klose et al., 2011). However, regarding the use of ARYL as a SH indicator, the observed sensitivity to detect changes in soil management is likely due to another mechanism. Although sulfatases are produced by both bacteria and fungi, these enzymes have been related to fungal biomass because fungi accumulate ester sulfate, which does not occur in bacteria and is the substrate for ARYL (Saggar et al., 1981). Thus, ARYL reflects the levels of recent fungal biomass production and turnover: the higher the fungal biomass production, the more the microbial community is compelled to produce sulfatases (Acosta-Martinez et al., 2021).

As mentioned by Acosta-Martinez et al. (2021), a conceptual model for enzyme assays as indicators of SQ was first articulated by Dick (1994). This model was based on the following criteria: (a) temporal sensitivity to negative or positive effects of external conditions within a reasonable time frame (on the order of a few years, not decades), which makes it useful for orientation of a sustainable soil management; (b) short-term stability and low in-season variability, which avoids obscuring the status of the sampled soil; (c) adequacy for commercial operations, in particular simplicity and high throughput capability; and (d) interpretability independent of soil type. Almost 30 years after the conceptualization of this model, current knowledge on soil enzymology has compiled evidence of the main advantages of using soil enzymes as SH indicators: short-term sensitivity to changes in management practices, the association of enzymes to soil functioning (nutrient cycling capacity), precision and coherence, and simple and inexpensive analytical determination procedures (Acosta Martinez et al., 2021; Bandick & Dick, 1999; Dick & Burns, 2011; Nannipieri et al., 2018; Ndiaye et al., 2000).

Several studies have shown that ARYL and GLU meet the criteria for SH indicators (Bandick & Dick, 1999; Balota et al., 2004; Mendes et al., 2018a, Lorenz et al., 2020; Passinato et al., 2021; Stott et al., 2010; Veum et al., 2014). These enzymes have been shown to detect effects of the following management or environmental conditions: (a) cover cropping (Bandick & Dick, 1999; Mendes et al., 1999; Miller & Dick, 1995; Ndiaye et al., 2000), (b) organic amendments (Carlson et al., 2015; Perucci, 1992), (c) heavy metals and herbicides (Hinojosa et al., 2004), (d) climate effects (Acosta-Martínez et al., 2014), (e) tillage (Balota et al., 2004, 2014; Dick, 1994), and/or (f) forest management (Vallejo et al., 2010, 2012).

Efforts to reach a global consensus became an integral part of the research associated with SH assessments worldwide. However, local realities must be taken into account. In Brazil, in all experiments/farms evaluated since 1999, ARYL and GLU, together or separately, in this order, were the indicators with consistently the highest sensitivity to detect soil changes associated with management practices (Balota et al., 2004, 2014; Mendes et al., 2003; Mendes et al., 2019a; Lisboa et al., 2012; Lopes et al., 2013, 2021; Peixoto et al., 2010). These two enzymes are closely related with SOM, the most important SH/SQ parameter, and grain yield, the critical economic parameter for farmers and the agribusiness (Lopes et al., 2013; Mendes et al., 2019b, 2021b; Passinato et al., 2021). Moreover, temporal variability of ARYL and GLU is low (Lopes et al., 2015), these enzymes are not influenced by fertilization and liming, they can be successfully used in the calculation of SQIs (Mendes et al., 2021b), and they can be measured directly in air-dried soil samples collected after harvest, with similar results to those obtained with field-moist soil (FFM) samples collected at flowering (Lopes et al., 2021; Mendes et al., 2019b).

Recent studies showed a linear relationship between ARYL and GLU and the diversity of the soil microbial community (fungi, bacteria, protists, and archaea) assessed by DNA characterization in subtropical Brazil (Passinato et al., 2021). At sites with higher crop yields, ARYL and GLU activities and biodiversity were also higher, with an abundance of beneficial soil microorganisms species (e.g., *Mortierella*, *Bacillus*, *Penicillium*, *Trichoderma*, *Pseudomonas*, *Bradyrhizobium*, and *Rhizobium*) and fewer plant pathogenic organisms (e.g., *Fusarium* and *Macrophomina*) (Passinato et al., 2021). The enzymes ARYL and GLU are also correlated with other microbiological properties (e.g., microbial biomass C [MBC], basal respiration, acid phosphatase, cellulase, and dehydrogenase) (Lopes et al., 2013; Mendes et al., 2019b). Considering their redundancy with other microbial indicators, all advantages listed above, and that, as suggested by Karlen (2021), at least two different indicators measurements should be used for each indicator group (i.e., biological, chemical, or physical), ARYL and GLU were selected as the two indicators that best expressed the biological functioning of the soil environment in Brazil. Less is more.

Finally, a SH indicator is only useful if its value can be unequivocally interpreted and reference values are available (Bunemann et al., 2018). This applies to ARYL and GLU, for which CLs can be determined, as will be discussed in the next section.

A Strategy to Interpret Soil Enzyme Activity Based on Crop Yield and Soil Organic Matter

To assist agricultural management decisions, target values of microbial indicators must be defined, which represents a major challenge for SH assessments (Biswas et al., 2017; Dick, 1992; Gil-Sotres et al., 2005; Gonzalez-Quiñones et al., 2011; Trasar-Cepeda et al., 1997). The establishment of CLs for microbial indicators aims to assist farmers, extension agents, agronomists, and soil scientists interested in SH assessments, making the interpretation of individual values of microbial indicators easier and independent of a comparative control or treatment (Lopes et al., 2013).

In contrast to chemical indicators of soil fertility, for which reference levels (low, medium, adequate, and high) are relatively well defined for each element and soil type (usually considering soil texture, soil use, or crop type), it is not a simple task to interpret a series of microbial indicators when data of comparative control treatments are not yet available.

In 2013, we proposed a framework that applies the principles of soil nutrient calibration tests to interpret individual values of several microbial parameters: MBC, basal respiration, and soil enzyme activities, including ARYL, GLU, cellulase, and acid phosphatase (Lopes et al., 2013). In that study, microbial indicators were interpreted as a function of long-term cumulative yields of maize and soybean and SOC in linear regression models. Soil samples (0–10 cm depth) of 24 treatments were collected in three long-term field experiments and analyzed to determine the microbial properties and SOC. Because of different P fertilizer management strategies (rates and placement methods), a gradient of grain yield was observed in the treatments, depending on the presence/absence of phosphate fertilizers (Figure 10.5). Over long periods (>14 years), this grain yield gradient resulted in a SOC gradient in the 0- to 10-cm soil layer. Due to this simultaneity, a set of microbial indicators could, for the first time, be interpreted as a function of (a) SOC (a parameter widely accepted as the best indicator of SQ) and (b) grain yield (a parameter that reflects the economical aspect of crops). Adequacy classes for each microbial indicator, as a function of the relative cumulative yield (RCY) and SOC, were established based on the following criteria: ≤40% RCY: low; 41%–80% RCY: moderate; and > 80% RCY: adequate (Figure 10.6).

The variable RCY is calculated to express the cumulative yield of corn and soybean as a percentage of the greatest cumulative yield obtained in each of the

Figure 10.5 Maize on low (left) and high (right) soil quality plots, side by side, in the long-term field experiments located at Embrapa Cerrados (Planaltina, Brazil). Different P fertilizer management strategies (rates and placement methods) resulted in simultaneous gradients of both grain yield and soil organic C (0- to 10-cm soil layer).

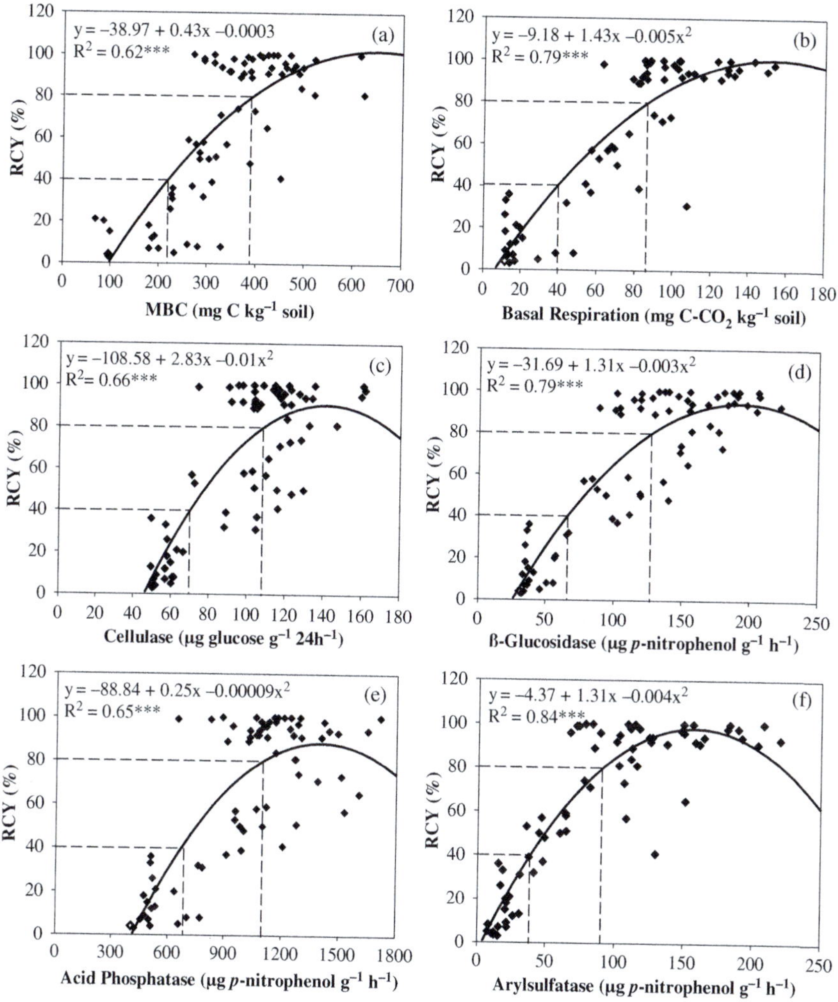

Figure 10.6 Relationships among (a) microbial biomass C, (b) basal respiration, (c) cellulase, (d) β-glucosidase, (e) acid phosphatase, and (f) arylsulfatase and the relative cumulative yield (RCY). The data points represent the three field replicates of the 24 selected treatments in the long-term P fertilization experiments at Embrapa Cerrados, Planaltina, Brazil. Dashed lines represent the limits of the interpretative classes: <40% RCY: low; 41%–80% RCY: moderate; and ≥ 80% RCY: adequate. ***Significant at $P < 0.001$. From: Lopes et al. (2013)/with permission of John Wiley & Sons.

three experiments. The use of RCY to generate the interpretation tables, rather than a single yield measurement, was key not only to minimize the influence of factors not related to SH (e.g., climatic adversities, plant genotypes, and the occurrence of pests and diseases) but also because it more accurately reflected all changes in soil throughout the long-term field experiments. However, considering that it is easier to assess the SOC content of a given area than the RCY, the interpretation strategy based on SOC is more practical (Lopes et al., 2013).

The reference values generated by this strategy are useful not only in the interpretation of microbial indicators independent of a comparative control or treatment but also in terms of establishing threshold limits for scoring functions to assess changes in SQ by SQIs, as will be shown in the next section.

Three nearby areas with native Cerrado vegetation were included as reference for the original soil conditions in the study by Lopes et al. (2013). The comparison with a native reference area allowed to conclude that the interpretative tables based on crop yields and SOC were more adequate and realistic to evaluate SQ. The biological functioning of native Cerrado soils, for instance, has some peculiarities related to the quality and quantity of the returned plant residues (Lopes et al., 2013; Mendes et al., 2019a; Peixoto et al., 2010) that, if not well understood, can lead to erroneous interpretations of some biological indicators, such as GLU and soil respiration (Lopes et al., 2013). In addition, the use of interpretative tables based on relationships among microbial indicators with RCY and SOC strongly reinforced the idea that the equilibrium MBC levels of native areas may not be attainable targets in agricultural soils, even under optimal management practices. In short, the use of native areas as the SQ reference criterion would penalize high-quality agricultural soils for their lower MBC levels than those found in native areas.

The interpretative framework based on crop yield and SOM proposed by Lopes et al. (2013) has been used to generate CLs for microbial indicators and SQIs in Canada (Lupwayi et al., 2015), India (Biswas et al., 2017), China (Mei et al., 2019; Tian et al., 2020), and Iran (Hemmati et al., 2022).

The FERTBIO Soil Sample Concept

As pointed out by Doran and Parkin (1994), once a suit of SH indicators has been selected, standard methods for collecting and handling samples in the field, processing, and analyzing them at the laboratory and interpreting the data are needed for monitoring and making appropriate comparisons.

In addition to interpretative algorithms for microbial indicators, another important achievement for implementing SoilBio in routine commercial soil analyses in Brazil was the development and adoption of the FERTBIO soil sample concept, which requires the collection of soil samples after crop harvest (post-harvest stage) and air-drying prior to laboratory tests (Mendes et al., 2019b). An advantage of the FERTBIO soil sample concept for farmers and commercial laboratories is the standardization of the sampling time and pretreatment procedures for soil chemical fertility (FERT) and biological (BIO) analyses, reducing operational costs. Mendes et al. (2019b) examined the effects of two soil sampling/handling strategies (flowering /field moist [FFM] and post-harvest/air-dried [PHAD] soil samples) on the CLs for MBC and soil enzymes (ARYL, GLU, acid phosphatase, and dehydrogenase). Soil samples (0- to 10-cm depth) were collected in 2013 and 2015 from the same 24 treatments in the long-term P fertilization experiments at Embrapa Cerrados (Figure 10.7). Reduced major axis regressions showed that PHAD sampling was representative of FFM. Despite significant reductions in MBC (11%), ARYL (49%), GLU (55%), acid phosphatase (54%), and dehydrogenase (46%), PHAD sampling preserved the ranking of treatments and did not alter the adjustment of regression models between SOC and microbial indicators. Specific CLs for PHAD soil samples were established for each microbial indicator (Table 10.3).

The development of interpretative tables using the FERTBIO soil sample concept represented an important advance toward the inclusion of ARYL and GLU in routine commercial soil analyses in Brazil. In practical terms, for farmers, performing soil sampling after harvest makes it easier to collect and send these samples to the laboratory because they coincide with soil fertility testing. In addition, air-dried soil samples facilitate the handling and storage of the material before measuring ARYL and GLU activity. Air-dried soil samples generally also result in less variation among replicates and facilitate the reanalysis of the original soil samples whenever necessary (Haney et al., 2004; Lorenz & Dick, 2011).

Taking soil texture into account is crucial for a meaningful SH assessment because it is the controlling factor in numerous soil physical, chemical, and biological properties, such as CEC, soil aggregation, and retention of organic matter (Karlen et al., 2017). By the FERTBIO soil sample concept, the interpretation strategy based on SOC and RCY was replicated in soils representing the most important textural classes found in Brazil (from sandy to clayey), and interpretative algorithms for ARYL and GLU were developed as a function of clay percentage.

Considering the clay content in the 0- to- 10-cm diagnostic soil layer in Brazil, SoilBio algorithms can be used in Oxisols, Ultisols, Alfisols, and Entisols (Dr. Helio do Prado, pedologist, personal communication). In the Cerrados region alone, these soil orders account for 75% of the soils (Correia et al., 2004).

As with soil fertility interpretative tables, the dynamic nature of the interpretative algorithms for SoilBio, in response to the improvement of the data bank, is worth mentioning. In comparison, in Rio Grande do Sul state, Southern Brazil, the first nutrient recommendation tables were established in 1967, which by 2004 had already been revised and improved 10 times (Nicolodi et al., 2004).

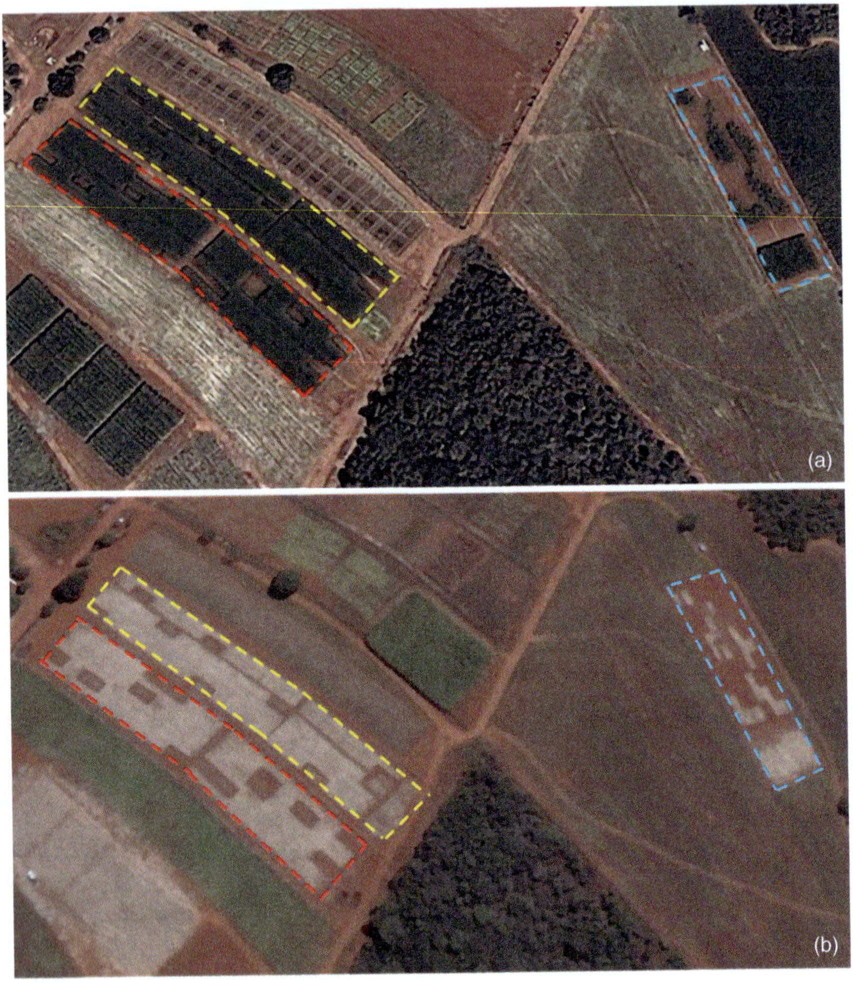

Figure 10.7 Google image of three long-term field experiments at Embrapa Cerrados, Planaltina, Brazil, showing maize plots at flowering (a) and post-harvest stage (b). Although the driving questions were related to P fertilizer management (rates and placement methods), these experiments were also used to develop a strategy to interpret individual values of microbial indicators and soil quality indices (SQIs), to validate the FERTBIO soil sample concept, and to develop the four-quadrant model to evaluate C trends in soil.

Table 10.3 Interpretative classes for microbial indicators, as a function of soil organic C, for clayey Oxisols (Latossolo Vermelho) of the Brazilian Cerrado (0- to 10-cm depth) determined at the flowering stage with field-moist soil samples and at the post-harvest stage with air-dried soil samples.

	Interpretative classes		
Microbial indicators[a)]	**Low**	**Moderate**	**Adequate**
		FFM	
Microbial biomass C	≤245	246–415	>416
β-Glucosidase	≤106	107–209	>210
Arylsulfatase	≤44	45–115	>116
Acid phosphatase	≤602	603–1056	>1057
Dehydrogenase	≤32	33–69	>70
		PHAD	
Microbial biomass C	≤152	153–324	>325
β-Glucosidase	≤66	67–115	>116
Arylsulfatase	≤30	31–70	>71
Acid phosphatase	≤263	264–494	>495
Dehydrogenase	≤19	20–40	>41

Abbreviations: FFM, Flowering field-moist soil samples; PHAD, post-harvest stage with air-dried soil samples.

a) Microbial biomass C expressed in mg C kg^{-1} soil; β-glucosidase, acid phosphatase, and arylsulfatase expressed in mg PNP kg^{-1} soil h^{-1}; dehydrogenase expressed in mg triphenylformazan kg^{-1} soil 24 h^{-1}.

Note. From Mendes et al. (2019b)/with permission of Elsevier.

Soil Quality Indices Based on Crop Yield and Soil Organic Matter

Soil quality indices enable the integration of complex information of different soil properties and processes in a simple tool that can be used to improve soil management decisions (Hussain et al., 1999). One of the most widely used approaches to determine SQIs was developed by Karlen and Stott (1994). Soil indicators are associated with the main soil functions, which are weighted according to their importance for the SQ management at a given location. An overall rating of SQ with respect to the pre-defined goal is obtained by summing the weighted soil functions.

By the SoilBio technology, SQ is evaluated combining chemical and biological indicators (Figure 10.8). A SQI based on chemical (FERT) and biological (BIO) properties ($SQI_{FERTBIO}$) is calculated from the interpretation of 11 chemical

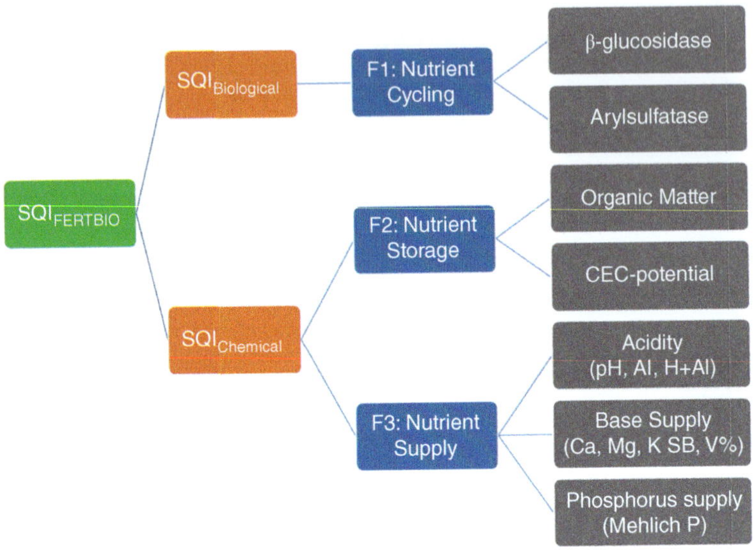

Figure 10.8 Schematic representation of the model used to define the soil quality indices (SQIs) of the SoilBio technology. SB, sum of bases; V%, base saturation.

indicators, used in routine fertility analysis, together with the two soil enzymes (ARYL and GLU). The SQI framework includes three soil functions: (F1) nutrient cycling (based on ARYL and GLU activities), (F2) nutrient storage (based on SOC and CEC), and (F3) nutrient supply (based on Ca^{2+}, Mg^{2+}, K^{+}, P, pH, H + Al; Al^{3+}, sum of bases, and base saturation [V%]). Scores for each of these soil functions are computed using measurements (indicators) that are normalized (converted in values between 0 and 1) with one of the three (more is better, optimum, or less is better) standardized scoring functions (Wymore, 1993). Nutrient cycling (F1) estimates the performance of biological activity and the processes directly or indirectly derived from it (e.g., nutrient cycling and SOM formation and decomposition). Nutrient storage (F2) assesses the size of the soil nutrient "reservoir," which is mainly related to the quantity and quality of clays and SOM content and quality. Nutrient supply (F3) determines the quality of the soil nutrient "reservoir," involving aspects related to soil acidity and the capacity to provide the main macronutrients.

The $SQI_{FERTBIO}$ index is computed by the weighted additive method (Karlen & Stott, 1994) given by the following equation:

$$SQI_{FERTBIO} = (SF1 \times W1) + (SF2 \times W2) + (SF3 \times W3)$$

where SF1, SF2, and SF3 are the scores calculated for the soil functions F1, F2, and F3, respectively, and W1, W2, and W3 are the weights attributed to the respective

Table 10.4 Soil functions, associated indicators (Levels 1 and 2), and respective weights used for calculations of soil quality indices by SoilBio technology.

		Level 1		Level 2	
Soil function	**Weight**	**Indicator**	**Weight**	**Indicator**	**Weight**
	%		%		%
F1: Nutrient cycling	33.3	GLU	50		
		ARYL	50		
F2: Nutrient storage	33.3	SOM	50		
		CEC	50		
F3: Nutrient supply	33.3	P supply	33.3	P-Mehlich	100
		Base supply	33.3	K^+	20
				Mg^{2+}	20
				Ca^{2+}	20
				Base saturation	20
				SB	20
		Soil acidity	33.3	pH (H_2O)	33.3
				H + Al	33.3
				Al^{3+}	33.3

Abbreviations: ARYL, arylsulfatase; CEC, cation exchange capacity; GLU, β-glucosidase; SB, sum of bases; SOM, soil organic matter.
Note. From Mendes et al. (2021b).

function in the model. For the $SQI_{FERTBIO}$ calculations, equal weights are assigned to F1, F2, and F3 (Table 10.4). As shown in Figure 10.8, the $SQI_{FERTBIO}$ is divided in two sub-indices: (a) the soil biological quality index (SQI_{BIO}), involving the performance of F1, and (b) the soil chemical quality index (SQI_{FERT}), consisting of the mean performances of F2 and F3. The three SQI and function scores vary from 0 to 1; the closer to 1, the better the SQ or the performance of the function.

A detailed description of the calculation of the SoilBio SQI is provided by Mendes et al. (2021b).

The application of the $SQI_{FERTBIO}$ of SoilBio technology to data from the above-mentioned experiments in Planaltina, Brazil, showed a strong relationship between the index with RCY and SOC (Figure 10.9). Therefore, P fertilization resulted in higher cumulative crop yields and SOC that in turn were associated with better $SQI_{FERTBIO}$ and SF scores. Relationships between the $SQI_{BIOLOGICAL}$, $SQI_{CHEMICAL}$, and soil functions scores with both RCY and SOC were also established. Critical limits to interpret scores of the SQI and soil functions

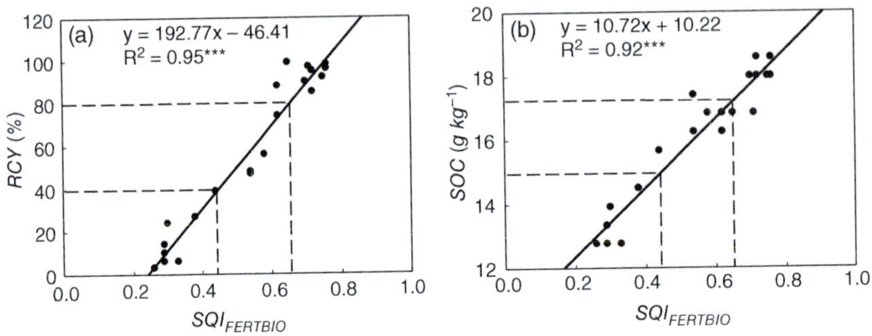

Figure 10.9 Relationships between relative cumulative yield (RCY) and the soil quality index $SQI_{FERTBIO}$ (a) and between the soil organic C (SOC) and the $SQI_{FERTBIO}$ (b). Plotted points represent the means of the 21 selected treatments in the 2013 and 2015 samplings in long-term P-fertilization experiments at Embrapa Cerrados, Brazil. Dashed lines represent the limits of the interpretative classes for RCY (<40% low, 41%–80% moderate, and >80% adequate) and SOC (< 14.7 g kg^{-1} soil: low; 14.7–17.4 g kg^{-1} soil: moderate; and >17.4 g kg^{-1} soil: adequate). ***Significant at $P < 0.001$. Mendes et al. (2021b)/with permission of Elsevier.

separately, based on RCY and SOC, were defined by linear regression analyses (Mendes et al., 2021b).

The combination of chemical and biological indicators of different soil functions in the $SQI_{FERTBIO}$ proved to be a simple, cost-effective, and useful strategy to assess SQ of clayey tropical Oxisols. Ideally, SQ/SH assessments must take chemical, biological, and physical indicators into consideration (Karlen et al., 1997). We acknowledge the absence of indicators associated to soil physical properties in the $SQI_{FERTBIO}$. However, because soil biology (represented by ARYL and GLU activity in SoilBio) plays a crucial role in the modifications of soil physical properties, being the first step in the process of soil regeneration (Hatfield et al., 2017; Lynch & Bragg, 1985), increased soil biological activity can be indirect good evidence (positively related) of better physical soil structure. In fact, good correlations between ARYL and GLU activity with soil physical properties have been reported by Anghinoni et al. (2021) and Passinato et al. (2021).

The SQIs of the SoilBio technology are used to help extensionists, agronomists, and soil scientists when assisting farmers in SH assessments by facilitating the explanation of the relevance of SQ for the economic performance of a farm. The SQ assessment provided by the SoilBio technology integrates chemical and biological indicators, which reflect regional Brazilian conditions and relate agronomic management practices to SQ, yield benefits, and SOM. As emphasized by Karlen et al. (2017), this is undoubtedly a win-win situation, not only because SQ/SH can be measured in the field but also because data can be used to improve crop yields and land management decisions, enhancing the sustainability of tropical agroecosystems.

The SoilBio Report: Beyond Deficiency/Excess of Nutrients

A traditional report of soil fertility analyses and a SoilBio report are presented in Tables 10.5 and 10.6, respectively. In both tables, lines represent different sites of a grain crop farm in the Cerrado region. From left to right, columns in the SoilBio report show ARYL and GLU activity levels, SOM, the three SQIs ($SQI_{FERTBIO}$, $SQI_{BIOLOGICAL}$, and $SQI_{CHEMICAL}$), and the scores of the three functions (F1: nutrient cycling, F2: storage, and F3: supply). The values of enzyme activity, SOM, SQI, and function scores are represented on a "traffic light" chromatic scale, where dark or light green represent adequate values, yellow represents intermediate values, and orange or red represent low values (Figure 10.10).

The comparison of the SoilBio report with a traditional soil fertility analysis report (Tables 10.5 and 10.6) highlights the more comprehensive view of the SoilBio technology, which goes beyond the detection of nutrient deficiency/excess issues.

Table 10.5 Example of a traditional soil fertility report of a farm in Mato Grosso State (soil type: Clayey Oxisol).

Soil sample	pH (H_2O)	Al	Ca	H + Al	Mg	CEC	SB	P	K	Base saturation	Clay
		–$mmol_c$ dm^{-3}–						–mg L^{-1}–		–%–	
1001 LV01	5.5	0.05	3.1	5.3	1.3	10	5	17	96	46	67
1002 09	5.6	0.01	7.7	2.6	4	15	12	39	114	82	70
1003 02/19 BCE	5.1	0.08	2.8	5.6	0.9	9	4	14	55	40	52
1004 12	5.7	0.03	6.6	5.2	1.8	14	9	27	75	62	71
1005 20	6.2	0.02	5.9	4	0.9	11	7	10	92	64	52
LV 101	5.6	0.01	5.5	5.3	1.1	12	7	35	36	56	73
LV 102	5.7	0.02	4	5.1	1.1	10	5	18	54	51	71
LV 103	6.1	0.01	3.7	3.9	2.2	10	6	13	39	60	66
LV 104	5.5	0.02	2.1	3.5	0.6	6	3	22	35	44	41
LV 105	5.9	0.02	5.3	4.1	0.8	10	6	17	61	60	69

Note. Al, Ca, and Mg determined by KCl (1.0 mol L^{-1}); K and P determined by Mehlich 1. Abbreviations: CEC, cation exchange capacity; SB, sum of bases.

Table 10.6 SoilBio Report Produced by the Embrapa Soil Quality Interpretation Module.

Soil sample	ARYL	GLU	SOM	SQI			Nutrient cycling	Nutrient storage	Nutrient supply
				Fertbio	Biological	Chemical			
1001 LV01	161	165	34	0.73	0.70	0.75	0.70	0.59	0.91
1002 09	219	151	45	0.72	0.71	0.72	0.71	0.61	0.83
1003 02/19 BC	78	96	32	0.82	0.61	0.93	0.61	0.88	0.97
1004 12	319	196	44	0.89	0.83	0.92	0.83	0.86	0.99
1005 20	71	60	34	0.79	0.51	0.93	0.51	0.88	0.98
LV 101	68	87	39	0.67	0.36	0.82	0.36	0.70	0.94
LV 102	73	94	35	0.63	0.39	0.76	0.39	0.58	0.93
LV 103	77	69	33	0.64	0.4	0.76	0.40	0.58	0.95
LV 104	23	45	19	0.54	0.42	0.61	0.42	0.35	0.86
LV 105	56	72	33	0.62	0.34	0.76	0.34	0.58	0.95

Abbreviations: ARYL, arylsulfatase; GLU, β-glucosidase (Tabatabai, 1994); SOM, soil organic matter (Walkley–Black); SQI, soil quality index.
Note. Data from the same farm as in Table 10.5.

To make SoilBio available to Brazilian producers, Embrapa has supported commercial soil analysis laboratories (Embrapa SoilBio Network). The basic principles of the network consist of the standardization of methods for soil sample collection and handling in the field, processing and analyzing them in the laboratory, and interpreting the data, along with appropriate proficiency testing, which makes nationwide comparisons of measurements possible. The network laboratories interact with Embrapa via SQIM, a Web platform developed by Embrapa Cerrados and Embrapa Agrobiologia. This platform provides separate interpretations of values of enzymes determinations and SOM levels and calculates the SQIs and scores for the three soil functions assessed by SoilBio: nutrient cycling (F1), storage (F2), and supply (F3).

The SoilBio reports show certain patterns that represent relevant indications about SQ and soil management in a given area by analyzing the color patterns assigned to functions F1 and F2. The four main types of SoilBio reports are

Class	Very low	Low	Moderate	High	Very high
Range	0 to 0.20	0.21 to 0.40	0.41 to 0.60	0.61 to 0.80	0.81 to 1.00

Figure 10.10 Color rating scale used to interpret soil enzymes, soil organic matter (SOM), soil quality indices (SQIs), and soil function scores of the SoilBio technology. Threshold values (defined arbitrarily as 0.2 intervals) apply to SQIs and soil function scores.

REDE EMBRAPA BioAS BIOANÁLISES DE SOLO

SoilBio Report Type	Enzymes (F1 Cycling)	Soil Organic Matter (Storage)
1- Healthy soil	High	High
2- Undergoing biological degradation	Low	High
3- Unhealthy soil	Low	Low
4- Regenerative processes	High	Low

Figure 10.11 Four main types of SoilBio reports. Each type is defined based on the combination of scores of nutrient cycling (Function 1 [F1]) and nutrient storage (Function 2 [F2]).

depicted in Figure 10.11 and reflect soil enzymes' ability to act as ecological sensors that respond more quickly to soil management changes than SOM. These patterns also constitute the basis of the four-quadrant model proposed by Chaer et al. (2023) to evaluate C trends, based on the relationship between SOC and the average ARYL and GLU activities per unit of SOC (average specific enzyme activity). Report Type 1 represents healthy, high-quality soils with high enzyme activity and high SOM (i.e., a soil that was well managed in the long term). Under management practices that tend to decrease overall SQ, the enzyme activity of a soil will initially decline quickly, which is not accompanied at the same rate by SOM. This situation (low enzyme activity status and high SOM) is observed in soils undergoing biological degradation and is represented in SoilBio Report Type 2. Soils with low enzyme activity levels as well as low SOM strongly indicate unhealthy/low-quality soils, as shown in SoilBio Report Type 3. In this case, biological degradation is associated with SOM loss. Finally, report Type 4 represents high enzyme activity levels coupled with low SOM contents. This indicates low quality/unhealthy soils in the process of restoration by conservation management practices. Due to the sensibility of soil enzymes, their activities shift to higher levels rapidly, whereas SOC remains relatively unchanged in the short term.

Figures 10.12–10.15 show examples of the four types of SoilBio reports of agricultural properties in the Brazilian Cerrado region. All these reports were elaborated by commercial soil laboratories of the Embrapa SoilBio Network.

Sample ID	ARYL	GLU	SOM	Clay %	SQI FERTBIO	SQI Biological	SQI Chemical	Nutrient Cycling	Nutrient Storage	Nutrient Supply
108043_Th. 2 4 - 01 0-10	156	202	40	56	0.90	0.88	0.91	0.88	0.93	0.89
108044_Th. 2 4 - 02 0-10	159	174	40	51	0.92	0.91	0.93	0.91	0.97	0.89
108045_Th. 05 - 01 0-10	187	233	45	51	0.94	0.97	0.93	0.97	0.98	0.88
108046_Th. 06 - 01 0-10	191	191	42	51	0.93	0.94	0.93	0.94	0.97	0.89
108047_Th. 06 - 02 0-10	204	201	48	49	0.95	0.97	0.94	0.97	0.99	0.89
108048_Th. 08 - 01 0-10	197	213	45	51	0.94	0.96	0.93	0.96	0.98	0.89
108049_Th. 08 - 02 0-10	211	219	47	54	0.94	0.96	0.94	0.96	0.98	0.89
108050_Th. 9 12 - 01 0-10	184	215	43	49	0.95	0.97	0.94	0.97	0.98	0.89
108053_Th. 14 - 01 0-10	189	167	43	56	0.90	0.87	0.91	0.87	0.94	0.89
108054_Th. 14 - 02 0-10	164	176	42	49	0.94	0.94	0.94	0.94	0.98	0.89
108055_Th. 15 - 01 0-10	124	181	41	44	0.94	0.95	0.93	0.95	0.99	0.88
108056_Th. 15 - 02 0-10	147	192	42	44	0.95	0.97	0.93	0.97	0.99	0.88
108057_Th. 16 - 01 0-10	161	166	47	44	0.94	0.96	0.94	0.96	0.99	0.88
108058_Th. 16 - 02 0-10	130	152	43	41	0.94	0.95	0.94	0.95	0.99	0.88
108059_Th. 17 - 01 0-10	122	131	43	41	0.93	0.92	0.94	0.92	0.99	0.88
108060_Th. 17 - 02 0-10	161	171	42	39	0.95	0.98	0.93	0.98	0.99	0.88
108061_Th. 18 19 - 01 0-10	199	190	46	51	0.94	0.95	0.94	0.95	0.99	0.89
108062_Th. 18 19 - 02 0-10	203	176	47	49	0.94	0.95	0.94	0.95	0.99	0.89
108063_Th. 20 - 01 0-10	185	132	46	54	0.90	0.84	0.93	0.84	0.96	0.89
108064_Th. 20 - 02 0-10	201	136	46	54	0.91	0.85	0.93	0.85	0.98	0.89
108065_Th. Furna 30 - 01 0-10	141	172	47	46	0.94	0.94	0.94	0.94	0.99	0.89
108066_Th. Furna 30 - 02 0-10	139	158	45	49	0.92	0.90	0.94	0.90	0.99	0.89
108067_Th. Furna 60 - 01 0-10	219	144	48	56	0.90	0.85	0.92	0.85	0.96	0.89
108068_Th. Furna 60 - 02 0-10	215	169	44	56	0.91	0.89	0.91	0.89	0.94	0.89

Figure 10.12 Example of a SoilBio report Type 1 of a farm in the state of Mato Grosso do Sul, Brazil. Predominantly green tones for nutrient cycling (Function 1 [F1]) and nutrient storage (Function 2 [F2]) in the columns indicate a healthy/high-quality soil environment. ARYL, arylsulfatase (mg *p*-nitrophenol kg^{-1} h^{-1}); GLU, β-glucosidase (mg *p*-nitrophenol kg^{-1} h^{-1}); SOM, soil organic matter (g kg^{-1}). Data obtained from the laboratories of the Embrapa SoilBio Network.

SAMPLE ID	ARYL	GLU	SOM	Clay %	SQI FERTBIO	SQI Biological	SQI Chemical	Nutrient Cycling	Nutrient Storage	Nutrient Supply
3258 2020_TH L01 - 01 - B	25	30	41	59	0.64	0.26	0.83	0.26	0.80	0.85
3259 2020_TH L01 - 02 - B	29	38	43	63	0.64	0.27	0.82	0.27	0.80	0.84
3260 2020_TH L01 - 03 - B	51	50	46	59	0.70	0.36	0.87	0.36	0.88	0.86
3261 2020_TH L02 - 01 - B	28	39	57	55	0.72	0.31	0.93	0.31	0.99	0.86
3262 2020_TH L02 - 02 - B	21	36	40	55	0.66	0.29	0.85	0.29	0.85	0.85
3263 2020_TH L02 - 03 - B	14	17	43	59	0.62	0.20	0.83	0.20	0.87	0.80
3264 2020_TH 05 - 01 - B	22	35	35	62	0.55	0.25	0.70	0.25	0.60	0.80
3265 2020_TH 05 - 02 - B	36	62	43	57	0.70	0.37	0.86	0.37	0.88	0.85
3266 2020_TH 05 - 03 - B	25	44	42	63	0.63	0.27	0.81	0.27	0.79	0.83
3267 2020_TH 12 - 01 - B	27	40	40	54	0.65	0.32	0.81	0.32	0.86	0.76
3268 2020_TH 12 - 02 - B	16	28	35	65	0.53	0.21	0.69	0.21	0.59	0.79
3269 2020_TH 12 - 03 - B	25	27	47	62	0.67	0.24	0.89	0.24	0.92	0.85
3270 2020_TH 15 - 01 - B	16	33	36	55	0.61	0.26	0.78	0.26	0.74	0.81
3271 2020_TH 15 - 02 - B	24	45	49	57	0.70	0.30	0.90	0.30	0.95	0.85
3272 2020_TH 15 - 03 - B	15	34	41	56	0.65	0.26	0.85	0.26	0.87	0.84
3273 2020_TH 16 - 01 - B	22	44	41	63	0.63	0.26	0.82	0.26	0.78	0.86
3274 2020_TH 16 - 02 - B	27	38	42	58	0.67	0.29	0.86	0.29	0.86	0.86
3275 2020_TH 16 - 03 - B	20	38	44	57	0.68	0.28	0.88	0.28	0.90	0.85
3276 2020_TH 19 - 01 - B	35	65	47	60	0.71	0.35	0.89	0.35	0.92	0.85
3277 2020_TH 19 - 02 - B	33	49	46	57	0.70	0.33	0.88	0.33	0.92	0.85
3278 2020_TH 19 - 03 - B	28	55	38	60	0.63	0.31	0.79	0.31	0.73	0.84
3279 2020_TH 25 - 01 - B	18	46	45	59	0.63	0.27	0.81	0.27	0.89	0.73
3280 2020_TH 25 - 02 - B	26	41	44	56	0.68	0.30	0.87	0.30	0.91	0.83
3281 2020_TH 33 - 01 - B	26	61	52	55	0.72	0.35	0.90	0.35	0.99	0.80
3282 2020_TH 33 - 02 - B	17	42	43	60	0.65	0.26	0.85	0.26	0.89	0.82

Figure 10.13 Example of a SoilBio report Type 2 of a farm in the state of Mato Grosso, Brazil. Predominantly light/dark red tones for nutrient cycling (F1) and green tones for nutrient storage (F2) indicate a soil undergoing biological degradation. ARYL, arylsulfatase (mg *p*-nitrophenol kg^{-1} h^{-1}); GLU, β-glucosidase (mg *p*-nitrophenol kg^{-1} h^{-1}); SOM, soil organic matter (g kg^{-1}). Data from laboratories of the Embrapa SoilBio Network.

Sample ID	ARYL	GLU	SOM	Clay %	SQI FERTBIO	SQI Biological	SQI Chemical	Nutrient Cycling	Nutrient Storage	Nutrient Supply
66273_Amt 103 Lav. 107	53	42	27	71	0.46	0.28	0.54	0.28	0.28	0.81
66274_Amt 104 Lav. 108	73	66	30	67	0.54	0.38	0.62	0.38	0.41	0.84
66275_Amt 105 Lav. 109	52	61	26	41	0.65	0.59	0.68	0.59	0.58	0.78
66276_Amt 106 Lav. 110	44	37	22	70	0.44	0.26	0.53	0.26	0.21	0.84
66277_Amt 107 Lav. 111	49	51	20	71	0.43	0.29	0.49	0.29	0.17	0.81
66278_Amt 108 Lav. 112	85	61	28	75	0.51	0.34	0.60	0.34	0.32	0.88
66279_Amt 109 Lav. 113	75	60	26	70	0.51	0.36	0.59	0.36	0.30	0.87
66280_Amt 110 Lav. 114	94	104	28	54	0.65	0.64	0.66	0.64	0.45	0.87
66281_Amt 111 Lav. 115	68	70	28	73	0.50	0.34	0.59	0.34	0.34	0.84
66282_Amt 112 Lav. 116	59	46	25	76	0.45	0.27	0.54	0.27	0.25	0.83
66283_Amt 113 Lav. 117	82	77	27	81	0.48	0.32	0.56	0.32	0.27	0.85
66284_Amt 114 Lav. 118	81	88	27	72	0.49	0.39	0.55	0.39	0.28	0.81
66285_Amt 115 Lav. 119	65	71	28	63	0.53	0.40	0.59	0.40	0.40	0.79
66286_Amt 116 Lav. 120	49	53	29	51	0.60	0.44	0.68	0.44	0.58	0.79
66287_Amt 117 Lav. 121	20	32	25	34	0.66	0.45	0.77	0.45	0.83	0.71
66288_Amt 118 Lav. 122	47	50	30	69	0.48	0.29	0.57	0.29	0.38	0.77
66289_Amt 119 Lav. 123	62	51	30	64	0.53	0.35	0.62	0.35	0.40	0.84
66290_Amt 120 Lav. 124	96	94	30	65	0.57	0.48	0.62	0.48	0.39	0.85
66291_Amt 121 Lav. 125	89	62	28	74	0.49	0.35	0.55	0.35	0.30	0.81
66292_Amt 122 Lav. 126	62	55	27	70	0.47	0.32	0.55	0.32	0.31	0.79
66293_Amt 123 Lav. 127	73	50	26	64	0.50	0.38	0.56	0.38	0.30	0.82
66294_Amt 124 Lav. 128	49	48	28	75	0.46	0.27	0.55	0.27	0.29	0.82
66295_Amt 125 Lav. 129	52	46	27	70	0.47	0.29	0.57	0.29	0.32	0.81
66296_Amt 126 Lav. 130	65	58	28	65	0.54	0.36	0.62	0.36	0.41	0.83
66297_Amt 127 Lav. 131	87	69	31	62	0.58	0.46	0.63	0.46	0.42	0.84

Figure 10.14 Example of a SoilBio report Type 3 of a farm in the state of Mato Grosso, Brazil. The predominantly yellow, light/dark red tones for nutrient cycling (F1) and nutrient storage (F2) in the columns indicate an unhealthy/low quality soil environment. ARYL, arylsulfatase (mg *p*-nitrophenol $kg^{-1} h^{-1}$); GLU, β-glucosidase (mg *p*-nitrophenol $kg^{-1} h^{-1}$); SOM, soil organic matter ($g kg^{-1}$). Data from the laboratories of the Embrapa SoilBio Network.

Sample ID	ARYL	GLU	SOM	Clay %	SQI FERTBIO	SQI Biological	SQI Chemical	Nutrient Cycling	Nutrient Storage	Nutrient Supply
16124_LOTE 6 N 1	19	86	9	22	0.73	0.81	0.69	0.81	0.58	0.81
16125_LOTE 6 N 5	28	143	11	22	0.76	0.92	0.68	0.92	0.55	0.80
16126_LOTE 6 N 11	20	86	8	24	0.65	0.78	0.59	0.78	0.39	0.79
16127_LOTE 6 N 13	20	93	8	24	0.58	0.79	0.47	0.79	0.25	0.70
16128_LOTE 6 N 15	25	124	9	22	0.65	0.89	0.53	0.89	0.42	0.64
16129_LOTE 6 N 23	19	69	8	26	0.55	0.69	0.48	0.69	0.19	0.76
16130_LOTE 6 N 25	26	105	12	24	0.70	0.86	0.63	0.86	0.46	0.79
16131_LOTE 6 N 27	19	85	9	24	0.63	0.77	0.56	0.77	0.33	0.79
16132_LOTE 6 N 31	20	105	9	24	0.60	0.80	0.50	0.80	0.24	0.76
16133_LOTE 6 N 33	20	79	9	26	0.62	0.73	0.57	0.73	0.34	0.80
16134_LOTE 6 N 35	21	95	9	24	0.63	0.80	0.55	0.80	0.31	0.79
16135_LOTE 6 N 45	19	77	9	26	0.60	0.71	0.54	0.71	0.31	0.77
16136_LOTE 6 N 47	13	67	9	26	0.55	0.63	0.51	0.63	0.24	0.77
16137_LOTE 6 N 49	11	54	8	26	0.53	0.57	0.51	0.57	0.23	0.78
16138_LOTE 6 N 52	26	93	9	24	0.63	0.84	0.53	0.84	0.27	0.78
16139_LOTE 6 N 54	15	59	8	24	0.51	0.66	0.43	0.66	0.23	0.64
16140_LOTE 6 N 56	18	88	9	24	0.58	0.76	0.48	0.76	0.20	0.77
16141_LOTE 6 N 65	15	76	8	24	0.49	0.71	0.38	0.71	0.13	0.64
16142_LOTE 6 N 66	21	100	10	24	0.64	0.81	0.56	0.81	0.34	0.78

Figure 10.15 Example of a SoilBio report Type 4 of a farm in the state of Bahia, Brazil. The predominantly green tones for nutrient cycling (F1) and orange, yellow, and light/dark red tones for nutrient storage (F2) indicate low-quality/unhealthy soil, recovering by conservation management practices. ARYL, arylsulfatase (mg *p*-nitrophenol $kg^{-1} h^{-1}$); GLU, β-glucosidase (mg *p*-nitrophenol $kg^{-1} h^{-1}$); SOM, soil organic matter ($g kg^{-1}$). Data from laboratories of the Embrapa SoilBio Network.

Usually, on-farm management decisions are largely influenced by operational and economic aspects, in detriment of the best agronomic practices. Therefore, the main purpose of SoilBio is to support management decisions in line with the best CA practices, capable of improving SH. In cases of good SH or of a regenerative process, SoilBio results should motivate farmers to continue with the applied management practices. On the contrary, in cases of incipient or advanced SH degradation, SoilBio represents an alert regarding the need for CA practices that will help to overcome or to interrupt these processes. Thus, the SH assessment by SoilBio is destined to improve the efficiency and sustainability of farming systems. This is possible by emphasizing the importance of healthy soils not only for increased crop yields but also for several benefits, such as better nutrient use efficiency, C sequestration, greenhouse gas mitigation, and yield stability under adverse conditions. Figure 10.16 shows the maps of the function nutrient supply and of SH (resulting from the combinations of the functions nutrient cycling and nutrient storage) of a 40,000-ha farm, located in Mato Grosso. As opposed to the nutrient supply function, where tones of dark and light green predominate—indicating the presence of soils with excellent chemical fertility—yellow and red colors on the SH map indicate the presence of unhealthy/low-quality soil environments. Not surprisingly, on the SH map, dark/light green represent either ICL systems or areas with deep-root brachiaria in recent years (see Chapter 2 for more information on these systems). These two maps demonstrate that, by providing information that would go unnoticed in soil fertility analyses, SoilBio anticipates SH problems that could negatively affect the economic performance of the farm. In fact, SoilBio can be considered as new ally of agronomists and rural technicians. Its inclusion in soil management decisions is challenging because sometimes a reassessment of the management practices adopted on the rural property may be required, which will result in customized, specific recommendations tailored for each farm.

Final Remarks

Future scenarios call for a more resilient, efficient, and multifunctional agriculture around the world. In addition to producing food, fiber, and fuel, science-based twenty-first century agriculture needs to be recognized and rewarded for its ability to provide important environmental services. As expressed in the famous Da Vinci quote at the head of this chapter, the "simplicity/sophistication" philosophy behind the SoilBio technology demonstrates that SH monitoring does not necessarily have to be an onerous/complicated task. As shown in Figures 10.12–10.15, in addition to on-farm SH assessments, the combination of the scores of the functions nutrient cycling and storage provides metrics to classify the different SH conditions (healthy, undergoing biological degradation, unhealthy, and recovering) of farms.

By November 2022, the SoilBio data bank contained 15,788 results of soil samples from 22 Brazilian states. This database underlies and feeds the Brazilian Soil

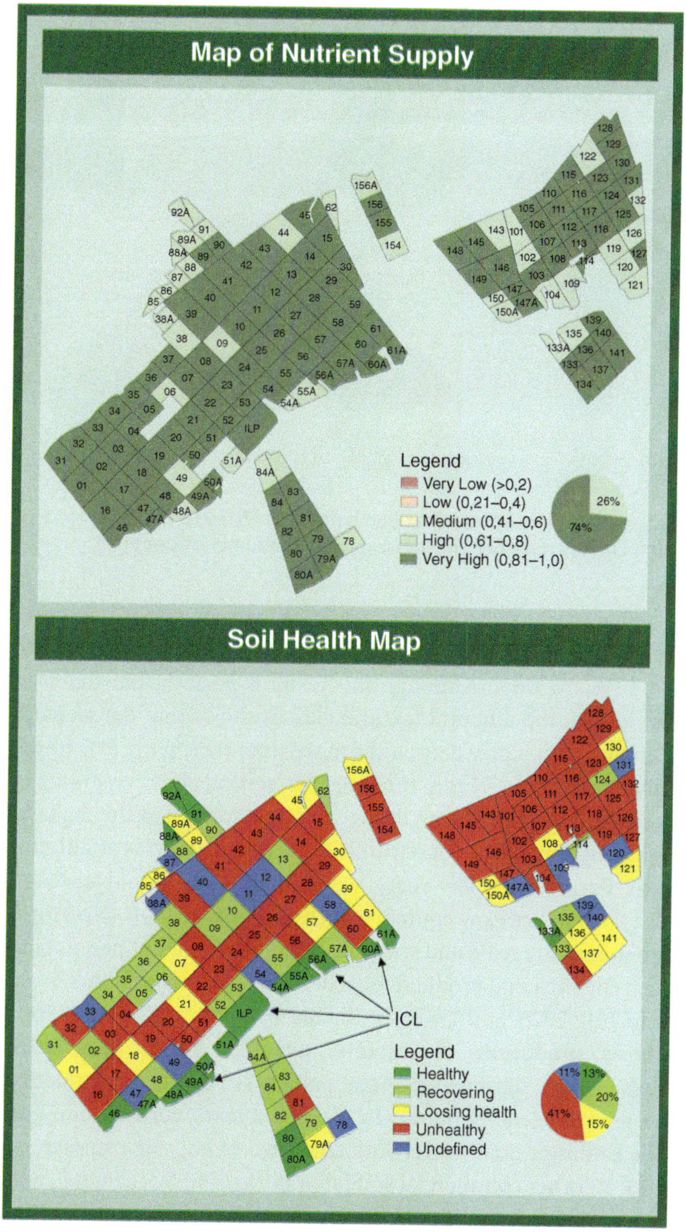

Figure 10.16 Maps of nutrient supply (a) and soil health (b) from a farm in Mato Grosso state, Brazil. On both maps, each square represents areas of on average 300 ha. Soil enzymes and chemical analyses were performed by one of the laboratories of the Embrapa SoilBio Network. On the soil health map, green highlights areas under integrated crop–livestock systems or brachiaria in the recent past.

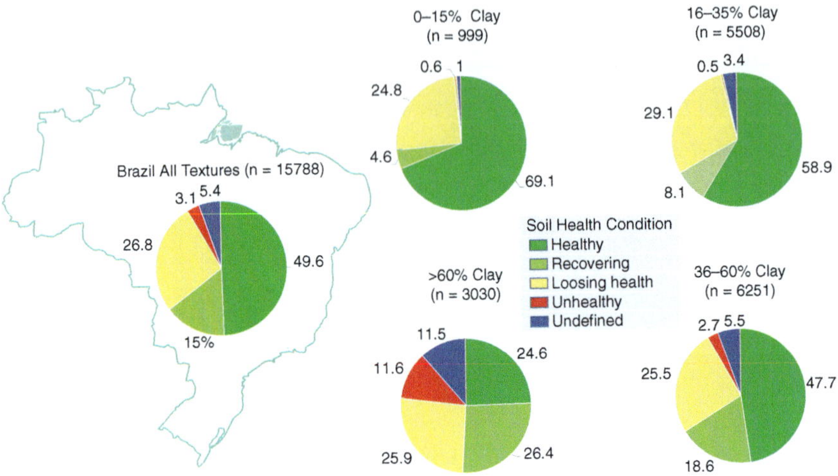

Figure 10.17 Scoreboard of the health of Brazilian soils (from 2020 to 2022), considering all textural classes together and separately (n = number of samples).

Health scoreboard (Figure 10.17). The combination of the scores of F1 (nutrient cycling) and F2 (nutrient storage) shows that 65% of all soil samples were classified as "healthy" and "recovering". Although this result is positive, the goal is to reach 100%. A close look at the different textural classes shows that the scenario changes for soil samples with a clay content of more than 60%, with 51% in the classes "healthy" and "recovering." From these results, several effective strategies can be applied to achieve high SQ. The analyses can be performed at the regional or national level, opening tremendous opportunities for public policies of soil conservation and health. For the next few years, we expect significant advances in SH in Brazil because farmers are now equipped with appropriate tools to perform on-farm SH assessments at any scale and with adequate management practices to maintain tropical soils healthy and productive, as described throughout this book.

Since its release in July 2020, SoilBio has attracted the attention of numerous stakeholders, such as producers, consultants, technical service providers, conservation groups, large agribusiness companies, researchers, students, policymakers, as well as the general public. In the first 2 years, corresponding to a training phase, the Embrapa SoilBio Network consisted of only six commercial laboratories. By 2023, 20 other commercial soil laboratories will be included, and over 40 will initiate proficiency testing. In December 2022, a complete SoilBio (i.e., ARYL and GLU + soil chemical parameters + soil texture analysis) cost around US$32.00.

The next steps in SoilBio research include its expansion into other agricultural regions, namely in the Northeast, and the development of interpretative algorithms for sugarcane, coffee, pasture, and forest plantations. A recent study by our group also has pointed out how ARYL and GLU can be used to evaluate C trends in soil (Chaer et al., 2023).

In the agricultural year 2021/2022, when soil enzyme measurements were transferred from research to commercial laboratories, there was a shortage of soil enzyme substrates due to the great demand for SoilBio analyses. It took months until the chemical companies were able to normalize the supply of PNS and PNG (ARYL and GLU substrates, respectively). For the SoilBio research team, this overwhelming response was thrilling not only because a new era for soil analyses in Brazil was about to begin but, most importantly, because it points toward the dawn of a new era in our relation with soils, based on the establishment of a biochemical dialogue mediated by GLU and ARYL. For us researchers, this is clear. Our wish is that it may soon also become clear to producers and trigger powerful transformations.

Acknowledgments

We dedicate this book chapter to our dear colleague Djalma Martinhão Gomes de Sousa. Djalma's research on soil fertility contributed to dramatic changes in the concepts of tropical agriculture. His passion for soil microbiology resulted in the development of the framework to interpret biological indicators and in the idealization of the FERTBIO soil sample concept. All these ideas were tested in his long-term field experiments related to P fertilizer management. Djalma will always be an inspiration for all of us. Many thanks Djalmão!

This work was financed by the Brazilian Agricultural research Corporation (Embrapa, projects 02.14.01.026.00, 20.20.03.017.00, and 22.14.01.026.00) and partially financed by the National Council for Scientific and Technological Development (CNPq), Edital Universal (grant number 404764/2016-9); by the Research Support Foundation of the Federal District (FAPDF), Edital Demanda Espontânea 2016 (grant number 1355/2016); and by the MCTI/CNPq/CAPES/FAPS (INCT-MPCPAgro-CNPq 465133/2014-4, Fundação Araucária-STI 043/2019, CAPES). I·C. Mendes, G. M. Chaer, M. A. Nogueira, and M. Hungria acknowledge research fellowships from CNPq. We thank Clodoaldo Sousa, Lucas Rolim, Osmar Oliveira, and Valmir Sousa (laboratory and field assistance); Samuel Teixeira Santos, Lucas Magalhães, Daniela Duarte (information technology team); and Chang Wilches, Cesar Araujo and Fabiola Araujo (innovation coordinators). Special thanks go to all the students who have worked with us during these past 22 years, especially Andre Alves de Castro Lopes and Leandro Sousa.

References

Acosta-Martinez, V., Moore-Kucera, J., Cotton, J., Gardner, T., & Wester, D. (2014). Soil enzyme activities during the 2011 Texas record drought/heat wave and implications to biogeochemical cycling and organic matter dynamics. *Applied Soil Ecology*, 75, 43–51.

Acosta-Martínez, V., Pérez-Guzmán, L., Veum, K. S., Nunes, M. R., & Dick, R. P. (2021). Metabolic activity—enzymes. In D. L. Karlen, D. E. Stott, & M. M. Mikha (Eds.), *Laboratory methods for soil health analysis* (Vol. 2, pp. 194–250). ASA. https://doi.org/10.1002/9780891189831.ch11

Anghinoni, G., Anghinoni, F., Tormena, C. A., Braccini, A. L., Mendes, I. C., Zancanaro, L., & Lal, R. (2021). Conservation agriculture strengthen sustainability of Brazilian grain production and food security. *Land Use Policy*, 108, 105591.

Anghinoni, I., & Vezzani, F. M. (2021). Systemic soil fertility as product of system self-organization resulting from management. *Revista Brasileira Ciencia Solo*, 45, e0210090. https://doi.org/10.36783/18069657rbcs20210090

Balota, E., Calegari, A., Nakatani, A. S., & Coyne, M. S. (2014). Benefits of winter cover crops and no-tillage for microbial parameters in a Brazilian Oxisol: A long-term study. *Agriculture Ecosystem Environment*, 197, 31–40.

Balota, E. L., Kanashiro, M., Colozzi Filho, A., Andrade, D. S., & Dick, R. P. (2004). Soil enzyme activities under long-term tillage and crop rotation systems in subtropical agro-ecosystems. *Brazilian Journal of Microbiology*, 35(4), 300–306. https://doi.org/10.1590/S1517-83822004000300006

Bandick, A. K., & Dick, R. P. (1999). Field management effects on soil enzyme activities. *Soil Biology Biochemistry*, 31(11), 1471–1479. https://doi.org/10.1016/S0038-0717(99)00051-6

Biswas, S., Hazra, G. C., Purakayastha, T. J., Saha, N., Mitran, T., Roy, S. S., & Mandal, B. (2017). Establishment of critical limits of indicators and indices of soil quality in rice-rice cropping systems under different soil orders. *Geoderma*, 292, 34–48. https://doi.org/10.1016/j.geoderma.2017.01.003

Bünemann, E. K., Bongiorno, G., Bai, Z., Creamer, R. E., Deyn, G., Goede, R., Fleskens, L., Geissen, V., Kuyper, T. W., Mäder, P., Pulleman, M., Sukkel, W., Groenigen, J. W. V., & Brussaard, L. (2018). Soil quality: A critical review. *Soil Biology and Biochemistry*, 120, 105–125. https://doi.org/10.1016/j.soilbio.2018.01.030

Carlson, J., Saxena, J., Basta, N., Hundal, L., Busalacchi, D., & Dick, R. P. (2015). Application of organic amendments to restore degraded soil: Effects on soil microbial properties. *Environmental Monitoring and Assessment*, 187, 109. https://doi.org/10.1007/s10661-015-4293-0

Castellano, S. D., & Dick, R. P. (1991). Cropping and sulfur fertilization influence on sulfur transformations in soil. *Soil Science Society of America Journal*, 55, 114–121. https://doi.org/10.2136/sssaj1991.03615995005500010020x

Chaer, G. M., Mendes, I. C., Dantas, O. D., Malaquias, J. V., Reis-Junior, F. B., & Oliveira, M. I. L. (2023). Evaluating C trends in clayey Cerrado Oxisols using a four-quadrant model based on specific arylsulfatase and β-glucosidase activities. *Applied Soil Ecology*, 183, 104742.

Correia, J. R., Reatto, A., & Spera, S. T. (2004). Solos e suas relações com o uso e o manejo. In D. M. G. Sousa & E. Lobato (Eds.), *Cerrado: Soil correction and fertilization* (2nd ed., pp. 29–61. (In Portuguese.)). Embrapa Cerrados. https://www.embrapa.br/busca-de-publicacoes/-/publicacao/555355/cerrado-correcao-do-solo-e-adubacao

Deng, S., & Popova, I. (2011). Carbohydrate hydrolases. In R. P. Dick (Ed.), *Methods of soil enzymology* (pp. 185–209). ASA and SSSA. https://doi.org/10.2136/sssabookser9.c9

Dick, R. P. (1992). A review: Long-term effects of agricultural systems on soil biochemical and microbial parameters. *Agriculture, Ecosystems & Environment*, 40(1–4), 25–36.

Dick, R. P. (1994). Soil enzyme activities as indicators of soil quality. In J. W. Doran, D. C. Coleman, D. F. Bezdicek, & B. A. Stewart (Eds.), *Defining soil quality for a sustainable environment*. https://doi.org/10.2136/sssaspecpub35.c7

Dick, R. P., & Burns, R. G. (2011). A brief history of soil enzymology research. In R. P. Dick (Ed.), *Methods of soil enzymology* (pp. 1–19). ASA and SSSA. https://doi.org/10.2136/sssabookser9.c1

Doran, J. W., & Parkin, T. B. (1994). Defining and assessing soil quality. In J. W. Doran, D. C. Coleman, D. F. Bezdicek, & B. A. Stewart (Eds.), *Defining soil quality for a sustainable environment* (pp. 1–21). ASA and SSSA. https://doi.org/10.2136/sssaspecpub35.c1

Gil-Sotres, F., Trasar-Cepeda, C., Leirós, M. C., & Seoane, S. (2005). Different approaches to evaluating soil quality using biochemical properties. *Soil Biology and Biochemistry*, 37(5), 877–887. https://doi.org/10.1016/j.soilbio.2004.10.003

Gonzalez-Quiñones, V., Stockdale, E. A., Banning, N. C., Hoyle, F. C., Sawada, Y., Wherrett, A. D., & Murphy, D. V. (2011). Soil microbial biomass—Interpretation and consideration for soil monitoring. *Soil Research*, 49(4), 287–304. https://doi.org/10.1071/SR10203

Haney, R. L., Franzluebbers, A. J., Porter, E. B., Hons, F. M., & Zuberer, D. A. (2004). Soil carbon and nitrogen mineralization: Influence of drying temperature. *Soil Science Society of America. Journal*, 68, 489–492. https://doi.org/10.2136/sssaj2004.0489

Hatfield, J. L., Sauer, T. J., & Cruse, R. M. (2017). Soil: The forgotten piece of the water, food, energy nexus. *Advances in Agronomy*, 143, 1–46. https://doi.org/10.1016/bs.agron.2017.02.001

Hemmati, S., Yaghmaeian, N., & Farhangi, M. B. (2022). Soil quality assessment of paddy fields (in northern Iran) with different productivities: Establishing the critical limits of minimum data set indicators. *Environmental Science and Pollution Research*, 30, 10286–10296. https://doi.org/10.1007/s11356-022-22846-8

Hinojosa, M. B., Carreira, J. A., García-Ruíz, R., & Dick, R. P. (2004). Soil moisture pre-treatment effects on enzyme activities as indicators of heavy metal-contaminated and reclaimed soils. *Soil Biology and Biochemistry*, 36, 1559–1568. https://doi.org/10.1016/j.soilbio.2004.07.003

Hussain, I., Olson, K. R., Wander, M. M., & Karlen, D. L. (1999). Adaptation of soil quality indices and application to three tillage systems in southern Illinois. *Soil and Tillage Research*, 50, 237–249.

Karlen, D. G., Goeser, N. J., Veum, K. S., & Yoist, M. A. (2017). On-farm soil health evaluations: Challenges and opportunities. *Journal of Soil and Water Conservation*, 72, 26–31. https://doi.org/10.2489/jswc.72.2.26A

Karlen, D. L. (2021). Preface. In D. L. Karlen, D. E. Stott, & M. M. Mikha (Eds.), *Approaches to soil health analysis* (Vol. 1, pp. xiii–xvii). ASA and SSSA.

Karlen, D. L., Mausbach, M. J., Doran, J. W., Cline, R. G., Harris, R. F., & Schuman, G. E. (1997). Soil quality: A concept, definition, and framework for evaluation (a guest editorial). *Soil Science Society of America Journal*, 61, 4–10. https://doi.org/10.2136/sssaj1997.03615995006100010001x

Karlen, D. L., & Stott, D. E. (1994). A framework for evaluating physical and chemical indicators of soil quality. In J. W. Doran, D. C. Coleman, D. F. Bezdicek, & B. A. Stewart (Eds.), *Defining soil quality for a sustainable environment* (pp. 53–72). ASA and SSSA. https://doi.org/10.2136/sssaspecpub35.

Klose, S., Bilen, S., Tabatabai, M. A., & Dick, W. A. (2011). Sulfur sycle enzymes. In R. P. Dick (Ed.), *Methods of soil enzymology* (Vol. 9, pp. 125–159). ASA and SSSA. https://doi.org/10.2136/sssabookser9.c7

Knight, T. R., & Dick, R. P. (2004). Differentiating microbial and stabilized β-glucosidase activity relative to soil quality. *Soil Biology and Biochemistry*, 36, 2089–2096. http://dx.doi.org/10.1016/j.soilbio.2004.06.007

Lehman, R. M., Cambardella, C. A., Stott, D. E., Acosta-Martinez, V., Manter, D. K., Buyer, J. S., Maul, J. E., Smith, J. L., Collins, H. P., Halvorson, J. J., Kremer, R. J., Lundgren, J. G., Ducey, T. F., Jin, V. L., & Karlen, D. L. (2015). Understanding and enhancing soil biological health: The solution for reversing soil degradation. *Sustainability*, 7(1), 988–1027. https://doi.org/10.3390/su7010988

Lisboa, B. B., Vargas, L. K., Silveira, A. O., Martins, A. F., & Selbach, P. A. (2012). Indicadores microbianos de qualidade do solo em diferentes manejos. *Revista Brasileira Ciencia Solo*, 36, 45–55.

Lopes, A. A. C., Bogiani, J. C., de Figueiredo, C. C., Reis Junior, F. B., Sousa, D. M. G., Malaquias, J. V., & Mendes, I. C. (2021). Enzyme activities in a sandy soil of Western Bahia under cotton production systems: Short-term effects, temporal variability, and the FERTBIO sample concept. *Brazilian Journal of Microbiology*, 52, 2193–2204. https://doi.org/10.1007/s42770-021-00606-z

Lopes, A. A. C., Sousa, D. M. G., Chaer, G. M., Reis Junior, F. B., Goedert, W. J., & Mendes, I. C. (2013). Interpretation of microbial soil indicators as a function of crop yield and organic carbon. *Soil Science Society of America Journal*, 77, 461–472. https://doi.org/10.2136/sssaj2012.0191

Lopes, A. A. C., Sousa, D. M. G., Reis Junior, F. B., & Mendes, I. C. (2015). Air-drying and long-term storage effects on β-glucosidase, acid phosphatase and arylsulfatase activities in a tropical Savannah Oxisol. *Applied Soil Ecology*, 93, 68–77. https://doi.org/10.1016/j.apsoil.2015.04.001

Lorenz, N., & Dick, R. P. (2011). Sampling and pretreatment of soil before enzyme analysis. In R. P. Dick (Ed.), *Methods of soil enzymology* (pp. 85–101). ASA and SSSA. https://doi.org/10.2136/sssabookser9.c5

Lorenz, N., Gardener, B. M. C. S., Lee, N. R., Ramsier, C., & Dick, R. P. (2020). Soil enzyme activities associated with differential outcomes of contrasting approaches to soil fertility management in corn and soybean fields. *Applied Ecology and Environmental Sciences*, 8, 517–525. https://doi.org/10.12691/aees-8-6-26

Lupwayi, N. Z., Harker, K. N., O'Donovan, J. T., Turkington, T. K., Blackshaw, R. E., Hall, L. M., Willenborg, C. J., Gan, Y., Lafond, G. P., May, W. E., & Grant, C. A. (2015). Relating soil microbial properties to yields of no-till canola on the Canadian prairies. *European Journal of Agronomy* 62, 110–119. https://doi.org/10.1016/j.eja.2014.10.004

Lynch, J. M., & Bragg, E. (1985). Microorganisms and soil aggregate stability. *Advances in Soil Science*, 2, 133–171. http://dx.doi.org/10.1007/978-1-4612-5088-3_3

Mei, N., Yang, B., & Tian, P. (2019). Using a modified soil quality index to evaluate densely tilled soils with different yields in Northeast China. *Environmental Science and Pollution Research*, 26, 13867–13877. https://doi.org/10.1007/s11356-018-3946-2

Mendes, I. C., Dick, R. P., Bandick, A. K., & Bottomley, P. J. (1999). Microbial biomass and activities in soil aggregates affected by winter cover crops. *Soil Science Society of America Journal*, 63, 873–881. https://doi.org/10.2136/sssaj1999.634873x

Mendes, I. C., Ono, F. B., Oliveira, M. I., Silva, R. G., Kappes, C., Reis Junior, F. B., & Zancanaro, L. (2020). Rotação de culturas, bioindicadores e saúde do solo. In P. A. Silva & L. C. Oliveira (Eds.), (Eds.)*Boletim de Pesquisa da Fundação MT* (pp. 102–110). Fundação MT.

Mendes, I. C., Sousa, D. M. G., Chaer, G. M., Reis Junior, F. B., Lopes, A. A. C., Dantas, O. D., & Souza, L. M. (2021a). Soil bioanalysis: A simple and effective tool to access and interpret soil health. In *Proceedings of the Global Symposium on Soil Biodiversity* (pp. 256–261). FAO. https://doi.org/10.4060/cb7374en

Mendes, I. C., Sousa, D. M. G., Dantas, O. D., Lopes, A. A. C., Reis Junior, F. B., Oliveira, M. I. L., & Chaer, G. M. (2021b). Soil quality and grain yield: A win-win combination in clayey tropical Oxisols. *Geoderma*, 388, 114880. https://doi.org/10.1016/j.geoderma.2020.114880

Mendes, I. C., Sousa, D. M. G., Reis Junior, F. B., Kappes, C., Ono, F. B., Semler, T. D., Zancanaro, L., & Lopes, A. A. C. (2017). Qualidade biológica do solo: por que e como avaliar. In *Boletim de Pesquisa da Fundação MT* (pp. 98–105). Fundação MT.

Mendes, I. C., Sousa, D. M. G., Reis Junior, F. B., Lopes, A. A. C., & Souza, L. M. (2019a). Bioanálise de solo: Aspectos teóricos e práticos. In E. C. Severiano, M. F. Morais, & A. M. Paula (Eds.), *Tópicos em ciência do solo* (Vol. X, pp. 399–462). SBCS.

Mendes, I. C., Sousa, D. M. G., Reis-Junior, F. B., & Lopes, A. A. C. (2018a). *Bioanálise de solo: como acessar e interpretar a saúde do solo*. In *Circular Técnica 38*. Cerrados.

Mendes, I. C., Souza, L. M., Sousa, D. M. G., Lopes, A. A. C., Reis Junior, F. B., Lacerda, M. P. C., & Malaquias, J. V. (2019b). Critical limits for microbial indicators in tropical Oxisols at post-harvest: The FERTBIO soil sample concept. *Applied Soil Ecology*, 139, 85–93. https://doi.org/10.1016/j.apsoil.2019.02.025

Mendes, I. C., Souza, L. V., Resck, D. V. S., & Gomes, A. C. (2003). Propriedades biológicas em agregados de um Latossolo Vermelho-Escuro sob plantio convencional e direto no Cerrado. *Revista Brasileira Ciencia Solo*, 27, 435–443. http://doi.org/10.1590/S0100-06832003000300005

Mendes, I. C., Tormena, C. A., Cherubin, M. R., & Karlen, D. L. (2018b). Soil health assessment and maintenance in central and south-central Brazil. In D. Reicosky (Ed.), *Managing soil health for sustainable agriculture: Monitoring and management* (1st ed., pp. 379–415). Burleigh Dodds. http://doi.org/10.19103/AS.2017.0033.35

Miller, M., & Dick, R. P. (1995). Thermal stability and activities of soil enzymes as influenced by crop rotations. *Soil Biology and Biochemistry*, 27, 1161–1666.

Nannipieri, P., Trasar-Cepeda, C., & Dick, R. (2018). Soil enzyme activity: A brief history and biochemistry as a basis for appropriate interpretations and meta-analysis. *Biology and Fertility of Soils*, 54(1), 11–19. https://doi.org/10.1007/s00374-017-1245-6

Ndiaye, E. L., Sandeno, J. M., McGrath, D., & Dick, R. P. (2000). Integrative biological indicators for detecting change in soil quality. *American Journal of Alternative Agriculture*, 15(1), 26–36. http://dx.doi.org/10.1017/S0889189300008432

Nicolodi, M., Gianello, C., & Anghinoni, I. (2004). Fertilidade: Uma propriedade emergente do sistema solo. In *Reunião Sul-brasileira de Ciência do Solo* (pp. 5). Embrapa.

Nicolodi, M., Gianello, C., Anghinoni, I., Marré, J., & Mielniczuk, J. (2008). Insuficiência do conceito mineralista para expressar a fertilidade do solo percebida pelas plantas cultivadas no sistema plantio direto. *Revista Brasileira Ciencia Solo*, 32, 2735–2744. https://doi.org/10.1590/S0100-06832008000700017

Passinato, J. H., Amado, T. J. C., Kassam, A., Acosta, J. A. A., & Amaral, L. P. (2021). Soil health check-up of conservation agriculture farming systems in Brazil. *Agronomy*, 11(12), 2410. https://doi.org/10.3390/agronomy11122410

Peixoto, R. S., Chaer, G. M., Franco, N., Reis Junior, F. B., Mendes, I. C., & Rosado, A. S. (2010). A decade of land use contributes to changes in the chemistry, biochemistry and bacterial community structures of soils in the Cerrado. *Antonie van Leeuwenhoek*, 98(3), 403–413. https://doi.org/10.1007/s10482-010-9454-0

Perucci, P. (1992). Enzyme activity and microbial biomass in a field soil amended with municipal refuse. *Biology and Fertility of Soils*, 14, 54–60.

Saggar, S., Bettany, J. R., & Stewart, J. W. B. B. (1981). Measurement of microbial S in soil. *Soil Biology and Biochemistry*, 13, 493–498. http://doi.org/10.1016/0038-0717(81)90040-7

Skujiņš, J. (1976). Extracellular enzymes in soil. *CRC Critical Review Microbiology*, 4, 383–421. http://doi.org/10.3109/10408417609102304

Stott, D. E., Andrews, S. S., Liebig, M. A., Wienhold, B. J., & Karlen, D. L. (2010). Evaluation of β-glucosidase activity as a soil quality indicator for the soil management assessment framework. *Soil Science Society of America Journal*, 74, 107–119. https://doi.org/10.2136/sssaj2009.0029

Tabatabai, M. A. (1994). Soil enzymes. In R. W. Weaver, S. Angle, P. Bottomley, D. Bezdicek, S. Smith, A. Tabatabai, & A. Wollum (Eds.), *Methods of soil analysis. Part 2. Microbiological and biochemical properties* (pp. 778–833). ASA and SSSA. https://doi.org/10.2136/sssabookser5.2

Tian, K., Zhang, B., Zhang, H., Huang, B., Darilek, J. L., Zhao, Y., & Yang, J. (2020). Evaluation of soil quality in major grain-producing region of the North China plain. Integrating minimum data set and established critical limits. *Ecological Indicators*, 117, 106613. https://doi.org/10.1016/j.ecolind.2020.10661

Trasar-Cepeda, C., Leiros, C., Gil-Sotres, F., & Seoane, S. (1997). Towards a biochemical quality index for soils: An expression relating several biological and biochemical properties. *Biology and Fertility of Soils*, 26(2), 100–106.

Vallejo, V. E., Arbeli, Z., Terán, W., Lorenz, N., Dick, R. P., & Roldan, F. (2012). Effect of land management and *Prosopis juliflora* (Sw.) DC trees on soil microbial community and enzymatic activities in intensive silvopastoral systems of Colombia. *Agriculture, Ecosystems & Environment*, 150, 139–148.

Vallejo, V. E., Roldan, F., & Dick, R. P. (2010). Soil enzymatic activities and microbial biomass in an integrated agroforestry chronosequence compared to monoculture and a native forest of Colombia. *Biology and Fertility of Soils*, 46, 577–587. https://doi.org/10.1007/s00374-010-0466-8

Veum, K. S., Goyne, K. W., Kremer, R. J., Miles, R. J., & Sudduth, K. A. (2014). Biological indicators of soil quality and soil organic matter characteristics in an agricultural management continuum. *Biogeochemistry*, 117(1), 81–99. https://doi.org/10.1007/s10533-013-9868-7

Wallenstein, M. D., & Burns, R. (2011). Ecology of extracellular enzyme activities and organic matter degradation in soil: A complex community-driven process. In R. P. Dick (Ed.), *Methods of soil enzymology* (pp. 35–55). SSSA. https://doi.org/10.2136/sssabookser9.c2

Wymore, A. W. (1993). *Model-based systems engineering. An Introduction to the Mathematical Theory of Discrete Systems and to the Tricotyledon Theory of Systems Design*. CRC.

11

Challenges to Managing Soil Health in the Newest Agricultural Frontier in Brazil

Rodrigo Estevam Munhoz de Almeida, Henrique Antunes de Souza, Balbino Antonio Evangelista, Alexandre Uhlmann, Michele Ribeiro Ramos, Edvaldo Sagrilo, Tais Souza dos Santos Dias, Laura Resplandes de Sousa Paz Oliveira, and Nídia Raquel Costa

Chapter Overview

The current agricultural frontier region in Brazil is located in the north and northeast of the country. This zone has specific characteristics, such as low altitudes, high temperatures, and, in some parts, restricted water regimes. The soils are highly weathered and acidic, with low buffering capacity, low organic matter (OM) content, and considerable occurrence of sandy or gravelly soils. The combination of these edaphoclimatic conditions results in low capacity to retain nutrients and water. Adoption of sustainable cropping systems that reduce soil disturbance and increase OM, crop diversity, and soil protection can further enhance soil health (SH) to face the adverse conditions of the tropical climate in this region. Management practices recommended to implement sustainable agriculture in this region start with understanding the climate risk, defining crops and sowing dates with adoption of practices to correct soil acidity, increasing fertility, preventing compaction and erosion, and ensuring high levels of biological activity in the soil. Inclusion of cover crops (CCs) for biomass inputs, such as oversowing grasses into soybean [*Glycine max* (L.) Merr.] and intercropping maize (*Zea mays* L.) with tropical forages, increases water retention and reduces soil temperature, leading to better nutrient cycling. This, consequently, promotes microbiological activity and SH.

Soil Health Series: Volume 3 Soil Health and Sustainable Agriculture in Brazil, First Edition.
Edited by Ieda Carvalho Mendes and Maurício Roberto Cherubin.

Introduction

An agricultural frontier region in Brazil is characterized by the geographical delimitation of areas; particularly, with the recent expansion of agricultural cultivation, this delimitation is done either through the conversion of degraded pastures or the clearing of areas with native vegetation. The development of agriculture in the central-northern region of the Brazilian territory began in the 1970s and 1980s, and it was made possible by advancements in technology and the adoption of agricultural management practices adapted to cultivation in tropical conditions. These practices include using limestone to correct soil acidity, fertilization to correct soil fertility, and genetic improvement of grain-producing crops suitable for the low latitude region.

The expansion of agriculture into new areas is often driven by low land prices in regions where agriculture is not yet an important activity. This trend is further accentuated during periods of high agricultural profitability, which is typically marked by factors such as (a) favorable climate in previous harvests, (b) abundant crops, (c) high grain prices, and (d) low input costs. In the case of central-northern Brazil, the initial expansion of agriculture prioritized areas with the greatest yield potential, which tended to be characterized by high rainfall, high altitudes, flat terrain, and clayey soils. Logistical considerations and proximity to major urban centers were also considered. However, because these areas have been occupied and cultivated for years, the agricultural frontier has moved to more challenging areas with lower altitudes, more restricted water regimes, sandy soils, undulating flat terrain, and less favorable logistics.

In Brazil, the officially designated frontier area for agricultural expansion is known as the MATOPIBA (the first syllables of the states of Maranhão, Tocantins, Piauí, and Bahia). This region includes 337 municipalities and spans 73 million ha (Lima et al., 2019). Across the four states, there are 4,803,471 ha of soybean cultivation (Lima et al., 2019), with an estimated yield of around 18.5 million tons in the 2022/2023 crop season, representing 12.3% of the 150.3 million tons estimated for Brazil (CONAB, 2022). The region is deemed strategic for the economic development of the entire country and has been attracting farmers from various regions (Pires et al., 2016).

However, the agricultural areas in Bahia that comprise the MATOPIBA region have already been mostly occupied. The production systems in these areas are already stabilized, with corrected soil fertility and high land prices. These areas are now undergoing vertical investment, with the implementation of irrigation systems and grain storage on farms. Bahia was the first area to be occupied in MATOPIBA due to its higher altitude and flat relief. However,

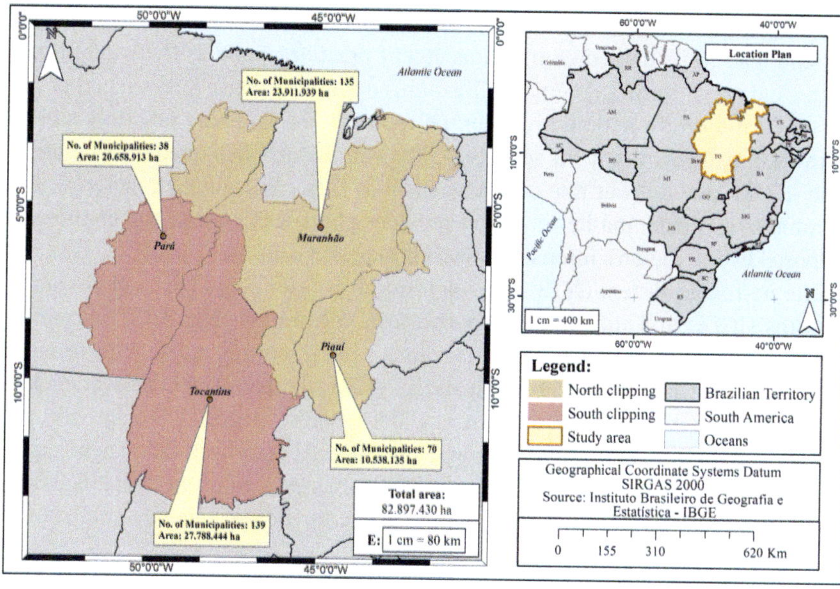

Figure 11.1 Location of the expansion region of the last agricultural frontier in Brazil covering the states of Maranhão, Piauí, Tocantins, and Pará. Prepared by the authors.

these areas have predominantly sandy or medium-textured soils and a restricted rainfall regime, which limits the cultivation of two crops in a same rainy season. Hence, the authors of this chapter have defined the new agricultural frontier area (Figure 11.1) by considering factors such as the increase in soybean cultivation area over the past 10 years and lower land prices compared with other regions. The proposed new frontier excludes the already consolidated areas of Bahia but includes the areas of Tocantins and the municipalities of Piauí and Maranhão, which were already part of MATOPIBA, and adds the eastern portion of Pará and some municipalities of Piauí that were not previously part of this frontier.

In the region of the newest Brazilian agricultural frontier discussed in this chapter, highly weathered, acidic soils with low buffering capacity are predominant, notably due to their low clay and OM content. Thus, these soils generally have a low capacity to retain nutrients and water. These limitations are exacerbated by "dry spells," which are common during the rainy season, causing severe drought stress and posing significant challenges to agricultural crop yield (Evangelista et al., 2017). In addition to these inherent characteristics of tropical soils, this agricultural frontier region significantly includes sandy and sandy/loam soils or soils with gravel presence, further complicating their management in the face of the tropical climate.

Continuous cultivation of crops under such conditions can impair soil quality and lead to its degradation. Management practices, such as the use of limestone and agricultural gypsum, fertilization, adoption of conservationist cropping systems, and crop rotation, are essential to improve SH. These practices aim at providing biomass input for soil cover, less soil disturbance, greater accumulation of OM, and a lower rate of soil water loss. This, consequently, increases the volume of soil explored by roots and provides plant resilience to the challenges imposed by the region's intense tropical climate and soil.

The no-tillage system (NTS) and, more recently, the integrated crop–livestock systems (ICLs) and integrated crop–livestock–forestry systems (ICLFs), which involve the inclusion of forage plants through intercropping maize with tropical grasses (primarily from the *Urochloa* and *Megathyrsus* genus), or even oversowing grasses on a soybean crop, are the most widely used strategies in the agricultural frontier region to increase C levels and promote SH improvement. These management practices effectively reduce erosion, increase nutrient use efficiency, and improve soil structure, resulting in higher crop yields (Almeida et al., 2020; Atwood et al., 2022; Guilherme et al., 2018).

Despite being very promising in the global agricultural scenario, the agricultural frontier region lacks systematic and holistic studies that address the management of soils and crops under such conditions. There is a lack of information that enables the restoration and maintenance of the functions of cultivation systems, aiming for greater sustainability in agriculture. This chapter focuses on the climate and edaphic characterization of the newest agricultural frontier in Brazil as well as on some of the best management practices that should be considered to promote SH in this region.

Climate Characterization

The agricultural frontier region focused on in this chapter (Figure 11.1) is a transition zone with specific characteristics, such as fragile soils and adverse climate, commonly found in the representative plant formations of the Caatinga, Cerrado, and Amazon biomes. Within the polygon being analyzed, the Cerrado biome, a tropical savanna characterized by a seasonal climate and vegetation consisting of a mosaic of grasslands, shrublands, and forestlands with different proportions, depending on the region, is predominant (Becerra et al., 2010).

According to Köppen's classification (1948), the entire studied area of the states of Tocantins, Maranhão, and Piauí is categorized as having a tropical climate (Aw), with variations of dry, hot, and humid weather. The average monthly temperature ranges from 25 to 27°C across most of the territory. Rainfall is irregularly distributed throughout the year, with a dry season from May to September and a rainy season from October to April. The average annual rainfall ranges from 800 to 2000 mm (Assad & Evangelista, 1994).

The portion of the territory that belongs to Maranhão has a high annual average temperature of around 26°C, annual rainfall volumes ranging from 1250 to 1500 mm, and a well-defined dry season. In Tocantins, the climate is characterized by a rainy season from October to April, followed by a dry period from May to September. The extensive latitudinal extension and relatively flat terrain influence the region's climate, with altitudes predominantly ranging from 200 to 600 m. To the north of the 6°S parallel, the climate is humid with typical Amazon biome vegetation, and no dry winter occurs. The average annual temperature is ~26°C, with 1500–2100 mm of rainfall. To the south of the 6°S parallel, the climate is predominantly sub-humid, with an equal balance of rainy and dry months, and average annual temperatures gradually decrease as altitude increases. The annual average temperature ranges between 23°C (at higher altitudes) and 27°C, and rainfall varies between 900 and 2100 mm. The state of Piauí experiences average temperatures ranging from 25 to 27°C, with variable rainfall of ~700 mm in the south and nearly 1200 mm annually in the north.

Using Köppen's (1948) climate classification method, Alvares et al. (2013) classified the region of this study as follows: the entire territory of the state of Tocantins as AW type, with a dry winter; almost all of the surface of the states of Maranhão and Piauí also as Aw, with a dry winter, with only the extreme north of these states having climate As, with a dry summer; the small longitudinal strip of the state of Pará that borders the states of Maranhão and Tocantins has an Aw climate, with a dry winter and an Am, with a monsoon climate, which occurs in most of the territory of Pará. The highest average annual precipitation rates, ranging from 1900 to 2200 mm, are observed in this portion of land in Pará. These rates decrease to 1600–1990 mm in the western half of Tocantins and to 1300–1660 mm in the eastern half of Tocantins, central-western Maranhão, and extreme northern part of Piaui. The lowest rainfall rates, between 1000 and 1300 mm, are observed in the southeast of Maranhão and in a small strip in the northeast of Piauí (Alvares et al., 2013).

In a land suitability assessment study in the MATOPIBA region, Lumbreras et al. (2015) used the Thornthwaite and Mather (1955) method to identify five climatic domains, denoted by the symbols C1–C5 (Figure 11.2), based on the volume of rainfall in millimeters (P), moisture index (Im), and number of dry months (dm). The five domains are as follows: C1: P = 1600–2000, Im = 20–30,

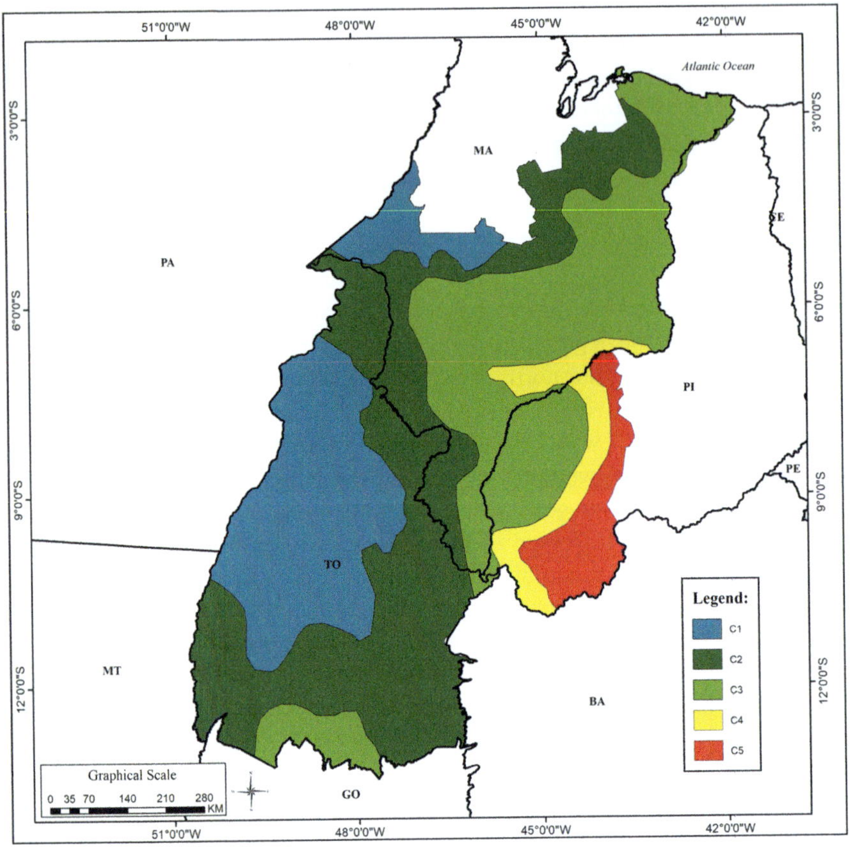

Figure 11.2 Climatic domains distribution of the Maranhão, Tocantins, and Piauí states: C1: P = 1600–2000, Im = 20–30, es = 3–4; C2: P = 1300–1600, Im = 0–20, es = 5–6; C3: P = 1100–1300, Im = 0–10, es = 6–7; C4: P = 900–1100, Im = −10 to −20, es = 6–7; and C5: P = 800–900 Im = −20 to −30, es = 7–8, where es is the number of dry months, Im is the moisture index, and P is precipitation (mm). Adapted from Lumbreras et al. (2015).

dm = 3–4; C2: P = 1300–1600, Im = 0–20, dm = 5–6; C3: P = 1100–1300, Im = 0–10, dm = 6–7; C4: P = 900–1100, Im = −10 to −20, dm = 6–7; and C5: P = 800–900, Im = −20 to −30, dm = 7–8. The spatial distribution of these domains indicates a decrease in moisture indices toward the south and, to a greater extent, from west to east. The areas near the Amazon biome, which fall under domains C1 and C2, experience the highest rainfall (>1,300 mm), shorter drought periods, and positive moisture indices. Climates C4 and C5 are characterized by severe water deficiency typical of semi-arid environments, with annual rainfall totals reaching ~600 mm. In summary, the inter-annual climatic variability in the region, especially concerning rainfall and water availability for crops, is quite significant (Lumbreras et al., 2015).

This agricultural frontier region is characterized by predominantly low altitudes of less than 400 m asl, associated with high temperatures and high rates of water loss due to evapotranspiration. Rainfall volumes increase gradually from the southeast, with 680–750 mm, to the northwest, with up to 2,500 mm. The greatest accumulation of rainfall is concentrated in the states of Pará and Tocantins, while the smallest indices are in the states of Piauí and Maranhão (Figure 11.3a). The average annual temperature ranges from 24 to 26°C in the southeastern parts of Tocantins, southern Maranhão, and small parts of Pará, whereas in most of the agricultural frontier territory, average temperatures generally exceed

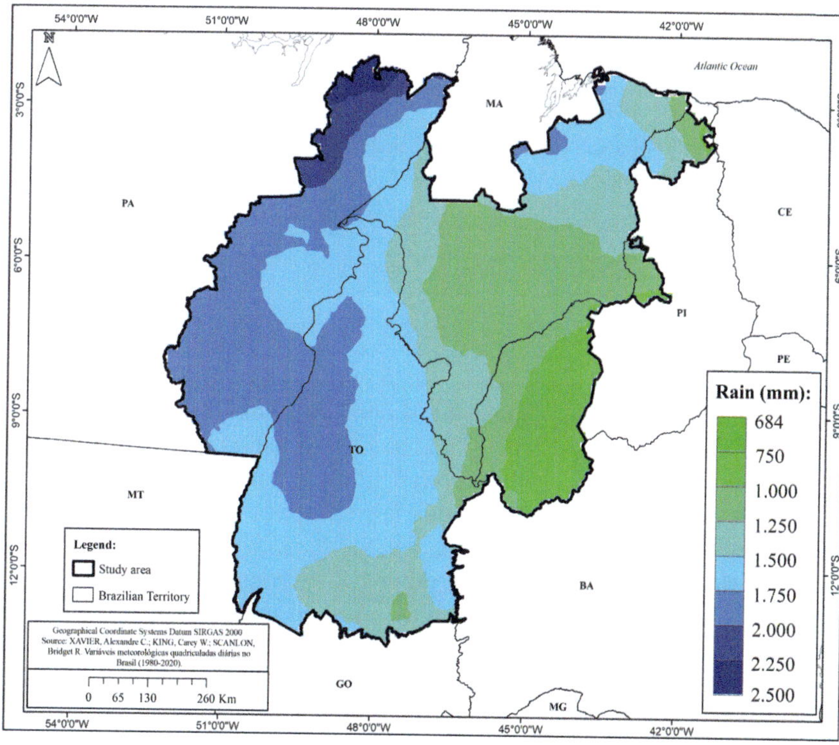

Figure 11.3 Average annual precipitation (a) and average annual temperature (b) in the agricultural frontier expansion region (Xavier et al, 2020).

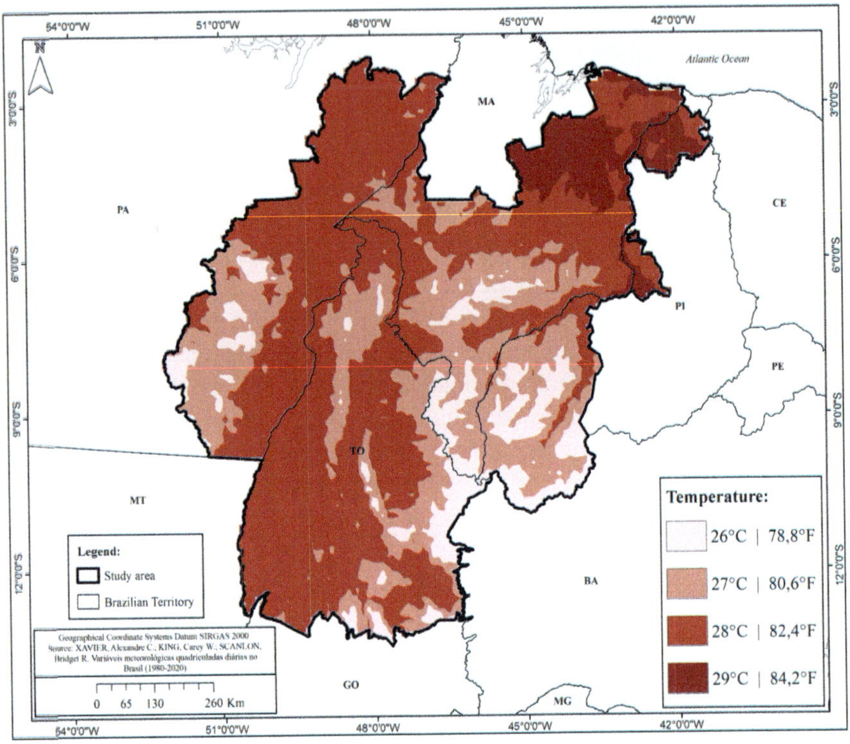

Figure 11.3 *(Continued)*

26°C (Figure 11.3b). The spatial distribution of the annual average temperature indicates a hot climate throughout the region, with little variation in average temperature, ranging from 26 to 29°C. The highest temperatures are found in the northeast of Maranhão and the lowest in the frontier region of Tocantins, Maranhão, Piauí, and Bahia. The region's largest area has average temperatures between 27 and 28°C (Figure 11.3b).

The maximum temperature indices exhibit little variability across the region (Figure 11.4a), ranging from 31°C in small zones in the southeast and west to 34°C in the southwest of Tocantins and northeast of Maranhão. However, there is a greater variability in minimum temperature (Figure 11.4b), ranging from 19°C in the high altitude regions of southern Piauí to 24°C in the central-north region of Maranhão.

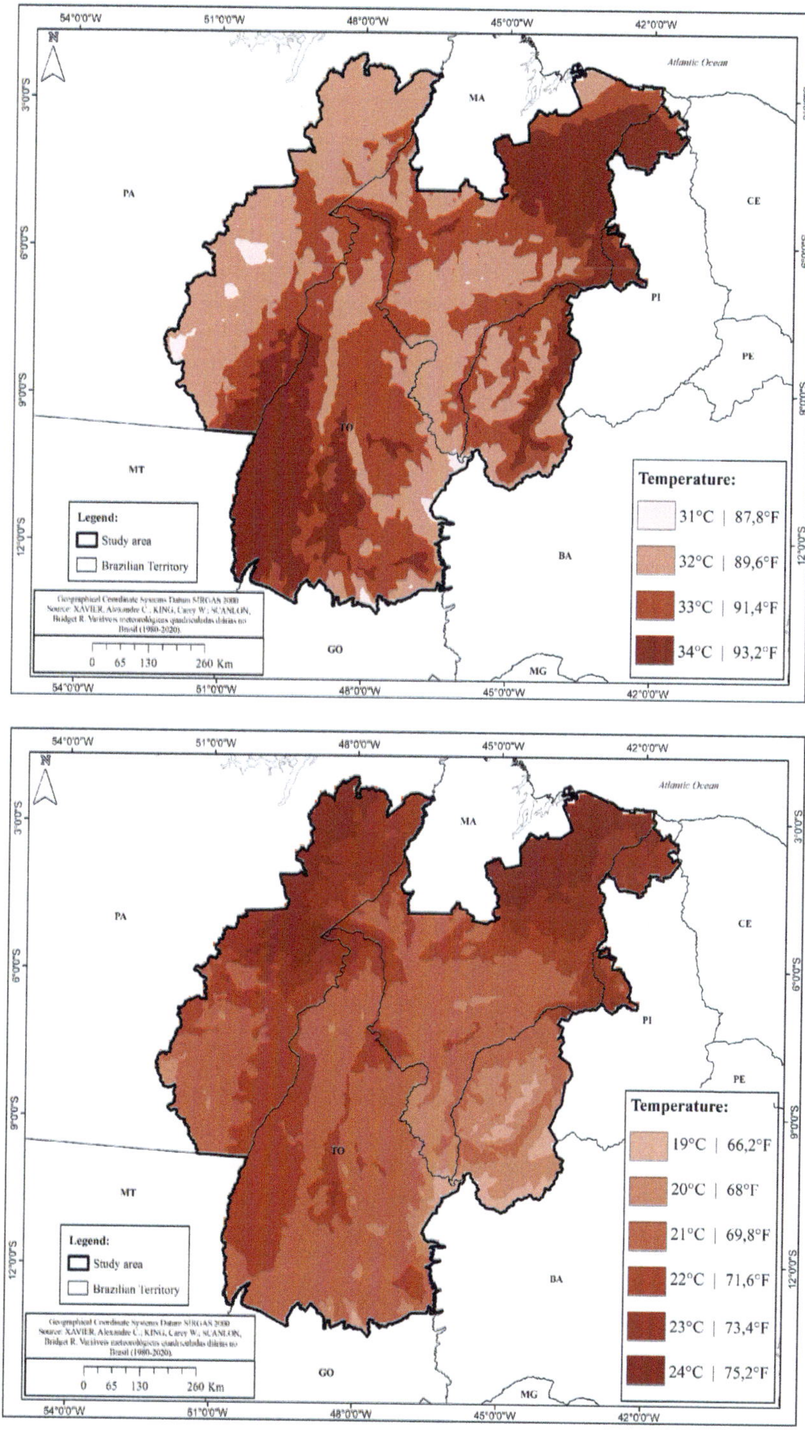

Figure 11.4 Annual average of maximum (a) and minimum temperatures (b) in the agricultural frontier expansion region (Xavier et al, 2020).

Main Soil Classes in the Newest Agricultural Frontier in Brazil

The soil types that occur in the agricultural frontier area, classified up to the second categorical level according to the Brazilian System of Soil Classification (Sistema Brasileiro de Classificação de Solos [SiBCS]) (Santos et al., 2018), are shown in Figure 11.5. The SiBCS is used in this chapter because the Soil Taxonomy was considered inadequate to represent soils in Brazil's large territory, with a complexity of tropical, dry, and subtropical environments and soil-forming processes (Schaefer et al., 2023). Despite that, Table 11.1 shows the approximate correspondence of soil classes between SiBCS, World Reference Base (WRB) (IUSS Working Group WRB, 2015), and Soil Taxonomy (USDA-NRCS, 1999, 2014).

The soil maps shown in this chapter were created by cutting out the soil map of Brazil, which is at a scale of 1:250,000 (IBGE, 2019). The area occupied by each soil was determined using the methodology of Araújo Filho et al. (2022). Each mapping unit area was divided according to the component soil types present in that unit. Units with only one component soil type were assigned 100% of the area for that soil; units with two component soil types were assigned 60% of the area to the main soil type and 40% to the secondary soil type; units with three component soil types were assigned 50%, 30%, and 20% of the area; and units with four component soil types were assigned 40%, 20%, 20%, and 20% of the area.

The soils in intertropical regions are deeply weathered, acidic, and nutrient-poor. Latossolos are the most prevalent soil type in the agricultural frontier polygon, covering 28% of the area. Argissolos also constitute a significant portion of the polygon (23%), with a higher concentration in the western edge of the area, in municipalities of Pará state. Neossolos make up a substantial portion of the soil types in the region, representing 21% of the total, with about half being Neossolos Litólicos (11%) and the rest being Neossolos Quartzarênicos (9%). Plintossolos are the fourth most prevalent soil type in the polygon, covering ~18% of the area. Plintossolos Pétricos comprise 12.7% of the total area and are the most common type covering large areas in the state of Tocantins and at the northern tip of the polygon between the states of Maranhão and Piauí (Figure 11.5).

Latossolos are tropical soils par excellence. They are deep and have a well-developed structure, resulting in good physical characteristics. However, they often have low chemical fertility (Ker, 1997; Oliveira et al., 2017). The sum of base values is often insufficient for successful cash crop production.

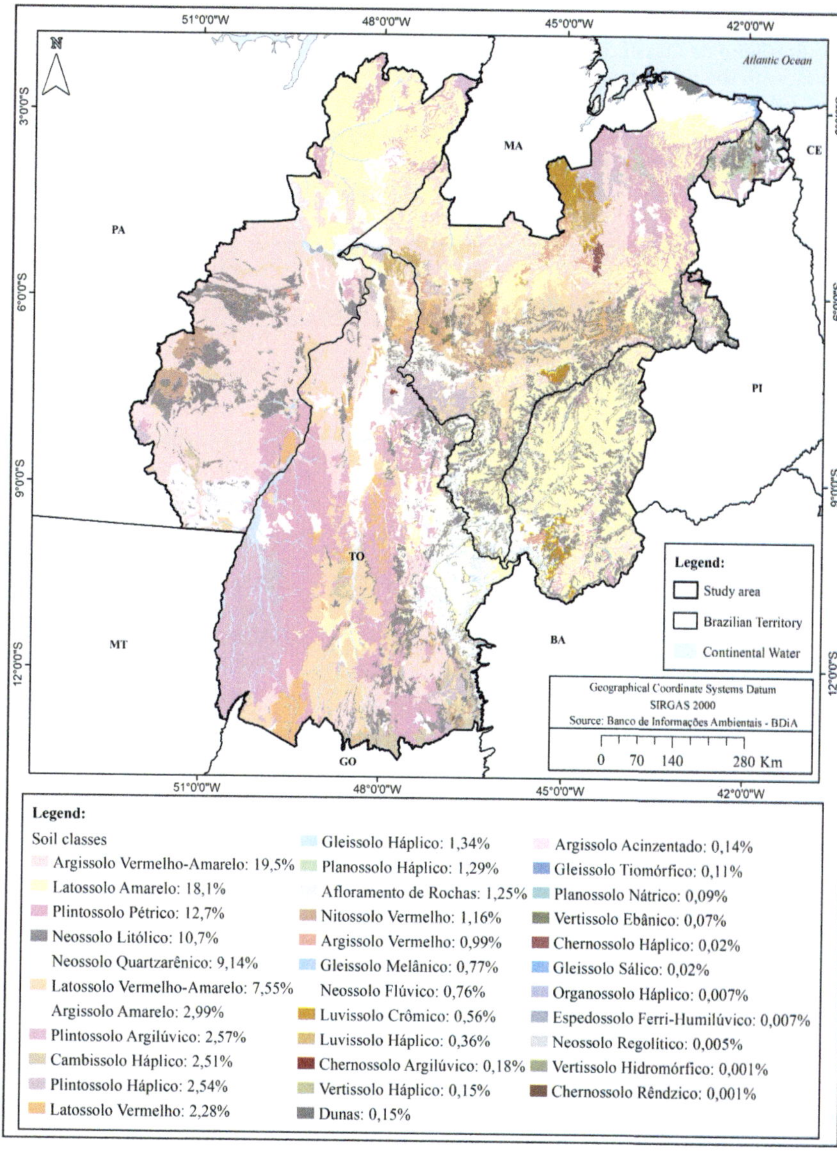

Figure 11.5 Spatial distribution of the main classes of soils classified up to the second categorical level according to Sistema Brasileiro de Classificação de Solos (SiBCS) (Santos et al., 2018) in the northern Brazilian agricultural frontier consisting of part of the states of Maranhão, Piauí, Pará, and the entire state of Tocantins. The approximate correlation of SiBCS soil classification with World Reference Base (WRB) and Soil Taxonomy can be seen in Table 11.1. Prepared by the authors.

Table 11.1 Approximate correspondence between high-level categorical soil classes in Sistema Brasileiro de Classificação de Solos (SiBCS), World Reference Base (WRB), and Soil Taxonomy, with the percentage of the agricultural frontier area (Figure 11.4) occupied by each soil type.

SiBCS[a]	Percentage of the area[b]	WRB[c]	Soil taxonomy[d]
Latossolos	28.0	Ferralsols	Oxisols
Argissolos	23.2	Acrisols, Lixisols, Alisols	Ultisols, some Oxisols (Kandic)
Neossolos	20.7	–	Entisols
(Neossolo Flúvicos)	(0.8)	Fluvisols	(Fluvents)
(Neossolos Litólicos)	(10.7)	Leptosols	(Lithic…Oerthents), (Lithic…Psamments)
(Neossolos Quartzarênicos)	(9.1)	Arenosols	(Quartzipsamments)
(Neossolos Regolíticos)	(0.1)	Regosols	(Psamments)
Plintossolos	17.9	Plinthosols	Subgroup Plinthic (many classes of Oxisols, Ultisols, Alfisols, Entisols, Inceptisols)
Cambissolos	2.5	Cambisols	Inceptisols
Gleissolos	2.3	Gleysols, some Stagnosols	Entisols (aqu-alf-and-ent-ept)
Planossolos	1.4	Planosols	Alfisols
Nitossolo	1.2	Nitisols, Lixisols, Alisols	Ultisols, Oxisols (kandic), Alfisols
Afloramento de rochas	1.2	Rock outcrops	Rock outcrops
Luvissolo	0.9	Luvisols	Alfisols, Aridisols (Argids)
Vertissolo	0.2	Vertisols	Vertisols
Chernossolos	0.2	Phaeozems, Kastanozems, Chernozems (some)	Molisols (only Ta)
Dunas	0.1	Dunes	Dunes
Organossolos	0.1	Histosols	Histosols
Espodossolos	0.1	Podzols	Spodosols

a) Soil classes at the first categorical level according to the Brazilian Soil Classification System (SiBCS) (Santos et al., 2018). Soils in parentheses are classified at the second categorical level.
b) Percentage of the area in Figure 11.4 occupied by each type of soil classified in the first categorical level. Values in parentheses correspond to soils classified at the second categorical level.
c) IUSS Working Group WRB (2015).
d) USDA-NRCS (1999, 2014).

Note. Adapted from Santos et al. (2018).

Other characteristics of Latossolos include low cation-exchange capacity (CEC), low base saturation, and high levels of H and Al.

Despite their chemical limitations, the acidity can be corrected using limestone, and proper fertilization can provide the necessary conditions for high yield levels. The main weaknesses of these soils are related to inadequate management, which can cause erosion (especially in soils with medium texture) and compaction from the intensive use of machinery. Because Latossolos are commonly associated with relatively flat terrain, high agricultural yield can be achieved by properly managing cash crops. Therefore, managing Latossolos is not a problem due to the wealth of successful experiences accumulated during the past few decades of agricultural development in the Brazilian Cerrado. However, the proportion of Latossolos in the agricultural frontier is lower than in other regions previously occupied in the southern part of the Brazilian Cerrado, where agriculture is well established, and is one of the reasons why agricultural expansion in the study area has been postponed until now.

The Argissolos have chemical and mineralogical characteristics similar to Latossolos but differ in having a higher clay content in subsurface horizons than surface horizons (similar to argillic horizons). Reduced surface clay content makes them more susceptible to erosion. However, the textural change allows for greater water accumulation in the transition zone, which is crucial for cultivating second crops at the end of the rainy season. These two soil classes together account for over 50% of the polygon area (Figure 11.5) and therefore are the main surfaces used for agriculture.

Neossolos Quartzarênicos are soils used for agriculture, but they present challenges for water storage due to their low soil CEC due to the small proportion of clay in the profile and the low levels of OM. In order to be classified as such, they must have a sandy texture, meaning that the sand content minus the clay content must be >700 $g\ kg^{-1}$ (Santos et al., 2018). Due to their pedological evolution, these soils are poor in primary minerals that are less resistant to weathering, or they have less than 4% of such minerals (Santos et al., 2018). Although these soils are deep, they are fragile and highly susceptible to erosion. Therefore, managing them requires integrated systems that prioritize increasing OM in the soil to promote aggregation and enhance the CEC (Ker, 1997; Oliveira et al., 2017). The vegetal cover is critical in sandy soils, especially for increasing water storage capacity. Many farmers have reported that during dry periods, soybean plants cultivated in integrated systems with management techniques for biomass input exhibit greater resilience to adverse conditions compared with cultivation without the addition of crop residues. This results in an improved crop establishment and significant yield gains that justify the technological investment.

The main impediment to using areas covered by Neossolos Quartzarênicos is related to the locations where they are most abundant. These soils are more frequent in the eastern region of Tocantins and the southern region of Maranhao. (Figure 11.5), except for the portions in the extreme north of Tocantins. These areas have lower levels of rainfall (Figure 11.3a), which exacerbates the severity of water deficits (Alvares et al., 2013). These soils have a low water retention capacity and are subject to seasonal weather conditions. However, the higher altitude in some areas can partially compensate for these factors by reducing temperatures and evapotranspiration. Nonetheless, most of the agricultural frontier areas considered in this chapter, excluding the western areas of Bahia, are located in low-lying areas.

Neossolos Litólicos, unlike Neossolos Quartzarênicos, are shallow and contain coarse fragments of the parent material (gravel, pebbles, and boulders, which can be found on the surface or subsurface). The primary constraint of Neossolos Litólicos is the limited volume of soil available for root exploration, leading to low water retention and creating significant challenges for farming in the hot climate of the agricultural frontier region. Additionally, these soils are typically found in areas with more rugged terrain, which poses challenges for mechanization. Because they are young soils, they are less weathered, and their chemical fertility is directly tied to the quality of the parent material. They inherit the CEC and nutrients that result from the weathering of the parent rock (Santos et al., 2018).

The agricultural sector in the state of Tocantins has been expanding mainly toward the western region, where precipitation is more abundant and the soils tend to have a higher clay content. However, this area is mainly dominated by Plintossolos Pétricos concrecionários (Figure 11.5), which are considered unsuitable for agricultural practices due to their excess gravel content (Ramalho-Filho & Beek, 1995). Despite this, these soils have been extensively used for agricultural purposes (Figure 11.6), resulting in crop yields comparable to those obtained from Latossolos (Almeida et al., 2020; Ramos, 2022).

According to SiBCS (Santos et al., 2018), Plintossolos are soils characterized by the formation of plinthite during their development, with or without the formation of ironstone. These mineral soils are typical of hot and humid areas with a well-defined dry season, which creates favorable conditions for the formation of plinthite. The Brazilian agricultural frontier region is an example of an area where these soils are found. At the second categorical level, Plintossolos are divided into Argilúvicos or Háplicos (soils with plinthite that occur more frequently in lowland areas) and Plintossolos Pétricos (soils with ironstone that are found in higher elevations compared to others). At the third categorical level, Plintossolos Pétricos, are further divided into two groups: Plintossolos Pétricos litoplíntico (soils with petroferric contact, which means a continuous layer of ironstone) and Plintossolos Pétricos concrecionários (Figure 11.7), which have a large volume of ironstone mixed with gravel and pebbles, allowing good drainage and deep root growth.

Figure 11.6 Soybean cultivated in a Plintossolo Pétrico concrecionário produces around 4200 kg ha^{-1} of grain yield. Paraíso do Tocantins, Tocantins, Brazil. Photo: Rodrigo Estevam Munhoz de Almeida.

Figure 11.7 Profile of a Plintossolo Pétrico concrecionário that occurs frequently in the agricultural frontier region. Palmas, Tocantins, Brazil. Photo: Rodrigo Estevam Munhoz de Almeida.

These soil characteristics make them less limiting for agriculture, and they are the soils on which rainfed grain agriculture expands in the Brazilian agricultural frontier region (Figure 11.6).

The Plintossolos Pétricos concrecionários (Figure 11.7) are soils in the tropics with limitations such as low natural fertility, high acidity, and low water retention due to a large proportion of their volume being taken up by coarse material (Lumbreras et al., 2015). An evident lack of knowledge about important aspects of their physical and hydraulic properties makes it difficult to predict the climatic risks for cash crops—an essential tool for the country's agricultural credit policy. Furthermore, cultivating crops on these Plintossolos Pétricos concrecionários (Figure 11.6) poses several disadvantages, including the need to remove stones during the preparation of the area for agriculture (Figure 11.8), the wear and tear on machinery and equipment due to friction with gravel, and the increase of surface temperature of the soil when exposed. This high temperature usually causes damage to emerging seedlings (Figure 11.9). Recent research has shown that soils with gravel in this region can reach temperatures above 40°C (Leite et al., 2022), which is extremely harmful to the establishment of the plant stand and to soil microbiology.

Figure 11.8 Heaps of discarded stones after manual collection in an area of Plintossolo Pétrico concrecionário during soil preparation operations in converting degraded pasture to farming in June 2022. Paraíso do Tocantins, Tocantins, Brazil. Photo: Rodrigo Estevam Munhoz de Almeida.

Figure 11.9 Soybean crop cultivated in Plintossolo Pétrico concrecionário in December 2022. Plant coverage is reduced due to the high amount of gravel on the soil surface. This restricts soil-seed contact leading to high plant mortality caused by scald due to the high temperature of the gravel when exposed to sunlight. Pium, Tocantins, Brazil. Photo: Rodrigo Estevam Munhoz de Almeida.

One of the greatest challenges in the agricultural frontier region is the management of Plintossolos Pétricos (Almeida et al., 2020; Leite et al., 2022; Ramos, 2022). These soil types have unique characteristics that significantly affect the production of grain crops, including (a) the presence of petroferric contact in the soil profile, (b) the presence of gravel on the surface, (c) the proportion of gravel and fine soil in the crop's root exploration layer, (d) the size of gravel fragments in the soil mechanical working layer, and (e) the particle size distribution of fine soil (percentage of clay). These soil characteristics, alone or in combination, affect several management practices, such as (a) land clearing, stone removal, and land preparation for agriculture (Figure 11.8); (b) soil preparation for deep incorporation of limestone and correction of soil acidity at depth; (c) water retention in the soil and water supply for crops, or the occurrence of flooding; (d) soil fertility management; and (e) soil-seed contact, which affects germination and seedling establishment (Figure 11.9).

The removal of stones (Figure 11.8) increases the initial preparation costs in these areas. However, with the use of management practices that address the challenges posed by gravel, it is possible to achieve good yield rates by using cropping systems that aim at improving SH. Among the recommended management actions for soils with gravel, the following stand out: deep incorporation of limestone, monitoring of the nutritional status of cultivated plants through foliar diagnosis, fertilization that is appropriate for the productive potential of each condition of Plintossolo Pétrico concrecionário, the use of agricultural systems that provide biomass input for soil protection (Figure 11.10), and the use of precision agriculture tools for mapping the variation in the amount of gravel and applying inputs at variable rates.

By using technologies developed through research focused on the sustainable management of Plintossolos Pétricos concrecionários and through years of successive cultivation in this region, either with living CCs or straw (Figure 11.10), it is possible to improve soil structure, increase OM content, enhance water retention, and decrease surface soil temperatures. This makes establishing a soybean production system combined with second-season maize cultivation in gravelly soils feasible, reducing risk, favoring high yields, and increasing profitability.

Figure 11.10 Soybean cultivation in Plintossolo Pétrico concrecionário protected by ruzigrass straw [*Urochloa ruziziensis* (Germ. & Evrard) Crins], with good plant population establishment in December 2022. Soybean yield in this area was 5080 kg ha^{-1}. Paraíso do Tocantins, Tocantins, Brazil. Photo: Rodrigo Estevam Munhoz de Almeida.

Characteristics and History of Cultivation on the Northern Edge of the Agricultural Frontier Area

The northern edge of the agricultural frontier discussed in this chapter encompasses 216 municipalities in Piauí, Maranhão, and the northeastern region of Pará, with a total area of 40,983,132 ha (Figure 11.11). In this region, the area under cultivation and grain production have both increased significantly over the past 5 years (IBGE, 2022). Specifically, soybean production increased by 33%, from 5,175,301 tons in 2017 to 6,882,146 tons in 2021. During the same period, the cultivated area expanded from 1,798,776 to 2,206,543 ha, reflecting a 23% increase.

For maize, there was a 44% increase in production, from 2,983,554 tons in 2017 to 4,308,891 tons in 2021, and an 18% increase in the planted area, which went up from 707,215 ha in 2017 to 836,013 ha in 2021.

The significant increase in production of cultivated areas is primarily due to yield improvements resulting from the increasing use of modern production technologies, which have become prevalent in the region and allow for agricultural intensification (Bolfe et al., 2016). In addition, the expansion of cultivated areas is driven by factors such as low land prices, favorable topography for mechanization, soils without physical limitations, and a well-defined rainy season. Notably, although the areas on the northern edge of the agricultural frontier share some characteristics with those on the southern edge, agricultural activity in the region as a whole has a set of unique characteristics.

The eastern portion, close to the states of Ceará, Pernambuco, and the northern region of Bahia (Figure 11.11), is considered a transition region (ecotone) between the Caatinga biome, where the semi-arid climate predominates, and the Cerrado, with a seasonal tropical climate. This ecotonal zone has two well-defined seasons: a shorter rainy season (4–5 months) and a longer dry season (6–7 months), which increases the risk of losses and makes it challenging to cultivate two crops in the same year.

The main crops grown in this region are soybean and maize, with the main focus being on cultivating oilseed and rotating with maize every 4 or 5 years. Generally, the short sowing seasons only allow for one-grain crop to be sowed, followed by a second straw crop usually of palisade grass [*Urochloa brizantha* (Hochst. ex A. Rich.) R. Webster], ruzigrass [*Urochloa ruziziensis* (Germ. & Evrard) Crins], or pearl-millet (*Pennisetum glaucum* L.). However, a small percentage of farmers (10–15%) have been growing earlier-maturing soybean genotypes, which allows for the possibility of a second grain crop (second crop), depending on the specific weather conditions of each year (i.e., on the influence of the phenomena El Niño or La Niña). Among the species most commonly used for secondary cultivation are those with shorter cycle and more tolerance to water deficit, such as sorghum

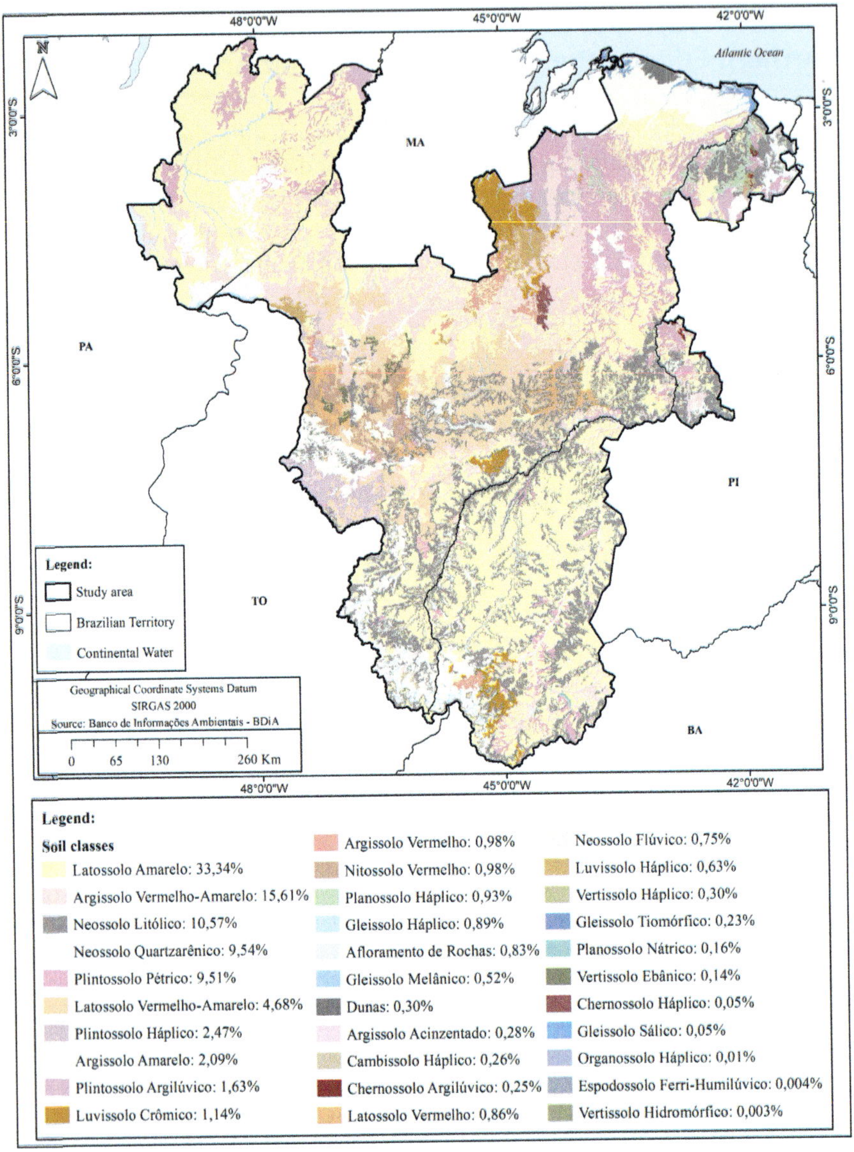

Figure 11.11 Spatial delimitation of the northern edge of the Brazilian agricultural frontier area consisting of part of the states of Maranhão, Piauí, and the northern part of Pará, and distribution of the main soil classes according to Sistema Brasileiro de Classificação de Solos (SiBCS) (Santos et al., 2018). The approximate correlation of SiBCS soil classification with World Reference Base (WRB) and Soil Taxonomy can be seen in Table 11.1. Prepared by the authors.

(*Sorghum bicolor* L.), cowpea [*Vigna unguiculata* (L.) Walp.], and, in some cases, even maize.

In the state of Piauí, there has been a significant expansion of cultivation, particularly of soybeans, in the central-northern region (middle and lower Parnaíba), despite grain crops being predominant in the southern part of the state for a longer period. Generally, the areas where grain crops are grown in Piauí are characterized by flat to gently undulating topography, which makes mechanization easier. Due to rainfall geography, soybean planting begins on October 20 in southwestern Piauí, in November in the extreme south of the state, in December in central Piauí, and in January in northern Piauí and northeastern Maranhão (Figure 11.12).

To the west of the northern edge (Western Maranhão and Eastern Pará; Figure 11.11), the climate changes and the rainfall volume is higher, which characterizes the transition between the Cerrado and the Amazon biomes. This region has a unique agricultural reality where water deficits do not impose severe restrictions on crops (Figure 11.3a). As a result, it has recently gained importance in the production of grains and soybeans, followed by millet as a CC. In wetter years, the cultivation of second-season maize, sesame (*Sesamum indicum* L.), and sorghum is becoming increasingly popular. Similarly to the eastern portion of the northern edge (Piauí), crop rotation with maize or sorghum occurs every 4 or 5 years during the crop season. The planting of soybeans is later compared with the southern edge of the agricultural frontier, starting at the end of November each year (Figure 11.12). However, the calendar for second-season crops is more extended in this region than in the eastern part (Figure 11.13).

The central portion of the northern edge, which includes the southern region of Maranhão and stretches to the border with Piauí state (Figure 11.11), is considered the most traditional area for grain cultivation, particularly soybean. This region has well-established enterprises with over 20 years of activity. Consequently, the soils in this region present better fertility (improved over many years), and there is a greater variety of cultivars adapted to these conditions. Additionally, this region has the highest number of producers who plant a second crop, mainly maize. This is due to a longer rainy season, which allows for a broader planting window. Soybean planting is recommended from the second half of October (Figure 11.12), whereas the ideal sowing date for the second crop of maize ends on February 28 or until March 10 (Figure 11.13). In the predominant soybean cultivation system that involves rotation with maize in the region, there has been a significant increase in the cultivation of maize intercropped with ruzigrass (Figure 11.19). This practice is viewed as a strategic management alternative for sustainable productivity. The maize–ruzigrass intercropping system promotes a conservationist approach to production by generating a significant amount of high-quality straw (Figure 11.20). Moreover, it enhances

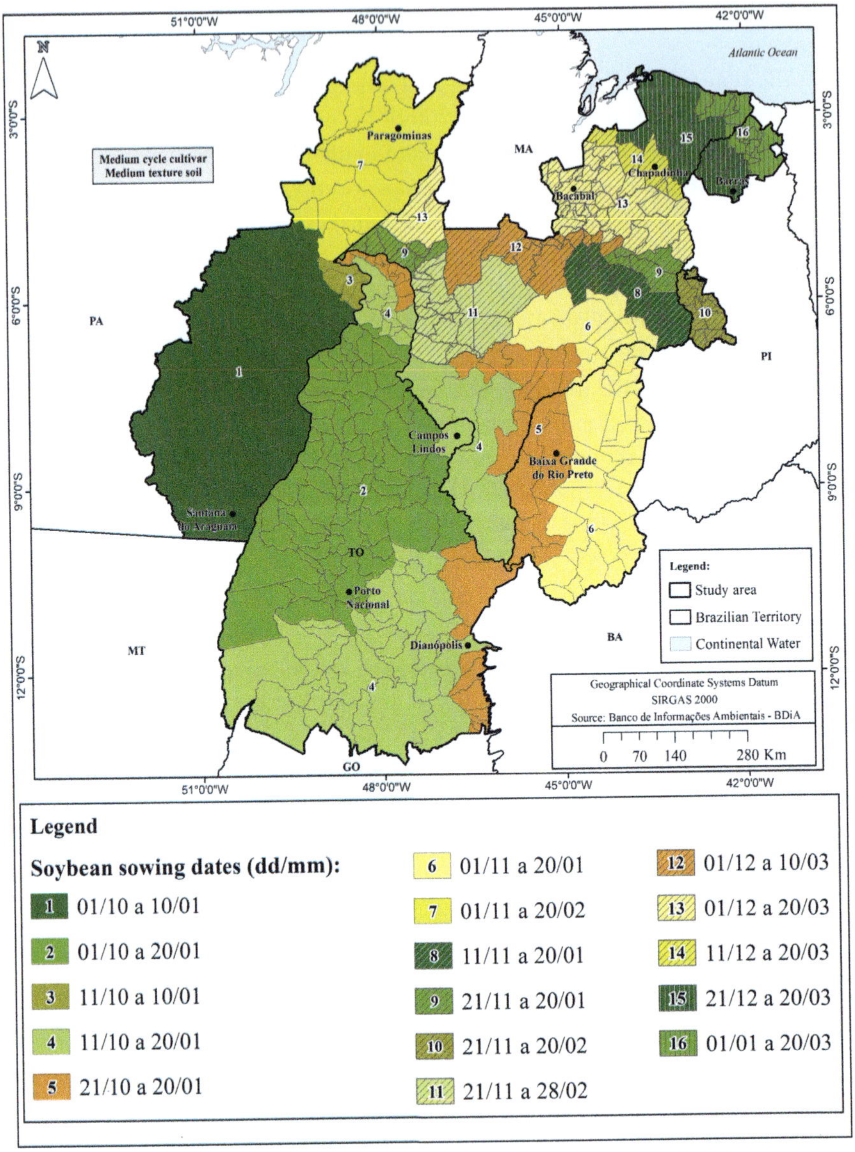

Figure 11.12 Soybean sowing times recommended by the Climate Risk Zoning (Zoneamento de Risco Climático [ZARC]) (Brasil, 2022a) for medium-cycle cultivars and medium-texture soils for the agricultural frontier region. The recommended dates may change depending on other soil textures and cultivar cycles. Prepared by the authors.

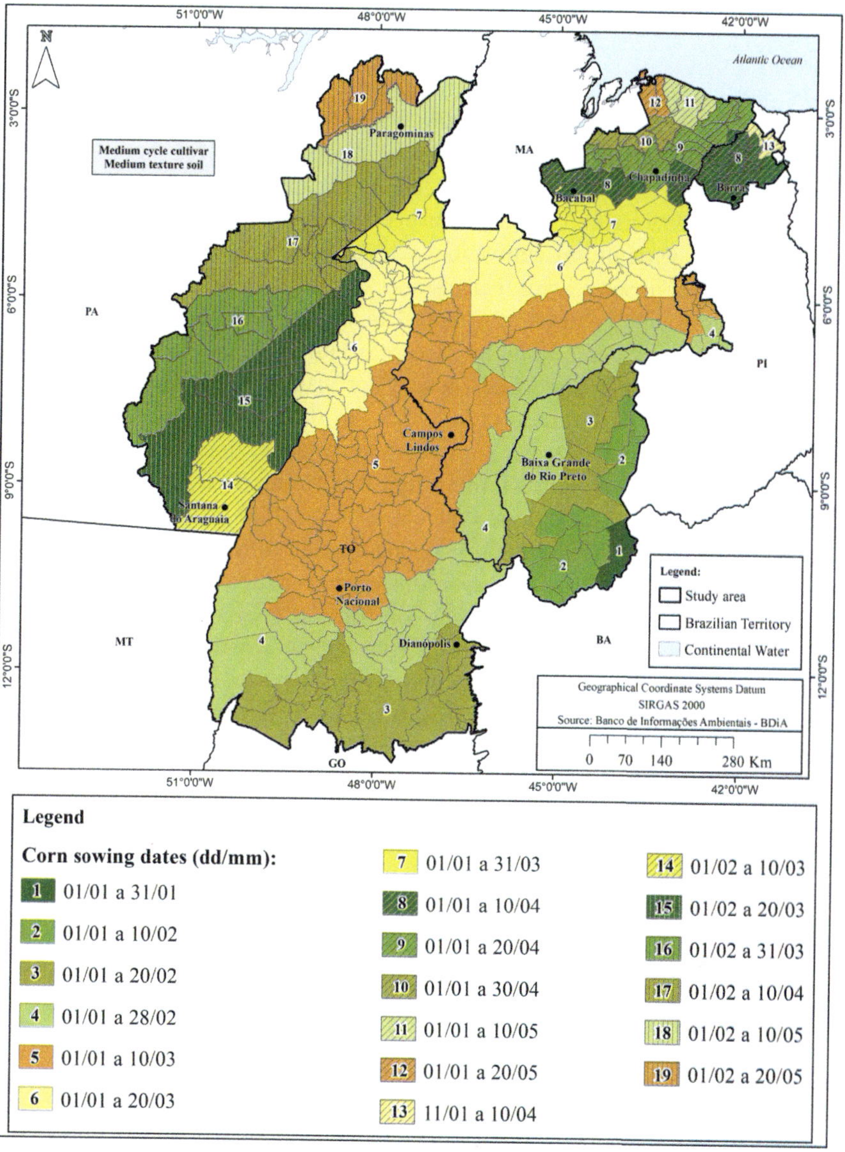

Figure 11.13 Sowing times for the second-season corn crop recommended by the Climate Risk Zoning (Zoneamento de Risco Climático [ZARC]) (Brasil, 2022b) for early-cycle cultivars and soils with medium texture for the agricultural frontier region. The recommended dates may change depending on other soil textures and cultivar cycles. Prepared by the authors.

the resilience of successive crops against dry periods that frequently occur in the region.

The northern edge has several prominent soil classes, such as Latossolo Amarelo, Argissolo Vermelho-Amarelo, Neossolo Litólico, Neossolo Quartzarênico, and Plintossolo Pétrico. These soil classes account for more than 78% of the soil types in the region (Figure 11.11). The main characteristic of these soil classes is their low natural fertility. However, some soil classes present physical restrictions on agricultural cultivation, which poses additional challenges for grain cultivation in some areas. For instance, Plintossolo Pétrico concrecionário contains a considerable amount of gravel in the soil volume, which hampers agricultural management. Similarly, Neossolo Litólico is widely present on the northern edge and has a reduced capacity for water storage, making it prone to water deficiency (Lumbreras et al., 2015). In the eastern part of Maranhão, some soils, especially the Argissolo Amarelo distrocoeso típico, have a naturally densified cohesive layer between the A and B horizons. The origin of this layer is widely debated in the literature (Fabiola et al., 2003; Moreau et al., 2006), but the fact remains that the cohesion is strong enough to hinder root penetration or even cause strangulation. However, this characteristic only manifests during the dry season, meaning that when the soil is moist, it becomes friable, making it difficult to diagnose the presence of cohesion at certain times of the year (Ramos et al., 2013). This characteristic does not pose any problems for soil use, especially for grain agriculture, because the cohesive character does not manifest during the wet season. However, this is not the case for perennial and semi-perennial crops, such as forestry, fruit-growing, and CCs. The management strategy for these crops includes the use of a subsoiler or the creation of larger pits than conventional ones, which increases the cost of preparing the areas for agricultural use.

Characteristics and History of Cultivation on the Southern Edge of the Agricultural Frontier Area

The southern part of the agricultural frontier area defined for the chapter of this book covers 166 municipalities in the Tocantins and southeastern Pará states, with an area of 41,914,299 ha (Figure 11.14). The planted area and grain production have increased considerably in the last 5 years in this region. Soybean production increased by 52% and its cultivated area by 43% during this period, going from 2,735,724 tons cultivated on 958,098 ha in 2017 to 4,165,140 tons cultivated on 1,374,821 ha in 2021. In the case of maize, the increase was also 52% in production and 44% in the planted area, going from 1,271,126 tons cultivated on 349,583 ha in 2017 to 1,937,412 tons cultivated on 502,925 ha in 2021 (IBGE, 2022).

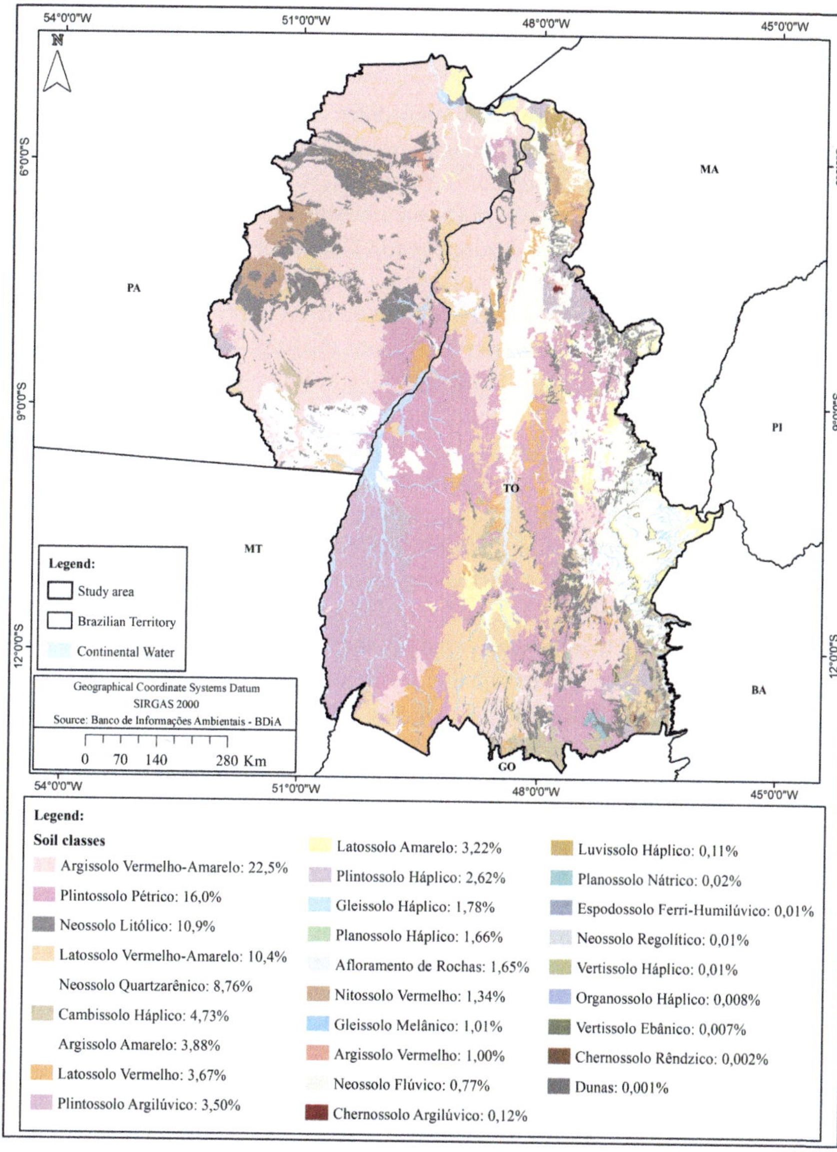

Figure 11.14 Spatial delimitation of the southern border of the Brazilian agricultural frontier area constituted by a portion of the state of Pará and the state of Tocantins. The map includes the distribution of the main soil classes according to Sistema Brasileiro de Classificação de Solos (SiBCS) (Santos et al., 2018). The approximate correlation of SiBCS soil classification with World Reference Base (WRB) and Soil Taxonomy can be seen in Table 11.1. Prepared by the authors.

The most representative soil classes in the southern part of the agricultural frontier area are Argissolos, occupying 27% of this region. The most common type is Argissolo Vermelho Amarelo, covering 22% of the area (Figure 11.14). These soils are mainly concentrated in Pará and the northern region of Tocantins state. The second most frequent class is Plintossolos, covering 22% of the area, with the Plintossolo Pétrico being the most frequent and covering 16% of the territory. Neossolos are also important, occupying 20% of the area, mainly divided between Neossolo Litólico (11%) and Neossolo Quartzarênico (9%). Latossolos represent 17% of the southern edge of the agricultural frontier, with Latossolo Vermelho Amarelo being the most common (10%). In total, these soil classes cover approximately 87% of the southern edge of the agricultural frontier (Figure 11.14).

Generally, this region is characterized by low altitude and a tropical climate typical of the Cerrado biome, with rainy and dry seasons. Large areas of the region have sandy or gravelly soils (Figure 11.14). These characteristics pose challenges for agriculture due to the high temperatures (Figure 11.4a) resulting from the low altitude and latitude of the region, which increase water consumption by crops due to increased evapotranspiration. Furthermore, water deficiency during extended dry spells (periods of >10 days without rain during the rainy season) is a common issue in the southern edge of the agricultural frontier, which can amplify crop damage. In soils with large proportions of gravel and/or sand and low water-holding capacity, management difficulties are even greater, and the sustainability of agriculture depends on the adoption of SH technologies to achieve high yields.

The annual rainfall volume in the southern edge of the agricultural frontier generally increases from southeast to northwest (Figure 11.3a). The beginning of the rainy season, and particularly the end of it, vary across the territory and affect the ideal planting times, with lower risks, for planting soybeans and second-season maize (Figures 11.12 and 11.13) and therefore increases or decreases the viability of two crops per season.

In the central region of Tocantins state, soybean planting is recommended from October 1 (Figure 11.12). However, in practice, it usually intensifies around October 15. Planting second-season maize with lower risk is recommended until March 10 (Figure 11.13), which is a difference of 146 days. After accounting for the soybean cycle (which depends on the cultivar), there are around 30 days left for soybean harvesting within the ideal planting period for the second crop. Hence, this region has a high percentage of areas with second-season crops, which depends on each rural property's planting and harvesting capacity.

In the southern and southeastern regions of Tocantins state, the rainy season begins later than in the central part of the same state. As a result, the recommended time for soybean planting starts on October 11 for the south and on October 21 for the eastern portion (Figure 11.12). However, in practice, planting in these regions usually intensifies in early November, whereas the recommendation for planting second crop maize with lower risk ends earlier, on February 20 (Figure 11.13). This creates a difference of 111 days, which is the period consumed by the soybean cycle, resulting in significant restrictions for the cultivation of the second crop of maize in these regions.

In the northernmost portion of Tocantins state, the recommended planting time for soybean begins later than in the central portion and at the same time as the southern and southeastern regions of the state; that is, on October 11 or 21 (Figure 11.12). However, the end of the rainy season occurs later than in the rest of the state, and therefore the second crop maize can be planted until March 20 (Figure 11.13), which facilitates the second crop maize cultivation in this region.

In the southeastern lands of Pará state, which make up the southern edge of the agricultural frontier, the planting of soybean is primarily recommended starting on October 1st (Figure 11.12). The planting schedule for second-crop maize varies, as the southernmost part of this area has a planting deadline of March 10 (Figure 11.13), following approximately the same planting and harvesting rules as the central portion of Tocantins state. Moving north of this area, the end of the rainy season is delayed, with April 10 being the ideal date for planting second-crop maize (Figure 11.13). This lengthy calendar provides ample time for cultivating two crops in one agricultural year. In practice, in these areas of Pará, soybean planting occurs later not due to a lack of rain in October but rather to avoid excessive rainfall during the harvest in January and February (exacerbated in years with the La Niña climate phenomenon), without compromising the second crops.

In the southern edge of the agricultural frontier area, there has been a substantial increase in the cultivation of second-season maize in recent years in locations where the climate conditions allow for such a crop. Farmers in these areas are increasingly adopting the practice of intercropping maize with grass (Figure 11.19) during both the main and second-season crops to provide biomass and establish more sustainable agricultural systems. In regions such as the south and southeast of Tocantins state, where the cultivation of second-season maize is restricted, or even in other regions with greater potential for second-season maize but in later areas of soybean cultivation where the harvest occurs after the ideal planting time for second-season maize, farmers are increasingly cultivating crops that require less water, such as sorghum, millet, and sesame. In more challenging soil conditions, such as Plintossolos Pétricos concrecionários and Neossolos Quartzarênicos, farmers prefer to plant later, after the beginning of November, due to the possibility of more frequent rains for the establishment of soybeans. However, this implies a reduction in the potential for second-season maize cultivation. In these cases, the use of less demanding crops in terms of water, and the cultivation of tropical grass species to provide biomass, is becoming increasingly common. This is a fundamental practice for managing these fragile soils.

Challenges in Conservation, Maintenance of Fertility, and Restoration of Soil Functionality in the Agricultural Frontier

Clearing of New Areas and Socio-environmental and Economic Impacts in the Region

The conversion of new areas for agricultural cultivation is an inherent reality in agricultural frontier areas, whether through legal deforestation or the conversion of degraded pastures into grain-producing crops. Specifically for the states of Maranhão and Piauí, the potential for Cerrado areas suitable for grain cultivation is around 5 million ha, with 2 million in Maranhão and 3 million in Piauí (Klepker, 2014). Of this projected total, around 50% of the available area has not yet been converted to grain production. In Tocantins, the area with agricultural potential is estimated at 13,825,070 ha, representing 50.25% of the total state territory. Of this amount, it is estimated that 7,498,250 ha are occupied with pastures, 1,000,000 ha are cultivated with grain crops, and 5,353,820 ha are still available to be converted to agricultural activity (Seagro, 2020). Therefore, the areas to be explored with grain crops can double or even triple in the coming years.

It is important to emphasize that the exploration of areas in the agricultural frontier region has promoted economic expansion and greater integration of the regional territory into national and global markets. This process has also driven the flux of goods and services, the provision of infrastructure and urbanization, as well as the generation of jobs, especially in sectors with greater productive specialization, aimed at directly and indirectly meeting the demands of the new economy (Alves, 2020).

Preparation, Acidity Correction, and Soil Fertilization in Newly Cleared Areas

Traditionally, first-year soybean yield is low due to problems on the establishment of a successful symbiosis with *Bradyrhizobium* strains, low nutrient concentrations, and soil acidity that has not been fully corrected (incomplete reaction of limestone due to lack of time), which hinders plant growth (Lustosa Filho et al., 2021). However, despite these limitations, the use of technologies associated with intensive management for high soybean yields has enabled yields >3000 kg ha^{-1} in the first year of cultivation (Lustosa Filho et al., 2021), making the activity economically viable in the region.

Soil preparation in the region involves removing or suppressing the native vegetation, removing stumps and rocks, burning plant residues, and using a plow

Figure 11.15 Deep incorporation of the first half of the limestone dose ($4\,t\,ha^{-1}$), with a 42″ grid in an area of Plintossolo Pétrico concrecionário. Paraíso do Tocantins, Tocantins, Brazil. Photo: Rodrigo Estevam Munhoz de Almeida.

(36″ or larger) and leveling harrow. The operation of removing stumps and rocks and leveling is repeated as necessary. Once this stage is completed, lime is applied, and the soil is further plowed (using a 36″ or larger plow) to ensure deep incorporation of the lime (as shown in Figure 11.15). Finally, the soil is leveled using a harrow. However, if the area has Plintossolos Pétricos concrecionários, the process of removing raised rocks due to soil movement can increase the cost of these stages.

The amount of limestone being applied in the newly cleared areas of the region has been exceeding the recommended values in official bulletins (Sousa & Lobato, 2004), with doses often being two or three times higher than the recommended values. One of the justifications for using high doses of limestone in these areas, when the goal is to prepare the land for more demanding crops, is the reaction time of this corrective agent in the soil, which is related to the predominant dry climate during the off-season in the region (Donagemma et al., 2016). Soil preparation is done during the year's dry season, and the reaction of the limestone begins with the start of the rains, concomitant with the period recommended for sowing soybeans. Thus, the use of high doses allows for a faster process of soil acidity correction because the reaction of only 20–40% of the limestone applied at high doses is equivalent to correcting acidity to a similar degree to recommendations obeying official management criteria (Lustosa Filho et al., 2021). Another justification for applying high doses of limestone with deep incorporation (Figure 11.15) is that this strategy makes it possible to increase soil fertility to the levels required by soybean crops, in addition to correcting acidity at greater depths and preparing a larger volume of soil for roots to explore, which ensures greater resilience to water stress inherent in these low-altitude agricultural frontier regions.

Obtaining economically viable yields in the first year of soybean cultivation is possible by combining the following management actions to achieve high productivity: (a) deep incorporation of high doses of limestone (>6 t ha^{-1}), which can exceed values of 10–11 t ha^{-1} (Donagemma et al., 2016); (b) use of agricultural gypsum to improve fertility in deeper layers of the soil, in addition to the layer corrected by limestone; (c) corrective fertilization with P, K, and micronutrients to raise their levels in the soil to adequate levels; (d) selection of high-yielding cultivars adapted for the region; (e) planting at the most appropriate time for high yields in the region, avoiding delays in soil clearing and preparation operations, which lead to planting delays; (f) efficient inoculation of soybean with high-quality *Bradyrhizobium* inoculants; (g) use of complementary seed and foliar nutrition with micronutrients, plant growth promoters, and bioinputs; and (h) control of pests, diseases, and weeds inherent in soybean management for high yields (Lustosa Filho et al., 2021).

Fertilization in clearing areas is carried out using the concept of corrective fertilization (Sousa & Lobato, 2004), which involves applying amounts of P and K that exceed the extraction and exportation of these nutrients by crops during harvest to increase their levels in the soil. In areas with greater financial resources and seeking high soybean yields in the first year, high doses of fertilizer are applied in the first year after the area is cleared, mainly P, which can reach values of 1 Mg ha^{-1} of single superphosphate (180–200 kg ha^{-1} of P_2O_5). The application of a corrective dose of P, incorporated to a depth of 20 cm in the soil, along with a maintenance dose in the planting groove, results in higher productivity and increased levels of available P in the soil in the first six harvests after converting pastureland to agriculture (Gotz et al., 2023).

Additional management options are available for newly cleared areas in the region that require lower investments. These options involve using smaller quantities of soil amendments and fertilizers or using rice (*Oryza sativa* L.) as the first crop, which is more tolerant to acidic soils. Another approach is to prepare and correct the soil over 2 or 3 years by applying lime and fertilizers in stages. In this approach, CCs such as *Brachiaria* alone or in association with pearl millet are planted in the first year (this method can be extended for up to 2 years). Only then soybean is introduced as the main crop. This allows sufficient time for the lime to react with the soil, increasing soil nutrient levels, and providing the benefits of CCs.

Cultivation Systems that Promote Soil Health in Agricultural Frontier Regions

The occurrence of highly weathered soils, which may eventually have physical limitations or an excess of gravel associated with low altitudes, are crucial factors to consider in the region. Additionally, the tropical climate with high temperatures, intense but seasonal rainfall, frequent dry spells, and variable length dry seasons, when combined with inadequate soil management, promotes the degradation of agricultural areas. These negative impacts are primarily associated with the loss of soil C, increased erosion, poor plant nutrition, water stress, soil compaction, and decrease microbial diversity. Reduced plant diversity in agroecosystems leads to increases in the occurrence of nematodes, pests, and plant diseases, which invariably affects productivity.

In this scenario, the adoption of conservationist cultivation systems that promote SH is essential for achieving high yield and greater agricultural sustainability in the frontier region, thereby avoiding degradation. To achieve these goals, it is recommended to sow CCs to protect the soil and recycle nutrients as well as to rotate crops to promote microbiota biodiversity and prevent the increase of pests and diseases. Additionally, using a subsoiler every 4 or 5 years can prevent soil compaction (Barbosa et al., 2022a). The implementation of cropping systems that reduce soil disturbance and increase OM, crop diversity, and soil protection through the supply of biomass, such as NTS, ICL, and ICLF, can further enhance SH.

Intercropping techniques for establishing CCs or tropical grasses are essential in the agricultural frontier region due to the prolonged dry season in the fall/winter period. The main techniques used in the region consist of (a) oversowing CCs in soybean fields or planting them after soybean harvest in areas that are not cultivated with a second crop and (b) intercropping maize with deep-root grasses in areas where maize is produced either as the main crop (summer crop) or as the second crop, after soybean harvest.

Oversowing tropical forage plants in soybean has become a widely adopted practice among farmers in the agricultural frontier region, especially for areas without a second crop (Figure 11.16). This technique involves sowing the forage plant seeds after the soybean crop has reached the beginning of the grain-filling stage (Andrade et al., 2017). After the soybean harvest, the forage plant grows during the offseason, producing straw for planting soybean in the following crop season using the NTS (Figure 11.17). According to Dias et al. (2020), the oversowing of soybean with tropical grasses produced significantly higher biomass than the control treatment of second-crop maize. Specifically, palisade grass, ruzigrass, Mombaça guineagras (*Megathyrsus maximus* 'Mombaça'), and Tamani guineagrass (*M. maximus* 'Tamani') produced 5.5, 3.1, 4.0, and 4.6 Mg ha^{-1} of biomass, respectively.

Figure 11.16 Soybean on harvest day with palisade grass [*Urochloa brizantha* (Hochst. ex A. Rich.) R. Webster] germinated and established (March 2022). The grass was implanted with the oversowing technique when the soybean was in the grain-filling stage (R5) in February 2022. Gurupi, Tocantins, Brazil. Photo: Rodrigo Estevam Munhoz de Almeida.

Figure 11.17 Mombasa guinea grass (*Megathyrsus maximus* 'Mombasa') at the end of the off-season in October 2022. Grass implanted in an oversowing system in the soybean crop of the previous harvest (February 2022). Gurupi, Tocantins, Brazil. Photo: Francelino Petenó de Camargo.

In contrast, the second-crop maize produced only 2.4 Mg ha^{-1} of biomass. Additionally, soybean grown in straw-covered plots of grasses produced an average of 4.3 Mg ha^{-1} of grains, which was higher than the 3.5 Mg ha^{-1} of grains produced in maize straw–covered plots. Alternatively, grass can be sown after the soybean harvest in areas with sufficient rainfall to establish the grass, which provides a more operationally straightforward approach with similar outcomes as oversowing (Figure 11.18).

The intercropping of maize with grass (Figure 11.19) aims at establishing a forage plant at the same time as maize grains are cultivated, bringing several benefits to the chemical, physical, and biological properties of the soil (Barbosa et al., 2022a, 2022b; Crusciol et al., 2015; Santos et al., 2021). This cultivation technique is well suited for agricultural frontier areas because it allows for the inclusion of biomass, which overcomes the prolonged dry season of the region that makes it impossible to grow CCs after maize cultivation.

When it is not possible to plant a second crop, intercropping can be done during the primary crop as an alternative for crop rotation with soybeans. The intercropped maize is harvested during the rainy season or very close to its

Figure 11.18 Soybean cultivated under grass straw Mombasa guineagrass (*Megathyrsus maximus* 'Mombasa') in November 2020. The soil was covered with straw throughout the soybean cycle, which produced around 5 Mg ha^{-1} of grains. The grass was implanted after the soybean harvest of the previous crop in March 2020. Paraíso do Tocantins, Tocantins, Brazil. Photo: Rodrigo Estevam Munhoz de Almeida.

Figure 11.19 Off-season corn intercropping with ruzigrass [*Urochloa ruziziensis* (Germ. & Evrard) Crins] 74 days after planting in May 2020. Maize yield around 10 Mg ha^{-1}. Paraíso do Tocantins, Tocantins, Brazil. Photo: Rodrigo Estevam Munhoz de Almeida.

end, resulting in a substantial amount of grass biomass. This situation is highly advantageous for farms that use the ICL system (see Chapter 2) because it enables the production of pasture for cattle during the off-season.

Intercropping is also adopted in regions where second-season maize cultivation is viable. In this case, implementation occurs after soybean harvesting, and maize harvesting occurs during the dry season, resulting in lower biomass accumulation in the grass. However, by cultivating two cash crops and a forage plant in the same year, the production system can be supplied with biomass to maintain or improve soil health and produce straw for the next crop under a NTS (Figure 11.20). For the eastern region of Maranhão state, Santos et al. (2021) observed that intercropping maize with tropical forage grasses led to less stressful conditions for soil microbial activity than sole maize cultivation. This improved condition for the soil microbiota resulted in a 33% increase in soybean grain yield when cultivated after intercropped maize, compared with soybean cultivated after sole maize.

Pearl millet remains the most widely used CC in the agricultural frontier area, primarily used in three ways. Option A, where it is a second crop after soybean, is used for grain production due to the high market value of millet grains in recent years. This option is predominantly used in regions where the narrow window of

Figure 11.20 Production of ruzigrass [*Urochloa ruziziensis* (Germ. & Evrard) Crins] biomass in October 2022, 3 months after the maize harvest, close to the planting of soybeans for the next crop. The grass was intercropped with off-season maize in February 2022. Paraíso do Tocantins, Tocantins, Brazil. Photo: Rodrigo Estevam Munhoz de Almeida.

the rainy season restricts other second crop cultivations or in areas where soybeans are harvested late, at the end of the rainy season. In this situation, the amount of biomass produced by the system decreases.

In Option B, pearl millet is used as CC after soybean cultivation, which is the most traditional method for biomass production at the beginning of the NTS adoption (Figure 11.21). However, because millet is an annual plant, it completes its life cycle before the end of the fallow period, thereby limiting the benefits for soil protection, nutrient cycling, and biomass input (Figure 11.22). Studies conducted in this frontier region have demonstrated that perennial tropical grasses have a greater capacity for biomass production than millet, as well as a higher amount of residual biomass over time (Leite et al., 2010). According to Andrade et al. (2017), in a comparison of CCs intercropped with soybean in the southern state of Tocantins, millet, palisade grass (*U. brizantha*), and Mombaça guineagrass (*M. maximus* Mombaça) produced 2044, 5854, and 9483 kg ha^{-1} of dry matter, respectively (Figure 11.22). Additionally, the soybean grown in straw-covered plots of these grasses produced 3332, 3658, and 4238 kg ha^{-1} of grains, respectively. According to Barbosa et al. (2022b), using the NTS alone in areas where soybean is cultivated and followed by millet for 14 years in eastern

Figure 11.21 Millet grown in succession to soybeans in a commercial grain production area in eastern Maranhão in September 2022. Brejo, Maranhão, Brazil. Photo: Henrique Antunes de Souza.

Maranhão resulted in total C stocks that were lower than those found in areas where the NTS was combined with a periodic rotation of ruzigrass intercropped with maize. Therefore, using perennial tropical grasses of the Urochloa or Megathyrsus genus appears to be more advantageous for maize cultivation because they grow throughout the offseason and produce a greater amount of biomass. These factors are crucial for improving SH in the agricultural frontier region.

Option C involves sowing millet at the onset of the first rains of the next rainy season (September/October) to generate a reasonable amount of straw before planting the soybean crop (Figure 11.23). This option is used when it is not possible to grow a CC after the previous harvest for any reason, such as late soybean harvesting during the dry season, the need for soil preparation during the off-season, or harvesting of second-crop grains without intercropping with any grass. In this scenario, the amount of biomass produced by millet will be greater the earlier the first rainfall in the new rainy season and the later the soybean planting in the area, which should not exceed the recommended ideal calendar (Figure 11.12). If the soybean planting is done too late, it will make it

Figure 11.22 Experimental plots with the inclusion of tropical forage plants using the soybean oversowing technique in October 2014 (Andrade et al., 2017). On the left, plot of Mombasa guinea grass with high biomass input, persisting throughout the off-season, and on the right, plot containing millet with a completed cycle and lower biomass input. Gurupi, Tocantins, Brazil. Photo: Rodrigo Estevam Munhoz de Almeida.

Figure 11.23 Millet with rapid biomass production on October 23, 2022, close to soybean planting. Millet was sown at the end of the off-season, before the first rain, which occurred in the last week of September 2022. Porto Nacional, Tocantins, Brazil. Photo: Rodrigo Estevam Munhoz de Almeida.

impossible to cultivate a second crop the following year. Using millet is a viable option because it grows faster than grasses of *Urochloa* or *Megathyrsus* genus and, therefore, can produce a greater amount of biomass in the short period before soybean sowing.

The evolution toward more sustainable cropping systems for maize cultivation, focusing on SH in the agricultural frontier region, involves combining different CCs from various plant families with distinct physiology and root systems. This combination of CCs creates a mix that provides numerous benefits, such as providing biomass, creating ground cover, fixing N, promoting more dynamic and efficient nutrient cycling, alleviating soil compaction, and controlling nematodes. Although this approach increases the operational complexity, using CC mixes has become a promising alternative for the region.

Notably, in some farms, more diversified and complex cropping systems are adopted with the presence of the forest component, associated with agricultural crop rotation practices and with forage grasses for cattle grazing (ICLF) (Figure 11.24). This additional stage in seeking balance and sustainability of agricultural systems in the region can provide greater production stability, with

Figure 11.24 Integrated crop–livestock–forestry (ICLF) system with maize intercropped with Tamani grass (*Megathyrsus maximus* 'Tamani') in between rows of eucalyptus in eastern Maranhão in June 2021. Brejo, Maranhão, Brazil. Photo: Fernando Devicari.

direct benefits to the soil through improving its chemical, physical, and biological quality. Under the conditions of northern Maranhão, Araújo et al. (2022) showed that intercropping maize with palisade grass between rows of eucalyptus and intercropping maize with Massai guineagrass resulted in negative CH_4 fluxes, denoting the potential of these systems to mitigate greenhouse gas (GHG) emissions. According to Araújo et al. (2022), degraded pasture areas resulted in high GHG emissions, confirming the positive effect of including integrated tropical forage species in agriculture and forestry.

As discussed in Chapter 2, the positive effects of the presence of trees in farming systems go beyond the fixation and storage of C in the soil. Cultivating forage grasses in crop areas generates a large amount of forage (Figure 11.25), which can be used for grazing cattle and sheep in the field. In these systems, the presence of trees creates shading conditions that ensure animal welfare in the face of high temperatures in the region (Figure 11.26), in addition to improving the nutritional quality of the pasture, increasing animal weight gain, and reducing enteric CH_4 emissions (Frota et al., 2017). Thus, the combined adoption of NTS, crop

Figure 11.25 Production of plant biomass by tropical forage grasses in an integrated crop–livestock–forestry system in eastern Maranhão in June 2022. Brejo, Maranhão, Brazil. Photo: Edvaldo Sagrilo.

Figure 11.26 Cattle grazing between rows of eucalyptus in an integrated crop–livestock–forestry system in December 2016. Campos Lindos, Tocantins, Brazil. Photo: Rodrigo Estevam Munhoz de Almeida.

rotation, intercropping, and ICL and ICLF systems has contributed to minimizing the limitations imposed on low-fertility, sandy, gravelly soils with low concentrations of OM in the agricultural frontier region.

Environmental Legislation in Agricultural Frontier Region

To overcome the challenging soil conditions in the agricultural frontier region, it is essential to implement agricultural management practices that utilize sustainable cropping systems to promote SH. The recent expansion of agriculture in this region has resulted in a significant advantage. In terms of environmental sustainability, there is an opportunity for agricultural expansion observing the premises of the new Forest Code (FC). Apart from the criticism made by both environmentalists (who focus on the generalized amnesty that reduced by 58% the environmental liabilities provided by 2012s FC) and landowners (who believe that the Law limits their production), the new Law introduces new devices. The environmental reserve quote (in Portuguese "CRA," cota de reserva ambiental) and the environmental services policy (Soares-Filho et al., 2014) are examples of such novelties.

Undoubtedly, implementing all the mechanisms devised by the law is challenging for state agencies, but no one can deny their critical role.

Brazilian environmental legislation has continuously progressed through legal provisions since the 1970s, particularly from the 1980s onward. This has resulted in the establishment of increasingly restrictive mechanisms for land use conversion (Franco, 2009). The Brazilian FC, established in 1965, is considered a strong pillar of the country's environmental policy (Soares-Filho et al., 2014). It is the central piece of legislation regulating land use and management on private properties and underwent significant changes due to a political debate that ultimately led to the approval, in 2012, of Law 12.651/12, commonly referred to as the "new forest code." Although this new law replaced the previous legislation from 1965 (Chamber of Deputies of Brazil, 2012; Franco, 2009), it preserved many of its key provisions, such as the Legal Reserve (LR) and Areas of Permanent Preservation (APPs). As of 2001, the FC required landowners to conserve native vegetation on their rural properties, setting aside a LR that occupies 80% of the property area in the Amazon and 20% in other biomes (Soares-Filho et al., 2014). The Law also designated environmentally sensitive areas as APPs, aiming to conserve water resources and prevent soil erosion. The APPs include riparian preservation areas, which protect riverside forest buffers, and hilltop preservation areas at hilltops, high elevations, and steep slopes.

Comparing the evolution of land anthropogenic occupation in the Cerrado biome (Brazilian neotropical savannas) by anthropic activities, it is possible to detect the process's speed. In 2002 around 60.5% of the Cerrado was occupied by native vegetation, having lost around 6% of this total after 13 years. Although the largest portions of anthropized land were concentrated in the southern parts of the biome, the MATOPIBA region showed a faster conversion rate (Victoria et al., 2020).

The agricultural expansion in the newest agricultural frontier in Brazil would be a model to be followed because there are still significant areas of native vegetation; moreover, recent deforestation is subject to the original rules of APPs and RLs, especially concerning large and medium-sized agricultural enterprises making the model a feasible approach. Although the new FC imposes relaxing norms on small landowners, the small landowners represent a smaller share of total land. In addition, the imposition of the Rural Environmental Registry (RER, or CAR in Portuguese) is a valuable instrument for monitoring and environmental management (Soares-Filho et al., 2014). However, the deforestation that does not comply with current regulations is subject to penalties as prescribed by law. Monitoring has been facilitated in recent years, mainly due to advances in geotechnology, the use of remote sensors, and the establishment of the RER. This registry delimits areas for environmental preservation within the property by the farmer, providing documentary evidence of intended use for farms. However, the repeated postponement of the deadline for the RER and the technical difficulties in validating the self-declared information submitted to the evaluation of state environmental agencies pose an additional risk to the Law implementation. It will probably harm rural producers and jeopardize the confidence in the State's capability and ultimately will lead to serious environmental damage.

The benefits of both RLs and APPs are undeniable (Metzger et al., 2019). The presence of forest and savannah formations strongly influences local weather conditions. Moreover, native tree formations contribute to maintaining soil permeability, plant and animal biodiversity, and increasing C in the soil (Soares-Filho et al., 2014). All these possibilities create opportunities to compensate farmers through environmental services by maintaining native formations. This path is still under development in Brazil (Metzger et al., 2019).

From a soil perspective, all of these mechanisms enable the integrated management of a farm so that natural resources can be optimized by planning the allocation of LRs and APPs. Hence, these processes can assist the cropping system to preventing erosion, increasing C fixation (and consequent payment for credits), increasing water infiltration in the soil, and achieving climate balance (Hissa et al., 2019; Metzger et al., 2019).

These factors present challenges for the region in answering questions about land use management. Although productive soils will understandably be primarily used for agricultural purposes, the definition of continuous areas of native vegetation on farms must be a criterion to maximize their benefits. Thus, although LRs may be located in less suitable soils for agriculture, they should be adjacent to APPs, forming ecological corridors that help to maintain biodiversity, soil quality, and water resources, and even contribute to filtering agricultural by-products.

In conclusion, it is important to note that the current Brazilian environmental policy, based on Law 12.651/2012, provides the necessary framework for cropping systems to adequately reconcile agricultural production with the conservation of natural resources.

Final Considerations

The agricultural frontier in Brazil presents the large expanses of native Cerrado vegetation (Sano et al., 2020) and is predominantly characterized by a tropical climate, low altitude, heavy rainfalls, frequent dry spells, prolonged dry seasons, and highly weathered soils (particularly sandy, medium-textured, and gravelly soils). These challenges impose significant difficulties for the sustainable management of agricultural production and the environment. Understanding the climate risk (i.e., the risk of yield loss due to extreme weather events) in each region of the agricultural frontier, defining crops (one or two harvests), and adopting management practices to correct soil acidity and increase fertility, along with the use of conservation practices that prevent compaction and erosion and ensure high levels of biological activity in the soil, are essential for an agricultural management strategy adapted to the edaphoclimatic conditions of this new Brazilian agricultural frontier.

The available data demonstrate that sustainable cropping systems, which use techniques to include CCs for biomass input, such as oversowing grasses into soybeans and intercropping maize with tropical forages, increase water retention and reduce soil temperature. This, consequently, promotes microbiological activity and SH. The greater the complexity of the cropping systems practiced in the region, with the integration of two or more plant species, together with the arboreal and animal components, the more resilient the crops will be in the face of the challenges of agricultural management in the agricultural frontier region.

However, the adoption of cropping systems aimed at SH in these areas remains a distant reality, even though it is increasingly prevalent in the region. Therefore, it is necessary to transfer technologies more efficiently to farmers, invest in research, and monitor these management systems and their long-term implications for soil quality.

Notably, given the prevailing soil and climate conditions in the agricultural frontier area, management practices aimed at improving SH are an immediate necessity to ensure the long-term yield sustainability of agricultural activity in the region.

References

Almeida, R. E. M., Uhlmann, A., Campos, L. J. M., & Costa, R. V. (2020). Expansão agrícola em áreas de difícil manejo: Cultivo em solos com cascalho. In *Anuário Brasileiro de tecnologia em nutrição vegetal* (pp. 71–75). Abisolo.

Alvares, C. A., Stape, J. L., Sentelhas, P. C., Gonçalves, J. L. M., & Sparovek, G. (2013). Köppen's climate classification map for Brasil. *Metereologische Zeishrift*, 22(6), 711–728. https://doi.org/10.1127/0941-2948/2013/0507

Alves, V. E. L. (2020). Expansão do agronegócio e os impactos socioambientais na região de Cerrados do centro-norte do Brasil (MATOPIBA). *Confins-Revista Franco-Brasileira Geografia,* https://journals.openedition.org/confins/28049#text

Andrade, C. A. O. D., Borghi, E., Bortolon, L., Bortolon, E. S. O., Camargo, F. P. D., Avanzi, J. C., & Fidelis, R. (2017). Straw production and agronomic performance of soybean intercropped with forage species in no-tillage system (in Portuguese, with English abstract). *Pesquisa Agropecuária Brasileira*, 52, 861–868.

Araújo Filho, J. C., Barros, A. H. C., Galvão, P. V. M., Teixeira, W. G., Lima, E. P., Victoria, D. C., Andrade Júnior, A. S., Xavier, J. P. S., Lumbreras, J. F., Coelho,

M. R., Baca, J. F. M., Monteiro, J. E. B. A., Oliveira, F. C. S. F., Silva Filho, A. D., & Barros, J. P. F. G. (2022). *Avaliação, predição e mapeamento de água disponível em solos do Brasil. Rio de Janeiro: Embrapa Solos (Boletim de pesquisa e desenvolvimento, 282).* E-book/ePub. https://ainfo.cnptia.embrapa.br/digital/bitstream/item/235187/1/CNPS-BPD-282-2022.epub

Araújo, M. D. M., Souza, H. A., Deon, D. S., Muniz, L. C., Costa, J. B., Souza, I. M., Reis, V. R. R., Brasil, E. P., & Pompeu, R. C. F. F. (2022). Integrated production systems in a Plinthosol: Greenhouse gas emissions and soil quality. *Australian Journal of Crop Science*, 16(02), 184–191. https://doi.org/10.21475/ajcs.22.16.02.3263

Assad, E. D., & Evangelista, B. A. (1994). Análise frequencial da precipitação pluviométrica. In E. D. Assad (Ed.), *Chuva nos cerrados. Análise e espacialização* (pp. 25–42). SPI/Embrapa.

Atwood, L. W., Racette, K. A., Diggelmann, M., Masala, C. A., Maund, S., Oliver, R., Screpanti, C., Wironen, M., & Wood, S. A. (2022). Soil health: New opportunities to innovate in crop protection research and development. *Frontiers in Environmental Science*, 10, 821742. https://doi.org/10.3389/fenvs.2022.821742

Barbosa, L. R., Souza, H. A., Oliveira, F. P., Nunes, L. A. P. L., & Leite, L. F. C. (2022a). Physical-hydraulic properties of an ultisol under no-tillage and crop-livestock integration in the cerrado. *Revista Caatinga*, 35(2), 460–469. https://doi.org/10.1590/1983-21252022v35n221rc

Barbosa, L. R., Souza, H. A., Teixeira Neto, M. L., & Leite, L. F. C. (2022b). Organic matter compartments in an Ultisol under integrated agricultural and livestock production systems in the Cerrado. *Ciência Rural*, 52, 10. https://doi.org/10.1590/0103-8478cr20200845

Becerra, J. A. B., Carvalho, S., Silva, E. A. D., Martins, I. A., & Randow, C. V. (2010). Variabilidade sazonal do clima e da vegetação no bioma Cerrado: I. Abordagem diagnóstico por sensoriamento remoto. In *XVI Congresso Brasileiro de Meteorologia – Anais*. https://doi.org/10.13140/RG.2.1.3121.0406. Sociedade Brasileira de Meteorologia (SBMET).

Bolfe, É. L., Victória, D. C., Contini, E., Bayma-Silva, G., Spinelli-Araujo, L., & Gomes, D. (2016). Matopiba in agriculture expansion: Territorial and socioeconomic aspects (In Portuguese, with English abstract). *Revista de Política Agrícola*, 25(4), 38–62.

Brasil. (2022a). Mapa – Ministério da Agricultura, Pecuária e Abastecimento. ZARC – Zoneamento Agrícola de Risco Climático. Portaria 254 de 05 de julho de 2022 – Soja 1ª Safra. Brasília. https://indicadores.agricultura.gov.br/zarc/index.htm

Brasil. (2022b). Mapa – Ministério da Agricultura, Pecuária e Abastecimento. ZARC – Zoneamento Agrícola de Risco Climático. Portaria 336 de 28 de setembro de 2022 – Milho 2ª Safra. Brasília. https://indicadores.agricultura.gov.br/zarc/index.htm

Chamber of Deputies of Brazil. (2012). *Lei no 12.651, de 25 de maio de 2012*. http://www2.camara.leg.br/legin/fed/lei/2012/lei-12651-25-maio-2012-613076-norma-pl.html

Companhia Nacional de Abastecimento (CONAB). (2022). Acompanhamento da safra Brasileira de grãos. In *Décimo segundo levantamento* (Brasília, v.9 – Safra 2021/22, n.12, pp. 1–88). https://www.conab.gov.br/info-agro/safras/graos

Crusciol, C. A. C., Nascente, A. S., Borghi, E., Soratto, R. P., & Martins, P. O. (2015). Improving soil fertility and crop yield in a tropical region with palisade grass cover crops. *Agronomy Journal*, 107, 2271–2280. https://doi.org/10.2134/agronj14.0603

Dias, M. B. D. C., Costa, K. A. P., Severiano, E. C., Bilego, U. O., Neto, A. E. F., Almeida, D. P., & Vilela, L. (2020). *Brachiaria* and *Panicum maximum* in an integrated crop–livestock system and a second-crop maize system in succession with soybean. *The Journal of Agricultural Science*, 158(3), 206–217.

Donagemma, G. K., Freitas, P. L., Balieiro, F. C., Fontana, A., Spera, S. T., Lumbreras, J. F., Viana, J. H. M., Araújo Filho, J. C., Santos, F. C., Albuquerque, M. R., Macedo, M. C. M., Teixeira, P. C., Amaral, A. J., Bortolon, E., & Bortolon, L. (2016). Characterization, agricultural potential, and perspectives for the management of light soils in Brazil. *Pesquisa Agropecuária Brasileira*, 51(9), 1003–1020. https://doi.org/10.1590/S0100-204X2016000900001

Evangelista, B. A., Silva, F. A. M., Simon, J., Campos, L. J. M., & Vale, T. M. (2017). Zoneamento de risco climático para determinação de épocas de semeadura da cultura da soja na região MATOPIBA. Embrapa Pesca e Aquicultura. In *Boletim de pesquisa e desenvolvimento*. Embrapa Pesca e Aquicultura.

Fabiola, N., Giarola, N., Silva, A. P., Imhoff, S., & Dexter, A. R. (2003). Contribution of natural soil compaction on hardsetting behavior. *Geoderma*, 13, 95–108. https://doi.org/10.1016/S0016-7061(02)00333-6

Franco, J. G. O. (2009). Legislação florestal brasileira: efetividade e implicações para a agricultura familiar (Doctoral Thesis, Universidade Federal do Paraná) https://www.conab.gov.br/info-agro/safras/graos.

Frota, M. N. P., Carneiro, M. S. S., Pereira, E. S., Berndt, A., Frighetto, R. T. S., Sakamoto, L. S., Moreira Filho, M. A., Cutrim Júnior, J. A. A., & Carvalho, G. M. C. (2017). Enteric methane in grazing beef cattle under full sun, and in a silvopastoral system in the Amazon. *Pesquisa Agropecuária Brasileira*, 52(11), 1099–1108. https://doi.org/10.1590/S0100-204X2017001100016

Gotz, L. F., Holzschuh, M. J., Vargas, V. P., Teles, A. P. B., Martins, M. M., & Pavinato, P. S. (2023). Phosphate management for high soybean and maize yields in expansion areas of Brazilian Cerrado. *Agronomy*, 13(1), 158. https://doi.org/10.3390/agronomy13010158

Guilherme, L. R. G., Lopes, A. S., & Corguinha, A. P. (2018). Challenges and opportunities for a sustainable agriculture in Brazil. *Acta Horticulturae*, 1224, 1–6. https://doi.org/10.17660/ActaHortic.2018.1224.1

Hissa, L. B. V., Aguiar, A. P. D., Camargo, R. R., Lima, L. S., Gollnow, F., & Lakes, T. (2019). Regrowing forests contribution to law compliance and carbon storage in private properties of the Brazilian Amazon. *Land Use Policy*, 88, 104163. https://doi.org/10.1016/j.landusepol.2019.104163

Instituto Brasileiro de Geografia e Estatística (IBGE). (2019). *BDiA – Banco de Informações Ambientais. Banco de dados cartográficos – Pedologia*. https://bdiaweb.ibge.gov.br/#/consulta/pedologia

Instituto Brasileiro de Geografia e Estatística (IBGE). (2022). *Sistema IBGE de Recuperação Automática – SIDRA: Produção agrícola municipal – Atualização 15/09/2022*. IBGE. https://sidra.ibge.gov.br/acervo#/S/Q

IUSS Working Group WRB. (2015). *World Reference Base for Soil Resources (2014, update 2015). International soil classification system for naming soils and creating legends for soil maps*. FAO. http://www.fao.org/3/i3794en/I3794en.pdf

Ker, J. C. (1997). Latossolos do Brasil: Uma Revisão. *Geonomos*, 5, 17–40. https://doi.org/10.18285/geonomos.v5i1.187

Klepker, D. (2014). Desafios para melhoria da qualidade do solo no Cerrado das novas fronteiras agrícolas. In L. F. C. Leite, G. A. Maciel, & A. S. F. Araújo (Eds.), *Agricultura conservacionista no Brasil* (pp. 217–230). Embrapa.

Köppen, W. (1948). *Climatologia: con un estudio de los climas de la tierra*. Fondo de Cultura Economica.

Leite, C., Oliveira, S., Henrique, J., Ribeiro, R., Fidelis, R. R., Tavares, R. D. C., Barilli, J., & Machado, Â. F. (2022). Liming in soils with plinthic materials of the Brazilian Savanna: Potentials and limitations. *Australian Journal of Crop Science*, 16, 488–494. https://doi.org/10.21475/ajcs.22.16.04.p3438

Leite, L. F. C., Freitas, R. C. A., Sagrilo, E., & Silva, S. R. S. G. (2010). Decomposição e liberação de nutrientes de resíduos vegetais depositados sobre Latossolo Amarelo no Cerrado Maranhense (In Portuguese, with English abstract). *Revista Ciência Agronômica*, 41(1), 29–35.

Lima, M., Silva Júnior, C. A., Rausch, L., & Gibbs, H. K. (2019). Demystifying sustainable soy in Brazil. *Land Use Policy*, 82, 349–352. https://doi.org/10.1016/j.landusepol.2018.12.016

Lumbreras, J. F., Carvalho Filho, A., Motta, P. E. F., Barros, A. H. C., Aglio, M. L. D., Dart, R. O., Silveira, H. L. F., Quartaroli, C. F., Almeida, R. E. M., & Freitas, P. L. (2015). *Aptidão agrícola das terras do MATOPIBA*. Embrapa Solos.

Lustosa Filho, J. F., Souza, H. A., Almeida, R. E. M., & Leite, L. F. C. (2021). Conservação e manejo da fertilidade do solo no Cerrado do Matopiba. In B. F. Iwata & I. L. Rocha (Eds.), *Cerrado: capital natural e serviços ambientais* (pp. 75–97). Paco Editorial.

Metzger, J. P., Bustamante, M. M. C., Ferreira, J., Fernandes, G. W., Librán-Embid, F., Pillar, V. D., Prist, P. R., Rodrigues, R. R., Vieira, I. C. G., & Overbeck, G. E. (2019).

Por que o Brasil precisa de suas Reservas Legais. *Perspectives in Ecology and Conservation*, 17, 104–116. https://doi.org/10.1016/j.pecon.2019.09.001

Moreau, A., Costa, L., Ker, J., & Gomes, F. (2006). Gênese de horizonte coeso, fragipã e duripã em solos do tabuleiro costeiro do sul da Bahia. *Revista Brasileira de Ciência do Solo*, 30, 1021–1030. https://doi.org/10.1590/S0100-06832006000600011

Oliveira, V., Jacomine, P., & Couto, E. (2017). Solos do Bioma Cerrado. In N. C. Curi, J. Ker, R. F. Novais, P. Vidal-Torrado, & C. R. G. R. Schaeffer (Eds.), *Pedologia dos biomas Brasileiros* (pp. 178–226). Sociedade Brasileira de Ciência do Solo.

Pires, G. F., Abrahão, G. M., Brumatti, L. M., Oliveira, L. J. C., Costa, M. H., Liddicoat, S., Kato, E., & Ladle, R. J. (2016). Increased climate risk in Brazilian double cropping agriculture systems: Implications for land use in northern Brazil. *Agricultural and Forest Meteorology*, 228–229, 286–298. https://doi.org/10.1016/j.agrformet.2016.07.005

Ramalho-Filho, A., & Beek, K. J. (1995). *Sistema de avaliação da aptidão agrícola das terras*. Empresa Brasileira de Pesquisa Agropecuária.

Ramos, M. R. (2022). A review of soybean cultivation on stony soils in Tocantins, Brazil. *International Journal of Science and Research*, 11, 367–371. https://doi.org/10.21275/SR22305001852

Ramos, M. R., Curcio, G. R., Dedecek, R. A., Melo, V. F., & Uhlmann, A. (2013). Influência da posição na encosta na manifestação do caráter coeso em solos da formação Macacu, no estado do Rio de Janeiro. *Revista Brasileira de Ciência do Solo*, 37, 837–845. https://doi.org/10.1590/S0100-06832013000400002

Sano, E. E., Bettiol, G. M., Martins, E. S., Couto Júnior, A. F., Vasconcelos, V., Bolfe, E. L., & Victoria, D. C. (2020). Características gerais da paisagem de Cerrado. In E. L. Bolfe, E. E. Sano, & S. K. Campos (Eds.), (pp. 21–37). *Dinâmica agrícola no Cerrado*. Embrapa.

Santos, H. G., Jacomine, P. K. T., Anjos, L. H. C., Oliveira, V. Á., Lumbreras, J. F., Coelho, M. R., Almeida, J. A., Cunha, T. J. F., & Oliveira, J. B. (2018). *Sistema Brasileiro de classificação de solos*. Embrapa.

Santos, S. F. C. B., Souza, H. A., Araújo Neto, R. B., Sagrilo, E., Ferreira, A. C. M., Carvalho, S. P., Brito, L. C. R., & Leite, L. F. C. (2021). Soil microbiological attributes and soybean grain yield in succession to corn intercropped with forage in the Maranhão Eastern Cerrado. *International Journal of Plant Production*, 15, 1–9. https://doi.org/10.1007/s42106-021-00167-z

Schaefer, C. E. G. R., Espindola, C. R., dos Anjos, L. H. C., Camargo, F. O., Ker, J. C., & Corrêa, G. R. (2023). A brief history of Brazilian soil science. In C. E. G. R. Schaefer (Ed.), *The soils of Brazil* (pp. 1–23). Springer.

Secretaria da Agricultura e Pecuária (Seagro). (2020). *Agricultura*. https://bit.ly/37vr7f3

Soares-Filho, B., Rajão, R., Macedo, M., Carneiro, A., Costa, W., Coe, M., Alencar, A. (2014). Cracking Brazil's Forest Code. *Science*, 344 (April), 363–364.

Sousa, D. M. G., & Lobato, E. (2004). *Cerrado: Correção do solo e adubação* (2nd ed.). Embrapa Cerrados.

Thornthwaite, C. W., & Mather, J. R. (1955). *The water balance* (Publications in Climatology, vol. VIII, no. 1). Drexel Institute of Technology – Laboratory of Climatology.

USDA-NRCS. (1999). *Soil taxonomy: a basic system of soil classification for making and interpreting soil surveys* (Agriculture handbook 436, 2nd ed.). USDA. https://www.nrcs.usda.gov/Internet/FSE_DOCUMENTS/nrcs142p2_051232.pdf

USDA-NRCS. (2014). *Keys to soil taxonomy* (12th ed.). https://www.nrcs.usda.gov/wps/portal/nrcs/detail/soils/survey/class/taxonomy/?cid=nrcs142p2_053580

Victoria, D. C., Bolfe, É. L., Sano, E. E., Assad, E. D., Andrade, R. G., Guimarães, D. P. & Landau, E. C. (2020). Potencialidades para expansão e diversificação agrícola sustentável do Cerrado. In E. L. Bolfe, E. E. Sano, & S. K. Campos (Eds.), *Dinâmica Agrícola no Cerrado: análises e projeções* (pp. 229–258). Brasilia, DF: Embrapa.

Xavier, A. C., King, C. W., & Scanlon, B. R. (2020). Variáveis meteorológicas quadriculadas diárias no Brasil (1980–2020). (In Portuguese, with English abstract). *Revista Internacional de Climatologia*, 36(6), 2644–2659.

12

Public Policies and Initiatives to Promote Soil Health and Carbon Sequestration in Brazil

Carlos Eduardo Pellegrino Cerri, Francisco Fujita de Castro Mello, Natália Braga Renteria, and Maurício Roberto Cherubin

Chapter Overview

Brazil is one of the world's largest producers and exporters of food, feed, fiber, and biofuel. Estimates indicate that Brazil needs to increase its food production by 40% to meet global demands by 2050. Meanwhile, tropical agricultural systems must contribute to sequester C, reduce greenhouse gas (GHG) emissions, and, consequently, mitigate climate change. Soil health (SH) is one the main pillars for achieving both challenges. Healthy soils are more productive and resilient, making agricultural systems less vulnerable to climate change in the next decades. However, to transform science and technology into action, it is necessary to have coordination of high-impact technical and policy interventions that meet the needs of all actors of agricultural sectors. Here, we provide examples of the public policies and initiatives adopted in different spatial scales that promote SH and C sequestration in Brazil. We briefly present the international climate agreement (Kyoto Protocol and Paris Agreement) and Brazil's commitments to mitigate climate change. We describe Living Soils of the Americas (LiSAm) program, a large initiative to enhance SH and C sequestration in agricultural lands across the continent. In Brazil, the Plan for Adaptation and Low Carbon Emission in Agriculture (ABC+ plan) is the main federal instrument for promoting sustainable agriculture in Brazil. Finally, two centers - the Research Center for Greenhouse Gas Innovation (RCGI) and the Center for Carbon Research in Tropical Agriculture (CCARBON) - are presented as examples of academic initiatives focused on developing solutions for enhancing SH and C sequestration to mitigate climate change and deliver other vital ecosystems services.

Soil Health Series: Volume 3 Soil Health and Sustainable Agriculture in Brazil, First Edition.
Edited by Ieda Carvalho Mendes and Maurício Roberto Cherubin.

Introduction

Agriculture and forestry have strategic roles in the face of a great world dilemma: global warming and the consequent climate change. Therefore, one of the most important challenges of the next decades is to reconcile food production to support population growth and environmental sustainability (Hong et al., 2022). In countries where agricultural activity is largely significant, such as Brazil, improving SH, enhancing C sequestration, and reducing GHG emissions are essential in the fight against climate change and food insecurity.

Brazilian agriculture has been substantially transformed over the last four decades, with incremental changes in the amount of cultivated area but also with substantial improvements in crop yields and the adoption of best management practices (BMPs). Among the different alternatives to reduce GHG emissions from cultivated areas, C sequestration in soil emerges as a natural solution (also called "nature-based solutions"). According to Bossio et al. (2020), global soil C sequestration can reach 24 Gt CO_2eq yr^{-1}, which represents 25% of the potential of natural climate solutions and comprises 47% for agriculture and grasslands. However, the adoption of sustainable management practices (i.e., conservation agriculture) can determine if agricultural soils will act as GHG sinks or GHG sources to the atmosphere. Sustainable management practices include no-tillage (NT); cover cropping; complex crop rotations; and integrated systems such as crop–livestock–forestry, irrigation, and organic amendments, among others (Amelung et al., 2020; Lal, 2013; Maia et al., 2022; Paustian et al., 2016). Most of these practices were described in the previous chapters of this book.

The adoption of NT in areas previously managed by conventional systems and pastures is a potential alternative for promoting C sequestration in agricultural soils in the different regions of Brazil. For example, after more than 20 years since the conversion from conventional tillage to NT, the sequestration rate for the 0- to 30-cm soil layer was 0.63 Mg C ha^{-1} yr^{-1}, or 17% soil organic C (SOC) (Maia et al., 2022). The time of the implementation is another factor that can boost soil C sequestration, where the long-term adoption of management practices that benefit soil C storage is recommended even after a new steady state is reached and no further GHG mitigation accrues (Smith, 2016). Integrated Crop Livestock Forest (ICLF) systems have emerged as sustainable strategies to intensify land productivity by combining annual crop, livestock, and/or forestry activities in the same area under different spatio-temporal arrangements (Landers et al., 2020). In Brazil, it is estimated that 17.4 million ha have been cultivated with some types of Integrated Agricultural Systems (IASs) (see Chapter 2). Among other

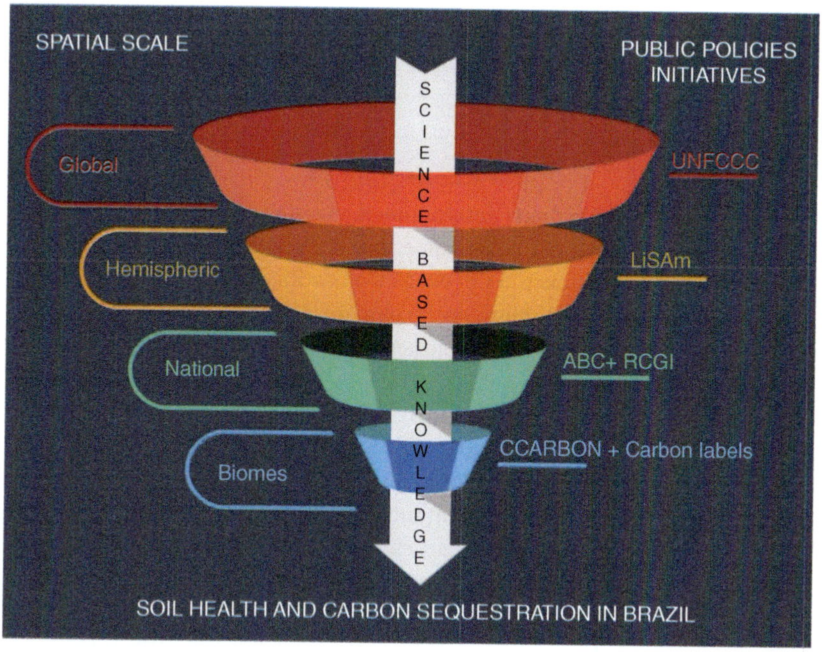

Figure 12.1 Public policies and initiatives, in different spatial scales, to promote soil health and C sequestration in agricultural systems in Brazil. ABC+, low-carbon emission in agriculture; CCARBON, Center for Carbon Studies in Tropical Agriculture; LiSAm, Living Soils of the Americas; RCGI, Research Centre for Greenhouse Gas Innovation; UNFCCC, United Nations Framework Convention on Climate Change. Carbon labels, Low-Carbon Brazilian Beef, Carbon-Neutral Brazilian Beef, Low-Carbon Soybean.

socioeconomic benefits, IASs have stood out as promising nature-based solutions for enhancing SH and C sequestration and for providing essential products to people (e.g., cereals, meat, and wood). Therefore, IAS can be a powerful solution to tackle global food and energy insecurity and climate change in the coming decades.

In this context, these examples of BMPs show that producing food, fiber, and energy for a growing world population and with a positive balance of C capture is a major challenge for modern agriculture. Thus, to transform science and technology into action, it is necessary to have better coordination of high-impact technical and policy interventions that meet the needs of farmers not only in Brazil but also in other developing countries of the world. Therefore, the present chapter presents examples of the main public policies and initiatives adopted in different spatial scales (Figure 12.1) that promote SH and C sequestration in the Brazilian agricultural sector.

Examples of Current Public Policies and Initiatives to Promote Soil Health and Carbon Sequestration in Brazil

United Nations Framework Convention on Climate Change and the Paris Agreement

Since Stockholm in 1972, many international meetings have addressed issues related to climate change and sustainable development. In 1992, Brazil hosted the United Nations Conference, ECO 92, as well as Rio +20 in 2012, placing the country in a leading role in discussions on climate change. It was during ECO 92 that discussions about a potential C market emerged as an initiative to curb the emission of GHGs.

Pathways to reverse the current scenario of global warming have been discussed and proposed in the last three decades by intergovernmental institutions, governments, and private and other sectors of society. One of the most important milestones was the "Kyoto Protocol," an international agreement created in 1997 during the Third Conference of the Parties held in Kyoto, Japan, which defines responsibilities and obligations for industrialized countries (the largest producers of GHGs) as part of the mission to reduce the percentage of gases emitted. The protocol went into effect in 2005 and lasted until 2015, when the Paris Agreement, signed in 2015 at COP 21 in Paris, replaced the Protocol (the Paris Agreement replaced the Kyoto Protocol, but its rules about C market were valid until 2020, which was the end of the second commitment period).

The subject must be treated globally, as agreed at the 21st Conference of the Parties (COP-21) in Paris (Paris Agreement) because all countries contribute to GHG emissions and therefore all countries must suffer consequences, to a greater or lesser extent. The 2015 Paris Agreement of the United Nations Framework Convention on Climate Change (UNFCCC) aims to hold the increase in global average temperatures by 2100 to well below 2°C above preindustrial levels and to pursue efforts to limit the temperature increase to 1.5°C above preindustrial levels (Richards et al., 2016; Rogelj et al., 2016). A surprisingly large number of countries (at least 119) voluntarily pledged to reduce their agricultural GHG emissions for the agreement in their statements of Intended Nationally Determined Contributions to the UNFCCC (Wollenberg et al., 2016). The main strategies to reduce GHG emissions from anthropogenic sources consist of reducing the burning of fossil fuels, minimizing deforestation and fires, promoting adequate soil management, and adopting strategies to maximize C sequestration in the soil. Conservationist practices are indisputable for the optimization of the last two strategies.

Individually, parties' contributions through Nationally Determined Contributions (NDCs) can be revised at any time if they "raise their level of ambition." National governments are responsible for determining the form and content

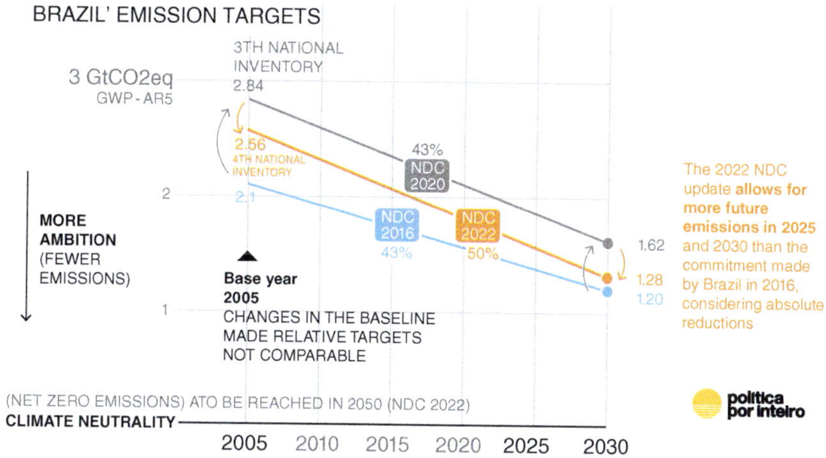

Figure 12.2 Brazilian emission targets as a contribution to the Paris Agreement. Unterstell and Martins (2022)/with permission of institutotalanoa.org.

of their contributions, taking into consideration the principles of the Paris Agreement (Unterstell & Martins, 2022). Given that the Brazil's NDC is an important instrument of foreign policy but also of national policy on climate change, Unterstell and Martins (2022) assessed its evolution, quantification, and quality, based on Brazil's submission to the UNFCCC of March 31, 2022 (Figure 12.2).

Brazil's NDC was submitted to UNFCCC in 2016. The base-year emissions were fixed as 2.1 $GtCO_2eq$ in 2005 and included the following commitments: (a) 1.30 $GtCO_2eq$ maximum reduction at 2005 levels (37% reduction); (b) 1.20 indicative maximum reduction $GtCO_2eq$ in 2030 at 2005 levels (43% reduction); (c) implementation of the national adaptation plan; and (d) implementation of sectoral mitigation actions, such as zero illegal deforestation in 2030, recovery of 15 million ha of degraded pastures, and others.

An update of the Brazilian target took place on 2020 and contained the following commitments: (a) 1.79 $GtCO_2eq$ maximum reduction at 2005 levels (37% reduction); (b) 1.62 indicative maximum reduction $GtCO_2eq$ in 2030 at 2005 levels (43% reduction); (c) climate neutrality (indicative) by 2060; (d) conditional target: US$10 billion/year support requested; (e) exclusion of reference to national adaptation plan; (f) exclusion of reference to the implementation of sectoral mitigation actions, such as zero illegal deforestation in 2030, restore 12 million ha of native vegetation, recover 30 million ha of degraded pastures, increase 5 million ha of ICLF, and others (Unterstell & Martins, 2022).

To enforce the commitments stated in COP21, the Brazilian government prepared some execution plans aimed at adapting and mitigating emissions from some sectors of the economy. To this end, the Sectorial Plan for Mitigation and

Adaptation to Climate Change (ABC Plan) created in 2012 (Brasil Ministério da Agricultura, Pecuária e Abastecimento, 2012) was updated in 2021 as the Plan for Adaptation and Low Carbon Emission in Agriculture: Strategic Vision for a New Cycle (ABC+ Plan) (Brasil Ministério da Agricultura, Pecuária e Abastecimento, 2021) in order to modernize national agriculture and make it more sustainable, based on the balance of GHG, as presented in other sections of this text.

Living Soils of the Americas

The LiSAm program is an initiative that is facilitating the collaboration among agricultural stakeholders, scientists, and donors. It addresses the challenge of demonstrating that SH restoration and soil C sequestration in agricultural lands are strategies that could be applied at large scales and potentially at low cost to be beneficial to farmers and at the same time contribute toward the goals set at the Conference of the Parties—UNFCCC negotiations in Paris 2015. Figure 12.3 captures the proposed conceptual framework of the initiative.

This initiative is being operationalized through public-private partnerships for the implementation of land management and C projects in different agroecosystems and regions of Latin America and the Caribbean (LAC). It is intended to respond to the needs and collaboration opportunities of main partner organizations and agricultural stakeholders.

A few outputs and expected results of the LiSAm program are illustrated in Table 12.1. These results are associated with the overall impact of the program,

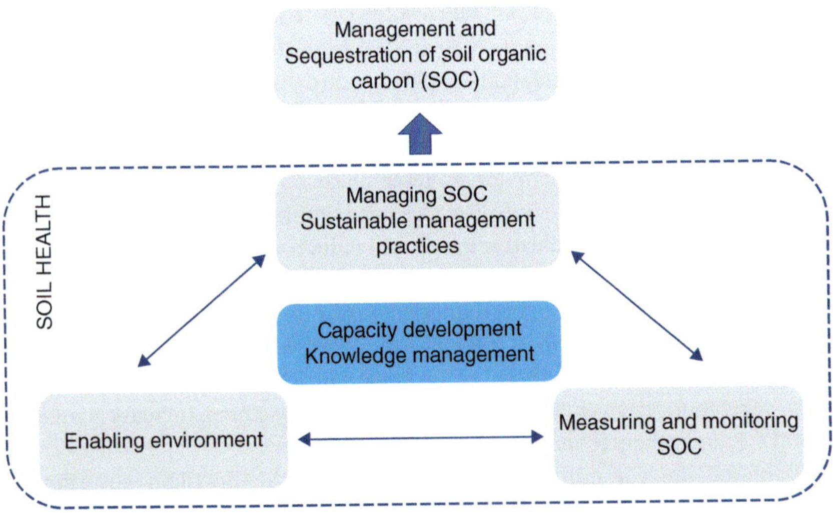

Figure 12.3 Conceptual framework of the initiative Living Soils of the Americas (LiSAm).

Table 12.1 Main expected results from the living soils of the Americas.

Strategic/ thematic areas	Outputs	Results
Measuring and monitoring soil organic C (SOC)	• Applied research agenda with academic partners institutions relating SOC with soil physical properties (bulk density), productivity, biological activity, vegetation cover, soil C sequestration rates, soil health. • SOC baselines of the main agricultural ecosystems based on the literature and ongoing work of main research actors • Spatial databases, remote sensing, and models development • Capacity development within key players in carbon science	• Countries use verified MRV protocols for soil C sequestration tested in the field and validated by CMASC • Easy to use GHG quantification tools for field projects • SOC stocks of main agricultural ecosystems are known. • Knowledge of soil structure and SOC interactions; relevance of surface vs. belowground biomass residues in SOM dynamics
Soil policies and regulations	• C PES, certification and green labelling schemes with farmers implementing BMPs • Promotion of compliance and voluntary markets (cap-and-trade systems) • Soil health policies	• Achieving SOC sequestration targets • Local governments implement effective land use regulations • Countries integrate soils into their NDCs and GHG programs to contribute to both adaptation and mitigation goals
Best management practices (BMPs)	• Assessment of what has been done in LAC and Systematization of Lessons Learned (reduced tillage, improved crop rotations, service crops, nutrient management, agro-forestry, organic amendments, etc.) • Evaluation of the BMPs using soil heath indicators and productivity indexes • Promotion of conservation agriculture practices and preventing land conversion from forest • Scaling (i.e., using multi-stakeholder platforms involving farmer associations, private sector, and state agencies) • Promoting land restoration/conservation projects	• Operational toolboxes • Databases and maps • Landscape/regional targets of area under BMPs • Emissions of SOC decrease due to mismanagement and unsuitable soil practices

Abbreviations: CMASC, Carbon Management and Sequestration Center at The Ohio State University; LAC, Latin America and the Caribbean; MRV, monitoring, reporting, and verification; PES, payment for ecosystem services.

especially related to the improved knowledge on GHG mitigation due to land use and conservation agriculture practices and the increase of land cultivated under management practices that increase C sequestration and SH.

A big challenge to achieving the above-mentioned goals, and for agriculture to participate in C projects, is the lack of dependable, credible, and cost-effective methods of monitoring changes in soil C in the large scale. Validating protocols for monitoring, reporting, and verification (MRV) in SOC; development of baselines; developing soil sampling capacities; guidelines for interpretation; and use of proxies or other indirect C assessments are among the crucial issues on the agenda of many organizations wishing to enter into C sequestration schemes. The measuring and monitoring line of action of this initiative will provide answers to these challenges as well as practical solutions to test and implement ongoing protocols within the projects developed by this initiative. The LAC region has a significant amount of knowledge and experience in land management practices. The application of MRV to assess C sequestration under specified practices and conditions will provide valuable information to policymakers to feed into other national climate change strategies. Innovative digital tools will be key for facilitating the measuring and monitoring of SOC at the scales necessary.

The best practices for the SOC management component aim to build the capacity of public and private extension agents, farmers, and other groups that are part of the initiative to improve SH in agriculture. One strategic line of action is fostering an environment that enables SOC management and sequestration. Capturing the benefits of healthy soil on the farm or at the landscape level is an important goal but alone would unlikely achieve the changes in the agricultural systems required to reach NDC targets. Policies and incentives, developing payment for ecosystem services schemes, promoting C offset markets, and developing C information networks are among the actions that will be required to engage agricultural stakeholders to reach the proposed mitigation and resilience goals. Moreover, to integrate SOC into the international climate change mechanisms, it is important to map the opportunities and to fully understand the action required from agricultural leaders.

This initiative offers a unique opportunity because it will count on the scientific and technical backstopping of the Carbon Management and Sequestration Center at The Ohio State University as well as the Inter-American Institute for Cooperation on Agriculture (IICA) network of 34 countries' representations that operate in close cooperation with the Ministries of Agriculture to respond to the most pressing agricultural challenges in the hemisphere. Brazil has actively contributed to this initiative. The IICA, founded in 1942, is the specialized agency of the Inter-American System that supports the efforts of its member states to achieve agricultural development and rural well-being. The IICA's medium-term plan establishes seven hemispheric programs as part of the technical cooperation agenda, including the Agricultural Climate Action and Sustainability, which fosters integrated solutions to achieve a more sustainable, climate-resilient, and low-C sector (IICA, 2022). The Agricultural Climate Action and Sustainability strengthens bridges between actors in the LAC region, thus promoting a common agenda that allows progress toward the achievement of the multiple agri-environmental goals of each country detailed in the NDCs developed in response to the Paris Agreement and the land degradation neutrality goals set under the UNCCD.

In this sense, one important outcome of the initiative is the publication "Soil carbon sequestration through adopting sustainable management practices: potential and opportunity for the American countries" (Cerri et al., 2021), where several of the issues presented in Table 12.1 were addressed, including methodologies for soil C evaluation at different scales, case studies based on evidence showing management practices that increase soil C stocks, and an estimate of the total soil C stocks for each region across the Americas, among others. Two main results are the indication of the land cover in the Americas and provide a soil C sequestration estimate for the entire region. The Americas region comprises a land extension of around 4 billion ha, where 2.2 billion ha are closed and open forests, 0.9 billion ha are pasture areas, 0.34 billion ha are dedicated to croplands, and the remaining are other land uses. The authors estimated a potential of soil C sequestration for the next 20 years of 9.8 Pg of CO_2eq (4.56–15.06 Pg CO_2eq) based in the adoption of two major land practices (i.e., pasture reclamation in 40% of the current land use and the expansion of 50% of the cropped area under NT system).

Low Carbon Emissions in Agriculture (ABC+ Plan)

To foster the commitments presented at the UNFCCC, the Brazilian government launched the Sectoral Plan for Mitigation and Adaptation to Climate Change for the Consolidation of a Low-Carbon Emission in Agriculture (ABC) (Brasil - Ministério da Agricultura, Pecuária e Abastecimento, 2012). The ABC Plan is one of the sectoral plans prepared in accordance with Article 3 of Decree No. 7.390/2010 and organizes and plans the actions to be carried out for the adoption of sustainable production technologies, selected with the objective of responding to the commitments to reduce GHG emissions in the agricultural sector assumed by the country.

During 2010–2020, the ABC Plan aimed to mitigate 133–162 MtCO_2eq through the adoption of seven technologies or practices, six of them referring to mitigation technologies and one with actions to adapt to climate change:

- Program 1: Recovery of degraded pastures
- Program 2: ICLF and agroforestry systems
- Program 3: NT systems

- Program 4: Biological N fixation
- Program 5: Planted forests
- Program 6: Animal waste treatment
- Program 7: Adaptation to climate change

The ABC Plan must be understood as an instrument for integrating government actions (federal, state, and municipal), the productive sector and civil society, to reduce GHG emissions from agricultural and livestock. For its effectiveness, an institutional arrangement was established involving representatives of the various entities involved. The participation of civil society is essential and reaffirms the democratic character in the conception and implementation of the programs. It is important to emphasize that all the technologies proposed by the ABC Plan were based in scientific evidence by publications such as Carvalho et al. (2009), Cerri et al. (2009), and Cerri et al. (2010), which presented specific data about the adoption of best agriculture practices and soil C sequestration at large scale in Brazil.

The ABC Plan governance structure is divided into three levels:

1. *Strategic National*: The ABC Plan will use the instances of the Interministerial Committee on Global Climate Change (CIM) and its Executive Group (GEx), established by Decree No. 6263, of November 21, 2007, with the purpose of evaluating the implementation of actions and to propose new measures that are necessary to reduce GHG emissions in agriculture.
2. *Tactical National*: The National Executive Committee of the ABC Plan, linked to the CIM/GEx, has the purpose of periodically monitoring and accompanying the implementation of the ABC Plan in addition to proposing measures to overcome eventual difficulties in this process. This Commission was coordinated by the Ministry of Agriculture, Livestock and Supply (MAPA) and the Ministry of Agrarian development (MDA), with the participation of representatives from the Civil House, the Ministries of Finance (MF) and the Environment (MMA), the Brazilian Agricultural Research Corporation (Empresa Brasileira de Pesquisa Agropecuária—Embrapa), and the Brazilian Forum on Climate Change.
3. *Operational State*: The State Management Groups (GEE) were constituted and are responsible for promoting the coordination and articulation of the Agriculture Sectorial Plan in the states. This group was coordinated by a representative of the State Secretary of Agriculture with the participation of MAPA, the MDA, the State Secretary for the Environment, Embrapa, and the State Agricultural Research Organizations, the official banks (Banco do Brasil, Banco da Amazônia, or Banco do Nordeste) and with the integration of civil society representatives (productive sector, workers, universities, research, cooperatives, agriculture federation, non-governmental organizations, etc.).

The consolidation of public-private partnerships is also essential to leverage the Plan's actions and replicate them at the state and municipal levels. It is possible to make them more efficient through the dissemination or adoption of sustainable practices with a fundamental role in the dissemination of this Plan aimed at reducing GHG emissions in agriculture and thus minimizing the possible negative impacts arising from climate change. The challenge is to address all the mentioned points through an integrated landscape approach, as illustrated in Figure 12.4, part of the updated version of the ABC Plan launched in 2021, called ABC+ (Brazil Ministry of Agriculture, Livestock and Food Supply, 2021).

The scope of the ABC Plan is national, and updates were scheduled for no longer than 2 years to readjust it to society's demands and new technologies and to incorporate new actions and goals, if necessary. To achieve the objectives outlined by the ABC Plan, in the period between 2011 and 2020, resources were made available to farmers to adopt the science-based practices incentivized by the program. According to the ABC Plan (Brasil Ministério da Agricultura, Pecuária e Abastecimento, 2012), the distribution of resources has not been uniform, with resources especially directed to the recuperation of degraded pastures and heavily concentrated in the Center-South of the country. It is worth noting that the volume of degraded pastures is an important environmental problem in the country; however, socioeconomically less developed regions with considerable

Figure 12.4 Conceptual bases of the Plan for Adaptation and Low Carbon Emission in Agriculture (ABC+ Plan) in Brazil. Adapted from MAPA (2022).

environmental problems have not been prioritized by the program (Gianetti & Ferreira Filho, 2021).

To face the adverse impacts of climate change and to increase the resilience and sustainability of Brazilian agriculture, ABC+ is based on the following three main pillars: (a) an integrated landscape approach; (b) synergy of adaptation and mitigation strategies; and (c) fostering adoption and maintenance of Sustainable Systems, Practices, Products and Production Processes (Brazil Ministry of Agriculture, Livestock and Food Supply, 2021). Based on these pillars, the strategies of ABC+ are listed below.

I. *Maintaining motivation for adoption of conservationist and sustainable farming systems, fostering increased productivity and income, resilience, and control of GHG emissions*
In this second round, the ABC+ is promoting the adoption of SPSABC, also present in the previous cycle, namely: (a) ICLF and NT systems, (b) biological N fixation, (c) planted forest, (d) restoration of degraded pastures, and (e) animal waste management. Other science-based Sustainable Systems, Practices, Products and Production Processes, proven to be effective in tackling climate change, are being developed and included, considering specificities such as soil, climate, water, and the ecosystems of Brazilian biomes and their geographic zones. This ensures greater yields and resilience of farming systems and provides effective control of GHG emissions from Brazilian agriculture (Brazil Ministry of Agriculture, Livestock and Food Supply, 2021).

II. *Strengthening initiatives for technology transfer and diffusion, training, and technical assistance*
Technical assistance, supported by training, is considered the main transformative instrument in the first cycle of the ABC Plan. Farm monitoring by trained professionals allows for not only the proper and correct adoption of recommended systems but also for the measurement of results being reaped from such systems. The so-called "reference units" (which are on-farm models for validation, demonstration, and technology transfer) have been an important catalyst for technical updating of professionals as well as for sources of feedback for research. To boost technical skills of professionals involved in technical assistance and rural extension, new ways of disseminating information are also being explored, enhanced by the increasing use of digital technologies in the field and connectivity in farming areas (Brazil Ministry of Agriculture, Livestock and Food Supply, 2021).

III. *Encouraging and supporting applied research for development or improvement of Sustainable Systems, Practices, Products and Production Processes*
Based on a robust national scientific framework and participation of the scientific community and with the aim of strengthening the ABC+, ideas for

innovation related to sustainable technologies for agricultural production are being tirelessly sought and permanently incorporated with a focus on increasing resilience, yields, and income and controlling GHG emissions (Brazil Ministry of Agriculture, Livestock and Food Supply, 2021).

IV. *Expanding mechanisms that recognize and reward farmers for adopting Sustainable Systems, Practices, Products and Production Processes*
Developing revenue streams for ecosystem services (markets for environmental services) represents a strong incentive for large-scale adoption and dissemination of Sustainable Systems, Practices, Products and Production Processes by farmers. These include a wide range of economic incentives and market instruments, such as certifications of different types and scopes, identification of origin, and traceability, among others. The main goal is to recognize effective efforts from the farming sector toward incorporating and maintaining sustainable production systems, promoting conservation of natural resources, and ensuring productivity and food supply. These instruments are also strategic for communication to Brazilian and international audiences regarding the efforts made and results achieved by the Brazilian farming sector in terms of sustainability (Brazil Ministry of Agriculture, Livestock and Food Supply, 2021).

V. *Fostering diversified financial and tax related instruments to support Sustainable Systems, Practices, Products and Production Processes*
Cross-cutting instruments for trading C credits will encourage the use of Sustainable Systems, Practices, Products and Production Processes in this new cycle in addition to those present in the previous cycle—the ABC Program and the "Plano Safra" (annual budget for farming support). This will enable the involvement of different financial agents, public and private, in a comprehensive process of effective promotion of sustainability in Brazilian agriculture (Brazil Ministry of Agriculture, Livestock and Food Supply, 2021).

VI. *Improving the ABC+ information management system for putting in place effective MRV mechanisms*
The integrated data management system (ABC Plan Information System [SINABC]) will be responsible for the systematization and consolidation of actions and results throughout the Plan's execution. SINABC will incorporate data from the ABC Plan Governance System, the System for Rural Credit Operations and Farming Insurance, and the multi-institutional platform for monitoring GHG reductions from agriculture. Information consolidated in SINABC, in its turn, will be monitored and validated by the Technical Committee for Monitoring the ABC Plan, which oversees defining guidelines for monitoring results from the ABC+ implementation. In addition, the National Executive Committee of the ABC+ will periodically monitor and follow up on implementation. Putting this new governance structure into practice

will allow for proper evaluation of effectiveness and efficiency of all efforts undertaken by the Brazilian agricultural sector to cope with climate change in a transparent manner, supported by a solid mechanism of evaluation, monitoring, and reporting (Brazil Ministry of Agriculture, Livestock and Food Supply, 2021).

Among all the modifications proposed from the previous plan to the updated ABC+ plan is the adoption technologies such as the Management Intensive Grazing in up to 5 million head, the increase of 3 million ha of irrigated area (use of intensification technologies), use of bio-inputs, and others. It is expected that the ABC+ technologies will affect a land area of 72 million ha and will mitigate approximately 1.11 billion t of CO_2eq by 2030. By jointly promoting actions for Brazilian agriculture to adapt to climate change events and mitigate GHG emissions, the ABC+ continues to be one of the most important national policies both for tackling climate change and for providing the world with food security within the context of sustainable development (Brazil Ministry of Agriculture, Livestock and Food Supply, 2021).

Research Centre for Greenhouse Gas Innovation

The RCGI was created in 2016 as a world center for advanced studies on energy transition for the sustainable use of natural gas, biogas, and H and for management, transport, storage, and usage of CO_2. The center, based at the University of São Paulo, is the result of The São Paulo Research Foundation (Fapesp) partnerships in support of high-level scientific research for the development of the energy sector. Its activities are based on three pillars: research, innovation, and diffusion of knowledge.

By 2021, the center encompassed 46 projects divided into five research programs: Engineering, Physical-Chemistry, Energy Policies & Economics, CO_2 Abatement, and Geophysics. In 2021, the center renewed its mission and began to develop five new programs. The RCGI aims to be a world-class center for advanced studies with a focus on innovation toward sustainability and mitigation of GHG emissions. The RCGI complements Fapesp's experiences in supporting high-level scientific research and technology development in these fields. In this venture, the RCGI has five distinct, but complementary, research programs: Nature Based Solutions (NBS), Carbon Capture and Utilization, Bioenergy Carbon Capture and Storage, GHG mitigation, and Advocacy (Figure 12.5). The integration of these five themes and all knowledge and innovation generated in the RCGI will directly support Brazil to achieve the NDCs through research and innovation. The RCGI is serving to unleash the Brazilian potential for a sustainable energy transition, targeting an increase of well below 2°C in the global scenario of climate change.

Figure 12.5 Research Centre for Greenhouse Gas Innovation (RCGI) programs and main questions.

The NBS program supports managing the soil–plant–atmosphere continuum to combat climate change and delivers other vital ecosystem services. The main goal of the NBS program is "to promote the development of sustainable solutions that help Brazil to fulfill its goals (NDCs) in the Paris Agreement aiming to mitigate global warming, as well as to support the provision of ecosystem services, encourage social well-being and support the elaboration of public policies." To address this ambitious goal, three national-scale projects are being developed, covering the three main forest and agriculture sectoral goals established in the Brazil's NDCs: restoration of native vegetation, expansion of ICLF, and reclamation of degraded pastures. In these projects, in addition to measuring the potential of this NBS to sequester C, SH and other components of sustainability are being measured in the field and modeled in space and time. More information about RCGI and its projects can be found at https://www.rcgi.poli.usp.br.

Center for Carbon Studies in Tropical Agriculture

Brazil is one of the world's largest producers and exporters of food, feed, fiber, and biofuel due to favorable soil and climate conditions associated with adapting technologies (e.g., plant and animal breeding, soil and water conservation management, diversified cropping systems, and forestry) to tropical conditions. Despite that, according to FAO's projection, Brazilian agriculture is expected to meet 40% of the food demand increase by 2050. Meanwhile, tropical agricultural systems must contribute to sequestering C and reducing GHG emissions. Therefore, the country has a tremendous potential to contribute not only to food security but also to sequestering C and mitigating climate change through implementing science-based, climate-smart agricultural solutions on a large scale.

In this context, the CCARBON, a center hosted by the President's Office of the University of São Paulo (USP) and physically located at Luiz de Queiroz College of Agriculture (ESALQ/USP), Piracicaba, São Paulo, was created in 2023. Recently funded by FAPESP, the São Paulo Research Foundation, CCARBON has the mission of developing innovative solutions and strategies for C-based sustainable tropical agriculture to mitigate climate change and improve living standards and conditions. The vision is to be recognized as a world-class leader on low-C tropical agricultural systems and qualification of human resources through research, innovation, and dissemination. For that, CCARBON will identify the main challenges and implement solutions to increase sustainable agri-food production of tropical agricultural systems (focusing on annual crops, such as soybean, maize, cotton, cover crops, and sugarcane; pastures; ICLF and agroforestry; and forestry/ecological restoration programs) by reducing GHG emissions and increasing C sequestration through climate-smart management practices.

The CCARBON will provide the necessary opportunity for the proper implementation of a variety of activities, including stocktaking and connecting existing research networks and projects to understand how international research cooperation is organized. The development of an international knowledge database will help researchers to increase international cooperation. The intended scientific contribution to sustainable C solutions on tropical agriculture is new, and, as part of a global problem, the research is complex and will surely require long-term inter- and multi-disciplinary approaches involving challenging topics such as climate change, food security, water security, SH and ecosystem services, ecological restoration, circular economy, and others directly associated with the UN Sustainable Development Goals. The overall CCARBON scope, including research, innovation, and dissemination actions, is schematically illustrated in Figure 12.6.

The CCARBON research goals will be achieved based on inter- and multi-disciplinary activities that collectively cover five major areas: soil, plant, animal, atmosphere, and digital tools. One of the main scientific and technological challenges of CCARBON is to integrate multiple disciplines and scales involved in the development of effective climate mitigation solutions for agriculture. There are several underlying mechanisms linking soil, microorganisms, plants, livestock, and the atmosphere, which are markedly affected by land use, agricultural management, and the inherent biophysical conditions of each agroecosystem. Unraveling such mechanisms is a major research challenge and will require important advances in each research field involved in CCARBON.

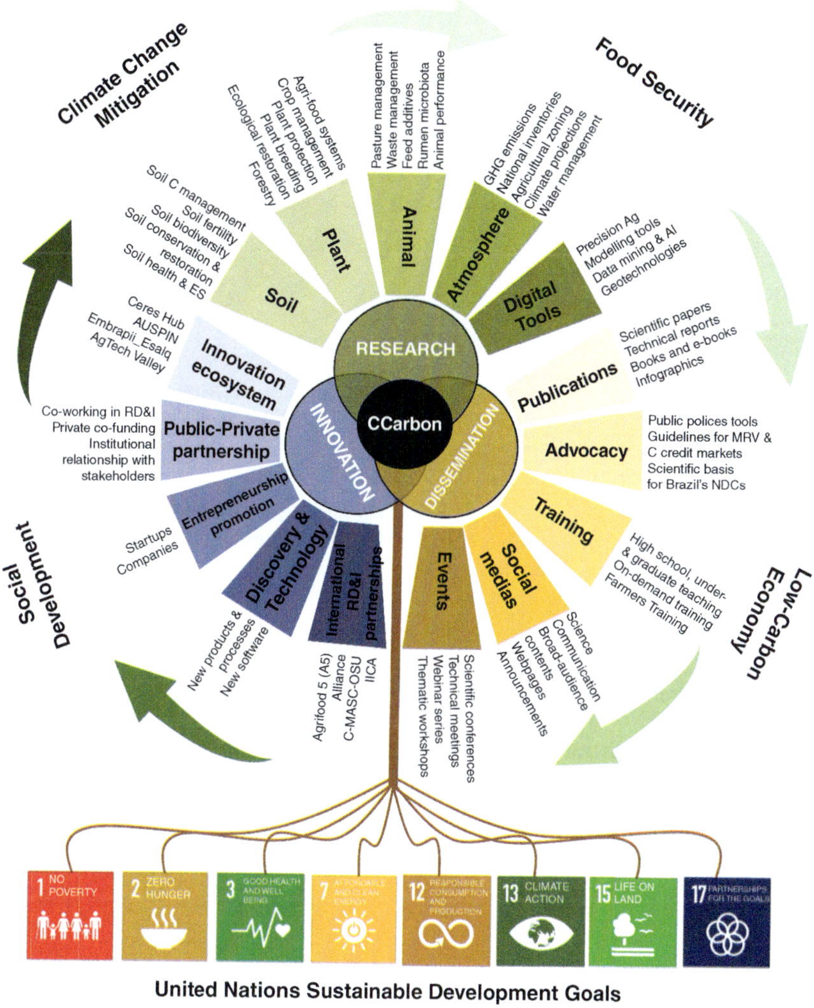

Figure 12.6 Schematic representation of general structure and purposes of the Center for Carbon Studies in Tropical Agriculture (CCARBON) Credit: United Nations.

In the first phase, the center will focus on investigating the dynamics of GHG capture and emissions by different agriculture, livestock, and forestry systems, aiming at mechanistically understanding the main processes driving these dynamics and exploring technological alternatives to improve the C balance. It will explore new avenues to convert agriculture from a C source to a C sink (Figure 12.7). This goal will be initially sought through disciplinary research, with different groups investigating this main topic under experimental conditions.

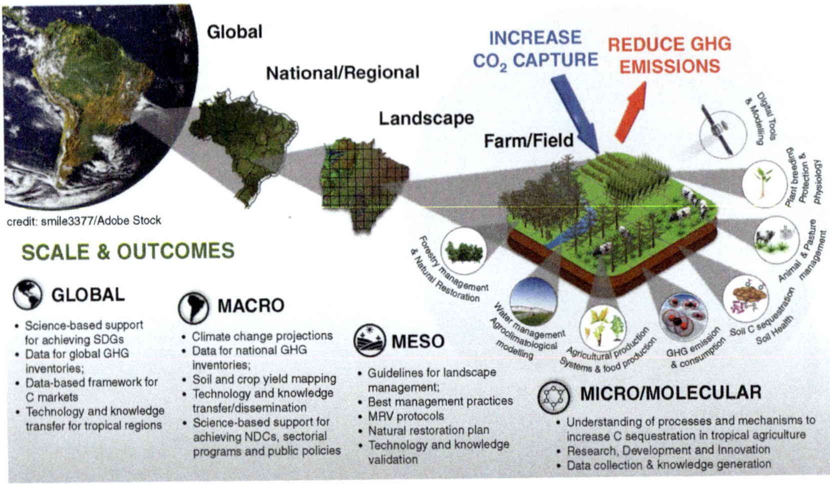

Figure 12.7 Center for Carbon Studies in Tropical Agriculture (CCARBON) strategic scales and outcomes to investigate and create C-related solutions in tropical agriculture using a multiscale approach from molecular to global.

Therefore, with the consolidation of CCARBON associated with public policies and other initiatives in this agenda, relevant advances in knowledge and technologies related to C and SH are expected in the next decade in Brazil. A new science-based transformation of Brazilian agriculture is expected for the near future, consolidating the country as a global leader in sustainable tropical agriculture.

Carbon Labels

Carbon-Neutral Brazilian Beef

In 2015, the Brazilian Agricultural Research Corporation (Embrapa) developed the concept "Carbon Neutral Brazilian Beef," which is represented by a label referring to beef cattle produced under integrated systems with the mandatory presence of a forestry component. This concept aims to support implementation of more sustainable cattle systems, especially regarding environment, through the introduction of trees that can neutralize CH_4 emitted by cattle. It ensures added value for beef produced under such systems. This concept aims also to spread the strategic importance of sustainability to the associated production chains (i.e., grains and forestry). It motivates farmers to integrate systems, optimizing the use of inputs and other production factors, resulting in synergistic positive effects (Alves et al., 2017).

In 2021, the traceability protocol finally certified the first Brazilian rural property as a producer of C-neutral meat: Fazenda Santa Verginia, located in Santa Rita do Pardo, in Mato Grosso do Sul. In the property, 30% of the area is occupied by trees and 70% by livestock activity (Brazilian Confederation of Agriculture and Livestock, 2023).

The label "Carbon Neutral Brazilian Beef" is a concept trademark, followed by a protocol with basic requirements, developed by Embrapa to enable a certification testifying that beef produced under given verifiable/certifiable parameters has its GHG emissions neutralized by the trees introduced through ICLF systems (Alves et al., 2017).

Low-Carbon Brazilian Beef

In 2015, Brazil presented intentions to expand its goals in the Paris Agreement because it already had an area of about 11.5 million ha with integration systems in place in half the time previously predicted. Of this total, 1.84 million ha, or about 2% of the area of cultivated pastures in Brazil, corresponded to the area used with of integration with the presence of the forest component, with the potential immediate production of Carbon Neutral Beef (Alves et al., 2017). The rest of the pasture areas did not include the forestry component; however, through adequate grazing management practices, fertilization, and consortium with legumes, the ICLF had a much greater potential to mitigate GHG emissions (Almeida & Alves, 2020).

Thus, the opportunity arises for the development of a new concept brand, the "Low Carbon Brazilian Beef," conceived and elaborated by Embrapa, whose objective to support sustainable livestock production systems capable of mitigating the CH_4 emitted by the herd during the production process in well-managed tropical pastures. Generally speaking, they are proposed for use in meat from animals whose CH_4 emissions were mitigated during the production process by increasing the C stock in the soil through the adoption of recovery and sustainable management of pastures and/or ICLF systems and whose production process is recognized, certifiable, and auditable (Almeida & Alves, 2020).

Low-Carbon Soybean

As a direct contribution to the global strategies for decarbonizing the economy and achieving the goals of governments and businesses to reduce GHG emissions, it was urgent and necessary to develop brand concepts, linked to a third-party certification system, based on internationally validated scientific criteria that ensure a differentiated product produced under conditions favorable to the mitigation of emissions of GHGs, with objective, measurable, and reportable criteria (Nepomuceno et al., 2021).

In this context, in April 2021, Embrapa Soja, a division of the Embrapa, presented the "Low Carbon Soja Program." The initiative intends to support sustainable agricultural practices through research and creation of production parameters. The positioning in defense of production with a focus on low C emissions in Brazil has been taking place for years on the part of the company. The country has different agricultural systems throughout its extension; therefore, to make this new model a reality, the entity is gathering information already published in scientific articles to establish the criteria that will be applied by farmers. In addition to optimizing costs for producers due to the reduction in the purchase of C credits, the initiative contributes to the recognition of Brazil as a more sustainable country in grain planting (Nepomuceno et al., 2021).

Carbon Market as a Potential Fostering Initiative to Promote Soil Health and Carbon Sequestration in Brazil

The idea of underlying C markets is intimately related to the structure of the climate change issue. One of the main features of climate change is that, regardless of the origin of the CO_2 emissions, the negative results affect everyone around the world. Local emissions will always cause a global effect, and this logic demands cooperation among countries to address the problem in an effective way. Although this basic feature poses a big challenge for international cooperations, it also offers a solution: because the effects are global, the solution can also be delocalized. Likewise, no matter where the C removal is done, the effects benefit everybody. Based on this view, the first C market mechanisms were developed in the Kyoto era of international climate governance. As mentioned, the Kyoto Protocol was the first mechanism to instrumentalize the UNFCCC. It set binding targets of emission reductions for developed countries and mechanisms to allow these countries to achieve the goal imposed.

The C markets were born then, within the UNFCCC, as flexibility mechanisms allowing countries with mandatory emission reduction targets to carry out their reductions in countries that did not have this obligation. The criterion of territoriality, hence, was pivotal in this strategy, helping to identify the additional action of the mechanism. This rationale underpinned the construction of the Clean Development Mechanism, the first regulated C market benefiting developing countries and which did not have mandatory CO_2 reduction targets. In this modality, the C projects were financed by developed countries within the developing countries' territory. The goal was to boost a financial influx to developing countries and to generate C credits for developed countries. Recognized by its dichotomy regarding following emissions obligations, adherence to the Kyoto logic declined after the Copenhagen Summit in 2009. The scientific evidence and the geopolitical order required engagement from all countries despite the historical responsibility from developed countries.

A new era emerged with the Paris Agreement, generating a huge impact on how international actors organize themselves regarding reduction pledges and C market features. In contrast to the binding dichotomy approach of the Kyoto Protocol, the Paris Agreements have brought about a voluntary approach that applied to all member countries. Thus, emission reduction goals have been adopted by all countries but on a voluntary basis, with each country presenting its reduction goals according to their local realities. This expansion of member countries with reduction objectives has also led to a larger number of actors in this market. As a result, the Paris Agreement called on the private sector to assume its responsibilities to reduce emissions. A new climate order had thus been born, with all countries and the private sectors having to take on, voluntarily, reduction commitments.

The C market has also been affected by these new demands. The Kyoto Protocol led to regulated C markets. Within UNFCCC, the Clean Development Mechanism has been formed, and outside it, Europe organized its own regional regulated market to address their climate commitments because voluntary markets were still scarce and are mainly present in the United States. With Paris, the reality of C markets changed, prompting an expansion of voluntary markets with companies' reduction pledges known as Net Zero commitments (to learn more about that see the Science Based Target Initiative, see https://sciencebasedtargets.org). Also, national regulated markets have gained importance as a C pricing instrument to set national reduction commitments. The Paris era also imposed new demands for C markets within UNFCCC. The territorial approach was no longer possible, and a new set of rules was negotiated in the period between the COP 21 and COP 26 in Glasgow. The Paris Rule Book was only finalized 7 years after the Paris Agreements, mainly due to the C market rules. Article 6 from the Paris Agreement, which addresses C market rules within the new structure, brought about new demands for C credits projects. If the territorial approach was the basis of the former protocol, in the new instrument, climate ambition is the key concept to identify the need for a C project.

In this sense, climate ambition consists of at least three elements: (a) exclusive finance (i.e., the finance source is determinant for the existence of the project), (b) social and biodiverse co-benefits (i.e., the project must generate a positive impact to local communities and preserve biodiversity), and (c) no double account (i.e., emission reductions can only be counted once, by one NDC). Considering these conditions, Brazil has emerged as a main C credit producer because it has unique conditions to provide high-integrity C assessments aligned with Paris Agreement Article 6 for the whole world. Moreover, Brazil's agricultural and forestry base can offer unique social and biodiverse co-benefits as few other countries can do (Griscom et al., 2020).

Initially, the Kyoto Protocol projects in Brazil were focused on renewable energies (mainly solar and wind), and the maturation of these technologies in the national market caused these projects to become less attractive. The development of new measurement and control technologies for NBS projects alongside the new global framework of the Paris Agreement has placed Brazil in a prime position for projected growth in the C market, enabling it to provide mitigation and removal of C while providing the best global cost-benefit. Presently, there are two C market schemes working simultaneously: the regulated market and the voluntary market. The regulated market is understood by all reduction targets submitted to binding rules, either at a national level or within UNFCCC Article 6 rules. The voluntary market or offset market refers to nonbinding pledges announced by companies. The C credits generated by C projects in this scenario can be used as the companies' pledges, after their insetting efforts, or by the UNFCCC market, with corresponding adjustments to avoid being double counted.

As for soil projects in agriculture, the potential to open new economic activity for producers is real. As mentioned before, agriculture is one of the main contributors of GHG emissions (>s10%); however, this sector has the capacity to sink C, which may offer great income potential for farmers as long as the technological and institutional requirements are met. Various initiatives designed to remunerate farmers who sink C (negative emissions) have sprung up around the world, although they are yet to be applied. Scalability and regulation questions must be addressed to support the development of soil C credits, thus allowing it to achieve its full potential.

Brazil is discussing its own C pricing strategy to develop a national regulated market. Despite a lack of binding mitigation rules, the voluntary C market is growing exponentially. The main economic sectors in the country are aware of the national agroforestry potential, and a structured debate is taking place to find the better blend between the regulated and the voluntary market. These initiatives will allow Brazil to take a relevant position in the new climate economy because this country offers unique high-quality C credits based on NBS projects.

Final Remarks

Brazil's voluntary pledge to decrease agricultural GHG emissions reflects the countries' interests and capacities and is limited to available technical options. Agricultural and agriculture-related emissions, including non-CO_2 emissions, will constitute the largest sector of surplus GHG emissions in the future because other sectors are projected to reduce their GHG emissions to the maximum extent by 2030; therefore, agriculture is critical to meeting global climate targets.

Excluding agricultural emissions from mitigation targets will increase the cost of mitigation in other sectors or reduce the feasibility of meeting the 2°C limit.

Therefore, agriculture will have to adapt to climate change, but it can also mitigate climate change by offsetting anthropogenic emissions through C sequestration in soils and biota. In this context, there are feasible pathways to enhance SH and C sequestration not only through ecological intensification and diversification of agricultural systems but also by restoring native ecosystems (Figure 12.8). Of course, there are several challenges to implementing those integrated agricultural systems because the level of complexity is higher than conventional production systems. Thus, international, national, and regional public policies are needed to incentivize the adoption of agricultural BMPs, including the intensification and the diversification of agricultural systems, in addition to bringing important social and economic benefits.

A positive agenda for SH and C sequestration depends on enhancing the awareness of policymakers and on the general society about the fragility of the soils and food systems. However, the enhanced awareness must be more intensively translated into political will and an effective action plan. More financial investments in BMPs could thereby hasten further sustainable agricultural development. Special efforts should be taken to ensure that new technologies are relevant, affordable,

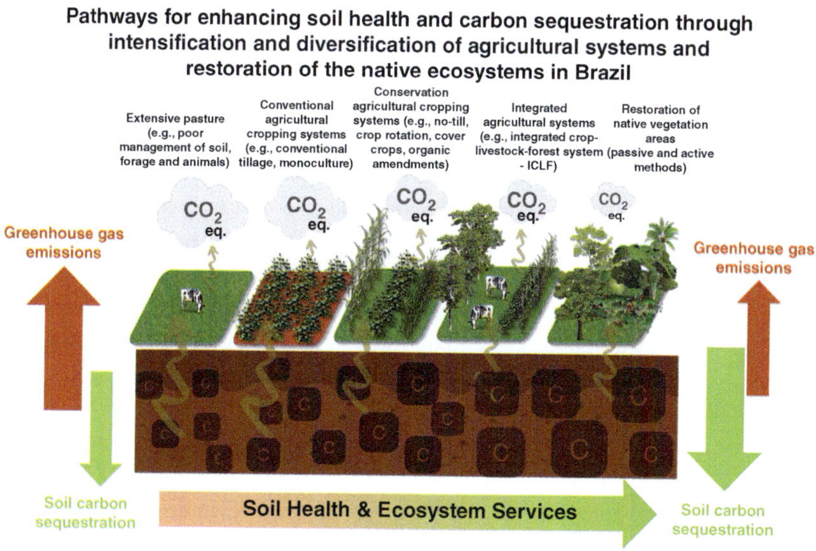

Figure 12.8 Pathways for enhancing soil health and C sequestration through intensification and diversification of agricultural systems and restoration native ecosystems in Brazil.

and accessible to many farmers. Now is the time to think beyond COP 27 and take concrete actions to address these issues of global significance.

Brazil has one of the largest potentials for soil C sequestration in the world, probably equivalent only to that of Indonesia and far above other countries, such as the United States and China. This C-based solution, in addition to being less costly and having greater potential for growth in the short term than purely technological solutions, brings additional potential benefits, such as the possibility to recover SH, biodiversity, water security, and socioeconomic development.

Acknowledgments

We gratefully acknowledge the support of the Center for Carbon Studies in Tropical Agriculture (CCARBON) sponsored by São Paulo Research Foundation (FAPESP) (2021/10573-4), Research Centre for Greenhouse Gas Innovation (RCGI), hosted by the University of São Paulo and FAPESP (2020/15230-5) and Shell Brasil, and the strategic importance of the support given by ANP (Brazil's National Oil, Natural Gas and Biofuels Agency) through the R&D levy regulation. M.R.C thanks CNPq for his Research Productivity Fellowship (311787/2021-5).

References

Almeida, R. G., & Alves, F.V. (2020). *Diretrizes técnicas para produção de carne com baixa emissão de carbono certificada em pastagens tropicais: carne baixo carbono (CBC)*. Embrapa Gado de Corte. https://www.embrapa.br/busca-de-publicacoes/-/publicacao/1120985/diretrizes-tecnicas-para-producao-de-carne-com-baixa-emissao-de-carbono-certificada-em-pastagens-tropicais-carne-baixo-carbono-cbc

Alves, F. V., Almeida, R. G., & Laura, V. A. (2017). *Carbon neutral brazilian beef: A new concept for sustainable beef production in the tropics*. Embrapa. https://www.embrapa.br/busca-de-publicacoes/-/publicacao/1120985/diretrizes-tecnicas-para-producao-de-carne-com-baixa-emissao-de-carbono-certificada-em-pastagens-tropicais-carne-baixo-carbono-cbc

Amelung, W., Bossio, W., Vries, W., Kögel-Knabner, I., Lehmann, J., Amundson, R., Bol, R., Collins, C., Lal, R., Leifeld, J., Minasny, B., Pan, G., Paustian, K., Rumpel, C., Sanderman, J., van Groenigen, J. W., Mooney, S., van Wesemael, B., Wander, M., & Chabbi, A. (2020). Towards a global-scale soil climate mitigation strategy. *Nature Communications*, 11, 5427. https://doi.org/10.1038/s41467-020-18887-7

Bossio, D. A., Cook-Patton, S. C., Ellis, P. W., Fargione, J., Sanderman, J., Smith, P., Wood, S., Zomer, R. J., von Unger, M., Emmer, M., & Griscom, B. W. (2020).

The role of soil carbon in natural climate solutions. *Nature Sustainability*, 3, 391–398. https://doi.org/10.1038/s41893-020-0491-z

Brasil Ministério da Agricultura, Pecuária e Abastecimento. (2012). *Plano setorial de mitigação e de adaptação às mudanças climáticas para a consolidação de uma economia de baixa emissão de carbono na agricultura: plano ABC (Agricultura de Baixa Emissão de Carbono)*/Ministério da Agricultura, Pecuária e Abastecimento, Ministério do Desenvolvimento Agrário, coordenação da Casa Civil da Presidência da República. 32p.

Brasil Ministério da Agricultura, Pecuária e Abastecimento. (2021). *Plano Setorial para Adaptação à Mudança do Clima e Baixa Emissão de Carbono na Agropecuária 2020-2030*: Plano Operacional/Ministério da Agricultura, Pecuária e Abastecimento. Secretaria de Inovação, Desenvolvimento Rural e Irrigação.

Brazil Ministry of Agriculture, Livestock and Food Supply. (2021). *Plan for adaptation and low carbon emission in agriculture strategic vision for a new cycle*. Secretariat for Innovation, Rural Development and Irrigation.

Brazilian Confederation of Agriculture and Livestock. (2023). *The taste of Brazil*. https://cnabrasil.org.br/agro.br-english

Carvalho, J. L. N., Cerri, C. E. P., Feigl, B. J., Piccolo, M. C., Godinho, V., & Cerri, C. C. (2009). Carbon sequestration in agricultural soils in the Cerrado region of the Brazilian Amazon. *Soil & Tillage Research*, 103, 342–349. https://doi.org/10.1016/j.still.2008.10.022

Cerri, C. C., Bernoux, M., Maia, S. M. F., Cerri, C. E. P., Costa Junior, C., Feigl, B. J., Frazão, L. A., Mello, F., Galdos, M. V., Moreira, C. S., & Carvalho, J. L. N. (2010). Greenhouse gas mitigation options in Brazil for land-use change, livestock and agriculture. *Scientia Agricola*, 67, 102–116. https://doi.org/10.1590/S0103-90162010000100015

Cerri, C. C., Maia, S. M. F., Galdos, M. V., Cerri, C. E. P., Feigl, B. J., & Bernoux, M. (2009). Brazilian greenhouse gas emissions: The importance of agriculture and livestock. *Scientia Agricola*, 66, 831–843. https://doi.org/10.1590/S0103-90162009000600017

Cerri, C. E. P., Cherubin, M. R., Damian, J. M., Mello, F. F. C., & Lal, R. (2021). *Soil carbon sequestration through adopting sustainable management practices: Potential and opportunity for the American countries*. IICA. https://repositorio.iica.int/handle/11324/19315

Gianetti, G. W., & de Souza Ferreira Filho, J. B. (2021). O plano e programa ABC: Uma análise da alocação dos recursos. *Revista de Economia e Sociologia Rural*, 59, 1–15. https://doi.org/10.1590/1806-9479.2021.216524

Griscom, W., Busch, J., Cook-Patton, S. C., Ellis, P. W., Funk, J., Leavitt, S. M., Lomax, G., Turner, W. R., Chapman, M., Engelmann, J., Gurwick, N. P., Landis, E., Lawrence, D., Malhi, Y., Schindler Murray, L., Navarrete, D., Roe, R., Scull, S., Smith, P., … Worthington, T. (2020). National mitigation potential from natural climate solutions in the tropics. *Philosophical Transactions of the Royal Society B*, 375, 20190126. https://doi.org/10.1098/rstb.2019.0126

Hong, C., Zhao, H., Qin, Y., Burney, J., Pongratz, J., Hartung, K., Liu, Y., Moore, F. C., Jackson, R. B., Zhang, Q., & Davis, S. J. (2022). Land-use emissions embodied in international trade. *Science*, 376, 597–603. https://doi.org/10.1126/science.abj1572

Inter-American Institute for Cooperation on Agriculture (IICA.). (2022). *Medium-term plan for the period 2022–2026*. https://repositorio.iica.int/handle/11324/20969

Lal, R. (2013). Soil carbon management and climate change. *Carbon Management*, 4, 439–462. https://doi.org/10.4155/cmt.13.31

Landers, J. N., de Freitas, P. L., Balbino, L. C., Júlio, C. S., & Marchão, R. L. (2020). Integration of crop-livestock in conservation agriculture systems. In A. Kassam (Ed.), *Advances in conservation agriculture volume 1: Systems and science*. Burleigh Dodds Science Publishing. https://doi.org/10.1201/9780429268724

Maia, S. M. F., Medeiros, A. S., Santos, T. C., Lyra, G. B., Lal, R., Assad, E. D., & Cerri, C. E. P. (2022). Potential of no-till agriculture as a nature-based solution for climate-change mitigation in Brazil. *Soil and Tillage Research*, 220, 105368. https://doi.org/10.1016/j.still.2022.105368

Ministério da Agricultura, Pecuária e Abastecimento (MAPA). (2022). *Plano ABC: Agricultura de Baixa Emissão de Carbono*. https://www.gov.br/agricultura/pt-br/assuntos/sustentabilidade/plano-abc/plano-abc-agricultura-de-baixa-emissao-de-carbono

Nepomuceno, A. L., Balbinot Junior, A. A., Rufino, C. F. G., Debiasi, H., Nogueira, M. A., Franchini, J. C., Alves, F. V., Almeida, R. G., Bungenstab, D. J. & Dall'Agnol, V. F. (2021). *Programa SBC -Soja Baixo Carbono: Um novo conceito de soja sustentável*. Embrapa.

Paustian, K., Lehmann, J., Ogle, S., Reay, D., Robertson, G. P., & Smith, P. (2016). Climate-smart soils. *Nature*, 532, 49–57. https://doi.org/10.1038/nature17174

Richards, M., Bruun, T., Campbell, B. M., Gregersen, L. E., Huyer, S., Kuntze, V., Madsen, S. T. N., Oldvig, M. B., & Vasileiou, I. (2016) *How countries plan to address agricultural adaptation and mitigation: An analysis of intended nationally determined contributions CCAFS dataset version 1.1*. CGIAR Research Program on Climate Change, Agriculture and Food Security. https://hdl.handle.net/10568/69115

Rogelj, J., den Elzen, M., Höhne, N., Fransen, T., Fekete, H., Winkler, H., Schaeffer, R., Sha, F., Riahi, K., & Meinshausen, M. (2016). Paris agreement climate proposals need a boost to keep warming well below 2 °C. *Nature*, 534, 631–639. https://doi.org/10.1038/nature18307

Smith, P. (2016). Soil carbon sequestration and biochar as negative emission technologies. *Global Change Biology*, 22, 1315–1324. https://doi.org/10.1111/gcb.13178

Unterstell, N. & Martins, N. (2022). *NDC: Analysis of the 2022 update submitted by the Government of Brazil. Analysis report*. TALANOA. 17p. https://www.politicaporinteiro.org/wp-content/uploads/2022/04/Brazils-NDC-2022-analysis_V0.pdf

Wollenberg, E., Richards, M., Smith, P., Havlík, P., Obersteiner, M., Tubiello, F. N., Herold, M., Gerber, P., Carter, S., Reisinger, A., van Vuuren, D. P., Dickie, A., Neufeldt, H., Sander, B. O., Wassmann, R., Sommer, R., Amonette, J. E., Falcucci, A., Herrero, M., … Campbell, B. M. (2016). Reducing emissions from agriculture to meet the 2 °C target. *Global Change Biology*, 22, 3859–3864. https://doi.org/10.1111/gcb.13340

Printed and bound by CPI Group (UK) Ltd, Croydon, CR0 4YY
16/09/2026
14963422-0001